VOLUME SIXTEEN

ADVANCES IN NEUROTOXICOLOGY

Nutrition and Neurotoxicology Part B

VOLUME SIXTEEN

ADVANCES IN NEUROTOXICOLOGY

Nutrition and Neurotoxicology Part B

Series Editor

MICHAEL ASCHNER
LUCIO G. COSTA

Edited by

ISAAC ADEGBOYEGA ADEDARA
Department of Food Science and Technology, Center of Rural Sciences, Federal University of Santa Maria, Camobi, Santa Maria, Rio Grande do Sul, Brazil; Molecular Drug Metabolism and Toxicology Research Laboratories, Department of Biochemistry, College of Medicine, University of Ibadan, Ibadan, Nigeria

TATIANA EMANUELLI
Department of Food Technology and Science, Center of Rural Sciences, Federal University of Santa Maria, Camobi, Rio Grande do Sul, Brazil

MICHAEL ASCHNER
Department of Molecular Pharmacology, Albert Einstein College of Medicine, Bronx, NY, United States

LUCIO G. COSTA
Department of Environmental and Occupational Health Sciences, University of Washington, Seattle, WA, United States

ELSEVIER

ACADEMIC PRESS
An imprint of Elsevier
elsevier.com/books-and-journals

Academic Press is an imprint of Elsevier
125 London Wall, London EC2Y 5AS, United Kingdom
50 Hampshire Street, 5th Floor, Cambridge, MA 02139, United States

ISBN: 978-0-443-34681-1
ISSN: 2468-7480

For Information on all Academic Press publications visit our website at https://www.elsevier.com/books-and-journals

Publisher: Zoe Kruze
Acquisitions Editor: Mariana Kuhl
Editorial Project Manager: Rehan Ismail
Production Project Manager: Abdulla Sait
Cover Designer: Muthukumaran Thangaraj

Typeset by Aptara, New Delhi, India

Contents

3 Influence of non-sugar sweeteners on neural health: The current state of knowledge

Adeboye A. Olaseni, Grazielle C.C. Weis, Luana H. Maurer, Isaac A. Adedara and Tatiana Emanuelli

4 Nutraceuticals and obsessive-compulsive disorder therapy

Amir Hossein Mohammadi, Negin Amin, Sayed Abolfazl Tavousi, Mahmood Khaksary Mahabady, Michael Aschner and Hamed Mirzaei

5 The nexus between the catabolism of tryptophan and neurotoxicity

Azubuike P. Ebokaiwe and Olusanya Olasehinde

Contributors

Amos Olalekan Abolaji
Drosophila Research and Training Centre, Basorun, Ibadan, Nigeria; Drosophila Laboratory, Drug Metabolism and Toxicology Unit, Department of Biochemistry, Faculty of Basic Medical Science, College of Medicine, University of Ibadan, Ibadan, Oyo State, Nigeria

Olusegun Gafar Adebayo
Department of Physiology, Faculty of Basic Medical Sciences, College of Medicine, University of Ibadan, Ibadan, Nigeria

Isaac A. Adedara
Nauphoeta cinerea Research Laboratory (NaciReL), Drug Metabolism and Toxicology Unit, Department of Biochemistry, College of Medicine, University of Ibadan, Ibadan, Nigeria; Department of Food Science and Technology, Center of Rural Sciences, Federal University of Santa Maria, Camobi, RS, Brazil

Lawrence Dayo Adedayo
Department of Human Physiology, Faculty of Basic Medical Sciences, College of Medicine, Federal University Wukari, Taraba, Nigeria

Ismaheel Akinwale Adeniyi
Department of Physiology, Faculty of Basic Medical Sciences, College of Medicine, University of Ibadan, Ibadan, Nigeria; Department of Physiology, Faculty of Basic Medical Sciences, College of Medicine, University of Ibadan, Ibadan, Nigeria

Olufunke Florence Ajeigbe
Department of Anatomy and Neurobiology, School of Medicine, University of California Irvine, Irvine, USA

Israel Akinrotimi Akinola
Department of Biological Sciences, East Tennessee State University, Johnson City, Tennessee, USA

Negin Amin
School of Health Science, College of Health, Massey University, Palmerston North, New Zealand

Kafilat Temidayo Aransi-ola
Department of Neuroscience and Cell Biology, Robert Wood Johnson Medical School, Rutgers University, Piscataway, NJ, United States of America

Michael Aschner
Department of Molecular Pharmacology, Albert Einstein College of Medicine, Bronx, NY, United States

Ifeoluwa Awogbindin
School of Medical Sciences, Faculty of Health, University of Victoria, Victoria, British Columbia, Canada

Adesina A. Babalola
Graduate Program in Biological Sciences, Department of Biochemistry and Molecular Biology, Federal University of Santa Maria, Santa Maria, RS, Brazil

Benneth Ben-Azu
DELSU Joint Canada-Israel Neuroscience and Biopsychiatry Laboratory, Department of Pharmacology, Faculty of Allied Health Sciences, College of Health Sciences, Delta State University, Abraka, Delta State, Nigeria; School of Medical Sciences, Faculty of Health, University of Victoria, Victoria, British Columbia, Canada

Emmanuel O. Chidebe
DELSU Joint Canada-Israel Neuroscience and Biopsychiatry Laboratory, Department of Pharmacology, Faculty of Allied Health Sciences, College of Health Sciences, Delta State University, Abraka, Delta State, Nigeria

Vanda Lopes de Andrade
Research Institute for Medicines (iMed.ULisboa), Faculty of Pharmacy, Universidade de Lisboa, Lisboa, Portugal; Life Quality Research Centre (CIEQV), IPSantarem/IPLeiria, Rio Maior, Portugal; Research Centre for Natural Resources, Environment and Society (CERNAS), Santarem Polytechnic University, Quinta do Galinheiro-S. Pedro, Santarém, Portugal

Rebecca P. Dickson
Nauphoeta cinerea Research Laboratory (NaciReL), Drug Metabolism and Toxicology Unit, Department of Biochemistry, College of Medicine, University of Ibadan, Ibadan, Nigeria

Ana Paula Marreilha dos Santos
Research Institute for Medicines (iMed.ULisboa), Faculty of Pharmacy, Universidade de Lisboa, Lisboa, Portugal

Azubuike P. Ebokaiwe
Toxicology and Immunotherapy Research Unit, Department of Medical Biochemistry, Delta State University Abraka, Nigeria

Tatiana Emanuelli
Department of Food Technology and Science, Center of Rural Sciences, Federal University of Santa Maria, Camobi, RS, Brazil

Daniel T. Esuku
DELSU Joint Canada-Israel Neuroscience and Biopsychiatry Laboratory, Department of Pharmacology, Faculty of Allied Health Sciences, College of Health Sciences, Delta State University, Abraka, Delta State, Nigeria

Hammed Olawale Faleke
Department of Chemistry and Biochemistry, Texas Tech University, Lubbock, Texas, USA

Oluwabukola Mary Farodoye
Programa de Pós-Graduação em Genética e Biologia Molecular, Universidade Federal do Rio Grande do Sul, Porto Alegre, RS, Brazil; Drosophila Research and Training Centre, Basorun, Ibadan, Nigeria

Opeyemi Babatunde Fasina
Department of Neuroscience, Lerner Research Institute, Cleveland Clinic, Ohio, USA

Omowumi FEMI-Akinlosotu
Developmental Neurobiology Unit, Department of Anatomy, College of Medicine, University of Ibadan, Nigeria

Olumayowa Igado
Neuroscience Unit, Department of Veterinary Anatomy, Faculty of Veterinary Medicine, University of Ibadan , Nigeria

Prosper Iwhiwhu
DELSU Joint Canada-Israel Neuroscience and Biopsychiatry Laboratory, Department of Pharmacology, Faculty of Allied Health Sciences, College of Health Sciences, Delta State University, Abraka, Delta State, Nigeria

Mahmood Khaksary Mahabady
Anatomical Sciences Research Center, Institute for Basic Sciences, Kashan University of Medical Sciences, Kashan, Iran

Luisa M. Mateus
Research Institute for Medicines (iMed.ULisboa), Faculty of Pharmacy, Universidade de Lisboa, Lisboa, Portugal

Luana H. Maurer
Department of Food Technology and Science, Center of Rural Sciences, Federal University of Santa Maria, Camobi, RS, Brazil

Hamed Mirzaei
Research Center for Biochemistry and Nutrition in Metabolic Diseases, Institute for Basic Sciences, Kashan University of Medical Sciences, Kashan, Iran

Amir Hossein Mohammadi
Research Center for Biochemistry and Nutrition in Metabolic Diseases, Institute for Basic Sciences, Kashan University of Medical Sciences, Kashan, Iran

Zainab Abiodun Molik
Department of Pharmacognosy, Faculty of Pharmacy, University of Ibadan, Ibadan, Nigeria

Marvellous A. Oghorodi
State University of Medical and Applied Sciences, Igbo-eno Enugu, Enugu State, Nigeria

Josephine Chigbolum Ohuche
Department of Veterinary Physiology and Biochemistry, Faculty of Veterinary Medicine, University of Ibadan, Ibadan, Nigeria

Abiola J. Oladepo
Nauphoeta cinerea Research Laboratory (NaciReL), Drug Metabolism and Toxicology Unit, Department of Biochemistry, College of Medicine, University of Ibadan, Ibadan, Nigeria

Memunat Oladepo
Department of Human Physiology, Amadu Bello University Zaria, Nigeria

Busayo Oladun
Exercise and Nutritional Sciences, College of Health Solutions, Arizona State University, Phoenix, United States of America

Halimat Olaniyan
Drosophila Research and Training Centre, Basorun, Ibadan, Nigeria

Olusanya Olasehinde
Department of Biochemistry, University of Benin

Adeboye A. Olaseni
Department of Food Technology and Science, Center of Rural Sciences, Federal University of Santa Maria, Camobi, RS, Brazil

Funmilayo Olopade
Developmental Neurobiology Unit, Department of Anatomy, College of Medicine, University of Ibadan, Nigeria

Samuel Adetunji Onasanwo
Department of Physiology, Faculty of Basic Medical Sciences, College of Medicine, University of Ibadan, Ibadan, Nigeria

Jackson E. Onuelu
DELSU Joint Canada-Israel Neuroscience and Biopsychiatry Laboratory, Department of Pharmacology, Faculty of Allied Health Sciences, College of Health Sciences, Delta State University, Abraka, Delta State, Nigeria

Benjamin Oritsemuelebi
DELSU Joint Canada-Israel Neuroscience and Biopsychiatry Laboratory, Department of Pharmacology, Faculty of Allied Health Sciences, College of Health Sciences, Delta State University, Abraka, Delta State, Nigeria

Titilayomi Ayomide Otenaike
Programa de Pós-Graduação em Genética e Biologia Molecular, Universidade Federal do Rio Grande do Sul, Porto Alegre, RS, Brazil; Drosophila Research and Training Centre, Basorun, Ibadan, Nigeria

Ademola Adetokunbo Oyagbemi
Department of Veterinary Physiology and Biochemistry, Faculty of Veterinary Medicine, University of Ibadan, Ibadan, Oyo State, Nigeria

Felix A.A. Soares
Graduate Program in Biological Sciences, Department of Biochemistry and Molecular Biology, Federal University of Santa Maria, Santa Maria, RS, Brazil

Sayed Abolfazl Tavousi
Student Research Committee, Kashan University of Medical Sciences, Kashan, Iran

John Oluwafemi Teibo
Department of Biochemistry and Immunology, Ribeirão Preto Medical School, University of São Paulo, Ribeirão Preto, São Paulo, Brazil

Grazielle C.C. Weis
Department of Food Technology and Science, Center of Rural Sciences, Federal University of Santa Maria, Camobi, RS, Brazil

Preface

The structure and function of the brain are directly impacted by nutrition. The consumption of high-quality balanced diets that provide essential vitamins, fatty acids, minerals and antioxidants enhances neurogenesis and protects against neuronal dysfunction. Empirical data have shown that simple dietary interventions are associated with neuroprotection and improvement of neurobehavioural function via several biochemical mechanisms at cellular and molecular levels in both animals and humans. Moreover, recent research in nutrition and neurotoxicology is unravelling the link between dietary factors and brain health. Several foods that are commonly consumed by people reportedly contain neurotoxicants, impacting the brain health of children, adolescents, adults and the elderly. The exact nature of interactions between nutrition and neurotoxicants may be difficult to fully delineate since one factor hardly changes unaided. Indeed, the overall expression of the complex interaction between nutrition and neurotoxins is reflected in the functional outcome of the nervous system.

This volume on Nutrition and Neurotoxicology - Part B provides a platform to synthesize various beneficial and detrimental constituents of foods, elucidate mechanistic signalling, and showcase *Caenorhabditis elegans* model in mycotoxin-mediated neurotoxicity. The volume commences with a chapter that discusses various neurotoxic and neuroprotective chemicals in cow milk and soy beverages by *de Andrade and colleagues.* The impact of dietary components on neuroinflammatory pathways is addressed in decoding the diet-inflammation axis: modulating neuroinflammation through nutritional interventions by *Fasina and colleagues*. The next chapter by *Olaseni and colleagues* discusses the current state of knowledge on the influence of non-caloric sweeteners on neural health. Moreover, a chapter by *Mohammadi and colleagues* provides an update on nutraceuticals and obsessive-compulsive disorder therapy, while the following chapter on the nexus between the catabolism of tryptophan and neurotoxicity is by *Ebokaiwe and Olasehinde.*

A few chapters are dedicated to the protective effects of bioactive compounds against environmental neurotoxicants and associated neurodegenerative diseases. Specifically, neuroprotective phytonutrients: mechanisms of action and evidence in neurotoxic and degenerative conditions by *Oritsemuelebi and colleagues*, nutritive agents ameliorating neurodegeneration by *Femi-Akinlosotu and colleagues*, and nutraceuticals and neurotoxicity by

Onasanwo and colleagues explain how environmental exposures to neurotoxicants contribute to neurological disorders and the potential beneficial health effects of dietary bioactive compounds seen in both animal models and humans. The penultimate chapter has been dedicated to neurotoxicity of heavy metals across the lifespan: the beneficial role of nutrition from foetus to the elderly by *Otenaike and colleagues*. The last chapter on dietary mycotoxin toxicity in *Caenorhabditis elegans* model: insights into neurotoxicological and neuroendocrine effects by *Oladepo and colleagues* advances our knowledge on the utility of *C. elegans* as an integrative and complementary model in understanding mechanisms associated with foodborne neurotoxins.

Isaac A. Adedara, Tatiana Emanuelli, Michael Aschner, Lucio G. Costa

CHAPTER ONE

Chemicals in cow milk and soy beverages: Neurotoxicity and neuroprotection

Vanda Lopes de Andrade[a,b,c], Luisa M. Mateus[a], Michael Aschner[d,*], Ana Paula Marreilha dos Santos[a]

[a] Research Institute for Medicines (iMed.ULisboa), Faculty of Pharmacy, Universidade de Lisboa, Lisboa, Portugal
[b] Life Quality Research Centre (CIEQV), IPSantarem/IPLeiria, Rio Maior, Portugal
[c] Research Centre for Natural Resources, Environment and Society (CERNAS), Santarem Polytechnic University, Quinta do Galinheiro-S. Pedro, Santarém, Portugal
[d] Department of Molecular Pharmacology, Forchheimer Building, Room 209, Albert Einstein College of Medicine, Bronx, NY, United States
*Corresponding author. e-mail address: Michael.aschner@einsteinmed.edu.

Contents

Advances in Neurotoxicology, Volume 16
ISSN 2468-7480
https://doi.org/10.1016/bs.ant.2026.02.006

Abstract

Cow milk constitutes a fundamental component of the human diet. Along with high nutritional value, contains neuroprotective components, caseins, lactoferrin, phospholipids and tryptophan. However, due to the use of pesticides in agriculture, veterinary drugs in livestock, and the presence of other contaminants, cow milk is frequently contaminated by these chemicals posing a risk to human health, namely through neurotoxicity. Strategies to decrease these contaminants include detoxification via microorganisms. Concomitantly, increased consumption of plant-based protein to substitute animal sources has been seen globally, being soy beverages (SB) utilized as an alternative to cow milk. Soy products are almost equivalent to animal sources in protein quality, and moreover, soy isoflavones like genistein, daidzein and glycitein and saponins, as well as soy peptides, may play neuroprotective roles. Conversely, neurotoxic risks associated with the consumption of soy beverages also exist, mainly due to the presence of excess of manganese, mycotoxins and consequences of human interventions during cultivation, resulting in pesticide residues, and pollution of soils, namely with metals. Due to the neurotoxicity and bioaccumulation of these contaminants, the fact that both cow milk and soy beverages are routinely ingested by all ages, and represent key ingredients in many processed foods, including baby foods, strict surveying of their levels is suggested.

Key words: Cow milk, Neuroprotection, Neurotoxic contaminants., Soy beverage

1. Introduction

Milk and its derivatives constitute one of the fundamental components of the human diet, particularly due to their protein and mineral content (Woźniak et al., 2022). Milk is widely consumed by children and adults and, consequently, is a very important food item. However, in the last few decades, the increased consumption of plant-based protein as a substitution for animal sources has been seen globally. In turn, soy beverages (SB) can be utilized as an alternative to cow milk by individuals who are vegan, lactose intolerant or allergic to cow milk, which is one of the most common food allergies in infancy. It can also be a substitute for cow milk where cow milk is not accessible in adequate amounts (Andrade et al., 2023). In this chapter, we will explore aspects related to general health, neuroprotection, and neurotoxicity associated with cow milk and soy beverages consumption, particularly focusing on contaminants found in these foods. We will also mention methods to reduce contamination in both cow milk and plant-based alternatives, such as and soy beverages.

2. Introduction to cow milk

Cow milk (CM) is considered a major part of human diet worldwide because of its overall nutritional value (Jadhav et al., 2019; Calahorrano-Moreno et al., 2022; Cimmino et al., 2023). Cow milk and its products are essential components of a balanced human diet as they have tremendous nutritional values regarding a good proportion of proteins, fats, and important minerals (Ćosić et al., 2021).

Milk is a fluid secreted by the mammary glands of female mammals and fulfils the nutritional requirements of the neonate, including the energetic part (provided by lipids, lactose, and in excess by proteins), essential amino acids, and amino groups necessary for the biosynthesis of non-essential amino acids (provided by proteins), essential fatty acids, vitamins, inorganic elements, and water (Raheem and Niamah, 2021; Calahorrano-Moreno et al., 2022).

Bovine milk is the most consumed food product representing about 48 percent of the total milk consumed globally. The European Union, Australia, and New Zealand are the most important producers, followed by the United States and India (Wiley, 2007; Calahorrano-Moreno et al., 2022). Global CM production has increased by about 20 percent in the last decade, from 694 million tons in 2008 to 843 million tons in 2018. One mechanism by which CM is introduced and promoted in countries throughout the world is school milk programs (Wiley, 2007).

Given the clear disparity in dairy consumption trends, and an overall increase in CM production, growth potential in the dairy market is the greatest in developing countries. Stable or declining CM consumption in western countries is related to demographic trends, notably the aging of the population, as well as increased competition with other beverages introduced in the market (Wiley, 2007). The consumption of CM and its derivatives belongs to the most ancient eating practices, since these products have been considered as the most balanced food found in nature, containing major sources of nutrients, especially for children, adults, and the elderly people (Nag, 2010; Boudebbouz et al., 2022). However, despite its nutritional value and high consumption, the use of pesticides in agricultural practices and antibiotics in livestock production, CM is frequently contaminated by these chemicals (Zhao and Wang, 2012; Akhtar and Karam, 2017; Welsh et al., 2019).

3. Chemical contamination in cow milk

Pesticide residues may appear in both CM and its products from direct or indirect sources, such as herbage, contaminated drinking water, and the application of various chemicals from preventing diverse diseases (Nag, 2010; Welsh et al., 2019; Cimmino et al., 2023). Thus, antibiotics and hormones are administered to cows orally, by injection, or as intramammary infusions to treat diseases, promote animal growth and increase milk production (Calahorrano-Moreno et al., 2022). Due to its nutritional and supplementary value, CM is being consumed by people of different age groups; therefore, those residues in CM represent an immediate attention of researchers and regulators (Akhtar and Karam, 2017).

Additionally, other common environmental contaminants are mycotoxins, and metals consumed by cattle through feed, forages, and water (Raheem and Niamah, 2021; Calahorrano-Moreno et al., 2022). In addition, direct contamination may occur during CM processing from milking, handling, storage and even pasteurization. During the industrialization process, CM may contain metals, residues of cleaning products, mycotoxins, among others (Calahorrano-Moreno et al., 2022). Thus, due to various direct and indirect contamination sources, different chemical and microbiological contaminants have been found in CM. Nevertheless, as population and industrial growth increased, new contaminants appeared, and with this, contamination of CM also increased not only by compounds of biological origin, but also by compounds of chemical origin (Calahorrano-Moreno et al., 2022).

Among the various chemical contaminants in CM, we will give special attention to chemicals with neurotoxic effects that may cause neurotoxicity on a direct or indirect way. Neurotoxic chemicals may target neuronal axons, myelin or neurotransmission causing neurotoxic effects. These effects depend on the chemical's potency and the repeatability of the exposure. The loss of neurons can happen affecting or losing some functions (Costa et al., 2008). As an example of chemical contaminants with neurotoxic effects we may include the group of pesticides and some veterinary drugs. Indeed, all the above referred cow milk contaminants, pesticides but also veterinary medicines residues, may derive in health risks to humans due to the consumption of contaminated CM and dairy products (Raheem and Niamah, 2021).

4. Pesticides residues in cow milk

Pesticides are defined by the Food and Agriculture Organization (FAO) of the United Nations as a substance or mixture of substances attended for controlling, preventing, destroying any pest, animal, or human disease-causing vectors, undesirable plants, or animal species affecting food production, managing, selling, storage, and transportation (WHO, 2020). Pesticides play a key role in agriculture and food production by controlling the insects, weeds, plant pathogens and microbial contaminations (Ćosić et al., 2021). These pesticides are used before harvest and during the storage period, and these pesticides transfer through the food chain from simple plants and animals to high-end animals which leads to the bioaccumulation of pesticides in animal products such as meat, milk, fats and eggs (Akhtar and Karam, 2017; Raheem and Niamah, 2021). A pesticide found in water is another source of residues in milk through drinking water (Akhtar and Karam, 2017; Ćosić et al., 2021).

The global consumption of pesticides in 2019 was approximately 4.19 million metric tons, where China was by far the largest pesticide-consuming country (1.76 million metric tons), followed by the United States (408 thousand tons), Brazil (377 thousand tons), and Argentina (204 thousand tons) (Fernández-González et al., 2015; Rajak et al., 2023; Kiruthika et al., 2024). Although pesticides are valuable in several aspects such as agriculture and public health, their inappropriate application may cause health problems, including carcinogenicity, reproductive toxicity, neurodegenerative diseases and disruption of endocrine system (Mohammadi et al., 2021; Schopf et al., 2022).

Milk and dairy products, although present considerable socioeconomic importance, their contamination with pesticide residues is a matter of a worldwide serious public health concern, since they are widely and repeatedly consumed by infants, children and elderly population

(Battu et al., 2004; Khaniki, 2007; Ćosić et al., 2021; Schopf et al., 2022). Infant exposure to these residues poses serious health risks, especially owing to the lack fully active enzymatic and metabolic systems having a detoxification role (Shaker and Elsharkawy, 2015). In addition to the great importance for human diet, milk has been studied as a bio-concentration process indicator of the environmentally persistent organic pollutants, such as some pesticides (Ćosić et al., 2021). Occurrence of pesticide residues in milk is not so rare, especially for CM (EFSA, 2019).

4.1 Different classes of pesticides

There are several different classes of pesticides, with different uses, mechanisms of action in target species, and toxic effects in nontarget organisms. Pesticides are usually classified according to their target species. The three major classes (and their target pests) are those of insecticides (insects), herbicides (weeds), fungicides (fungi, molds) (Costa et al., 2008; Richardson et al., 2019). Within each class, several subclasses exist, with substantially different chemical and toxicological characteristics (Costa et al., 2008).

Depending on climate and the type of crops, countries may preferentially apply greater amounts of insecticides, herbicides or fungicides (Costa et al., 2008). Pesticides are one of the most found contaminants, not only in raw CM but also after the pasteurization and Ultra Heat Treatment (UHT) process. Their presence in CM, even below the maximum permitted levels, represents a health risk to the consumer (Calahorrano-Moreno et al., 2022). Some studies report that animal origin food is responsible for 90 percent pesticides entry in the human body (Akhtar and Karam, 2017). Pesticides source in dietary rout (products like milk, meat, fat and eggs) is main way of chronic exposure to these substances (Abd-Rabo et al., 2016; Akhtar and Karam, 2017). Organochlorine, organophosphates, synthetic pyrethroid and/or triazine were all pesticides' chemical classes found in fluid CM, powder products, yogurts, cheese, butter, and sour cream. (Schopf et al., 2022).

Many pesticides and their residues that enter the human body through the food chains are recognized as causative agents or primarily contributory factors in several serious human diseases, such as disruption of the endocrine system, neurological and immune system disorders, sterility, different cancer types (breast, lung, cervix, and prostate), heart diseases, Alzheimer's and Parkinsonism (Nath et al., 2013; Bedi et al., 2015). Insecticides play a relevant role in the control of insect pests. All the chemical insecticides in use today are neurotoxicants, and act by poisoning the nervous systems of the target organisms. The target sites for insecticides in insects are also found in mammals, hence insecticides are, for the most part, not species-selective regarding targets of toxicity, and mammals, including humans, are highly sensitive to their toxicity (Costa et al., 2008).

As noted, insecticides have higher acute toxicity toward nontarget species compared to other pesticides. There are several classes of insecticides; the most widely used are the cholinesterase inhibitors (organophosphates and carbamates), followed by the pyrethroids, and by other more recently developed compounds, such the neonicotinoids. The organochlorine compounds (e.g., DDT) (also insecticides) were widely used until most of them were banned in the mid-1970s; however, some are still used in certain countries, and exposure of the general population still occurs, given their high environmental persistence (Costa et al., 2008).

4.2 Organophosphorus

Organophosphorus are a commonly used group of pesticides and constitute roughly 38 percent of the total pesticide utilized universally (Boudebbouz et al., 2022). Organophosphorus compounds were developed in the early 1940s, while carbamates were introduced as insecticides in the 1950s. Compounds of both classes although having a common target of toxicity, the enzyme acetylcholinesterase (AChE), display different toxicological features.

Exposure to organophosphorus can cause known effects—acute (short) and chronic (longer-term)—to humans as well as to animals (Richardson et al., 2019; Perry et al., 2020; Ajiboye et al., 2022). The primary target for organophosphorus is AChE, whose physiological role is to hydrolyse acetylcholine, a major neurotransmitter in the central and peripheral (autonomic and motor-somatic) nervous systems. Inhibition of AChE by organophosphates causes accumulation of acetylcholine at cholinergic synapses, with overstimulation of cholinergic receptors of the muscarinic and nicotinic type, with symptoms such as weakness or paralysis of the muscles (Costa et al., 2008; Richardson et al., 2019; Perry et al., 2020). Two additional areas of concern regarding the neurotoxicity of organophosphates are possible long-term CNS effects in adults, and developmental neurotoxicity (Costa et al., 2008).

Abundant evidence exists in support of pre- and/or post-natal exposure to organophosphates-induced developmental neurotoxicity (Costa et al., 2008; Richardson et al., 2019). Organophosphates can disrupt various cellular processes (e.g., DNA replication, neuronal survival, neurite outgrowth), alter non-cholinergic pathways, induce oxidative str, ess, and cause various behavioral abnormalities (Costa, 2006; Costa et al., 2008). The intensive use of organophosphates is mainly due to their broad spectrum of action and to their lesser persistence and accumulation in the environment in comparison with organochloride pesticides (Soares et al., 2019; Mulla et al., 2020). However, despite this relatively low stability, organophosphates can accumulate in the food chain including CM (Bolles et al., 1999). This is due to the ability of organophosphates to covalently link to proteins in milk; in fact, it has been proved that casein, mainly with its seryl and phosphoseryl residues, can bind with organophosphates (Deiana and Fatichenti, 1992; Pagliuca et al., 2006).

Since CM can be one of the human dietary sources of organophosphates, the detection of these residues in this food is important to correctly evaluate the overall consumer risk with particular attention to the safety of infants and children (Pagliuca et al., 2006). Nonetheless, despite their wide use, there is a relatively low number of studies (12 publications) reporting organophosphate pesticides analysis in raw CM compared to other related pesticides such as organochlorine (Boudebbouz et al., 2022).

4.3 Carbamate

Carbamate insecticides (carbaryl, carbofuran, and aminocarb) are organic compounds derived from carbamic acid, having different degrees of acute oral toxicity, ranging from moderate to low toxicity (carbaryl), to extremely high toxicity (e.g., aldicarb) (Costa et al., 2008). Carbamate insecticides are used abundantly by households and for agriculture purposes (Blodgett and Means, 2013). Milk contamination with carbamate pesticides could cause a serious risk to humans and animals since these compounds have been linked with cancer, reproduction toxicity and neurotoxic effects (Vidair, 2004; Herbert et al., 2021; Boudebbouz et al., 2022).

The mechanism of toxicity of carbamates is similar to organophosphates, as they also inhibit AChE, although in case of carbamates the inhibition of the enzyme is transient and rapidly reversible. Nevertheless, the sign and symptoms of carbamate poisoning are similar to those of organophosphates (Costa et al., 2008).

4.4 Organochlorine pesticides

Organochlorine pesticides, polychlorinated biphenyls, dibenzo-*p*-dioxins and dibenzofurans are widely distributed halogenated aromatic compounds which persistently contaminate the environment (Mrema et al., 2013). These compounds are chemically stable with long biological half-life which leads to high biomagnification in the food chain across a wide range of trophic levels (Serrano et al., 2008; Boudebbouz et al., 2022).

Milk-producing animals accumulate pesticides from contaminated feed and by inhaling contaminated air. Organochlorine pesticides due to their lipophilic properties are initially stored in fat-rich tissues and subsequently translocated and excreted in milk (Deti et al., 2014). The organochlorine insecticides are categorized into three groups namely dichlorodiphenylethane (include DDT, DDD and dicofol), chlorinated cyclodienes (Includes aldrin, dieldrin and heptachlor) and hexachlorocyclohexanes (include lindane only). These compounds exist in the form of residues in all food chains from soil to animals, but these may vary in toxic doses, skin absorption levels, metabolism and even elimination from the body (Hayes and Laws, 1982; Muhammad et al., 2021).

Organochlorine pesticides, although known endocrine disrupting chemicals (Martins et al., 2019), can also cause impaired immune function, a wide variety of cancers and neurobehavioral disorders (Singh, 2016). In addition, exposure to these chemicals can induce a wide range of adverse health effects including possible long-term effects on intellectual function in infants and negative effects on central nervous system functioning (Zhao et al., 2007).

Most of the organochlorine insecticides employ their hazardous effects by injuring the insect nervous system. DDT exerts its insecticidal activity by causing voltage-gated sodium channels (Na_V) to remain open, leading to persistent depolarization and hyperactivity in the nervous system (Richardson et al., 2019). These chemicals follow the same mechanism for humans by attacking the nervous system, although the lethal doses may differ for insects and humans. Due to the dependence of the organism on the dose for strong toxic effects, care should always be taken when these pesticides encounter humans (Muhammad et al., 2021). It was suggested that the chronic exposure to organochlorine insecticides may be related to the increase of Alzheimer's Disease (AD). Recently, Richardson and co-workers (2019) reported that serum levels of DDE (a DDT metabolite) were approximately four-fold higher in AD patients, and this was associated with increased risk of AD diagnosis.

It is noteworthy that organochlorine pesticides such as hexachlorocyclohexane, dichlorodiphenyl trichloroethane, and endosulfan are still present despite having been banned since the 1970s (between 1970 and 1980). They are commonly detected in CM because of their high persistence in the environment. This indicates that they are still used in agriculture and animal husbandry. In addition, and internationally, DDT has been recommended by the World Health Organization (WHO) for malaria eradication, which has increased its usage in recent years (van den Berg et al., 2017). Additionally, with a few exceptions (cyhalothrin, cypermethrin, fenvalerate, deltamethrin, permethrin, and diazinonella), most pesticides found in CM are not regulated by Codex and the EU (Raheem and Niamah, 2021). This demonstrates the low efficiency of the regulatory controls of these contaminants in the unprocessed and post-processed product, leading to an inefficient safety of CM (Calahorrano-Moreno et al., 2022).

The presence of organochlorine insecticides (such as HCB, Lindane, Aldrin and DDT) in CM has been verified with values exceeding the Minimal Risk Level (MRL), according to the WHO, which may eventually cause a large risk on human health due to its biological accumulation, and the consequent negative effects already mentioned (Rêgo, 2019). Notably some pesticide residues disappeared with skim milk and buttermilk during the production process. Yogurt has low level of pesticides due to heat treatment (it will be mentioned ahead) (Raheem and Niamah, 2021). Among the organochlorine pesticides reported in 32 studies in raw CM, DDTs have the highest monitoring frequency (27 publications), followed by endosulfan (20 publications), hexachlorocyclohexane (HCH) (19 publications), drins (16 publications), heptachlor (11 publications), and endrin (8 publications) (Boudebbouz et al., 2022).

Endosulfan and HCH with its isomers, were also considered persistent organic pollutants in 2011 by the Stockholm Convention. Although their use has been restricted by several countries, it is still used in developing countries due to the absence of other organic choices and/or regulation lacking (Sathishkumar et al., 2021). Heptachlor, another chlorinated hydrocarbon insecticide besides being considered carcinogenic to humans, can attack the central nervous system and causes neurological disorders such as Lewy pathology (Ross et al., 2019).

4.5 Pyrethroids

Pyrethroids are synthetic organic insecticides derived from the naturally occurring flowers of pyrethrums (*Chrysanthemum coccineum* and *Chrysanthemum cinerariae folium*) (Zacharia, 2010). They are used worldwide since 1980 and are considered as the safest compounds for use in food due to their photo degradation, effectiveness against various insects, and the low toxicity compared to other classes of pesticides (Yoo et al., 2016). The mechanism of action of pyrethroids is the same in insects and in mammals, as they bind to the sodium channels and slow its activation, as well as the rate of inactivation, leading to a stable hyperexcitable state (Costa et al., 2008).

However, despite low toxicity, recent evidence suggests that exposure to pyrethroids during pregnancy or childhood may be associated with childhood behavioral problems and could affect the nervous system (Furlong et al., 2017). Exposure to these compounds *in utero* or during childhood can contribute to a variety of neurodevelopmental and neurological disorders. Developmental neurotoxicants may also cause silent damage, which would manifest itself only as the individual ages, and may contribute to neurodegenerative diseases such as PD or AD (Giordano and Costa, 2012).

5. Strategies to decay pesticide residues in milk

As it was mentioned before, milk products might contain undesired organophosphate or organochlorine pesticides when dairy cows are fed with pesticides-polluted forages or drinking water (Battu et al., 2004; Schopf et al., 2022).

Several processing technologies can be utilized to destroy pesticide residues or reduce its concentration in food and dairy products through its degradation (Gafour, 2018). These technologies reducing the pesticide residues in CM, may indirectly contribute to the decrease of eventual neurotoxic effects caused by the pesticides. Recently, detoxification via microorganisms such as lactic acid bacteria (LAB) and probiotics has been extensively studied and degradation of pesticides through hydrolytic enzymes has been introduced as the possible mechanism, and it has been highlighted that some probiotics harbor pesticide-degrading genes (Mohammadi et al., 2021). In fact, microorganisms play a significant role in the breakdown of pesticides and have been recognized for their influence and many uses in human welfare (Kafilzadeh et al., 2015).

Lactic acid bacteria are gaining global attention, especially due to their role as a probiotic. Besides their role in food processing, LAB also have a significant role in degrading insecticide residues in the environment (Kiruthika et al., 2024). Lactic acid bacteria are a group of Gram-positive, non-spore forming, immobile and catalase negative bacteria, which excretes lactic acid as major end product and generally recognized as safe (Kiruthika et al., 2024). The key enzymes produced by LAB, such as phosphatase and esterase, play a vital role in breaking down insecticide molecules (Kim et al., 2005; Kiruthika et al., 2024).

Biosorption methods employing the use of microorganisms prove to be efficient in the removal of pesticides, metals, and mycotoxins. Biosorption with LAB managed to eliminate organophosphate pesticides from CM, being more effective for chlorpyrifos, fenitrothion, and malathion (Calahorrano-Moreno et al., 2022). According to Schopf et al. (2022) although the application of LAB was shown to be an effective method for removing diazinon, dimethoate, and methyl parathion in CM it was not very selective since it cannot eliminate all the organophosphates studied. However, Zhou and Zhao (2014) observed that such LAB species present in CM (*Lactobacillus casei*, *L. bulgaricus*, *L. rhamnosus*, *L. acidophilus* and *Streptococcus thermophilus*) could degrade organophosphorus insecticides, namely, chlorpyriphos, chlorpyriphos-methyl, diazinon, dichlorvos, fenthion, malathion, phorate, pirimiphos-methyl and trichlorphon.

Ayoub et al. (2012) showed that the sterilization process of raw CM is the most effective method for reduction the pesticide residues to minimum value. Concerning the study of organochlorine pesticides residues in CM, Ayoub as well as Abou-Zied before (1994) concluded that sterilization of CM is the most effective method for reduction of p,p'-DDE residues in milk followed by boiling and finally pasteurization. Raheem and Niamah (2021) also noted that the concentration of residues of pesticides in CM decay after heat process. The heat treatment had affected highly the residue in CM sample and the sterilization and boiling provided effective degradation of HCH isomers (organochlorine insecticide). Analysis of organophosphorus pesticides results also indicated that the pesticides degraded faster at higher temperature in CM during heat treatment; and among the pesticides methyl parathion was the most unstable one. Heat treatment and lactic acid fermentation could reduce the level of organophosphorus pesticide in dairy products (Bo, 2010). Gafour (2018) found that in the dairy industry field, the heat treatments such as pasteurization, boiling, and sterilization had a significant positive impact on malathion reduction.

Moreover, ultrasound technology has the potential to increase microbial metabolic activity; so, Yuan et al. (2020) explored the stimulatory effect of ultrasound pre-treatment on the degradation of four common organophosphate pesticides (fenitrothion, chlorpyrifos, profenofos, and dimethoate) during milk fermentation by *Lactobacillus plantarum*. The authors observed that the ultrasonic pretreatment significantly accelerated the degradation of the four pesticides studied during yogurt fermentation (Yuan et al., 2020). The enhancement of pesticide degradation by ultrasound may be promoted by the activity of phosphatase in CM, which was the main mechanism for L. *plantarum* to degrade organophosphorus pesticides (Yuan et al., 2020).

6. Veterinary drug residues in cow milk

The common use of veterinary medicines by livestock breeders are utilized mainly to enhance the production of CM and its products besides treating diseases such as mastitis and parasites. The failure to know the adequate period for disposal of these medicines by the animal, is one of the causes that leads to their presence or residues in animal-derived foods, including CM, with potential dietary exposure to humans (Piñeiro et al., 2021; Raheem and Niamah, 2021).

The presence of drug residues in food products has become a growing concern because of the adverse health risks and regulatory implications. Drug residues in food refer to the presence of pharmaceutical compounds or their metabolites in products such as meat, fish, eggs, milk, poultry and ready-to-eat foods, which are intended for human consumption. These residues can come from the use of drugs in the field of veterinary medicine, such as antibiotics, antiparasitic agents, growth promoters and other veterinary drugs given to livestock with the aim of providing them as prophylaxis, therapy and for promoting growth (Beyene, 2016; Pratiwi et al., 2023). Compliance with the maximum residue limit (MRL) is very important for food manufacturers according to the Food and Drug Administration (FDA) or European Union (EU) regulations (Pratiwi et al., 2023). The residues of these drugs in food animals can have various potential adverse effects on humans, including carcinogenic, allergic, toxic, neurologic disorders and microbiological effects (Babapour et al., 2012; Rana et al., 2019; Pratiwi et al., 2023).

6.1 Antibiotics

The presence of antibiotics in human food has been linked to several negative public health outcomes, including gastrointestinal disturbance, kidney damage, and neurological diseases (Apata, 2009). Due to their availability and low cost, antibiotics have been widely used in the dairy, livestock, poultry, aquaculture and honey production industries in various countries around the world. Currently, approximately 80 percent of food animals receive antibiotics for part or most of their lifetime, which may lead to residue accumulation in animal products (Rana et al., 2019).

Antibiotics are natural products of a micro-organism, identical synthetic products or similar semi synthetic product that inhibit the growth of or destroy microorganisms. In veterinary medicine antibiotics are used for therapeutic, prophylactic, metaphylactic and nutritive purposes (Babapour et al., 2012; Beyene, 2016; Pratiwi et al., 2023). Cows eliminate antibiotics and their metabolites through milk, depending on the dose and route of application, level of milk production, type and degree of mammary disease, and time between treatment and milking (Calahorrano-Moreno et al., 2022).

The presence of antibiotics residues in human food is associated with several adverse public health effects including hypersensitivity, tissue damage, gastrointestinal disturbance like perturbation of the intestinal microbiome and modification of the antimicrobial resistance profile of the human intestinal microbiota and neurological disorders (Babapour et al., 2012; Piñeiro et al., 2021). These antimicrobial drug classes are currently used and licensed in many jurisdictions for therapeutic indications in food animal veterinary medicine. However, some regulatory agencies no longer approve their use as growth promoters in food-producing animals (Beyene, 2016). Antimicrobials used according to label directions should not result in residues at slaughter. However, reasons for residue presence in edible tissues include non-adherence to recommended label directions or dosage (extra-label usage); non observance of recommended withdrawal periods; administration of too large a volume at a single injection site; use of antibiotic-contaminated equipment, or failure to properly clean equipment used to mix or administer drugs, and also mixing errors (Okocha et al., 2018).

While the process varies among national authorities, all antimicrobial new animal drugs are assessed for toxicological and pharmacological hazards as appropriate to the drug class and to the specific drug under consideration. These can include, but are not limited to carcinogenicity, haemotoxicity, genotoxicity, immunotoxicity, neurotoxicity, hepatotoxicity, nephrotoxicity, developmental toxicity, and reproductive effects (Cerniglia et al., 2016).

The natural intestinal bacteria serve as a protective barrier, preventing incoming pathogens from taking hold and causing illnesses; hence, intestinal microorganisms play a key role in maintaining the body's homeostasis. Recent experimental findings suggest that dysbiosis (reduced diversity) of the gut microbiome is linked to systemic inflammatory diseases including AD (Khalifa et al., 2024). Consequently, maintaining the homeostasis of the intestinal microbiota is essential. Broad-spectrum antibiotics are frequently implicated in disrupting intestinal flora, leading to gastrointestinal disorders (Yang et al., 2021; Khalifa et al., 2024). Tetracycline, oxytetracycline, penicillin, and amoxicillin are prevalent antibiotics found in milk (Calahorrano-Moreno et al., 2022).

Antibiotic treatment can disrupt both short- and long-term conditions of microbial balance in the gut (Yang et al., 2021). In turn, antibiotics residues in CM can induce neurotoxicity owing to the disruption of the microbiota in the gut (Helaly et al., 2019; Yang et al., 2021). To reiterate the previous point, Ke et al. (2024) report that the intestinal microbiota helps maintain the integrity of the intestinal barrier, which may provide protection against the neurotoxic effects induced by methylmercury (MeHg).. The balance of the microbial population has been considered as an independent biological factor that affects the hypothalamic-pituitary axis. Antibiotic abuse caused the animals to show anxiety-related disorders (Golubeva et al., 2015).

The neurotoxic effect result from abuse of antibiotics works by both local and more prominent systemic mechanisms (Helaly et al., 2019; Yang et al., 2021). It can be said that antibiotic abuse is a cofactor behind the rise of neuropsychiatric diseases (Helaly et al., 2019). Corroborating these observations, it was observed that long-term antibiotic therapy for tuberculosis has been reported to induce clear neurotoxicity including cognitive disorders and delusional thinking (Helaly et al., 2019).

Excess use of antibiotics like Amoxicillin Clavulanic combination induced neurotoxic effects demonstrated by neurobehavioral tests and histopathology *in vivo* studies. Neomycin induced hippocampal excitability that reflected calcium binding protein activity. The neurobehavioral tests in mice showed potential impaired memory, cognition and motor function in a remote way (Helaly et al., 2019). A study performed by Vuong et al. (2020) using mice demonstrated that antibiotic maternal exposure could influence fetal neurodevelopment (Vuong et al. 2020; Yang et al., 2021; Patangia et al. 2022). This study is relevant concerning the exposure of fetus to antibiotic residues in milk consumed by the mothers. However, concerning Rahman et al. opinion (2021) that although most of the milk samples of the study area possess veterinary antibiotics residue above the MRL value, it would not be detrimental to humans' health following consumption. This statement may be very debatable because the consumption of CM varies very much between each one, depends also on the age and health status of the consumer as well as on the type of diet and medicines that are taken.

6.2 Antiparasitics

6.2.1 Coccidiostats

Coccidiostats, also referred to as antiprotozoal agents, are compounds extensively used as additives or drugs for the prevention and treatment of livestock coccidiosis (Peek and Landman, 2011; Kant, 2013). Specifically, these drugs are used to treat or control the disease caused by parasites, organisms of the genus *Eimeria*, known to damage the intestinal mucosa of the host animal (Zhao et al., 2018). They primarily impact poultry and ruminant animals (Tuck et al., 2016). However, coccidiostats may cause residue accumulation in edible animal products and, consequently, posing serious health risks to the consumer (Kadykalo et al., 2018; Rana et al., 2019). Excessive anticoccidial residues in animal-derived food have been associated with toxicological effects such as teratogenicity, hepatotoxicity, or neurotoxicity (Tuck et al., 2016; Pratiwi et al., 2023).

6.2.2 Avermectins

Avermectins (AVM)s have widespread global application in veterinary and human medicine to control parasitic infections. The AVMs are the most widely used anti-parasitic drugs in veterinary field with sales exceeding one billion US dollars annually (Salman et al., 2022). Among the AVMs, ivermectin was one of the first, broad-spectrum antiparasitic agents extensively used for the prevention and treatment of parasites in food-producing animals. It has been widely used for over 25 years and is listed in the WHO essential medicines for several parasitic diseases (Campillo et al., 2021; Suvarna, 2023). It is indicated for use in the treatment of strongyloidiasis (*Strongyloides stercoralis*) and onchocerciasis (*Onchocerca volvulus*), diseases caused by the referred parasites.

All AVMs including ivermectin are highly lipophilic, implying a long withdrawal period and presence (accumulation) in the edible parts of the treated animals, specifically those with high fat content (Rana et al., 2019; Campillo et al., 2021). According to Campillo et al. (2021) Ivermectin application at the label dose for beef cattle will cause detectable CM residues that may persist for an extended period.

From a human safety perspective, the administration of ivermectin to food producing animal species entails potential risks related to the presence of drug residues in edible tissues, CM, and other derived products (Escribano et al., 2012). The AVMs may pose a serious toxicity threat to animals (Salman et al., 2022). Ivermectin has been regarded as potential neurotoxin;

it acts mainly by binding to the glutamate-dependent chloride channels of invertebrate nerve and muscle cells, causing an increase in membrane permeability leading ultimately to neuromuscular paralysis and death of certain parasites (Salman et al., 2022). Generally, the use of AVMs in animals has a wide safety margin as far as the nervous system is concerned (Lankas et al., 1997; Salman et al., 2022). However, in some species like rats and dogs the AVMs exhibit nervous effects by damaging the cerebral cortex and inducing diffused cerebellar dysfunction (Salman et al., 2022).

In neurotoxicity studies various signs are observed like depression, tremors, salivation, ataxia, convulsions, mydriasis, coma, and ultimately death in animals like dogs, cattle, and lions (Lankas et al., 1997; Campillo et al., 2021; Salman et al., 2022). Serious neurological adverse events were also reported in public health programs in Africa to eliminate onchocerciasis through community-based ivermectin treatment. Therefore, it has been considered free of the potential to cause neurological adverse drug reactions, except in situations of overdose (Campillo et al., 2017; Chandler et al., 2018). Various regulatory authorities including the Joint Expert Committee of Food Additives and Contaminants (JECFA) and European Union have fixed the MRL of ivermectin in CM at 0.01 mg/kg. Therefore, monitoring of ivermectin residue levels in the various food products of animal origin is of great importance to ensure the safety of consumers (Suvarna, 2023).

7. Cow milk—General protection

Milk proteins have also been reported for their biological activities, with peptides derived from their gastrointestinal digestion playing an influential role. These activities include nutrient uptake, immunoregulation, and postprandial hormone secretion. In addition milk has antioxidant protective effects (Erukainure et al., 2020).

Several studies have analyzed the anti-inflammatory effects of various CM-derived molecules including protein isolates, fractions of bio-active lipids and calcium. The milk fat globule membrane (MFGM) which chaperones encases CM lipids, contains an estimated 120 proteins, some of which have been demonstrated to potently modulate immune function. In one study, lactadherin, a protein naturally enriched in CM, markedly suppressed inflammation in mice induced with a fluid percussion injury. Similarly, mucins have been shown to play a role in the innate immune system and may attenuate bacterial infections (D'Alonzo et al., 2020).

8. Cow milk neuroprotection

8.1 Caseins

Caseins are amongst the milk proteins and abundant in most mammalian species, making 80 percent of CMs proteins (Kunz et al., 1999). They are regarded as micelles owing to their assembly in supra-macromolecular structures (Erukainure et al., 2020). The results from Erukainure et al. (2020) study indicate a neuroprotective effect of casein micelles from CM. There have been increasing interests on cognitive impairment as it has been recognized as a major age-related neuropathology (Berr, 2000). Oxidative stress has been implicated in the pathogenesis and progression of the two major types of cognitive impairment: AD and vascular dementia (VaD). Treatment of oxidative brain injury *in vivo* using isolated casein micelles resulted in a significant enhancement of the brain's antioxidant defense systems. Specifically, levels of key enzymatic antioxidants including glutathione (GSH), superoxide dismutase (SOD), and catalase (CAT), were markedly increased, indicating a reinforced capacity to neutralize reactive oxygen species (ROS) and mitigate oxidative stress. Moreover, the treatment also led to elevated expression of ecto-nucleoside triphosphate diphosphohydrolase (E-NTPDase), an enzyme involved in the hydrolysis of extracellular nucleotides, which plays a crucial role in modulating neuroinflammation and facilitating tissue repair. Additionally, levels of 5'-nucleoside-triphosphatase, another enzyme implicated in nucleotide metabolism, were also increased, further supporting the therapeutic potential of casein micelles in restoring homeostasis in oxidatively stressed neural tissues. These findings suggest a multifaceted neuroprotective mechanism by which casein micelles not only counteract oxidative damage but also support inflammatory resolution and regenerative pathways in the brain (Feldbrügge et al., 2018). Treatment with the isolated casein micelles led to complete depletion of oxidative stress generated lipid metabolites, while deactivating oxidative-activated lipid metabolic pathways. These results indicate the microstructural properties, particle size, elemental composition, and antioxidant neuroprotective effect of casein micelles from CM (Erukainure et al., 2020).

8.2 Lactoferrin (Lf)

Lactoferrin is a multifunctional glycoprotein with approximately 80 kDa molecular mass and 700 amino acids, that is usually known for its presence in CM. lactoferrin's is found in different regions of the body, especially in the brain, exerting antioxidant, antibacterial, antiviral and anti-inflammatory effects (Le Parc et al., 2017; Eker et al., 2023). To no surprise, lactoferrin can decrease the number of proinflammatory cytokines that are secreted from glial cells under certain circumstances (Xu et al., 2019).

8.3 Phospholipids and tryptophan

Bovine dairy milk phospholipids reduce endoplasmic reticulum stress in neuronal cells, implicated in development of neurodegenerative disorders. Moreover, relevant to brain function, CM phospholipids such as phosphatidylcholine and sphingomyelin, were found to promote cell structural integrity (D'Alonzo et al., 2020). In addition, tryptophan, which is abundant in CM, is a precursor of the neuroactive compounds, serotonin and melatonin (Joung et al., 2021).

9. Soy and soy beverages consumption

Soybean [*Glycine* max (L.) Merr.] is originated from China since over 5000 years ago; in 1830, travelled a great distance to Europe, and from there spread throughout the world to North America, South America, India, and other regions (Cao et al., 2022; Yip et al., 2023). Nowadays, is the most important oil plant worldwide and the major oil crop, which also provides plant protein to the food industry, with a global oil plant cultivation area of 46 percent in 2017 (Heydebreck, 2021; Cao et al., 2022; Yip et al., 2023). In this same year, 14.4 million tons and 20.8 million tons of soybeans and soybean meal, respectively, were imported to the European Union (EU), being the majority of soybeans imported from the USA and Brazil (>70 percent), whereas about 87 percent of soybean meal was originated from Brazil and Argentina (Heydebreck, 2021). In 2018, the average *per capita* per year of soybean food consumption was 7.94 Kg in Japan, 3.61 kg in China, 0.21 kg in the USA and 0.28 kg in Europe (Cao et al., 2022; Yip et al., 2023).

Noteworthy is the fact that the average global warming potential of soybean production (0.58 kg CO_2-eq/Kg) is much lower than beef (28.73 Kg CO_2-eq/Kg), lamb (27.91 Kg CO_2-eq/Kg), pork (5.85 Kg CO_2-eq/Kg) or chicken (4.12 Kg CO_2-eq/Kg) meat. Therefore, soy products have been increasingly recognized as an alternative to livestock production to reduce climate changes (Cao et al., 2022; Yip et al., 2023). The use of soy in foodstuffs is quite diverse with unfermented foods including tofu, edamame, soy nuts and sprouts, while fermented soy products comprise miso, tempeh, natto and soy sauce.

Currently, soy-based beverages are one of the most consumed plant-based beverages, which is a global trend in the food market, representing an alternative for cow milk (Pérez-Rodríguez et al., 2023). Although the consumption of milk and dairy products are recommended in numerous national dietary guidelines, their inclusion in the daily diet is declining in both Europe and USA. Such is due to lactose intolerance and milk protein allergies, as well as to a growing tendency for vegan or flexitarian diets in the belief of improved health and lowered environmental impact (Sterup Moore et al., 2024). For vegans and vegetarians, soy and soy products such as SB successfully took its place due to its high protein content, which is estimated to be 36.49 mg/100*g* according to United States Department of Agriculture Food Composition Databases; in the case of SB, the product contains 7 g of protein per 236.6 mL which is similar to the protein content of dairy milk. Moreover, SB is inexpensive, refreshing, nutritional and contains different functionally active components responsible for beneficial roles in the body (Paul et al., 2020).

The SB is made by triturating soybeans with water, filtering the aqueous extract, and heat processing the aqueous extract to inactivate enzymes, particularly trypsin inhibitors, and pasteurizing the product (Haisman, 2011). Therefore, the main soy sources in SB are the beans, the hydro soluble extract, and the isolate protein. The isolate protein is an end-product of the soy processing and contains more than 90 percent of protein, in dry base (McArdle and Hamill, 2011). With regards to the terminology of SB, although soy-based drinks are known as soy milk all over the world, the European regulation prohibits the use of the word milk for drinks that are not made from mammary secretions. In a different way, the Food and Drug Administration (FDA) considers the legality of using the term "milk" in plant-based dairy alternatives. Nowadays, soy-based drinks are referred to as "soy milk," "soy drinks," or "soy beverages" (Olías et al., 2023).

10. Nutritional value of soy beverages and general effects on health

The nutritional content of plant-based beverages varies markedly depending on their plant origin, fortification, and industrial processing, and consequently, the commercial brand. In truth the largely variable composition of the plant-based beverages available on the market impairs a fair nutrient calculation (Sterup Moore et al., 2024). In the case of SB, since the food industry continually innovates to enhance nutritional and sensory qualities to make it more attractive to consumers, the market is full of different products (Olías et al., 2023). A study in Spain based on the labeling information, leaded to consider that the average soybean content was approximately 11 percent (w/v) in these beverages (Olías et al., 2023). In turn, soybeans contain around twice the amount of protein compared to other commonly consumed beans and legumes, which is 35–40 percent of protein on a dry-weight basis and is the only plant source containing all the nine essential amino acids; moreover, the leucine content of soybeans is comparable to levels found in fish and eggs. Soy products are almost equivalent to animal sources in protein quality, with the additional benefit of containing less saturated fat and no cholesterol (Jones et al., 2008; Chatterjee et al., 2018; Belobrajdic et al., 2023). They also contain higher amounts of poly unsaturated fatty acids, including both omega-6 and omega-3 essential fatty acids, than other common legumes. Soybeans have a similar dietary fiber content to other beans and pulses, mainly in the form of oligosaccharides, and contain non-starch polysaccharides which include pectic polysaccharides, xyloglucan and cellulose. They are also a good source of several micronutrients, often limited in plant-based diets, such as iron, zinc and calcium; the high calcium content of soybeans, coupled with the fact that this calcium is efficiently absorbed, makes them particularly amenable to be used as a dairy alternative (Belobrajdic et al., 2023).

Despite the already mentioned nutritional variations among SB brands, SB is considered an excellent source of high-quality protein, vitamin B and unsaturated fatty acids, and the fortification these beverages by the industry turn the nutritional gap with cow milk successfully bridged (Olías et al., 2023). Of brief reference are the infant soy formulas as well as milk-based formulas, that fed millions of infants worldwide and promoted growth and development (Badger et al., 2009). Soybeans and derived foods (which includes SB) have been considered healthy for humans, with their phytochemicals such as isoflavones, saponins, proteins or peptides, phospholipids, phenolic acids, trypsin inhibitors, and phytic acid, believed to be responsible for several health benefits (Chatterjee et al., 2018; Chen and Chen, 2021; Kim et al., 2021). Actually, more than 100 phytochemicals were identified in soybeans and soy food (Chatterjee et al., 2018) and in this chapter, general health effects due to the presence of isoflavones, saponins, proteins or peptides will be described.

Epidemiological studies have associated soy consumption with reduction of the risk for chronic diseases such as obesity, cardiovascular disease, insulin-resistance/type II diabetes, immune disorders, certain type of cancers, such as breast and prostate cancers, as well as better bone health, and relief of menopausal symptoms (Jones et al., 2008; Chatterjee et al., 2018). Other authors mention antioxidative properties, antiaging, antithrombotic effects, anti-renal failure, diuretic action, prevention of gallstone formation, anti-obesity, relief from constipation, prevention of osteoporosis, promotion of liver function, lower blood pressure and boosting of immunity; soy phytochemicals can also inhibit HIV (Kim et al., 2021).

The major bioactive compounds in soybeans are the isoflavones and both soy foods and soy-based infant formulas are rich sources of isoflavones, containing 1–4.2 mg isoflavones/g, which are associated with soy proteins (Jones et al., 2008; Chatterjee et al., 2018; Messina et al., 2022). Jackson et al. (2002) describes mean recoveries of isoflavones in SB, in relation to their initial concentration in raw soybeans, of 54 percent. Biologically inactive glucoside isoflavones are broken down by intestinal bacteria into the biologically active aglycones genistein, diadzein and glycitein, which account for approximately 50, 40 and 10 percent, respectively, of the total isoflavone content of soybeans (Soni et al., 2014; Chatterjee et al., 2018). Limited research exists regarding the bioavailability and pharmacodynamics of isoflavones, but it is achieved that gender and form of the soy product has the potential to influence those processes. Peak concentrations of daidzein after soy consumption have higher levels in women, while in SB, isoflavones are absorbed faster and peak levels are attained earlier than with other soy products (Soni et al., 2014).

Isoflavones are often referred to as phytoestrogens because they structurally resemble 17β-estradiol and can affect estrogen-mediated processes through binding to both estrogen receptors (ER) α and β (ERα/β), although their affinity for Erβ is higher (Lee et al., 2005; Chatterjee et al., 2018). ERα is expressed in various tissues and organs, and, ERβ have been found to be expressed in bone, cardiovascular system, and in the brain (Lee et al., 2005). Genistein is the most potent isoflavone, whereas daidzein and glycitein have affinities to ERs approximately 100–500 times lower than genistein (Soni et al., 2014). The levels of genistein, diadzein and glycitein vary considerably across soy food products (Lee et al., 2005). A study in which SB was subjected to various heat treatments at 95, 121 and 140°C for various lengths of time, showed that heating may cause an increase or decrease in the genistein content of SB which depends on the temperature and time used. Even so, genistein showed greater stability to heat treatment than daidzein and glycitein, with both daidzein and glycitein contents decreasing rapidly during the early stage of heating, despite on continued heating the rates of decrease were much slower (Huang et al., 2006). An interesting study with postmenopausal women, aimed to compare the pharmacokinetics of genistein and daidzein after ingestion of SB with the ingestion soy extract capsules; the results showed that the bioavailability of daidzein following a single oral administration of SB was significantly less than that of soy extract capsules, whereas the bioavailability of genistein from both soy preparations were comparable (Anupongsanugool et al., 2005).

The estrogenic activity of isoflavones raise some concern regarding their potential to act as endocrine disruptors, and such is truth in the case of infants and small children because their hormone-sensitive organs, including the brain and reproductive system, are still undergoing sexual differentiation and maturation; their susceptibility to the endocrine-disrupting activities of soy phytoestrogens may be especially high (Patisaul, 2017). However, whether soy phytoestrogens are beneficial or harmful to human health is neither straightforward nor universally applicable to all groups (Patisaul, 2017). In male rats, continuous exposure to low and high doses of isoflavones is described as promoting hormonal imbalance, resulting in a decrease in circulating and testicular androgen and an increase in estrogen levels, associated reduced sperm quality parameters and testicular weight (Caceres et al., 2023). Contradictorily a review from Messina et al. (2022) concludes that in adult humans, soy isoflavone intake do not negatively affect testosterone or estrogen levels, sperm or semen parameters in men, as well as thyroid function, breast, endometrial tissue or estrogen levels in women, in which although the menstrual cycle length may be slightly increased, ovulation is not prevented (Messina et al., 2022). It is described that when endogenous levels of circulating estrogen are high (e.g., during the ovulatory menstrual cycle phase in pre-menopausal women) phytoestrogens can act as ER antagonists by competing with the activity of natural estrogens, but with a lower potency; when circulating estrogen is low (e.g., in men or post-menopausal women), phytoestrogens can operate as ER agonists (Soni et al., 2014). The most effective way of relieving menopausal syndrome is hormone replacement therapy. However, such treatment has also been associated with increased risks for coronary heart disease, stroke and breast cancer. As an alternative, the estrogenic activity of isoflavones affords efficacy in post-menopausal women, with epidemiological data showing a connection between the consumption of soy and reduced vasomotor symptoms associated with menopause, including hot flush, sweating and palpitations (Chen and Chen, 2021).

Isoflavones has also beneficial impacts on the risk factors for cardiovascular disease by lowering liver or blood triglycerides, total and LDL cholesterol levels, and by increasing the HDL cholesterol and the HDL/LDL cholesterol ratio (Jones et al., 2008). Other studies suggested that isoflavones may be protective against age-related diseases such as osteoporosis. Furthermore, estrogen receptor-independent actions of isoflavones include antioxidant action, inhibition of DNA topoisomerases, inhibition of cell cycle progression and inhibition of angiogenesis and tumor invasiveness (Wong et al., 2008). Other benefic compounds found in soy are saponins, due to their potential anti-colitic, anti-inflammatory, antimicrobial, hepato- and cardio-protective, estrogen-like and anti-carcinogenic effects. Soyasapogenol B, a metabolite of soyasaponin I, has been reported to inhibit the proliferation of human breast cancer cells (Hong et al., 2013; Chatterjee et al., 2018).

With respect to soy proteins, they can be partially hydrolyzed enzymatically to physiologically active soy peptides (Kim et al., 2021). Several soybean peptides that have been discovered over the last decade, some of them with antihypertensive, antithrombotic, hypocholesterolemic, anti-diabetic, immunostimulating, anti-inflammatory, chemopreventive, antimicrobial, opiate-like, mineral-binding and antioxidative properties, due to their structure, composition, and amino acid sequence (Cabanos et al., 2021; Kim et al., 2021).

11. Neuroprotection

11.1 Isoflavones

Lately, there is a growing interest in the physiological functions of soy isoflavones, especially in whether they affect positively cognitive functions and have beneficial effects on neurodegenerative diseases (Lee et al., 2005). In several studies it was observed that soy isoflavones can improve cognitive function in humans (Cui et al., 2020). However, while soy isoflavones appear to improve cognitive function in postmenopausal women, very few works have involved men; one study demonstrated that soy isoflavones appear to be harmful to the cognitive function in middle-aged men. In another study (Lee et al., 2005), participants received a daily intake of 100 mg of soy isoflavones, a dosage considerably higher than that typically consumed in Western diets. Over the course of the intervention, both short-term and long-term memory performance demonstrated statistically significant improvements. In addition to memory, executive functions, particularly mental flexibility, the ability to adapt cognitive strategies in response to changing stimuli or demands were also enhanced. Importantly, these cognitive benefits were observed in both male and female participants, suggesting a broad, non-sex-specific neurocognitive effect of isoflavone intake. The findings support the hypothesis that soy isoflavones, which exhibit phytoestrogenic and antioxidant properties, may modulate cognitive function through mechanisms such as estrogen receptor-mediated neuroprotection, enhanced cerebral blood flow, or improved cholinergic signaling in brain regions responsible for memory and executive processing. These results add to a growing body of evidence that dietary flavonoids can positively influence brain function even in healthy, non-elderly populations.

In recent years, the interest in soybean as a neuroprotective nutrient in the management of Alzheimer disease (AD) has increased, with clinical studies demonstrating that soy isoflavones can have beneficial effects on the cognition of older adults, especially older adults with AD (Lu et al., 2018). While the underlying mechanisms of improved brain function have yet to be fully characterized, the administration of soy isoflavones in mice with chronic ethanol-induced dementia significantly improved their cognitive performance, regulated the acetylcholinesterase (AChE) activity and acetylcholine (Ach) levels, elevated synaptic plasticity-related protein expressions and inhibited neuron apoptosis-related protein expressions in the cortex and in the hippocampus (Lu et al., 2020). Being achieved the estrogenic effect of isoflavones, it is also reported that they may improve the cognitive functions by mimicking the effects of estrogen in the brain; in fact, the mammalian 17β-estradiol has been shown to protect neurons and positively affect the cognitive function. Some studies suggest that estrogen affects neuron survival and growth, synaptic plasticity and brain function via an ER-mediated pathway and/or antioxidant properties. The effects of estrogen on ERs can also modulate brain functions by regulating gene transcription and second messenger systems (Lee et al., 2005).

11.2 Genistein

Genistein can pass through the blood–brain barrier (BBB) and induce neuroprotective effects through antioxidant, anti-inflammation, and estrogenic properties (Li et al., 2022). It is broadly known that the amyloid β protein (Aβ) is the main component of neuritic plaques in AD, produced from the amyloid precursor protein (APP) by sequential proteolytic cleavages by β-secretase and γ-secretase; since its accumulation is the molecular driver of ADs pathogenesis and progression, Aβ is currently the prime target for the development of AD therapy (Hur, 2022; Zhang et al., 2023). Genistein might be a promising isoflavone in the management of AD, since has been shown to reduce neuronal loss, ameliorate Aβ-induced cell death *in vivo*, and even downregulate Aβ generation possibly by serving as an inhibitor of γ-secretase (Soni et al., 2014; Li et al., 2022). Furthermore, while the initial neurologic change in AD patients is the activation of microglia around the Aβ deposition site, thereby increasing proinflammatory cytokine levels, genistein possess (as mentioned) the capability of mediating anti-inflammatory effects. One of the mechanisms is the inhibition of inflammatory factor levels, with proven decrease of proinflammatory chemokines, monocyte chemoattractant protein-1, which is a mediator that recruit monocytes in various inflammation models and diseases. It also induces an interleukin-6 mRNA decrease, being interleukin 6 produced in response to tissue injuries, having its dysregulated continual synthesis a pathological effect on chronic inflammation (Tanaka et al., 2014; Hosaka et al., 2017; Duan et al., 2021).

Parkinson's disease (PD) is the second most common degenerative disorder of the central nervous system, and it has been suggested that genistein may have therapeutic or preventive value on PD due to their anti-inflammatory properties. Antioxidant capabilities through ROS-scavenging via redox-sensitive signaling pathways is another cited mechanism (Uddin and Kabir, 2019; Li et al., 2022). In agreement, it was already observed a significant dose-dependent decrease in oxidative stress and an increase in dopamine content in genistein-exposed PD flies, more concretely transgenic *Drosophila* flies with key PD features such as age-dependent loss of dopaminergic neurons (Li et al., 2022).

Since most neurodegenerative diseases share some common etiologies such as oxidative stress, and neuroinflammation, beneficial effects of genistein are also described by Li et al. (2022) on Huntington's disease, spinal and bulbar muscular atrophy, multiple sclerosis, and amyotrophic lateral sclerosis.

11.3 Daidzein

In the same way as genistein, daidzein can pass through the BBB exerting antioxidant, anti-inflammation, and estrogenic effects (Soni et al., 2014; Wang et al., 2022). *In vitro* study (Liu et al., 2009) have demonstrated that daidzein, a major soy isoflavone, exerts notable anti-inflammatory effects, which may contribute to reducing the risk or slowing the progression of Alzheimer's disease (AD). Specifically, daidzein has been shown to down-regulate pro-inflammatory cytokines such as tumor necrosis factor-alpha (TNF-α), interleukin-1 beta (IL-1β), and interleukin-6 (IL-6) in microglial and astrocytic cultures activated by lipopolysaccharide (LPS) or amyloid-beta (Aβ) peptides. These cytokines are key mediators of neuroinflammation, a central feature of AD pathogenesis. Daidzein also appears to inhibit activation of the NF-κB signaling pathway, which plays a critical role in the transcription of inflammatory genes. Furthermore, it has been shown to reduce inducible nitric oxide synthase (iNOS) **and** cyclooxygenase-2 (COX-2) expression, thereby limiting the production of neurotoxic reactive nitrogen and oxygen species. Through these mechanisms, daidzein not only curbs inflammation but may also help preserve synaptic integrity and neuronal viability in models of neurodegeneration. Although these effects have yet to be fully validated *in vivo* or in clinical settings, the mechanistic data suggest that daidzein's anti-inflammatory actions could offer neuroprotective benefits and potentially mitigate the inflammatory cascades that drive AD progression. Moreover, in the same way as genistein, diadzein was already associated with a reduction in tau protein phosphorylation associated with neurofibrillary tangle formation (Soni et al., 2014).

In vivo studies demonstrated that daidzein partly improves the symptoms of PD by restoring dopamine levels and regulating antioxidant systems, with reduced loss of dopaminergic neurons and improvement of movement disorders (Wang et al., 2022). Daidzein was shown to confer neuroprotective effects in Parkinson's disease (PD) models by modulating both dopaminergic signaling and oxidative stress responses. Specifically, daidzein treatment was associated with a partial restoration of striatal dopamine levels, which are typically depleted due to the progressive degeneration of dopaminergic neurons in the substantia nigra pars compacta. This preservation of dopamine was accompanied by a reduction in the loss of tyrosine hydroxylase-positive neurons, indicating a protective effect on the nigrostriatal pathway. In parallel, daidzein significantly improved motor deficits, as assessed by behavioral assays such as the rotarod and open-field tests, reflecting a functional recovery of motor coordination and balance. Mechanistically, these benefits were linked to daidzein's ability to enhance endogenous antioxidant defenses, including upregulation of superoxide dismutase (SOD), glutathione peroxidase (GPx), and catalase (CAT), thereby reducing lipid peroxidation and reactive oxygen species (ROS) in the brain. Collectively, these findings suggest that daidzein may ameliorate PD-related pathology by targeting both dopaminergic neurodegeneration and oxidative damage, offering promise as a complementary or preventive therapeutic strategy (Wang et al., 2022). Additionally, genetic studies have advocated the role of improper protein phosphorylation caused by a dysfunction in kinases from the mitogen-activated protein (MAP) kinase family, as an important player in the progression and pathogenesis of PD. The MAP kinases, which is responsible for intracellular signaling, comprises four main groups: the extracellular signal-regulated kinase 1 and 2 (ERK1/2), the ERK5, the c-Jun N-terminal kinase (JNKs) and the p38 mitogen activated protein kinase (p38) (Bohush et al., 2018). The dysregulation of ERK1/2, JNK, and p38 is associated with various stages of inflammatory processes, contributing to age-related neurodegenerative diseases (Ahmed et al., 2020). Hence, controlling inflammation through the regulation of MAP kinase could be a worthwhile approach against neurodegenerative diseases and 6,7,4′-trihydroxyisoflavone, a major metabolite of daidzein, can reduce *in vitro* the phosphorylation of JNKs, p38 and ERK1/2; these results are suggestive that this isoflavone might be an attractive choice for the prevention of PD (Ahmed et al., 2020; Wang et al., 2022).

11.4 Glycitein

Glycitein can only be obtained via soy in the diet and is the least studied of the three mentioned estrogenic aglycones (Soni et al., 2014). Still, glycitein (as well as genistein and daidzein) is known to possess significant neuroprotective activity in humans, via suppression of the activities of monoamine oxidases, β-secretase, AChE and butyrylcholinesterase (BChE), which dysregulation is involved in the neurodegenerative process of AD and PD. Glycitein might be regarded a beneficial phytochemical for alleviating depression and dementia associated with neurodegenerative diseases (Prajapati et al., 2021). It is also described that glycitein have the ability to trigger increased activity of ROS scavenging enzymes, like superoxide dismutase (SOD) and catalase (CAT), being these properties pointed by some authors as potentially useful in the management of neurodegenerative diseases, such as PD (Dong and Yang, 2022).

11.5 Saponins and soy peptides

Two major saponins from soybeans were already isolated to investigate their learning- and memory-enhancing effects. Memory impairment was induced in rats using intracerebroventricular injection of amyloid-beta ($A\beta_{1-42}$) to mimic Alzheimer's disease-like pathology, and behavioral performance was assessed using the Morris water maze and passive avoidance tests. Rats were then administered soyasaponin I orally over a set period to evaluate **its therapeutic efficacy.** Such events were proposed to occur via the promotion of proliferation and differentiation of neural precursor cells in the hippocampus facilitating neuronal regeneration, and by minimizing neuroinflammation (Hong et al., 2013).

As regards to soy peptides, a study involving subjects with mild cognitive impairment which received soy peptide supplementation demonstrated that they significantly improved the delayed memory score and tended to improve immediate memory, compared with placebo; in addition, the soy peptide intake significantly increased serum dopamine, known to be associated with memory function (Maebuchi et al., 2013). *In vivo* experiments support these findings through the proposal that some soy peptides improve long-term memory and neurogenesis by activating cAMP response element-binding protein (CREB)-mediated downstream proteins as a molecular memory pathway. These peptides have also been reported as having positive effects on brain wave and neurotransmitter control, in patients with mild cognitive impairment; their role in preventing aging-related cognitive decline and neurodegenerative diseases, such as AD are also referred (Kim et al., 2021).

12. Neurotoxicity

The risks associated with the consumption of soy and their products, including SB, are mainly due to the presence of Manganese (Mn), mycotoxins and human interventions during cultivation; soybean crops are often treated with large amounts of pesticides, and pollution of soils, namely with metals can result in high concentrations of potentially toxic elements in food (Gil-Serna et al., 2020; Rubio et al., 2021). However, it should be mentioned in brief that despite the previously described neuroprotective properties genistein and daidzein, when these compounds are present in high dose, they can have toxic effects on neurons *in vitro*, possibly through interference with GABA (Jin et al., 2007).

12.1 Manganese

Despite Mn is a metal, it will be considered here in the context of its natural occurrence in soy, and not as a contaminant; effectively, this essential mineral nutrient is found at high levels in plants, such as soy (Cockell et al., 2004). As regards to SB, high levels of Mn are reported in soy baby formulas, leading to elevated serum levels in the consumers (Mitchell et al., 2020). In another work the calculated mean Mn intakes from SB for children with 1–3 years old, when assuming complete substitution of cow milk by these products (0.78 L/day), approached the Tolerable Upper Intake Level (UL) for this age range; no UL for Mn is available for infants under 1 year of age (Cockell et al., 2004).

Excessive Mn exposure increases the risk of adverse neurological effects, being demonstrated that mitochondrial impairment, oxidative stress, inflammation and excitotoxicity are involved in Mn-induced neurotoxicity (Cockell et al., 2004; Dobson et al., 2004; Pajarillo et al., 2021). In particular, the accumulation of Mn in the brain may cause a neurodegenerative disorder known as manganism, which has Parkinson-like symptoms. The similarities between the two disorders can be partially explained by the fact that the basal ganglia accumulate most of the excess Mn compared with other brain regions, and dysfunction in the basal ganglia is also the aetiology of PD (Dobson et al., 2004). Mn accumulates in several other brain regions, apart from the basal ganglia, such as the frontal cortex, hippocampus and cerebellum. Exposure to Mn may also be a risk factor for AD, with relatively recent studies beginning to show links between brain Mn concentration and the development of AD or AD-like symptoms; increases in plasma Aβ peptides were showed to be associated with increased concentrations of Mn. Additionally, overexposure to Mn increase the risk of the abnormal neurodevelopment, abnormal behavior, cognitive decline and learning deficits in infants and children (Martins et al., 2019; Pajarillo et al., 2021).

12.2 Mycotoxins

Mycotoxins are fungal secondary metabolites that are a considerable threat to food safety, since are toxic to humans. More than 300 mycotoxins are identified, although ochratoxin A (OTA), aflatoxins (AFT), trichothecenes (T2), zearalenone (ZEN) and fumonisins (FUM) are considered among the most important due to their high toxicity and frequent occurrence in food products (Gil-Serna et al., 2020). The complete eradication of mycotoxins is difficult as they are formed both pre- and post-harvest under natural environmental conditions (Niyibituronsa et al., 2018). Soy is susceptible to the growth of moulds that produce mycotoxins and in accordance, the

Dsm-Firmenich World Mycotoxin Survey performed from January to March, in 2024, have indicated the presence of all the mentioned mycotoxins in soybean meal; in SB, T-2 have been detected (Niyibituronsa et al., 2018; Azam et al., 2021; Dsm, 2024). The occurrence of emerging mycotoxins, such as enniatins and beauvericin in plant-based milks, was also evaluated by leading to found that 12.5 percent of SB, were contaminated by at least one of these mycotoxins (Gil-Serna et al., 2020). Quite concerning is the fact that the European Commission has not yet set guidelines for the maximum mycotoxin concentrations in plant-based beverages, nor has the European Food Safety Authority conducted a risk assessment (Pavlenko et al., 2024). According to the authors of the mentioned Survey, monitoring mycotoxins in all soybean meals every year is always advisable (Dsm, 2024).

Pertaining to the neurotoxicity of mycotoxins, several of them can cross the BBB and accumulate in the brain disrupting its normal function (Pei et al., 2021). Although research on OTA mediated neurotoxicity has increased greatly in recent years, detail knowledge about their induced neurotoxic effects in humans is currently inadequate. Nevertheless, it is already known that OTA, namely OTAA, has a high affinity with the brain where induces ROS generation, lipid peroxidation and apoptosis. Neuroinflammatory response associated with repeated exposure to OTA A is also demonstrated in three-dimensional whole-brain aggregate *in vitro* cultures (Obafemi et al., 2023). *In vivo*, OTA can cause male-specific autism by combining microRNA and X-chromosomes (Pei et al., 2021; Obafemi et al., 2023). The toxicity of OTA toxicity can lead to AD progression, with *in vivo* studies showing that upon OTA administration, moderately high levels of OTA can accumulate in the hippocampus, which is an essential site of neurodegeneration in AD. Among several proposed mechanisms for this neurodegeneration, an interesting one is the induction of diminished expression of the brain-derived neurotrophic factor (BDNF) in a dose-dependent way. Indeed, BDNF abnormal expression in the brain leads to leads to oxidative stress and apoptotic assault, being associated with AD and even with other disorders, which include schizophrenia, Huntington's disease and PD (Niaz et al., 2020). Indeed, the association of OTA A sub chronic exposure with PD is demonstrated by both *in vitro* and *in vivo* studies; OTA can trigger some of the key pathological features of this disease, such as the loss of striatal dopaminergic innervation and dopaminergic cell dysfunction, accompanied with motor impairments, and increased phosphorylated alpha-synuclein levels (Arce-López et al., 2021).

Among AFT, aflatoxin B1 (AFB1) is the most toxic congener of 21 identified AFTs, with *in vivo* studies indicating that AFB1 exposure can provoke anxiety, lethargy disorders, depression-like behavior, cognitive, learning and memory defects, and decreased feeding behavior. Mechanistically, AFB1 exposure has been associated with lipid peroxidation, ablation of non-enzymatic and enzymatic antioxidant defence systems and decreased neurotransmitter levels, as well as neuroinflammation; DNA damage, apoptosis and mitochondrial dysfunction in the brain tissue, is also demonstrated (Dai et al., 2024; Wang et al., 2023). This mycotoxin can additionally indirectly affect *in vivo* neurons, by disturbing the function of glial cells, and potentially by activating immune cells in the bloodstream, causing neurodevelopmental and function abnormalities (Pei et al., 2021). After observing a growing trend of neuronal deterioration *in vivo* with continuous increasing time of AFT administration, few authors considered that chronic exposure to aflatoxin or dose accumulation could lead to neurodegeneration events like Huntington disease or AD (Alsayyah et al., 2019). Additionally, the observation that oral exposure to AFB1 trigger α-synuclein pathology and cause dopaminergic neurotoxicity, leaded also to propose a contributing role of AFB1 exposure to PD aetiology (Wang et al., 2023).

During the last decade, the neurotoxicity of T2 toxin has been a major concern. This toxin crosses the BBB by altering permeability and induces oxidative stress responses, including ROS generation, lipid peroxidation, and protein carbonyl formation. Exposure to T2 can interfere with certain signal transduction pathways, such as the inhibition of Nrf2/hemeoxygenase-1 (HO-1) and the activation of the p53 pathway, which leads to oxidative stress and apoptosis; changes in brain dopamine levels and decreases in ganglion area are also reported (Zhang et al., 2020). It has been reported that the concentration of T2 in the serum of subjects with a history of mycotoxin exposure was significantly increased, concomitantly with hindered neurodevelopment, and weakened exercise balance, visual function, reaction speed, cognitive ability, and emotional management ability; a decrease in frontal cortex activity was also detected by quantitative electroencephalogram (Pei et al., 2021). The T2 toxins exert also mitochondrial dysfunction, being of notice that mitochondrial dysfunction dependent diseases include neurobehavioral and psychiatric diseases, such as autism spectrum disorders, schizophrenia, and bipolar and mood disorders; important changes occur also in mitochondrial dynamics in neurodegenerative diseases including Huntington's, Amyotrophic Lateral Sclerosis, and in AD and PD (Islam et al., 2018; Alqahtani et al., 2023). In the case of AD, Aß and tau problems both have effects on mitochondria, which leads to mitochondrial malfunction and, ultimately, the onset of the disease, while in PD disease, α-synuclein aggregation, and inflammation, results from reduced brain mitochondria activity (Alqahtani et al., 2023).

The available information on the neurotoxic effects of ZEN is very much limited (Venkataramana et al., 2014). However, being achieved that even at low doses, ZEN causes estrogenic effects, negative effects of this mycotoxin on the nervous system are reported, and these effects are related to the presence of estrogen receptors in the brain (Makowska et al., 2017). The underlying mechanisms behind sex hormonal modulation of neural circuits are not well understood, but sex differences are apparent in AD and PD, with so far, ERs ascribed has having important roles in the risk and progression of these neurodegenerative diseases (Maioli et al., 2021). To investigate the neurotoxic effects of zearalenone (ZEN), differentiated SH-SY5Y human neuroblastoma cells were treated with varying concentrations of ZEN. Following exposure, several biomarkers of cellular stress and damage were assessed. Changes in gene expression related to oxidative stress pathways were quantified, alongside measurements of reactive oxygen species (ROS) formation and malondialdehyde (MDA) levels, a marker of lipid peroxidation. Mitochondrial integrity was evaluated through assessment of mitochondrial membrane potential, and DNA damage was quantified using appropriate assays. All observed effects—including ROS generation, MDA elevation, mitochondrial dysfunction, and DNA damage—were found to be dose-dependent, consistent with findings reported by Pei et al. (2021) and Venkataramana et al. (2014).

Several studies on experimental animals or on cell cultures of neural origin have established that FUM has a neurodegenerative potential, although the mechanism of its neurotoxicity is still vague (Domijan, 2012). However, it should be mentioned that FUM B1 can inhibit sphingolipid biosynthesis in cultured cerebellar neurons, affecting the formation of sphingomyelin; actually, sphingolipids are especially abundant in central nervous system compared with other parts of the organism, where play different signaling roles (Merrill et al., 1993; Mei et al., 2023).

12.3 Pesticides

It is estimated that >899 million litters of pesticides has been sprayed in soybean, corn and sugar cane crops since 2015. Within the pesticides used in soybean crops, the most frequently applied are glyphosate, fipronil, and imidacloprid. Of particular concern, is the demonstration *in vitro* that this mixture of pesticides elicits significant synergistic effects regarding cytotoxicity, which were greater than the expected by both addictive predictions, as ahead will be described (Conte et al., 2022).

12.4 Glyphosate

Glyphosate is a controversial herbicide due to concerns regarding public health risks. Hence, there is an intense debate, namely in Europe, around glyphosate, with a vote in 2023 on renewing its marketing authorization (Bloem and Boonstra, 2023). Studies on glyphosate residues in food are scarce, existing a study performed in USA showing that glyphosate residues above the limit of quantification were not found in SB (Guo, 2014). Most of the studies on nervous systems effects used concentrations of glyphosate-based herbicides or glyphosate that are much higher than the ones to which most humans are likely to be exposed (Madani and Carpenter, 2022). Even so, chronic exposure to glyphosate scenarios should not be relaxed since several scientific works demonstrated that glyphosate can cross and affect the BBB and cause various types of short-term or long-term disturbances in the human nervous system. Exposure to glyphosate appears to affect neuronal development, altering the expression of molecules involved in the growth and maturation of neurons, induce oxidative stress, potentiates inflammatory responses by upregulating the expression of the proinflammatory cytokine IL-6 genes, interferes with neurotransmission, namely cholinergic and dopaminergic transmission, and with the glutamatergic system. In fact, the glutamatergic system is the one of most affected, exhibiting neuronal death due to autophagy, necrosis, or apoptosis, with the appearance of behavioral and motor disorders (Faria et al., 2021; Costas-Ferreira et al., 2022; de Castro Vieira Carneiro et al., 2023; Costas-Ferreira et al., 2024). A study in a rural area leaded to suggest a possible relationship between exposure to glyphosate during prenatal and childhood stages and an increased risk of developing autism (Costas-Ferreira et al., 2022). In a publication by Winstone et al. (2022). It is mentioned that the recent rise in glyphosate application to soy crops correlates positively with increased death rates due to AD and other neurodegenerative disorders (Winstone et al., 2022). There is also evidence that exposure to glyphosate is a risk factor for PD, since individuals who attempted suicide by glyphosate-based herbicides ingestion, but survived to this acute exposure, developed PD several months later (Madani and Carpenter, 2022).

12.5 Fipronil

Fipronil is an insecticide of the phenylpyrazoles class, for which very little information exist regarding it neurotoxicity. However, it is demonstrated that fipronil elicits neurotoxicity in mammals by inhibition of $GABA_A$-gated chloride channels, producing hyperexcitability of the central nervous system (Gupta and Anadón, 2018). In turn, while the status of fipronil as a potential developmental neurotoxicant is still under debate, as several *in vivo* and *in vitro* studies suggests it, other *in vitro* studies do not confirm this concern (Schmitz et al., 2021).

12.6 Imidacloprid

Among the neonicotinoid's insecticides, imidacloprid is the most used in soy crops. In general, neonicotinoids are antagonists of the nicotinic acetylcholine receptor (nAChR), and therefore they cause constant nerve stimulation, generating lethal effects in insects (Conte et al., 2022). Although neonicotinoids are thought to have minimal mammalian toxicity, several studies suggested that prolonged imidacloprid exposure can induce various health problems in humans (Hassanen et al., 2023). A potent neurotoxic effect of imidacloprid is reported *in vivo* by neurobehavioral and neuropathological alterations and a reduction of both acetylcholine esterase activity, and dopamine, serotonin, and γ-aminobutyric acid levels in the brain. In this organ it was also observed elevated levels of MDA and reduced antioxidant capacity, as well as upregulation of the transcription levels of nuclear factor-κB, interleukin-1 β, and tumor necrosis factor genes, as well as strong caspase-3 and inducible nitric oxide synthase expression (Hassanen et al., 2023). Imidacloprid can affect mammals brain function during development, affecting proliferation and neural formation, and memory and cognitive functions (Kara et al., 2015; Conte et al., 2022).

12.7 Pesticide mixtures

As mentioned, in soybean crops it is common to use glyphosate, fipronil, and imidacloprid in combination. A study performed to evaluate the individual and the mixture effects of fipronil, glyphosate and imidacloprid in human HepG2 cells, showed that this pesticides mixture elicited significant synergistic effects. Decreased mitochondrial membrane potential were observed, justifying the need for a more realistic approach in safety evaluations to better predict the effects to human health (Conte et al., 2022). Given such information, studies on the neurotoxicity of this mixture are deemed necessary.

12.8 Metals

Numerous human activities including mining, smelting, agriculture, waste disposal, and traffic and transportation have resulted in the release of significant quantities of metals to the environment. Soy belongs to a group of crops that strongly accumulate metals, and these include, neurotoxic metals, such as lead (Pb), cadmium (Cd) and copper (Cu) (Zhuang et al., 2013). In addition, soy consumption can also lead indirectly to metal toxicity because soy foods contribute to low levels of cysteine, zinc, pantothenic acid and other factors needed for detoxification. This can result in a buildup of mercury, cadmium, arsenic, aluminum and other toxic metals in the body, ultimately affecting every metabolic process (Daniel, 2009).

12.9 Lead

Lead primarily affects the central nervous system, particularly the developing brain, making children to be at a greater risk than adults. The ability of Pb to pass through the BBB is due in large part, to its ability to substitute for calcium ions; once in the brain, Pb induce damage in the prefrontal cerebral cortex, hippocampus, and cerebellum, may leading to a variety of neurological disorders, such as mental retardation, behavioral problems, nerve damage, and possibly AD, PD, and schizophrenia (Polido et al., 2009; Sanders et al., 2009; Bakulski et al., 2012; Rubio et al., 2021).

12.10 Cadmium

Once ingested, Cd exerts toxic effects that pose a significant threat to human health, being the nervous system particularly vulnerable to prolonged, low-dose Cd exposure. Within the nervous system, Cd disrupts mitochondrial respiration by decreasing ATP synthesis and increasing the production of ROS. The metal can also impair the BBB, neurotransmission, and change the regulation of glycogen metabolism, and such cumulative nervous system damage can increase the risk for neurological and neurodegenerative disorders (Cheng et al., 2021; Arruebarrena et al., 2023; Ojo et al., 2023). A Cd-dependent neurotoxicity has been related to neurodegenerative diseases such as AD and PD, amyotrophic lateral sclerosis, and multiple sclerosis (Branca et al., 2018).

12.11 Copper

Practically any data is available regarding Cu levels in SB, but it is known that although women are a hundred times more likely to become copper toxic than men, anyone who uses soy as meat or dairy replacements or who adheres to a vegetarian diet is at risk for copper toxicity (Daniel, 2009). Excessive Cu intake cause Cu accumulation in the brain, leading to apoptosis or cuproptosis in neuronal cells, while neuronal loss leads to memory deficits and cognitive impairment. On the other hand, ROS production, which attack biological macromolecules, leading to oxidative damage to the proteome, lipids, and DNA, inducing apoptosis, exacerbates neuronal toxicity and dysfunction; elevated glutamate, malondialdehyde, and caspase-3 in the frontal cortex and hippocampus of Cu-exposed mice is also described (Duan et al., 2023).

13. Strategies to decay contaminants in soy

Most of the available work has been performed in the reduction of microbiological risk and indeed decrease of metal levels but under a perspective o heavy metal stress induced in the plants rather than public health concerns. Therefore, described approaches are focused on heavy metal tolerance mediated by the plants (Shaffique et al., 2023). Even so, a work developed in China mention the production of low Cd crop varieties through genotypic selection as a means to reduce the risk of Cd migration into human diet, since physical, chemical and biological methods for remediation of heavy metal contaminated soil, have been shown to be insufficient in effectiveness, sustainability and economy (Zhi et al., 2020), Another work mention the need of reducing the chlorpyrifos content essential for food safety, with methods such as gamma irradiation, pulsed electric field, high-pressure processing and ultrasound pointed as some of the effective non-thermal treatments used in pesticide degradation. However, operational safety (irradiation), limited application to solids foods (PEF) and freshness modification (HPP & US) are some of the constraints associated with these technologies (Anbarasan et al., 2022). Regarding mycotoxins prevention methods, including pre-harvest field management and post-harvest storage management, are still the most effective strategies, since mycotoxins are hardly to be eliminated once they are present in the ingredients. Methodologies involving physical, thermal, chemical methods, and mycotoxin controlling feed additives so far, show evident mycotoxin reducing effects, but of different consistencies and limited by various practical conditions such as costs of high grain mass loss (Peng et al., 2018).

14. Conclusions

Nowadays, cow milk and soy beverages are the two main nutritive beverages ingested by humans at birth, during childhood, and for a lifetime, and they are also important ingredients in many processed foods including baby foods (Vitoria, 2017; Andrade et al., 2023; Bakshi, 2023). Therefore, the consumption of dairy products together with other contaminated food may expose consumers to unexpected levels of chemicals with emphasis on some pesticides and metals (Waliszewski et al., 2003; Armendáriz et al., 2004).

The effects on human health, particularly children to the exposure of these contaminants in cow milk or soy beverages, even below the permissible levels, are unpredictable because human beings are not only exposed to these food contaminants; they are also exposed to contaminants from other foods, exposed to chemicals in the environment, medicines, as well as chemicals at homes. Many of these chemicals are neurotoxic and their final consequences of the exposure to these mixtures may be very harmful mainly after chronic exposure.

Nevertheless, due to the toxicity and neurotoxicity and bioaccumulation of some of these chemical contaminants, the fact that these foods are routinely ingested by all ages, mainly children, and represent key ingredients in many processed foods, including baby foods, we suggest strict surveying of these chemical residue levels in cow milk and soy beverages.

Michael Aschner was supported in part by a grant from the National Institute of Environmental Health Sciences (NIEHS) R01ES10563.

References

Abd-Rabo, F., Elsalamony, H., Sakr, S., 2016. Reduction of pesticide residues in Egyptian buffalo milk by some processing treatments. Int J. Dairy Sci. 11, 75–80.

Ahmed, T., Zulfiqar, A., Arguelles, S., Rasekhian, M., Nabavi, S.F., Silva, A.S., Nabavi, S.M., 2020. Map kinase signaling as therapeutic target for neurodegeneration. Pharmacol. Res. 160, 105090.

Ajiboye, T.O., Oladoye, P.O., Olanrewaju, C.A., Akinsola, G.O., 2022. Organophosphorus pesticides: impacts, detection and removal strategies. Environ. Nanotechnol. Monit. Manag. 17, 100655.

Akhtar, S., Karam, A., 2017. Pesticides residue in milk and milk products: mini review. Pak. J. Anal. Environ. Chem. 18 (1), 37–45.

Alqahtani, T., Deore, S.L., Kide, A.A., Shende, B.A., Sharma, R., Dadarao Chakole, R., Nemade, L.S., Kishor Kale, N., Borah, S., Shrikant Deokar, S., Behera, A., Dhawal Bhandari, D., Gaikwad, N., Kalam Azad, A., Ghosh, A., 2023. Mitochondrial dysfunction and oxidative stress in Alzheimer's disease, and Parkinson's disease, Huntington's disease and Amyotrophic Lateral Sclerosis—an updated review. Mitochondrion 71, 83–92.

Alsayyah, A., ElMazoudy, R., Al-Namshan, M., Al-Jafary, M., Alaqeel, N., 2019. Chronic neurodegeneration by aflatoxin B1 depends on alterations of brain enzyme activity and immunoexpression of astrocyte in male rats. Ecotoxicol. Environ. Saf. 182, 109407.

Anbarasan, R., Jaspin, S., Bhavadharini, B., Pare, A., Pandiselvam, R., Mahendran, R., 2022. Chlorpyrifos pesticide reduction in soybean using cold plasma and ozone treatments. LWT 159, 113193.

Andrade, V.L., Ribeiro, I., Dos Santos, A.P.M., Aschner, M., Mateus, M.L., 2023. Metals in cow milk and soy beverages: is there a concern? Toxics 11 (12), 1013.

Anupongsanugool, E., Teekachunhatean, S., Rojanasthien, N., Pongsatha, S., Sangdee, C., 2005. Pharmacokinetics of isoflavones, daidzein and genistein, after ingestion of soy beverage compared with soy extract capsules in postmenopausal Thai women. BMC Clin. Pharmacol. 5, 2.

Apata, D., 2009. Antibiotic resistance in poultry. Int. J. Poult. Sci. 8, 404–408.

Arce-López, B., Alvarez-Erviti, L., De Santis, B., Izco, M., López-Calvo, S., Marzo-Sola, M.E., Debegnach, F., Lizarraga, E., de Cerain, A.L., González-Peñas, E., Vettorazzi, A., 2021. Biomonitoring of mycotoxins in plasma of patients with Alzheimer's and Parkinson's disease. Toxins 13 (7), 477.

Armendáriz, C., Pérez de Ciriza, J.A., Farré, R., 2004. Gas chromatographic determination of organochlorine pesticides in cow milk. Int. J. Food Sci. Nutr. 55 (3), 215–221.

Arruebarrena, M.A., Hawe, C.T., Lee, Y.M., Branco, R.C., 2023. Mechanisms of cadmium neurotoxicity. Int. J. Mol. Sci. 24 (23), 16558.

Ayoub, M., Desoki, M., Hassanin, A., Thabet, M., Mansour, M., Loutfy, N.M., Raslan, A., 2012. Detection of pesticide residues in milk and some dairy products. J. Plant Prot. and Path. Mansoura Univ. 3 (8), 865–880.

Azam, M.S., Ahmed, S., Islam, M.N., Maitra, P., Islam, M.M., Yu, D., 2021. Critical assessment of mycotoxins in beverages and their control measures. Toxins 13 (5), 323.

Babapour, A., Azami, L., Fartashmehr, J., 2012. Overview of antibiotic residues in beef and mutton in Ardebil, North West of Iran. World Appl. Sci. J. 19 (10), 1417–1422.

Badger, T.M., Gilchrist, J.M., Pivik, R.T., Andres, A., Shankar, K., Chen, J.R., Ronis, M.J., 2009. The health implications of soy infant formula. Am. J. Clin. Nutr. 89 (5), 1668S–1672S.

Bakshi, S., Paswan, V.K., Yadav, S.P., Bhinchhar, B.K., Kharkwal, S., Rose, H., Kanetkar, P., Kumar, V., Al-Zamani, Z.A.S., Bunkar, D.S., 2023. A comprehensive review on infant formula: nutritional and functional constituents, recent trends in processing and its impact on infants' gut microbiota. Front. Nutr. 10, 1194679.

Bakulski, K.M., Rozek, L.S., Dolinoy, D.C., Paulson, H.L., Hu, H., Alzheimer, C., Author, R., 2012. Alzheimer's disease and environmental exposure to lead: the epidemiologic evidence and potential role of epigenetics. Curr. Alzheimer Res. 9 (5), 563–573.

Battu, R.S., Singh, B., Kang, B.K., 2004. Contamination of liquid milk and butter with pesticide residues in the Ludhiana district of Punjab state, India. Ecotoxicol. Environ. Saf. 59 (3), 324–331.

Bedi, J.S., Gill, J.P., Aulakh, R.S., Kaur, P., 2015. Pesticide residues in bovine milk in Punjab, India: spatial variation and risk assessment to human health. Arch. Environ. Contam. Toxicol. 69 (2), 230–240.

Belobrajdic, D.P., James-Martin, G., Jones, D., Tran, C.D., 2023. Soy and gastrointestinal health: a review. Nutrients 15 (8), 1959.

Beyene, T., 2016. Veterinary drug residues in food-animal products: its risk factors and potential effects on public health. J. Vet. Sci. Technol. 7 (1), 1–7.

Berr, C., 2000. Cognitive impairment and oxidative stress in the elderly: results of epidemiological studies. Biofactors 13, 205–209.

Blodgett, D.J., Means, C., 2013. Organophosphate and carbamate insecticides. In: Peterson, M.E., Talcott, P.A. (Eds.), Small Animal Toxicology. Saunders, Philadelphia, pp. 715–724.

Bloem, B.R., Boonstra, T.A., 2023. The inadequacy of current pesticide regulations for protecting brain health: the case of glyphosate and Parkinson's disease. Lancet Plan. Health 7 (12), e948–e949.

Bo, L.Y., Zhao, X.H., 2010. Preliminary study on the degradation of seven organophosphorus pesticides in bovine milk during lactic acid fermentation or heat treatment. Afr. J. Microbiol. Res. 4, 1171–1179.

Bohush, A., Niewiadomska, G., Filipek, A., 2018. Role of mitogen activated protein kinase signaling in Parkinson's disease. Int. J. Mol. Sci. 19 (10), 2973.

Bolles, H.G., Dixon-White, H.E., Peterson, R.K., Tomerlin, J.R., Day Jr., E.W., Oliver, G.R., 1999. U.S. market basket study to determine residues of the insecticide chlorpyrifos. J. Agr. Food Chem. 47, 1817–1822.

Boudebbouz, A., Boudalia, S., Boussadia, M.I., Gueroui, Y., Habila, S., Bousbia, A., Symeon, G.K., 2022. Pesticide residues levels in raw cow's milk and health risk assessment across the globe: a systematic review. Environ. Adv. 9, 100266.

Branca, J.J.V., Morucci, G., Pacini, A., 2018. Cadmium-induced neurotoxicity: still much ado. Neural Regen. Res. 13 (11), 1879–1882.

Cabanos, C., Matsuoka, Y., Maruyama, N., 2021. Soybean proteins/peptides: a review on their importance, biosynthesis, vacuolar sorting, and accumulation in seeds. Peptides 143, 170598.

Caceres, S., Crespo, B., Alonso-Diez, A., de Andrés, P.J., Millan, P., Silván, G., Illera, M.J., Illera, J.C., 2023. Long-term exposure to isoflavones alters the hormonal steroid homeostasis-impairing reproductive function in adult male wistar rats. Nutrients 15 (5), 1261.

Calahorrano-Moreno, M.B., Ordoñez-Bailon, J.J., Baquerizo-Crespo, R.J., Dueñas-Rivadeneira, A.A, B. S. M. Montenegro, M.C., Rodríguez-Díaz, J.M., 2022. Contaminants in the cow's milk we consume? Pasteurization and other technologies in the elimination of contaminants. F1000Res 11, 91.

Campillo, J.T., Boussinesq, M., Bertout, S., Faillie, J.L., Chesnais, C.B., 2021. Serious adverse reactions associated with ivermectin: a systematic pharmacovigilance study in sub-Saharan Africa and in the rest of the world. PLoS Negl. Trop. Dis. 15 (4), e0009354.

Cao, P., Zhao, Y., Wu, F., Xin, D., Liu, C., Wu, X., Lv, J., Chen, Q., Qi, Z., 2022. Multi-omics techniques for soybean molecular breeding. Int. J. Mol. Sci. 23 (9), 4994.

Cerniglia, C.E., Pineiro, S.A., Kotarski, S.F., 2016. An update discussion on the current assessment of the safety of veterinary antimicrobial drug residues in food with regard to their impact on the human intestinal microbiome. Drug Test. Anal. 8, 539–548.

Chandler, R.E., 2018. Serious neurological adverse events after ivermectin-do they occur beyond the indication of onchocerciasis? Am. J. Trop. Med. Hyg. 98 (2), 382–388.

Chatterjee, C., Gleddie, S., Xiao, C.W., 2018. Soybean bioactive peptides and their functional properties. Nutrients 10 (9), 1211.

Chen, L.R., Chen, K.H., 2021. Utilization of isoflavones in soybeans for women with menopausal syndrome: an overview. Int. J. Mol. Sci. 22 (6), 1–23.

Cheng, H., Yang, B., Ke, T., Li, S., Yang, X., Aschner, M., Chen, P., Carmona, A., Lein, P., 2021. Mechanisms of metal-induced mitochondrial dysfunction in neurological disorders. Toxics 9 (6), 142.

Cimmino, F., Catapano, A., Villano, I., Di Maio, G., Petrella, L., Traina, G., Pizzella, A., Tudisco, R., Cavaliere, G., 2023. Human, cow, and donkey milk comparison: focus on metabolic effects. J. Dairy Sci. 106 (5), 3072–3085.

Cockell, K.A., Bonacci, G., Belonje, B., 2004. Manganese content of soy or rice beverages is high in comparison to infant formulas. J. Am. Coll. Nutr. 23 (2), 124–130.

Conte, F.M., Cestonaro, L.V., Piton, Y.V., Guimarães, N., Garcia, S.C., Dias da Silva, D., Arbo, M.D., 2022. Toxicity of pesticides widely applied on soybean cultivation: synergistic effects of fipronil, glyphosate and imidacloprid in HepG2 cells. Toxicol. *In Vitro* 84, 105446.

Ćosić, M., Špirović-Trifunović, B., Petrovic, A., Tasić, S., Puvača, N., Đurić, S., Vuković, G., Konstantinović, B., Marinković, D., Buršić, V., 2021. Pesticide residues in cow's milk. Mljekarstvo 71 (2), 165–174.

Costa, L.G., 2006. Current issues in organophosphate toxicology. Clin. Chim. Acta 366, 1–13.

Costa, L., Giordano, G., Guizzetti, M., Vitalone, A., 2008. Neurotoxicity of pesticides: a brief review. Front. Biosci. 13, 1240–1249.

Costas-Ferreira, C., Durán, R., Faro, L.R.F., 2022. Toxic effects of glyphosate on the nervous system: a systematic review. Int. J. Mol. Sci. 23 (9), 4605.

Costas-Ferreira, C., Durán, R., Faro, L.R.F., 2024. Evaluation of the potential role of glutamatergic, cholinergic, and nitrergic systems in the dopamine release induced by the pesticide glyphosate in rat striatum. J. Appl. Toxicol. 44 (10), 1489–1503.

Cui, C., Birru, R.L., Snitz, B.E., Ihara, M., Kakuta, C., Lopresti, B.J., Aizenstein, H.J., Lopez, O.L., Mathis, C.A., Miyamoto, Y., Kuller, L.H., Sekikawa, A., 2020. Effects of soy isoflavones on cognitive function: a systematic review and meta-analysis of randomized controlled trials. Nutr. Rev. 78 (2), 134–144.

Dai, C., Tian, E., Li, H., Gupta, S.Das, Hao, Z., Wang, Z., Velkov, T., Shen, J., 2024. Molecular mechanisms of aflatoxin neurotoxicity and potential neuroprotective agents. Food Sci. Hum. Well. 13 (5), 2445–2455. doi:10.26599/FSHW.2022.9250201.

D'Alonzo, Z., Lam, V., Nesbit, M., Graneri, L., Takechi, R., Mamo, J.C.L., 2020. Chronic consumption of bovine dairy milk attenuates dietary saturated fatty acid-induced blood-brain barrier dysfunction. Front. Nutr. 7, 58.

Daniel, K. T., 2009. Wise traditions in food, farming and the healing arts, the quarterly journal of the Weston A. Price Foundation. https://www.westonaprice.org/health-topics/soy-alert/soy-recovery-the-toxic-metal-component/#gsc.tab=0 (Assessed 9 October 2024).

de Castro Vieira Carneiro, C.L., Chaves, E.M.C., Neves, K.R.T., Braga, M.D.M., Assreuy, A.M.S., de Moraes, M.E.A., Aragão, G.F., 2023. Behavioral and neuroinflammatory changes caused by glyphosate: base herbicide in mice offspring. Birth Defects Res. 115 (4), 488–497.

Deiana, P., Fatichenti, F., 1992. Pesticide residues in milk processing. Ital. J. Food Sci. 24, 229–245.

Deti, H., Hymete, A., Bekhit, A.A., Mohamed, A.M., Bekhit, A.D., 2014. Persistent organochlorine pesticides residues in cow and goat milks collected from different regions of Ethiopia. Chemosphere 106, 70–74.

Dobson, A.W., Erikson, K.M., Aschner, M., 2004. Manganese neurotoxicity. Ann. N. Y. Acad. Sci. 1012, 115–128.

Domijan, A.M., 2012. Fumonisin B(1): a neurotoxic mycotoxin. Arh. Hig. Rada Toksikol. 63 (4), 531–544.

Dong, N., Yang, Z., 2022. Glycitein exerts neuroprotective effects in Rotenone-triggered oxidative stress and apoptotic cell death in the cellular model of Parkinson's disease. Acta Biochim. Pol. 69 (2), 447–452.

dsm-firmenich. World Mycotoxin survey Assessed 23 September 2024.

Duan, F., Yu, Z., Wei, L., Xuechao, L., Jixiang, W., Fangjun, W., 2023. Copper neurotoxicity: induction of cognitive dysfunction: a review. Medicine 102 (48), e36375.

Duan, X., Li, Y., Xu, F., Ding, H., 2021. Study on the neuroprotective effects of Genistein on Alzheimer's disease. Brain Behav. 11 (5), e02100.

EFSA, 2019. The 2017 European Union report on pesticide residues in food. EFSA J. 17, 6.

Eker, F., Bolat, E., Pekdemir, B., Duman, H., Karav, S., 2023. Lactoferrin: neuroprotection against Parkinson's disease and secondary molecule for potential treatment. Front. Aging Neurosci. 15, 1204149.

Erukainure, O.L., Salau, V.F., Alabi, O.O., Ebuehi, O.A.T., Koorbanally, N.A., Islam, M.S., 2020. Casein micelles from bovine milk exerts Neuroprotection by stalling metabolic complications linked to oxidative brain injury. Metab. Brain Dis. 35, 1417–1428.

Escribano, M., San Andrés, M.I., de Lucas, J.J., González-Canga, A., 2012. Ivermectin residue depletion in food producing species and its presence in animal foodstuffs with a view to human safety. Curr. Pharm. Biotechnol. 13 (6), 987–998.

Faria, M., Bedrossiantz, J., Ramírez, J.R.R., Mayol, M., García, G.H., Bellot, M., Prats, E., Garcia-Reyero, N., Gómez-Canela, C., Gómez-Oliván, L.M., Raldúa, D., 2021. Glyphosate targets fish monoaminergic systems leading to oxidative stress and anxiety. Environ. Int. 146, 106253.

Feldbrügge, L., Jiang, Z.G., Csizmadia, E., Mitsuhashi, S., Tran, S., Yee, E.U., Rothweiler, S., Vaid, K.A., Sévigny, J., Schmelzle, M., Popov, Y.V., Robson, S.C., 2018. Distinct roles of ecto-nucleoside triphosphate diphosphohydrolase-2 (NTPDase2) in liver regeneration and fibrosis. Purinergic Signal 14 (1), 37–46.

Fernández-González, R., Yebra-Pimentel, I., Martínez-Carballo, E., Simal-Gándara, J., 2015. A critical review about human exposure to polychlorinated Dibenzo-p-Dioxins (PCDDs), Polychlorinated Dibenzofurans (PCDFs) and Polychlorinated Biphenyls (PCBs) through foods. Crit. Rev. Food Sci. Nutr. 55 (11), 1590–1617.

Furlong, M.A., Barr, D.B., Wolff, M.S., Engel, S.M., 2017. Prenatal exposure to pyrethroid pesticides and childhood behavior and executive functioning. NeuroToxicology 62, 231–238.

Gafour, W.A.S., 2018. Impact of cow milk manufacturing processes on the degradation of malathion pesticide residues. Am. J. Food Sci. Technol. 6 (6), 290–294.

Gil-Serna, J., Vázquez, C., Patiño, B., 2020. Mycotoxins in functional beverages: a review. Beverages 6 (3), 1–11.

Giordano, G., Costa, L.G., 2012. Developmental neurotoxicity: some old and new issues. ISRN Toxicol. 2012, 814795.

Golubeva, A.V., Crampton, S., Desbonnet, L., Edge, D., O'Sullivan, O., Lomasney, K.W., Zhdanov, A.V., Crispie, F., Moloney, R.D., Borre, Y.E., Cotter, P.D., Hyland, N.P., O'Halloran, K.D., Dinan, T.G., O'Keeffe, G.W., Cryan, J.F., 2015. Prenatal stressinduced alterations in major physiological systems correlate with gut microbiota composition in adulthood. Psychoneuroendocrinology 60, 58–74.

Guo, E.F.R., 2014. Survey of glyphosate residues in honey, corn and soy products. J. Environ. Anal. Toxicol. 5 (1), 1–8.

Gupta, R.C., Anadón, A., 2018. Fipronil. In: Gupta, R.C. (Ed.), Veterinary Toxicology. Academic Press, Cambridge, pp. 533–538.

Haisman, D., 2011. Imitation dairy products. In: Fuquay, J.W. (Ed.), Encyclopedia of Dairy Sciences. Academic Press, Cambridge, pp. 913–916.

Hassanen, E.I., Issa, M.Y., Hassan, N.H., Ibrahim, M.A., Fawzy, I.M., Fahmy, S.A., Mehanna, S., 2023. Potential mechanisms of imidacloprid-induced neurotoxicity in adult rats with attempts on protection using Origanum majorana L. Oil/Extract: *in vivo* and in silico studies. ACS Omega 8 (21), 18491–18508.

Hayes, W., Laws, E., 1982. Chlorinated hydrocarbon insecticides. In: Hayes, W.J. (Ed.), Pesticides Studied in Man. Williams & Wilkins, Pennsylvania Furnace.

Helaly, A.M.N., El-Attar, Y.A., Khalilm, M., Ahmed Ghorab, D.S.E., El-Mansoury, A.M., 2019. Antibiotic abuse induced histopathological and neurobehavioral disorders in mice. Curr. Drug Saf. 14 (3), 199–208.

Herbert, L.T., Cossi, P.F., Painefilú, J.C., Goñalons, C.M., Luquet, C.M., Kristoff, G., 2021. Acute neurotoxicity evaluation of two anticholinesterasic insecticides, independently and in mixtures, and a neonicotinoid on a freshwater gastropod. Chemosphere 265, 129107.

Heydebreck, F., 2021. Monitoring of paraquat in soya products intended for animal feed. Int. J. Food Contam. 8 (1), 4.

Hong, S.W., Heo, H., Yang, J.H., Han, M., Kim, D.H., Kwon, Y.K., 2013. Soyasaponin I improved neuroprotection and regeneration in memory deficient model rats. PLoS One 8 (12), e81556.

Hosaka, K., Rojas, K., Fazal, H.Z., Schneider, M.B., Shores, J., Federico, V., McCord, M., Lin, L., Hoh, B., 2017. Monocyte chemotactic protein-1-interleukin-6-osteopontin pathway of intra-aneurysmal tissue healing. Stroke 48 (4), 1052–1060.

Huang, H., Liang, H., Kwok, K.C., 2006. Effect of thermal processing on genistein, daidzein and glycitein content in soymilk. J. Sci. Food Agric. 86 (7), 1110–1114.

Hur, J.Y., 2022. γ-Secretase in Alzheimer's disease. Exp. Mol. Med. 54 (4), 433–446.

Islam, M.T., Mishra, S.K., Tripathi, S., de Alencar, M.V.O.B., e Sousa, J.M.de C., Rolim, H.M.L., de Medeiros, M.das G.F., Ferreira, P.M.P., Rouf, R., Uddin, S.J., Mubarak, M.S., de C. Melo-Cavalcante, A.A., 2018. Mycotoxin-assisted mitochondrial dysfunction and cytotoxicity: unexploited tools against proliferative disorders. IUBMB Life 70 (11), 1084–1092.

Jadhav, M.R., Pudale, Anjali, Raut, Prakash, Utture, Sagar, Shabeer, T.P.Ahammed, Banerjee, K., 2019. A unified approach for high-throughput quantitative analysis of the residues of multi-class veterinary drugs and pesticides in bovine milk using LC-MS/MS and GC–MS/MS. Food Chem. 272, 292–305. doi:10.1016/j.foodchem.2018.08.033.

Jackson, C.-J.C, Dini, J.P., Lavandier, C., Rupasinghe, H.P.V, Faulkner, H., Poysa, V., Buzzell, D., DeGrandis, S., 2002. Effects of processing on the content and composition of isoflavones during manufacturing of soy beverage and tofu. Process Biochem. 37 (10), 1117–1123.

Jin, Y., Wu, H., Cohen, E.M., Wei, J., Jin, H., Prentice, H., Wu, J.Y., 2007. Genistein and daidzein induce neurotoxicity at high concentrations in primary rat neuronal cultures. J. Biomed. Sci. 14 (2), 275–284.

Jones, P.J., Asp, N.G., Silva, P., 2008. Evidence for health claims on foods: how much is enough? Introduction and general remarks. J Nutr. 138 (6), 1189S–1191S.

Joung, J.Y., Song, J.G., Kim, H.W., Oh, N.S., 2021. Protective effects of milk casein on the brain function and behavior in a mouse model of chronic stress. J. Agr. Food Chem. 69 (6), 1936–1941.

Kadykalo, S., Roberts, T., Thompson, M., Wilson, J., Lang, M., Espeisse, O., 2018. The value of anticoccidials for sustainable global poultry production. Int. J. Antimicrob. Ag. 51, 304–310.

Kafilzadeh, F., Ebrahimnezhad, M., Tahery, Y., 2015. Isolation and identification of endosulfan-degrading bacteria and evaluation of their bioremediation in Kor River, Iran. Osong Pub. Health Res. Perspect. 6, 39–46.

Kant, V., Singh, P., Verma, P.K., Bais, I., Parmar, M.S., Gopal, A., Gupta, V., 2013. Anticoccidial drugs used in the poultry: an overview. Sci. Int. 1, 261–265.

Kara, M., Yumrutas, O., Demir, C.F., Ozdemir, H.H., Bozgeyik, I., Coskun, S., Eraslan, E., Bal, R., 2015. Insecticide imidacloprid influences cognitive functions and alters learning performance and related gene expression in a rat model. Int. J. Exp. Pathol. 96 (5), 332–337.

Ke, T., Rajoo, A., Tinkov, A.A., Skalny, A.V., Tizabi, Y., Rocha, J.B.T., Bowman, A., Aschner, M., 2024. Intestinal microbiota protects against methylmercury-induced neurotoxicity. Biometals 37 (3), 561–576.

Khalifa, H.O., Shikoray, L., Mohamed, M.I., Habib, I., Matsumoto, T., 2024. Veterinary drug residues in the food chain as an emerging public health threat: sources, analytical methods, health impacts, and preventive measures. Foods 13 (11), 1629.
Khaniki, G.R.J., 2007. Chemical contaminants in milk and public health concerns: a review. Int. J. Dairy Sci. 2, 104–115.
Kim, I.S., Yang, W.S., Kim, C.H., 2021. Beneficial effects of soybean-derived bioactive peptides. Int. J. Mol. Sci. 22 (16), 8570.
Kim, Y.J., Choi, G.S., Kim, S.B., Yoon, G.S., Kim, Y.S., Ryu, Y.W., 2005. Screening and characterization of a novel esterase from a metagenomic library. Protein Expre. Purificat. 45 (2), 315–323.
Kiruthika, K., Suganthi, A., Edward, Y.S., Anandham, R., Renukadevi, P., Murugan, M., Sahoo, B.K., Ikram, M., Kavitha, P.G., Jayakanthan, M., 2024. Role of lactic acid bacteria in insecticide residue degradation. Probiotics Antimicrob. Proteins 17 (1), 81–102.
Kunz, C., Rodriquez-Palmero, M., Koletzko, B., Jensen, R., 1999. Nutritional and biochemical properties of human milk, part I: general aspects, proteins, and carbohydrates. Clin. Perinatol. 26, 307–333.
Lankas, G.R., Cartwright, M.E., Umbenhauer, D., 1997. P-glycoprotein deficiency in a subpopulation of CF-1 mice enhances avermectin-induced neurotoxicity. Toxicol. Appl. Pharmacol. 143, 357–365.
Le Parc, A., Karav, S., Rouquié, C., Maga, E.A., Bunyatratchata, A., Barile, D., 2017. Characterization of recombinant human lactoferrin N-glycans expressed in the milk of transgenic cows. PLoS One 12, 0171477.
Lee, Y.B., Lee, H.J., Heon, S.S., 2005. Soy isoflavones and cognitive function. J. Nutr. Biochem. 16 (11), 641–649.
Li, R., Robinson, M., Ding, X., Geetha, T., Al-Nakkash, L., Broderick, T.L., Babu, J.R., 2022. Genistein: a focus on several neurodegenerative diseases. J. Food Biochem. 46 (7), e14155.
Liu, M.H., Lin, Y.S., Sheu, S.Y., Sun, J.S., 2009. Anti-inflammatory effects of daidzein on primary astroglial cell culture. Nutr. Neurosci. 12 (3), 123–134.
Lu, C., Gao, R., Lv, J., Chen, Y., Li, S., Zhang, L., Zhang, N., Wang, Y., Fan, B., Liu, X., Wang, F., 2020. Neuroprotective effects of soy isoflavones on chronic ethanol-induced dementia in male ICR mice. Food Funct. 11 (11), 10011–10021.
Lu, C., Wang, Y., Wang, D., Zhang, L., Lv, J., Jiang, N., Fan, B., Liu, X., Wang, F., 2018. Neuroprotective Effects of Soy Isoflavones on Scopolamine-Induced Amnesia in Mice. Nutrients 10 (7), 853.
Madani, N.A., Carpenter, D.O., 2022. Effects of glyphosate and glyphosate-based herbicides like RoundupTM on the mammalian nervous system: a review. Environ Res. 17 (1), 163–174.
Maebuchi, M., Kishi, Y., Koikeda, T., Furuya, S., 2013. Soy peptide dietary supplementation increases serum dopamine level and improves cognitive dysfunction in subjects with mild cognitive impairment. JPT 41 (1), 67–73.
Maioli, S., Leander, K., Nilsson, P., Nalvarte, I., 2021. Estrogen receptors and the aging brain. Ess. Biochem. 65 (6), 913–925.
Makowska, K., Obremski, K., Zielonka, L., Gonkowski, S., 2017. The influence of low doses of zearalenone and t-2 toxin on calcitonin gene related peptide-like immunoreactive (CGRP-LI) neurons in the ENS of the porcine descending colon. Toxins 9 (3), 98.
Martins, A.C., Morcillo, P., Ijomone, O.M., Venkataramani, V., Harrison, F.E., Lee, E., Bowman, A.B., Aschner, M., 2019. New insights on the role of manganese in Alzheimer's disease and Parkinson's disease. Int. J. Environ. Res. Pub. Health 16 (19), 3546.
McArdle, R., Hamill, R., 2011. Utilisation of hydrocolloids in processed meat systems. In: Kerry, J.P., Kerry, J.F. (Eds.), In: Series in Food Science, Technology and Nutrition, Processed Meats. Woodhead Publishing, Swatson, UK, pp. 243–269.

Mei, M., Liu, M., Mei, Y., Zhao, J., Li, Y., 2023. Sphingolipid metabolism in brain insulin resistance and neurological diseases. Front. Endocrinol. 14, 1243132.
Merrill, A.H., van Echtenll, G., WangS, E., Sandhoffl, K., 1993. Fumonisin B1 inhibits sphingosine (Sphinganine) N-acyltransferase and de Novo sphingolipid biosynthesis in cultured neurons in situ. J. Biol. Chem. 268 (36), 27299–27306.
Messina, M., Mejia, S.B., Cassidy, A., Duncan, A., Kurzer, M., Nagato, C., Ronis, M., Rowland, I., Sievenpiper, J., Barnes, S., 2022. Neither soyfoods nor isoflavones warrant classification as endocrine disruptors: a technical review of the observational and clinical data. Crit. Rev. Food Sci. Nutr. 62 (21), 5824–5885.
Mitchell, E.J., Frisbie, S.H., Roudeau, S., Carmona, A., Ortega, R., 2020. Estimating daily intakes of manganese due to breast milk, infant formulas, or young child nutritional beverages in the United States and France: comparison to sufficiency and toxicity thresholds. J. Trace Elem. Med. Biol. 62, 126607.
Mohammadi, M., Shadnoush, M., Sohrabvandi, S., Yousefi, M., Khorshidian, N., Mortazavian, A.M., 2021. Probiotics as potential detoxification tools for mitigation of pesticides: a mini review. Int. J. Food Sci. Technol. 56 (5), 2078–2087.
Mrema, E.J., Rubino, F.M., Brambilla, G., Moretto, A., Tsatsakis, A.M., Colosio, C., 2013. Persistent organochlorinated pesticides and mechanisms of their toxicity. Toxicology 307, 74–88.
Muhammad, A.A., Javed, I., Ayaz, M., Abdullah, M., Imran, M., Shahbaz, M., Aslam Gondal, T., Ali, M., Iqbal, Z., Iqbal, Z., Salehi, B., Sharifi-Rad, J., Martins, N., 2021. Organochlorine pesticide residues in raw milk samples collected from dairy farms and urban areas of Lahore district, Pakistan. J. Food Sci. Technol. 58 (1), 129–137.
Mulla, S.I., Ameen, F., Talwar, M.P., Eqani, S.A.M.A.S., Bharagava, R.N., Saxena, G., Tallur, P.N., Ninnekar, H.Z., 2020. Organophosphate pesticides: impact on environment, toxicity, and their degradation. In: Saxena, G., Bharagava, R. (Eds.), Bioremediation of Industrial Waste for Environmental Safety. Springer, Singapore, pp. 265–290.
Nag, S.K., 2010. Pesticides, veterinary residues and other contaminants in milk. In: Griffiths, M.W. (Ed.), Series in Food Science, Technology and Nutrition, Improving the Safety and Quality of Milk. Woodhead Publishing, Cambridge, pp. 113–145.
Nath, A., Vendan, S.E., Kashyap, P., Singh, J.K., Singh, C.K., Kumar, S., 2013. Carcinogenic pesticides residue detection in cow milk and water samples from Patna, India. Curr. Trends Biotechnol. Chem. Res. 3 (1), 0265–2321.
Niaz, K., Zahid, S., Shah, A., Khan, F., Bule, M., 2020. Ochratoxin A-induced genotoxic and epigenetic mechanisms lead to Alzheimer disease: its modulation with strategies. Environ. Sci. Pollut. Res Int. 27 (36), 44673–44700.
Niyibituronsa, M., Onyango, A.N., Gaidashova, S., Imathiu, S.M., Uwizerwa, M., Wanjuki, I., Nganga, F., Muhutu, J.C., Birungi, J., Ghimire, S., Raes, K., De Boevre, M., De Saeger, S., Harvey, J., 2018. Evaluation of mycotoxin content in soybean (*Glycine max L.*) grown in Rwanda. Afr. J. Food Agric. Nutr. Dev. 18 (3), 13808–13824.
Obafemi, B.A., Adedara, I.A., Rocha, J.B.T., 2023. Neurotoxicity of ochratoxin A: molecular mechanisms and neurotherapeutic strategies. Toxicology 497–498, 153630.
Ojo, O.A., Rotimi, D.E., Ojo, A.B., Ogunlakin, A.D., Ajiboye, B.O., 2023. Gallic acid abates cadmium chloride toxicity via alteration of neurotransmitters and modulation of inflammatory markers in Wistar rats. Sci. Rep. 13 (1), 1577.
Okocha, R.C., Olatoye, I.O., Adedeji, O.B., 2018. Food safety impacts of antimicrobial use and their residues in aquaculture. Public Health Rev. 39, 21.
Olías, R., Delgado-Andrade, C., Padial, M., Marín-Manzano, M.C., Clemente, A., 2023. An updated review of soy-derived beverages: nutrition, processing, and bioactivity. Foods 12 (14), 2665.
Pagliuca, G., Serraino, A., Gazzotti, T., Zironi, E., Borsari, A., Rosmini, R., 2006. Organophosphorus pesticides residues inItalian raw milk. J. Dairy Res. 73, 340–344.

Pajarillo, E., Nyarko-Danquah, I., Adinew, G., Rizor, A., Aschner, M., Lee, E., 2021. Neurotoxicity mechanisms of manganese in the central nervous system. Adv. Neurotoxicol. 5, 215–238.
Patangia, D.V., Ryan, C.Anthony, Dempsey, Eugene, Stanton, Catherine, Ross, R.Paul, 2022. Vertical transfer of antibiotics and antibiotic resistant strains across the mother/baby axis. Trends Microbiol. 30 (1), 47–56. doi:10.1016/j.tim.2021.05.006.
Patisaul, H.B., 2017. Endocrine disruption by dietary phyto-oestrogens: impact on dimorphic sexual systems and behaviours. Proc. Nutr. Soc. 76 (2), 130–144.
Paul, A.A., Kumar, S., Kumar, V., Sharma, R., 2020. Milk analog: plant based alternatives to conventional milk, production, potential and health concerns. Crit. Rev. Food Sci. Nutr. 60, 3005–3023.
Pavlenko, R., Berzina, Z., Reinholds, I., Bartkiene, E., Bartkevics, V., 2024. An occurrence study of mycotoxins in plant-based beverages using liquid chromatography–mass spectrometry. Toxins 16 (1), 53.
Peek, H.W., Landman, W.J.M., 2011. Coccidiosis in poultry: anticoccidial products, vaccines and other prevention strategies. Vet. Q. 31 (3), 143–161.
Pei, X., Zhang, W., Jiang, H., Liu, D., Liu, X., Li, L., Li, C., Xiao, X., Tang, S., Li, D., 2021. Food-origin mycotoxin-induced neurotoxicity: intend to break the rules of neuroglia cells. Oxid. Med. Cell Longev. 2021, 9967334.
Peng, W.-X., Marchal, J.L.M., van der Poel, A.F.B., 2018. Strategies to prevent and reduce mycotoxins for compound feed manufacturing. Anim. Feed Sci. Technol. 237, 129–153.
Pérez-Rodríguez, M.L., Serrano-Carretero, A., García-Herrera, P., Cámara-Hurtado, M., Sánchez-Mata, M.C., 2023. Plant-based beverages as milk alternatives? Nutritional and functional approach through food labelling. Food Res. Int. 173 (1), 113244.
Perry, J., Cotton, J., Rahman, M.A., Brumby, S.A., 2020. Organophosphate exposure and the chronic effects on farmers: a narrative review. Rural Remote Health 20 (1), 4508.
Piñeiro, S.A., Cerniglia, C.E., 2021. Antimicrobial drug residues in animal-derived foods: potential impact on the human intestinal microbiome. J. Vet. Pharmacol. Ther. 44, 215–222.
Polido, K., Olympio, K., Gonçalves, C., Risso Günther, W.M., José, E., Bechara, H., 2009. Neurotoxicity and aggressiveness triggered by low-level lead in children: a review. Rev. Panam. Salud. Publica. 26 (3), 266–275.
Prajapati, R., Park, S.E., Park, H.J., Jung, H.A., Choi, J.S., 2021. Identification of a potent and selective human monoamine oxidase-A inhibitor, glycitein, an isoflavone isolated from Pueraria lobata flowers. ACS Food Sci. Technol. 1 (4), 538–550.
Pratiwi, R., Ramadhanti, S.P., Amatulloh, A., Megantara, S., Subra, L., 2023. Recent advances in the determination of veterinary drug residues in food. Foods 12 (18), 3422.
Raheem, W., Niamah, A., 2021. Contamination methods of milk with pesticides residues and veterinary drugs. IOP Conf. Series Earth Environ. Sci. 877, 012003.
Rahman, M.S., Hassan, M.M., Chowdhury, S., 2021. Determination of antibiotic residues in milk and assessment of human health risk in Bangladesh. Heliyon 7 (8), e07739.
Rajak, P., Roy, S., Ganguly, A., Mandi, M., Dutta, A., Das, K., Nanda, S., Ghanty, S., Biswas, G., 2023. Agricultural pesticides—friends or foes to biosphere? J. Haz. Mat. Adv. 10, 100264.
Rana, M.S., Lee, S.Y., Kang, H.J., Hur, S.J., 2019. Reducing veterinary drug residues in animal products: a review. Food Sci. Anim. Resour. 39 (5), 687–703.
Rêgo, I.C.V., dos Santos, G.N.V., Ribeiro, J.S., Lopes, R.B., dos Santos, S.B., et al., 2019. Organochlorine pesticides residues in commercial milk: a systematic review. Acta Agron. 68 (2), 99–107.
Richardson, J.R., Fitsanakis, V., Westerink, R.H.S., Kanthasamy, A.G., 2019. Neurotoxicity of pesticides. Acta Neuropathol. 138 (3), 343–362.

Ross, G.W., Abbott, R.D., Petrovitch, H., Duda, J.E., Tanner, C.M., Zarow, C., Uyehara-Lock, J.H., Masaki, K.H., Launer, L.J., Studabaker, W.B., White, L.R., 2019. Association of brain heptachlor epoxide and other organochlorine compounds with lewy pathology. Mov. Disord. 34, 228–235.

Rubio, C., Paz, S., Gutiérrez, Á.J., González-Weller, D., Martín, R., Hardisson, A., 2021. Human exposure to potentially toxic elements from the consumption of soybean beverages commercialized in Spain. J. Food Prot. 84 (6), 932–937.

Salman, M., Abbas, R.Z., Mehmood, K., Hussain, R., Shah, S., Faheem, M., Zaheer, T., Abbas, A., Morales, B., Aneva, I., Martínez, J.L., 2022. Assessment of avermectins-induced toxicity in animals. Pharmceuticals (Basel) 15 (3), 332.

Sanders, T., Liu, Y., Buchner, V., Tchounwou, P.B., 2009. Neurotoxic effects and biomarkers of lead exposure: a review. Rev. Environ. Health 24 (1), 15–45.

Sathishkumar, P., Mohan, K., Ganesan, A.R., Govarthanan, M., Yusoff, A.R.M., Gu, F.L., 2021. Persistence, toxicological effect and ecological issues of endosulfan—a review. J. Haz. Mat. 416, 125779.

Schmitz, A., Dempewolf, S., Tan, S., Bicker, G., Stern, M., 2021. Developmental neurotoxicity of fipronil and rotenone on a human neuronal *in vitro* test system. Neurotox. Res. 39 (4), 1189–1202.

Schopf, M.F., Pierezan, M.D., Rocha, R., Pimentel, T.C., Esmerino, E.A., Marsico, E.T., Lindner, J.D., da Cruz, A.G., Verruck, S., 2022. Pesticide residues in milk and dairy products: an overview of processing degradation and trends in mitigating approaches. Crit. Rev. Food Sci. Nutr. 63 (33), 12610–12624.

Serrano, R., Blanes, M.A., López, F.J., 2008. Biomagnification of organochlorine pollutants in farmed and wild gilthead sea bream (*Sparus aurata*) and stable isotope characterization of the trophic chains. Sci. Total Environ. 389 (2), 340–349.

Shaffique, S., Hussain, S., Kang, S.M., Imran, M., Kwon, E.H., Khan, M.A., Lee, I.J., 2023. Recent progress on the microbial mitigation of heavy metal stress in soybean: overview and implications. Front. Plant Sci. 14, 1188856.

Shaker, E.M., Elsharkawy, E.E., 2015. Organochlorine and organophosphorus pesticide residues in raw buffalo milk from agroindustrial areas in Assiut, Egypt. Environ. Toxicol. Pharmacol. 39 (1), 433–440.

Singh, Z., Kaur, J., Kaur, R., Hundal, S.S., 2016. Toxic effects of organochlorine pesticides: a review. Am. J. BioSci. 4 (3–1), 11–18. doi:10.11648/j.ajbio.s.2016040301.13. http://www.sciencepublishinggroup.com/j/ajbio.

Soares, S., Rosado, T., Barroso, M., Vieira, D.N., Gallardo, E., 2019. Organophosphorus pesticide determination in biological specimens: bioanalytical and toxicological aspects. Int. J. Legal Med. 133 (6), 1763–1784.

Soni, M., Rahardjo, T.B.W., Soekardi, R., Sulistyowati, Y., Lestariningsih, Yesufu-Udechuku, A., Irsan, A., Hogervorst, E., 2014. Phytoestrogens and cognitive function: a review. Maturitas 77 (3), 209–220.

Sterup Moore, S., Costa, A., Pozza, M., Weaver, C.M., De Marchi, M., 2024. Nutritional scores of milk and plant-based alternatives and their difference in contribution to human nutrition. LWT Food Sci. Technol. 191, 115688.

Suvarna, V., 2023. Ivermectin: a critical review on characteristics, properties, and analytical methods. J. AOAC Int. 106 (3), 534–557.

Tanaka, T., Narazaki, M., Kishimoto, T., 2014. Il-6 in inflammation, immunity, and disease. Cold Spring Harb. Perspect. Biol. 6 (10), a016295.

Tuck, S., Furey, A., Danaher, M., 2016. Analysis of anthelmintic and anticoccidial drug residues in animal-derived foods. In: Kay, J.F., MacNeil, J.D., Wang, J. (Eds.), Chemical Analysis of Non-antimicrobial Veterinary Drug Residues in Food. Wiley, Hoboken, NJ, USA, pp. 245–309.

Uddin, M.S., Kabir, M.T., 2019. Emerging signal regulating potential of genistein against Alzheimer's disease: a promising molecule of interest. Front. Cell Develop. Biol. 7, 197.
van den Berg, H., Manuweera, G., Konradsen, F., 2017. Global trends in the production and use of DDT for control of malaria and other vector-borne diseases. Malar. J. 16 (1), 401.
Venkataramana, M., Chandra Nayaka, S., Anand, T., Rajesh, R., Aiyaz, M., Divakara, S.T., Murali, H.S., Prakash, H.S., Lakshmana Rao, P.V., 2014. Zearalenone induced toxicity in SHSY-5Y cells: the role of oxidative stress evidenced by N-acetyl cysteine. Food Chem. Toxicol. 65, 335–342.
Vidair, C.A., 2004. Age dependence of organophosphate and carbamate neurotoxicity in the postnatal rat: extrapolation to the human. Toxicol. Appl Pharmacol. 196 (2), 287–302. doi:10.1016/j.taap.2003.12.016.
Vitoria, I., 2017. The nutritional limitations of plant-based beverages in infancy and childhood. Nutr. Hosp. 34, 1205–1214.
Vuong, H.E., Pronovost, G.N., Williams, D.W., Coley, E.J.L., Siegler, E.L., Qiu, A., Kazantsev, M., Wilson, C.J., Rendon, T., Hsiao, E.Y., 2020. The maternal microbiome modulates fetal neurodevelopment in mice. Nature 586 (7828), 281–286.
Waliszewski, S.M., Villalobos-Pietrini, R., Gomez-Arroyo, S., Infanzon, R.M., 2003. Persistent organochlorine pesticide levels in cow's milk samples from tropical regions of Mexico. Food Addit. Contam. 20, 270–275.
Wang, W., Wang, Y., Wagner, K.M., Lee, R.D., Hwang, S.H., Morisseau, C., Wulff, H., Hammock, B.D., 2023. Aflatoxin B1 increases soluble epoxide hydrolase in the brain and induces neuroinflammation and dopaminergic neurotoxicity. Int. J. Mol. Sci. 24 (12), 9938.
Wang, X., Yin, Z., Meng, X., Yang, D., Meng, H., Liao, C., Wei, L., Chen, Y., Yang, X., Han, J., Duan, Y., Zhang, S., 2022. Daidzein alleviates neuronal damage and oxidative stress via GSK3β/Nrf2 pathway in mice. J. Funct. Foods 92, 105060.
Welsh, J.A., Braun, H., Brown, N., Um, C., Ehret, K., Figueroa, J., Boyd Barr, D., 2019. Production-related contaminants (pesticides, antibiotics and hormones) in organic and conventionally produced milk samples sold in the USA. Public. Health Nutr. 22 (16), 2972–2980.
World Health Organization [WHO], 2020. Chemical Safety: Pesticides. https://www.who.int/news-room/questions-and-answers/item/chemical-safety-pesticides Assessed 4 August 2024.
Wiley, A., 2007. The globalization of cow's milk production and and consumption: biocultural perspectives. Ecol. Food Nutr. 46 (3–4), 281–312.
Winstone, J.K., Pathak, K.V., Winslow, W., Piras, I.S., White, J., Sharma, R., Huentelman, M.J., Pirrotte, P., Velazquez, R., 2022. Glyphosate infiltrates the brain and increases pro-inflammatory cytokine TNFα: implications for neurodegenerative disorders. J. Neuroinflammation 19 (1), 193.
Woźniak, D., Cichy, W., Dobrzyńska, M., Przysławski, J., Drzymała-Czyż, S., 2022. Reasonableness of enriching cow's milk with vitamins and minerals. Foods 11, 1079.
Wong, M.C.Y., Emery, P.W., Preedy, V.R., Wiseman, H., 2008. Health benefits of isoflavones in functional foods? Proteomic and metabonomic advances. Inflammopharmacology 16 (5), 235–239.
Xu, S.F., Zhang, Y.H., Wang, S., Pang, Z.Q., Fan, Y.G., Li, J.Y., Wang, Z.-Y., Guoa, C., 2019. Lactoferrin ameliorates dopaminergic neurodegeneration and motor deficits in MPTP-treated mice. Redox Biol. 21, 101090.
Yang, L., Bajinka, O., Jarju, P.O., Tan, Y., Taal, A.M., Ozdemir, G., 2021. The varying effects of antibiotics on gut microbiota. AMB Express 11 (1), 116.
Yip, C.S.C., Yip, Y.C., Chan, W., 2023. The associations of soya intakes with non-communicable diseases: a scoping review of meta-analyses. Br. J. Nutr. 129 (1), 135–146.

Yuan, S., Li, C., Zhang, Y., Yu, H., Xie, Y., Guo, Y., 2020. Degradation of parathion methyl in bovine milk by high-intensity ultrasound: degradation kinetics, products and their corresponding toxicity. Food Chem. 327, 127103.

Yoo, M., Lim, Y.H., Kim, T., Lee, D., Hong, Y.C., 2016. Association between urinary 3-phenoxybenzoic acid and body mass index in Korean adults: 1st Korean National Environmental Health Survey. Ann. Occup. Environ. Med. 28 (1), 1–8.

Zacharia, J.T., 2010. Identity, physical and chemical properties of pesticides. In: Stoytcheva, M. (Ed.), Pesticides in the Modern World—Trends in Pesticides Analysis. InTech, London.

Zhang, J., You, L., Wu, W., Wang, X., Chrienova, Z., Nepovimova, E., Wu, Q., Kuca, K., 2020. The neurotoxicity of trichothecenes T-2 toxin and deoxynivalenol (DON): current status and future perspectives. Food Chem. Toxicol. 145, 111676.

Zhang, Y., Chen, H., Li, R., Sterling, K., Song, W., 2023. Amyloid β-based therapy for Alzheimer's disease: challenges, successes and future. Signal Transduct. Target Ther. 8 (1), 248.

Zhao, X., Wang, B., Xie, K., Liu, J., Zhang, Y., Wang, Y., Wang, J., 2018. Development and comparison of HPLC-MS/MS and UPLC-MS/MS methods for determining eight coccidiostats in beef. J Chromatogr. B 1087–1088, 98–107.

Zhao, X.-H., Wang, J., 2012. A brief study on the degradation kinetics of seven organophosphorus pesticides in skimmed milk cultured with *Lactobacillus spp.* at 42°C. Food Chem. 131 (1), 300–304.

Zhao, Y., Yang, L., Wang, Q., 2007. Pulsed large volume injection gas chromatography coupled with electron-capture negative ionization quadrupole mass spectrometry for simultaneous determination of typical halogenated persistent organic pollutants. J. Am. Soc. Mass Spectrom. 18 (2), 1375–1390.

Zhi, Y., Sun, T., Zhou, Q., Leng, X., 2020. Screening of safe soybean cultivars for cadmium contaminated fields. Sci. Rep. 10 (1), 12965.

Zhou, X.W., Zhao, X.H., 2014. Susceptibility of nine organophosphorus pesticides in skimmed milk towards inoculated lactic acid bacteria and yogurt starters. J. Sci. Food Agric. 95, 260–266.

Zhuang, P., Li, Z.A., Zou, B., Xia, H.P., Wang, G., 2013. Heavy metal contamination in soil and soybean near the Dabaoshan Mine, South China. Pedosphere 23 (3), 298–304.

CHAPTER TWO

Decoding the diet-inflammation axis: Modulating neuroinflammation through nutritional interventions

Opeyemi Babatunde Fasina, PhD[a,*,†],
Olufunke Florence Ajeigbe, PhD[b,†], **Memunat Oladepo**[c],
Isaac A. Adedara, PhD[d]

[a] Department of Neuroscience, Lerner Research Institute, Cleveland Clinic, Ohio, USA
[b] Department of Anatomy and Neurobiology, School of Medicine, University of California Irvine, Irvine, USA
[c] Department of Human Physiology, Amadu Bello University Zaria, Nigeria
[d] Department of Food Technology and Science, Federal University of Santa Maria, Santa Maria, Brazil
*Corresponding author. e-mail address: fasinao@ccf.org.

Contents

Abstract

Neuroinflammation, evidence-based research identifies the essential protective response of the central nervous system (CNS), as it involves dynamic interactions among resident immune cells, neurons, and infiltrating peripheral immune cells. In acute phases, neuroinflammation fosters tissue repair and pathogen defense. However, when prolonged, it underlies several neurodegenerative conditions, including Alzheimer's disease, Parkinson's disease, and multiple sclerosis to name a few. Evidence underscores the significant impact of dietary components on neuroinflammatory pathways. Excessive intake of simple sugars and saturated fats promotes systemic inflammation, alters gut

†Co-first author:

Advances in Neurotoxicology, Volume 16
ISSN 2468-7480
https://doi.org/10.1016/bs.ant.2026.02.001

microbiota composition, and compromises the blood-brain barrier (BBB), exacerbating CNS pathologies. By contrast, specific dietary carbohydrates and fatty acids exhibit anti-inflammatory and neuroprotective effects, improving microbiota balance, enhancing antioxidant defenses, and modulating immune cell function. Vitamins and minerals also shape the neuroinflammatory landscape by regulating cytokine release, oxidative stress, and neuronal survival. Furthermore, amino acids and proteins influence CNS immune mechanisms via metabolic pathways governing excitotoxicity, neurotransmitter balance, and antioxidant production. This chapter explores the multifactorial nature of neuroinflammation, highlighting the roles of astrocytes and microglia as key regulators of immune responses, cytokine production, and reactive oxygen species (ROS) generation and emphasizes the importance of diet on neuroinflammation. The role and mechanisms of nutritional interventions in the modulation of neuroinflammation were discussed.

Keywords: Blood-brain barrier, Central nervous system, Neuroinflammation, Nutrition, Oxidative damage

1. Introduction

Neuroinflammation represents the reactive response of central nervous system (CNS) components to disruptions in homeostasis, whether of endogenous or exogenous origin. Under normal physiological conditions, neuroinflammation plays a pivotal role in maintaining CNS homeostasis, defending against pathogens, and facilitating tissue repair (Perry et al., 2010). This process involves both localized and recruited immune cells and signaling mechanisms. Neuroinflammatory responses can be triggered by intrinsic (e.g., neurodegenerative diseases) or extrinsic (e.g., infections) factors (Kölliker-Frers et al., 2021). The CNS, due to its unique architecture and function, is particularly susceptible to immune-inflammatory responses, underscoring its vulnerability to sustained or excessive inflammation. Historically, the CNS was considered an immune-privileged organ due to the absence of conventional lymphatic drainage and the protective role of the blood-brain barrier (BBB). However, emerging evidence has challenged this paradigm, revealing that the brain is intricately connected to systemic immune processes. The constitutive and reactive elements of the CNS including the BBB, astrocytes, oligodendrocytes, microglia, and infiltrating myeloid and lymphoid cells play essential roles in orchestrating immune responses within the brain. Astrocytes are critical contributors to neuroinflammation. In healthy CNS, astrocytes maintain the integrity of the BBB, regulate neurotransmitter levels, and provide metabolic support to neurons. Upon activation by injury or disease, astrocytes undergo astrogliosis, a process characterized by hypertrophy and the secretion of cytokines and chemokines

(Liddelow and Barres, 2017). These molecules attract peripheral immune cells to the site of injury while astrocytes themselves secrete neurotrophic factors that support tissue repair and regeneration (Sofroniew, 2009). Moreover, astrocytes help shield the CNS from T-cell-mediated neuroinflammation (Sanmarco et al., 2021). Furthermore, microglia, the resident immune cells of the CNS, continuously monitors the brain's microenvironment. Upon detecting harmful stimuli such as pathogens, toxic molecules, or tissue damage, microglia transition from a surveillance state to an activated state. This activation is marked by morphological changes and the release of inflammatory mediators (Nimmerjahn et al., 2005).

Activated microglia are crucial for clearing pathogens, apoptotic cells, and debris, thereby promoting tissue repair and neuroprotection (Lucas et al., 2006). Moreover, neuroinflammation is implicated in both direct and indirect CNS pathologies, including traumatic, neoplastic, ischemic, metabolic, toxic, infectious, autoimmune, developmental, and degenerative conditions (Kölliker-Frers et al., 2021). Neurodegenerative diseases, which primarily affect neurons, represent a subset of these conditions. These diseases, including Alzheimer's disease (AD), Parkinson's disease (PD), frontotemporal dementia, amyotrophic lateral sclerosis (ALS), and Huntington's disease, target distinct neuronal populations in specific CNS regions, resulting in diverse clinical and pathological manifestations. Chronic neuroinflammation is a hallmark of neurodegenerative disorders, despite variations in their underlying pathogenetic pathways, such as protein aggregation and genetic predispositions (Mayne et al., 2020). In diseases like AD and PD, persistent activation of microglia leads to the sustained release of pro-inflammatory cytokines, including tumor necrosis factor-alpha (TNF-α), interleukin-1 beta (IL-1β), and interleukin-6 (IL-6) (Heneka et al., 2015). While these cytokines are initially protective, their prolonged expression exacerbates neuronal damage through mechanisms such as oxidative stress, excitotoxicity, and apoptosis (Glass et al., 2010). Additionally, chemokines such as C–C motif chemokine ligand 2 (CCL2) and C-X-C motif chemokine ligand 10 (CXCL10) recruit peripheral immune cells to the CNS, amplifying the inflammatory response and further contributing to neurodegeneration (Ransohoff and Brown, 2012).

Understanding the complex interplay between neuroinflammation and CNS pathologies is critical for developing effective therapeutic strategies to mitigate neurodegenerative diseases. Additionally, microglia and invading myeloid cells are key immune players in CNS inflammation. Microglia, the CNSs resident immune cells, become activated in response to trauma or disease and release pro-inflammatory cytokines that modulate inflammation and immunity (Borst et al., 2021). Other CNS constituents, such as astrocytes, oligodendrocytes, and the BBB, are integral to the inflammatory milieu. Chronic activation of microglia and astrocytes can lead to an overproduction of pro-inflammatory mediators, such as cytokines and chemokines, contributing to neurodegeneration (Heneka et al., 2015). This dysregulated neuroinflammatory response is implicated in various neurological disorders, including Alzheimer's disease (AD), Parkinson's disease (PD), multiple sclerosis (MS), and stroke (Perry et al., 2010). According to Ransohoff et al. (2015), neuroinflammation is an essential CNS response to various disruptions, including infections, trauma, and neurodegenerative diseases. This response is mediated by pattern-recognition receptors (PRRs), which serve as the immune system's initial defense by detecting pathogens and damage-associated signals. PRR activation triggers inflammatory pathways that recruit immune cells to the site of injury or infection. Adaptive immune components, including T and B cells, also contribute to neuroinflammation. T cells infiltrate the CNS to regulate immune responses, while B cells produce antibodies, both of which can influence the neuroinflammatory environment. Cytokines play a dual role in CNS immune responses, exhibiting both protective and detrimental effects depending on the context and duration of exposure. Acute neuroinflammation can facilitate repair processes following injury, while chronic neuroinflammation is associated with neurodegenerative diseases, where persistent inflammation exacerbates neuronal damage and disease progression (Tohidpour et al., 2017). Pro-inflammatory cytokines such as IL-1β, IL-6, and TNF-α amplify inflammatory responses, whereas anti-inflammatory cytokines like interleukin-10 (IL-10) and transforming growth factor-beta (TGF-β) mitigate inflammation and promote tissue repair (Ransohoff and Brown, 2012). The balance between these opposing cytokines is critical for determining the outcome of neuroinflammatory processes (Glass et al., 2010). Chemokines, such as CCL2, CXCL10, and CXCL12, play a pivotal role in regulating immune cell migration to CNS sites of injury or infection (Ransohoff and Brown, 2012). These molecules facilitate the recruitment of peripheral immune cells, such as monocytes and lymphocytes, across the BBB. Depending on the context, this immune cell infiltration can either aid in pathogen clearance or exacerbate neuroinflammation (Heneka et al., 2015). Activated microglia and astrocytes also produce reactive oxygen species (ROS) and nitric oxide (NO), key inflammatory mediators in the CNS (Lucas et al., 2006). While

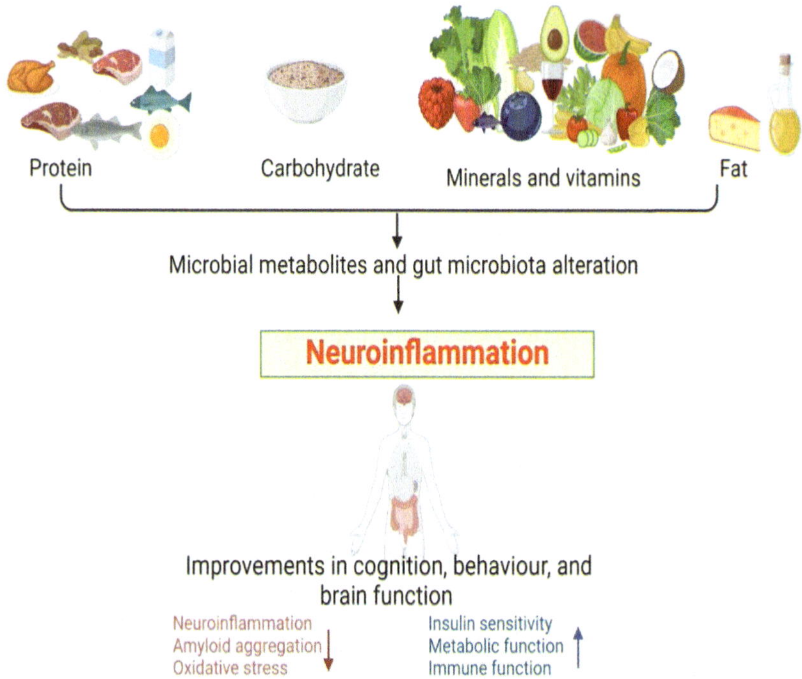

Fig. 1 Effect of main diets in neuroinflammation condition and ameliorative systemic pathways affected owing to diet intervention. The Fig. was created with BioRender.com.

ROS and NO are vital for pathogen defense and cellular signaling, excessive production can induce oxidative stress, damaging neurons and glial cells. This oxidative damage is a significant contributor to neurodegenerative disease progression (Perry et al., 2010). This review thus identifies the role of nutrients, carbohydrates, protein, fats, minerals and vitamins in managing neuroinflammation and its mediated pathogenesis while looking closely at the complexities related to its onset (Fig. 1).

2. The role carbohydrates plays in neuroinflammation

Carbohydrates play a complex and multifaceted role in neuroinflammation, which is a critical component in the pathogenesis of neurodegenerative diseases such as Alzheimer's disease, Parkinson's disease, multiple sclerosis, traumatic brain injury, and age-related cognitive decline (Marra et al., 2024). Increasingly, diet, particularly carbohydrate intake, has been shown to significantly modulate these inflammatory responses, influencing brain health either positively or negatively depending on the type and amount of carbohydrate consumed (Lee et al., 2021). Carbohydrates, ranging from simple sugars like glucose and fructose to complex polysaccharides, have diverse effects on neuroinflammation. Neuroinflammation involves the activation of immune cells within the central nervous system (CNS), especially microglia and astrocytes, in response to various stimuli like oxidative stress, metabolic imbalances, and immune dysregulation. Some carbohydrates, especially those prevalent in Western diets, may exacerbate neuroinflammation, accelerating cognitive decline and contributing to the onset of neurodegenerative conditions (Papadakis et al., 2024).

Conversely, other carbohydrates, particularly prebiotics and polysaccharides, show promise in reducing inflammation and oxidative stress, supporting neuronal health, and mitigating cognitive dysfunction. A key mechanism through which carbohydrates influence neuroinflammation is via the gut-brain axis, a bi-directional communication system between the gut and the brain (Haase et al., 2024). The gut microbiota, composed of trillions of microorganisms, plays a crucial role in maintaining immune homeostasis and regulating inflammation, including neuroinflammation. Carbohydrates, particularly those that serve as prebiotics, can influence the composition of the gut microbiota, promoting the growth of beneficial bacteria that produce anti-inflammatory metabolites such as short-chain fatty acids (SCFAs) (Chao et al., 2024). One of the most studied prebiotic carbohydrates is the polysaccharide derived from *Lycium barbarum* (LBPs), which has been shown to protect against cognitive dysfunction induced by a high-fat, high-fructose diet (HFFD) (Tian et al., 2024). In mice, the consumption of 200 mg/kg/day LBPs prevented learning and memory deficits commonly seen with HFFD, which is known to induce gut dysbiosis and cognitive decline after 14 weeks (Tian et al., 2024). LBPs were found to alter the gut microbiota by increasing the abundance of beneficial bacteria like *Allobaculum* and *Lactococcus*, while reducing the levels of harmful *Proteobacteria* (Tian et al., 2024). These changes in the gut microbiota were linked to an increase in SCFAs, which have been shown to reduce neuroinflammation by modulating immune responses and supporting neuronal health (Tian et al., 2024). Similarly, xylo-oligosaccharides (XOS), another prebiotic fiber, has demonstrated its capacity to alleviate neuroinflammation (Teng et al., 2024). In mice fed a high-fat diet (HFD), two doses of 30 g/kg diet and 60 g/kg XOS supplementation reversed cognitive decline, improved spatial memory, and reduced amyloid plaque accumulation in the hippocampus for 12 weeks (Teng et al., 2024). The beneficial effects of XOS were associated with a shift in gut microbiota composition, with an increase in the abundance of beneficial bacteria like Faecalibacterium (Teng et al., 2024). This gut microbiota modulation was linked to reduced neuroinflammatory markers, including decreased levels of pro-inflammatory cytokines in the brain (Teng et al., 2024). These findings highlight the potential of prebiotic fibers in preventing or mitigating neuroinflammation through their influence on gut microbiota and subsequent impact on brain health (Teng et al., 2024).

Lactulose and trehalose are two neuroprotective disaccharides. Lactulose, a synthetic sugar used primarily as a laxative, is also known for its prebiotic properties (Lee et al., 2021). It serves as a substrate for beneficial gut bacteria, promoting the production of SCFAs and reducing gut-derived inflammation (Lee et al., 2021). In a rat model of hepatic encephalopathy, a condition characterized by both liver dysfunction and neuroinflammation, lactulose (5.4 g/kg/day) combined with cinnamaldehyde (CA) (40 mg/kg/day) significantly reduced markers of neuroinflammation (Abd El Salam et al., 2024). The combination of lactulose and CA decreased levels of pro-inflammatory cytokines such as interleukin-1β (IL-1β) and attenuated the activation of P2X purinoceptor 7 and nucleotide-binding domain, leucine-rich repeat (P2 $\times$ 7R/NLRP3) inflammasome pathway in the brain, which is a critical driver of neuroinflammation (Abd El Salam et al., 2024). Additionally, lactulose improved cognitive function in these rats, further supporting the role of prebiotic carbohydrates in mitigating neuroinflammation and promoting brain health (Abd El Salam et al., 2024).

Trehalose, another disaccharide with prebiotic properties, has also shown promise in reducing neuroinflammation. In mouse models of Alzheimer's disease, 2 percent trehalose for 14 days was found to enhance autophagy, a process by which cells degrade and recycle damaged components, and reduce microglial activation (Zozulya et al., 2024). Trehalose also improved synaptic protein expression, suggesting that it supports synaptic health while reducing inflammation (Sinha et al., 2021). The ability of trehalose to stimulate autophagy may be particularly beneficial in neurodegenerative diseases where the accumulation of damaged proteins and cellular debris exacerbates neuroinflammation (Sinha et al., 2021). Trehalose's dual action of promoting cellular repair processes and reducing inflammatory responses makes it a promising candidate for therapeutic strategies targeting neuroinflammatory conditions (Sinha et al., 2021). For instance, trehalose was shown to ameliorate gliosis, neuroinflammation, and endoplasmic reticulum stress in a Lafora disease (LD) mouse model, significantly reducing the susceptibility to seizures (Sinha et al., 2021). However, it did not affect the formation of Lafora bodies, suggesting that the epileptic phenotype in LD could be secondary to or independent of these structures (Sinha et al., 2021). This highlights the potential of trehalose as a therapeutic agent, particularly in neurodegenerative diseases where neuroinflammation plays a pivotal role (Sinha et al., 2021). Furthermore, oxidative stress, which occurs when the production of ROS exceeds the body's ability to neutralize them, is closely linked to neuroinflammation (Gong et al., 2022). The brain, with its high metabolic rate and oxygen consumption, is particularly susceptible to oxidative damage. Excessive ROS can damage neuronal cells, triggering inflammatory responses and contributing to the progression of neurodegenerative diseases (Zhang et al., 2024). Certain carbohydrates, especially polysaccharides, have been found to exert antioxidant effects, thereby reducing oxidative stress and neuroinflammation. For example, polysaccharides from Armillariella tabescens (ATEP) have demonstrated potent neuroprotective effects in animal models of Alzheimer's disease (Zhang et al., 2024). ATEP (200 mg/kg/day) administration enhanced the activity of antioxidant enzymes such as superoxide dismutase (SOD) and glutathione peroxidase (GPX4), which play critical roles in neutralizing ROS (Zhang et al., 2024). ATEP also modulated the Keap1/Nrf2 signaling pathway, a major regulator of the antioxidant response, leading to reduced oxidative damage and neuroinflammation in the brain. Furthermore, ATEP reduced the expression of pro-inflammatory cytokines, showing its ability to simultaneously reduce oxidative stress and neuroinflammation.

Similarly, a neutral polysaccharide extracted from *Gastrodia elata* (NPGE) demonstrated significant neuroprotective effects in ischemic stroke models, where oxidative stress and neuroinflammation are key drivers of neuronal damage (Zhang et al., 2024). NPGE (1mg/kg) reduced neuronal ferroptosis, a form of iron-dependent cell death associated with lipid peroxidation, by upregulating the nuclear factor erythroid 2-related factor 2 and heme oxygenase 1 NRF2/HO-1 antioxidant pathway (Zhang et al., 2024) 3 days after middle cerebral artery occlusion (MCAO). This led to decreased ROS levels and reduced neuroinflammatory markers, suggesting that NPGE could be a valuable therapeutic agent for protecting against neurodegenerative diseases by targeting both oxidative stress and inflammation (Zhang et al., 2024). While complex carbohydrates like polysaccharides can reduce neuroinflammation, simple sugars such as glucose and fructose often have the opposite effect, exacerbating inflammation and contributing to the development of neurodegenerative diseases (Papadakis et al., 2024). This is particularly evident in the context of a Western-style diet, which is characterized by high consumption of sugars and saturated fats (Papadakis et al., 2024). The Western diet has been associated with increased risk of cognitive decline, largely due to its negative impact on gut health, systemic inflammation, and the integrity of the blood-brain barrier (BBB) (Więckowska-Gacek et al., 2021). Studies have shown that a Western diet promotes gut microbiota dysbiosis, driving systemic inflammation and impairing the BBB, allowing harmful substances to enter the brain and trigger neuroinflammatory responses (Więckowska-Gacek et al., 2021).

Galactose, another simple sugar commonly found in the Western diet, has also been shown to exacerbate neuroinflammation. In models of experimental autoimmune encephalomyelitis (EAE), a model for multiple sclerosis, galactose promoted the formation of advanced glycation end products (AGEs), which are toxic compounds that arise from the reaction between sugars and proteins or lipids (Haase et al., 2024). AGEs contribute to oxidative stress and inflammation by activating receptors for advanced glycation end products (RAGE) on immune cells, including microglia, further exacerbating neuroinflammation and neuronal damage (Haase et al., 2024). The immune cells of the CNS, particularly microglia and astrocytes, are central to the process of neuroinflammation. Microglia act as the brain's resident macrophages, responding to injury and disease by clearing debris and releasing cytokines to coordinate immune responses (Kong et al., 2020). However, chronic activation of microglia can lead to sustained neuroinflammation, contributing to neurodegenerative diseases. Several carbohydrates have been shown to modulate the activity of these immune cells, either promoting or suppressing neuroinflammation depending on the context. D-mannose (20 percent (w/v) in drinking water), a glucose epimer, has gained attention for its anti-inflammatory properties in the CNS (Hwang et al., 2022) in a 27-day experimental autoimmune encephalomyelitis (EAE) model. D-mannose treatment delayed disease onset and reduced neuroinflammation by increasing the frequency of regulatory T cells (Tregs; CD4+FoxP3+ regulatory T cells) in the brain. Tregs play a crucial role in maintaining immune tolerance and suppressing pro-inflammatory responses (Hwang et al., 2022). By enhancing Treg activity, D-mannose reduced the activation and infiltration of pro-inflammatory microglia and macrophages into the CNS, leading to a decrease in neuroinflammation (Wang et al., 2021).

In contrast, 25 mM high glucose levels have been shown to exacerbate neuroinflammation by enhancing microglial activation (Hung et al., 2022). In vitro, studies on microglial cells exposed to high glucose concentrations revealed prolonged activation of the NF-κB (Nuclear factor kappa-light-chain-enhancer of activated B cells) signaling pathway, a key regulator of inflammation (Hung et al., 2022). High glucose also increased the expression of toll-like receptor 4 (TLR4), a receptor that recognizes pathogens and activates immune responses (Hung et al., 20222). When activated by lipopolysaccharide (LPS), TLR4 promotes the production of pro-inflammatory cytokines such as IL-1β and TNF-α, which contribute to neuroinflammation (Hung et al., 20222). These findings suggest that hyperglycemia, commonly seen in diabetes, can worsen neuroinflammation and increase the risk of neurodegenerative diseases (Hung et al., 2022). Traumatic brain injury (TBI) and spinal cord injury (SCI) are associated with severe neuroinflammatory responses, which can exacerbate neuronal damage and impair recovery (Gong et al., 2022). Carbohydrates, particularly in the context of hyperglycemia, have been shown to influence trauma-induced neuroinflammation (Zhang et al., 2021). Following TBI, hyperglycemia worsened neuroinflammation and neuronal apoptosis, contributing to cognitive deficits and impaired recovery (Zhang et al., 2021). Hyperglycemia also inhibited the MEK5 (MAPK/ERK kinase 5) and ERK5 (Extracellular signal-regulated kinase 5) signaling pathway, which is critical for neuronal survival and recovery, further worsening neuroinflammation and neuronal death (Zhang et al., 2021). In contrast, trehalose at 0.1 mL/10 g body weight by gavage daily for 3 days has shown potential in reducing neuroinflammation and promoting recovery following spinal cord injury (SCI) (Gong et al., 2022). Trehalose inhibited ferroptosis by activating the NRF2/HO-1 antioxidant pathway, reducing oxidative stress and improving motor function recovery in SCI models.

The diverse effects of carbohydrates on neuroinflammation offer several therapeutic opportunities for managing neurodegenerative diseases. Prebiotic fibers, complex polysaccharides, and specific sugars like D-mannose and trehalose have shown potential in reducing neuroinflammation and supporting brain health (Wang et al., 2021; Zozulya et al., 2024). Future therapeutic strategies could focus on optimizing the intake of beneficial carbohydrates while minimizing the harmful effects of simple sugars like glucose and fructose, particularly in the context of Western diets (Papadakis et al., 2024). Prebiotic fibers such as lactulose, trehalose, LBPs, and XOS could be incorporated into dietary regimens to promote gut health and reduce neuroinflammation as shown in Fig. 2(Tian et al., 2024; Lee et al., 2021; Teng et al., 2024). These fibers modulate the gut microbiota, increasing the production of SCFAs, which have anti-inflammatory effects in the brain. Additionally, polysaccharides like ATEP and NPGE could be developed

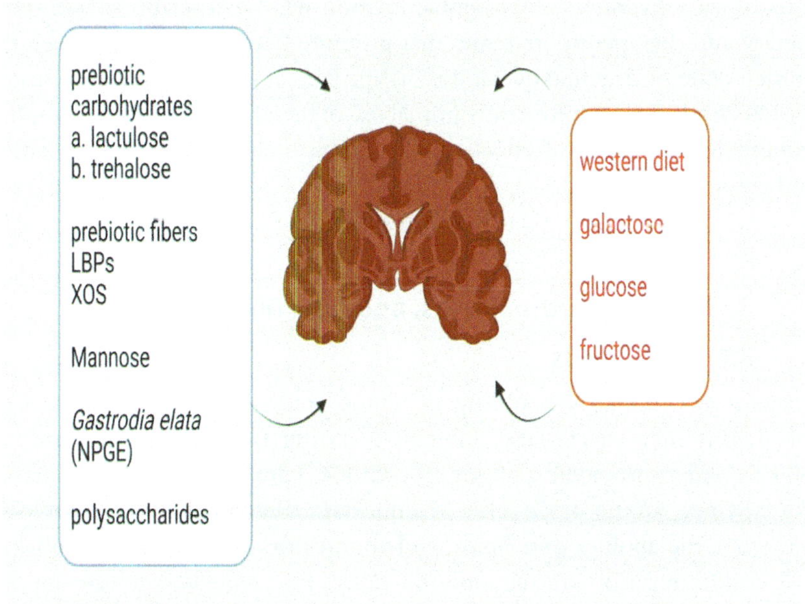

Fig. 2 Certain carbohydrates involved in the neuroprotection of the brain and sugar rich diets which promote neuroinflammation and cognitive decline. XOS- xylo-oligosaccharides, LBPs- *Lycium barbarum*. The Fig. was created with BioRender.com.

into therapeutic agents that target oxidative stress and inflammation, offering neuroprotection in conditions like Alzheimer's disease and ischemic stroke (Zhang et al., 2024;). D-mannose, with its ability to enhance regulatory T cell activity and reduce pro-inflammatory microglial activation, could be explored as a low-cost treatment for neuroinflammatory diseases like multiple sclerosis (Hwang et al., 2022). Similarly, trehalose, with its ability to enhance autophagy and reduce ferroptosis, holds promise as a therapeutic agent for both neurodegenerative diseases and trauma-induced neuroinflammation (Sinha et al., 2021, Gong et al., 2022). Carbohydrates exert profound effects on neuroinflammation, playing both harmful and beneficial roles depending on their structure and context of consumption (Papadakis et al., 2024). Simple sugars, particularly those consumed in excess as part of Western diets, can exacerbate inflammation and accelerate neurodegeneration (Papadakis et al., 2024). In contrast, complex carbohydrates, especially prebiotics and polysaccharides, offer significant therapeutic potential by modulating the gut-brain axis, reducing oxidative stress, and suppressing immune activation in the CNS (Hwang et al., 2022). Understanding the dual roles of carbohydrates in neuroinflammation will be crucial for developing dietary and therapeutic strategies that promote brain health and reduce the burden of neurodegenerative diseases (Papadakis et al., 2024). Future research should continue to explore the optimal use of beneficial carbohydrates while minimizing the harmful effects of sugar-rich diets on neuroinflammation and cognitive decline.

3. Proteins, amino acids, and neuroinflammation

3.1 Amino acids

Proteins are primarily made up of amino acids, as is most often understood. But in a strict sense, substances having both an amine group (-NH^{3+}) and a carboxylic acid group (-COO^{-}) are considered amino acids. For instance, amino acids serve as important metabolites, building blocks for other metabolites and lipids, and modulators of gene expression and cell communication throughout the body. The regulated release of tiny molecule neurotransmitters, many of which are amino acids, is necessary for neurotransmission between neurons, which may happen in a matter of milliseconds (Li et al., 2021). Several amino acids, most notably glutamate, are known to function as neuromodulators, small molecule neurotransmitters, or precursors of other small molecule neurotransmitters just in the nervous system (Dalangin et al., 2020). Synapses are specialized spaces where neurons interact with one other by releasing chemicals that are stored in synaptic vesicles. By attaching to their target receptors on nearby neurons, these substances diffuse across the synapse and modify the physiological states of those neurons. There is sometimes misunderstanding about the distinction between neurotransmitters and neuromodulators, even though these messenger molecules are all referred to as neurotransmitters (Nimgampalle et al., 2023). Some recognized neurotransmitters whose main function is to engage G protein-coupled receptors (GPCRs) in order to initiate a longer-lasting second messenger have also been referred to as "neuromodulators."

According to Dalangin et al. (2020), glutamatergic synapses comprise 80–90 percent of brain synapses and at least 60 percent of all central nervous system synapses. L-glutamate has been identified as the primary excitatory neurotransmitter in the peripheral nervous system. It is commonly accepted that L-aspartate, along with L-glutamate, is the secondary excitatory neurotransmitter in the central nervous system. However, some research indicates that aspartate and glutamate may be released simultaneously (Dalangin., 2021).

Most of the glutamate produced in glutamatergic synapses is absorbed by astrocytes and transformed into glutamine-by-glutamine synthetase. Following its export to the extracellular space, glutamine is absorbed by neurons, where phosphate-activated glutaminase transforms it back into glutamate before it is stored in vesicles. Aspartate may also be produced by some of the generated glutamate being digested. Cysteine and its derivatives have distinct chemical features that allow them to perform certain tasks within cells because they include a nucleophilic thiol group. Most people know that L-cysteine is a precursor to glutathione, which is the body's primary antioxidant (for in-depth talks on glutathione's functions in the neurological system, see Saso et al., 2023). According to Schulz et al. (2018), L-proline is a well-known neuromodulator in the brain that satisfies many of the requirements for a traditional neurotransmitter—possibly even more so than L-aspartate, which is also widely accepted to be a neurotransmitter. For instance, sodium-dependent neurotransmitter transporter NTT4, a vesicular transporter for L-proline, has been found, in contrast to L-aspartate (Das, 2022).

Serine has neurologically active enantiomers. In addition to serving as a precursor for neuroactive chemicals, L–serine is a crucial developmental and signaling component. Four distinct routes are used by astrocytes in the brain to generate serine, and deficits in this nutrient have been related to a variety of neuropathies and developmental disorders (Maugard et al., 2021). Tyrosine, an aromatic amino acid that is either absorbed by food or produced by phenylalanine hydroxylation, is the starting point for the sequential synthesis of dopamine, norepinephrine, and epinephrine and tyrosine hydroxylase catalyzes the first step, which is rate-limiting (Kandasamy et al., 2018). Tyrosine hydroxylase in the brain is another enzyme that may catalyze the hydroxylation of phenylalanine (Kandasamy et al., 2018). The temporary reduction of dopamine synthesis through acute phenylalanine and tyrosine depletion has been shown to have some effects on mood and cognition. Meanwhile, histidine decarboxylase converts histidine to histamine, a neurotransmitter that is primarily known for controlling sleep and wakefulness but is also involved in other critical processes such as endocrine regulation, motivation, arousal, and feeding (Scammell et al., 2019). Tryptophan, branched chain amino acids (leucine, isoleucine, valine), tyrosine, and phenylalanine are the amino acids that are most strongly associated with the onset of neuroinflammation. Quinolinic acid and kynurenic acid, two metabolites that are produced when tryptophan is metabolized via the kynurenine pathway, can affect neuroinflammation. Its metabolite quinolinic acid is an agonist that has been shown to have neurotoxic effects on the N-methyl-D-aspartate (NMDA) receptor. Higher levels of quinolinic acid have been linked to increased neuroinflammation and oxidative stress in several neurodegenerative illnesses, including Parkinson's and Alzheimer's diseases (Savonije and Weaver, 2023).

Yet by blocking NMDA receptor activation, its second metabolite, kynurenic acid, demonstrates neuroprotective qualities and anti-inflammatory benefits, perhaps reducing excitotoxicity linked to neuroinflammation (Sorgdrager et al., 2019). Additionally, serotonin, a byproduct of tryptophan with anti-inflammatory properties, has a role in the pathophysiology of neurodegenerative illnesses and mood disorders as well as the aggravation of inflammation. Furthermore, tryptophan is a well-known antioxidant that effectively suppresses the generation of proinflammatory T-helper-1 cytokines, superoxide (O_2^-), and inflammatory cytokines while also avoiding autoimmune diseases, reducing neuroinflammation, and boosting immunity. According to Yoo et al. (2022), branch-chain amino acids are known to cause excitotoxicity, hyperexcitability, inflammation, and oxidative stress in the brain. Variations in these parameters have been observed in neurodegenerative diseases such as Alzheimer's, Parkinson's, and ALS. Additionally, a study found that administering branch-chain amino acids supplements to obese rats affected neuroinflammatory parameters such as interleukin (IL-1) and astrocyte reactivity but did not improve memory function as observed with zinc supplementation (Feijó et al., 2022). Phenyl alanine levels control the production of neurotransmitters that control cell metabolism and brain function, as well as the synthesis of tetrahydrobiopterin, a cofactor for nitric oxide (NO) formation (Vasquez-Vivar et al., 2022). Neuromelanin, which is produced by tyrosine, dopamine, and melanin, has been identified as a neurotransmitter in the control of immunological response and as an antioxidant in the suppression of the production of inflammatory cytokines and superoxide (Moreno-García et al., 2021).

3.2 Proteins

Long-term negative neuronal (neuropsychiatric/neurodegenerative) effects, such as attention deficit hyperactivity disorder (ADHD) or even Alzheimer's disease, have been linked to neuroinflammation (Ghazy et al., 2022). It is interesting to note that a variety of proteins and their variations have been connected to neurodegenerative illnesses. Depending on the protein's constituents, there are several neurodegenerative disorders, including Parkinson's disease, sepsis-induced neuroinflammation, and several neurodegenerative diseases. Sepsis-induced neuroinflammation has been associated with the downregulation of extracellular matrix (ECM) homeostasis and axonal and synaptic network development (Jiang et al., 2022). The onset of neuroinflammation in Alzheimer's disease is linked to several proteins, including triggering of receptor expressed on myeloid cells 2 (TREM2), myeloid cell surface antigen CD-33 (CD-33), membrane-spanning 4-domains subfamily A (MS4A), clusterin (CLU), ATP-binding cassette subfamily A member 7 (ABCA-7), paired immunoglobin-like type 2 receptor alpha (PILRA), complement receptor type 1 (CR-1), and ephrin type-A receptor 1 (EPHA-1), as described by O'Day and Huber (2022). Neuroinflammatory biomarkers associated with Alzheimer's disease include changes in ATP-binding cassette subfamily A member 1 (ABCA1), chitinase-3-like protein I (CH3L-1/YKL-40), and NACHT, LRR, and PYD domain-containing protein 3 (NLRP-3). It is noteworthy that a multitude of proteins are still being researched as biomarkers in neurodegenerative illnesses, including Huntington's disease, Lewy body dementia (LBD), Parkinson's disease with Toll-like Receptor Cascades and Interleukins 4, 10, and 13 signaling as the most significantly enriched processes (Heidari et al., 2022), and frontotemporal dementia (FTD).

Additional proteins include protein clumps such as tau and amyloid-beta, which cause neurodegeneration by inducing neuroinflammation (Didonna, 2020). Neuroinflammation can also be triggered by mutations in mitochondrial-associated proteins, such as LRRK-2 (Leucine-Rich Repeat Kinase 2), PINK-1 (PTEN-Induced Kinase 1), PARK-7 (DJ-1, a multifunctional protein involved in protecting cells from oxidative stress and maintaining mitochondrial function), and PRKN (Parkin, an E3 ubiquitin ligase), which are also present in familial Parkinson's disease (Castelo Rueda and Maria, 2021) as seen in Fig. 3. DAMPs, or danger-associated molecular patterns, are released from damaged brain cells, mitochondrial DNA, HMGB-1, S100 proteins, chromogranin A, ATP, and uric acid. These DAMPs activate pattern recognition receptors (PRRs), causing neuroinflammation (Patel, 2018). In amyotrophic lateral sclerosis (ALS), aberrant accumulation of misfolded proteins can function as danger-associated molecular patterns (Sonninen et al., 2020), and inflammasomes are involved in identifying and removing danger-associated molecular patterns that exacerbate the chronic inflammatory state (Murao et al., 2021). There is proof that proteins play a significant role in neuroinflammation from two case-control studies that involved older people with mental illness by van de Rest et al. (2013), stating that elderly individuals may be more susceptible to low protein intake

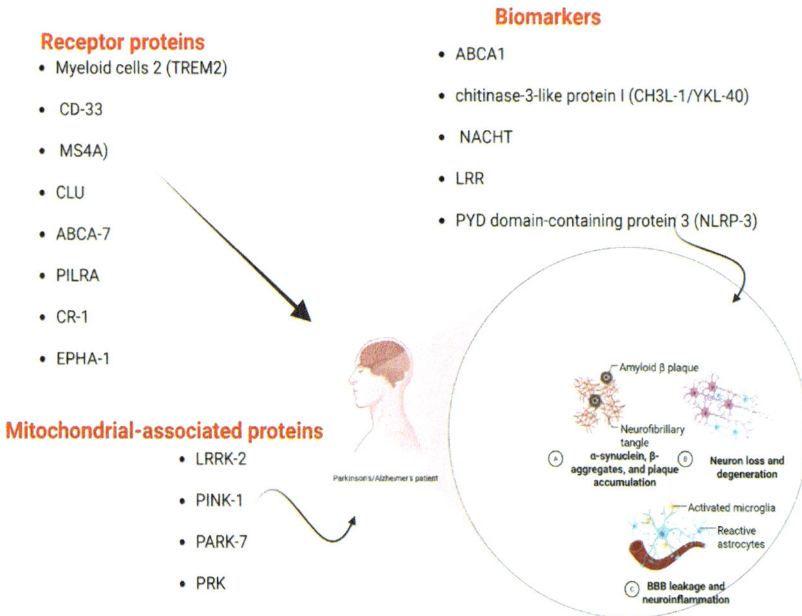

Fig. 3 Involvement of endogenous proteins, including receptor proteins and mitochondria-associated proteins, in the onset of neuroinflammation in neurodegenerative diseases, Parkinson's disease, and Alzheimer's disease. These proteins are also identified by protein biomarkers that reflect neuroinflammatory responses and how they impact the protein expressions observed in these conditions. LRRK-2- Leucine-Rich Repeat Kinase 2, PINK-1- PTEN-Induced Kinase 1, PARK-7-DJ-1 and PRKN-Parkin, ABCA1-ATP-binding cassette subfamily A member 1, CH3L-1/YKL-40- chitinase-3-like protein I, LRR- C-terminal leucine-rich repeat, and PYD domain-containing protein 3 (NLRP-3), myeloid cell surface antigen CD-33 (CD-33), membrane-spanning 4-domains subfamily A (MS4A), clusterin (CLU), ATP-binding cassette subfamily A member 7 (ABCA-7), paired immunoglobin-like type 2 receptor alpha (PILRA), complement receptor type 1 (CR-1), and ephrin type-A receptor 1 (EPHA-1). The Fig. was created with BioRender.com.

because of their decreased skeletal muscle mass, which results in lower whole body protein synthesis. This, in turn, may cause a reduction in the ability of the elderly to adjust to lower dietary protein intake and a decrease in the availability of neurotransmitter precursors (Glenn et al., 2019).

4. Minerals, vitamins and neuroinflammation

4.1 Minerals

Minerals serve a multitude of purposes in the immune system's proper operation. They have several purposes, including affecting muscle and nerve function, forming the building blocks for our bones, and controlling the body's water balance (Weyh et al., 2022). They are also found in hormones, enzymes, and other substances that are physiologically active. Additionally crucial to the immune system's efficient operation are a few minerals. This pertains to the adaptive immune response as well as the innate defensive mechanism. Notably, zinc, iron, copper, and magnesium are among the minerals that modulate neuroinflammation; they primarily mediate oxidative stress and inflammation pathways (Leyane et al., 2022). An imbalance in copper causes neuroinflammation and oxidative stress in Alzheimer's patients. Immune competence is largely dependent on the availability of copper. On the other hand, decreased humoral and cellular immune activity results from a copper shortage. Copper -deficient mice have been demonstrated to have an enlarged spleen and a smaller thymus. Additionally, there is a decrease in T cell counts and neutropenia (Stefanache et al., 2023). The function of B cells and NK cells is also compromised, as is mitogen-induced T-cell proliferation. Consequently, raising plasma copper levels may benefit human innate and adaptive immunity. Conversely, dietary copper consumption reduces the risk of Alzheimer's disease, according to Hsu et al. (2018).

Zinc has been researched for its impact on neuroinflammation and is well recognized for its immunomodulatory qualities. In models of obesity-induced neuroinflammation, zinc supplementation has demonstrated potential in lowering neuroinflammatory markers, such as IL-1β (Feijó et al., 2022). Sufficient absorption of zinc is necessary for the innate and adaptive immune systems to operate properly. Zinc is essential for the action of neutrophil granulocytes' nicotinamide adenine dinucleotide phosphate (NADPH) oxidase within the framework of the innate immune response (Gammoh and Rink, 2017). Consequently, a zinc deficit may lead to a decrease in the production of ROS, which have a lower capacity to cause harm. Studies conducted in vivo have also demonstrated that a zinc shortage results in decreased monocyte and neutrophil granulocyte adhesion and chemotaxis, as well as decreased macrophage maturation and activity. Zinc is also important for natural killer (NK) cells. A zinc shortage can result in decreased peripheral blood NK cell counts and compromised function. In this regard, it has been shown that decreased chemotaxis and lysis of virus-infected cells or tumor cells disclose this (Gammoh and Rink, 2019).

Increased neuroinflammation has been linked to a deficiency in magnesium. Magnesium supplements may lessen the neuroinflammation connected to a variety of neurological disorders. Magnesium has a variety of roles in the control of immune responses, especially with regard to the functioning of various immune cell subpopulations (DiNicolantonio et al., 2021). The nature of the innate immune system, the adaptive immune system, and the control of acute and chronic inflammatory processes can all serve as especially clear examples of the varied consequences. Magnesium is essential for controlling metabolism and preserving brain equilibrium. Lack of it is associated with neuroinflammation and systemic low-grade inflammation, especially in neurodegenerative illnesses including Parkinson's, Alzheimer's, and multiple sclerosis (Maier et al., 2022). When glutamate acts only on the α-amino-3–hydroxyl-5-methyl-4-isoxazole-propionate (AMPA) receptor, it inhibits the glutamate N-methyl-D-aspartate receptor (NMDA-R) at the physiological membrane potential, which is approximately −70 mV. This prevents the NMDA-R from being sustainedly stimulated, which ultimately results in neuronal death (Yamanaka et al., 2019). The capacity of magnesium to prevent the opening of the mitochondrial permeability transition pore and the consequent release of cytochrome c, which leads to apoptosis, is another factor contributing to its protective effect. It also has a role in maintaining ionic homeostasis, GABA-R agonism, and calcium antagonism, maintenance of ionic homeostasis, protection of blood brain barrier homeostasis (Maier et al., 2022) and Mg deficiency can lead to the upregulation of inflammatory cytokines and activation of microglia, contributing to neuroinflammation (Tjalkens et al., 2017).

4.2 Vitamins

Vitamins are regarded as organic substances that are necessary for the body's growth and regular operation. Organisms do not synthesize vitamins entirely or synthesize insufficient amounts. Therefore, dieting is the only way to receive them. Crucially, vitamins frequently serve as enzymatic cofactors or antioxidants. There are two primary types of vitamins: water-soluble vitamins, which need a constant external supply every day, and fat-soluble vitamins, which are stored in the body until your body requires them (Stevens, 2021). Vitamin supplementation has avoided the toxicity of alpha-synuclein (Yang et al., 2024). Furthermore, vitamins also shield the dopamine transporter from harm. A larger vitamin dosage also progressively prevents diseases caused by tau and amyloid. The B complex, (Vitamin B12: 0.5 mg/day, Vitamin B6: 20 mg/day) as well as folic acid 90.8 mg/day), vitamins A, C, E, and D_3, are essential for controlling neuroinflammation and shielding the brain system from inflammatory damage. All-trans retinoic acid (ATRA), a metabolite of vitamin A, offers protection against the neurotoxic effects of alcohol. Numerous essential processes are regulated by vitamin A, including the regulation of neural cell differentiation, patterning in the creation of the neural tube, and neuroinflammatory alterations in the brain (Rai et al., 2021). Oxidative stress is one of the main causes of Parkinson disease development, and vitamin A and its derivatives, including retinoic acid (RA), have potent antioxidative action (Huang et al., 2018). In animal models of Alzheimer's disease, vitamin E (tocopherol) prevents progressive neurodegeneration and exhibits synergistic effects with antioxidants. Meanwhile, vitamin C functions as an antioxidant in the brain, aids in the maintenance of vitamin E between 2.5 - 375 mg/kg for 6 months have been reported to contribute to the production of neurotransmitters and brain development (Gugliandolo et al., 2017) in Ttpa-/- mice. According to Holton (2021), vitamin C functions as an antioxidant, aids in lowering microglial activation, and decreases the synthesis of pro-inflammatory cytokines including TNF- and IL-1. It also helps the brain produce neurotransmitters and myelinate, and it guards against excitotoxicity brought on by glutamate. In a ubiquitin C-terminal hydrolase L1 (UCH-L1) gene knockdown Drosophila Parkinson's disease model, vitamin C treatments have been seen to reduce the Parkinson's disease-like phenotype of dopaminergic neuron degradation and locomotor impairments (Man Anh et al., 2019). The role of vitamin C in Huntington's disease has been studied extensively. Ascorbate dysregulation is seen in the cortical and striatal pathways of transgenic mice produced by Huntingtins (Rai et al., 2021). In the pathophysiology of Huntingtin, these pathways use the neurotransmitter glutamate to control ascorbate action. Additionally, vitamin E therapy in the Parkinson's disease rat model appears to have mitigated the course of the illness caused by intra-striatal injection of 6-OHDA (6-hydroxydopamine), according to biochemical and behavioral findings (Rai et al., 2021). In the brain, vitamin B_{12} controls oxidative stress, immunological response, and energy consumption.

It inhibits excitotoxicity, protects neurons from homocysteine-induced apoptosis, and modifies the expression of genes involved in inflammation (Alam., 2022). One important observation is that vitamin B_{12} prevents tau aggregation, which in turn prevents the formation of neurofibrillary tangles, by inhibiting tau fibrillization and tangle formation (Rafiee et al., 2017). Increased homocysteine levels, a sulfur-containing byproduct of methionine biosynthesis, have been reported to have numerous neurotoxic consequences in a few neurodegenerative illnesses, especially Parkinson's disease (Moretti and Caruso, 2019). Therefore, there was a high association between homocysteine level and Parkinson's disease pathogenesis in individuals with elevated homocysteine levels (Saadat et al., 2018). In Parkinson's disease patients, the loss of nigrostriatal dopaminergic neurons was caused by elevated homocysteine levels. Therefore, controlling homocysteine levels may stop Parkinson's disease from getting worse. Increased homocysteine levels, a sulfur-containing byproduct of methionine biosynthesis, have been reported to have numerous neurotoxic consequences in a few neurodegenerative illnesses, including Parkinson's disease (Moretti and Caruso, 2019). Therefore, there was a high association between homocysteine level and Parkinson's disease pathogenesis in individuals with elevated homocysteine levels (Saadat et al., 2018).

Vitamin D_3 reduces neuroinflammation and oxidative stress linked to neurodegenerative disorders including Alzheimer's and Parkinson's, depending on the stage and severity of Alzheimer's disease (Calou et al., 2023). The majority of vitamin D_3's biological effects are mediated via target gene regulation via nuclear vitamin D_3 Receptors. As a ligand-inducible transcription factor that is a member of the nuclear hormone receptor superfamily, vitamin D_3 has anti-inflammatory and antioxidative properties that may be seen at both peripheral and central locations. Patients with Parkinson's disease had significantly lower blood plasma concentrations of calcitriol than age-matched healthy controls, which may have implications for bone health and fracture risk (Anjum et al., 2018). Remarkably, the CNS expresses the calcitriol receptor as well (Lv et al., 2020). So, calcitriol, the active form of vitamin D, and related receptors in the CNS mediate vitamin Ds actions in Parkinson's disease (Anjum et al., 2018). Vitamins that are soluble in lipids or water both significantly protect Parkinson's and Alzheimer's disease. The disorders Huntington's, Parkinson's, and multiple sclerosis also demonstrate the therapeutic benefits of vitamins (Rai et al., 2021). Amyotrophic lateral sclerosis is a fatal form of motor neuron disease (MND) characterized by progressive degeneration of motor neurons in the brain and spinal cord. The loss of motor neurons leads to the deterioration of whole-body muscle mass (Martin et al., 2017). Reversible Parkinsonism symptoms have also been linked to vitamin D poisoning, according to another study (Chatterjee et al., 2017) and according to Rai et al. (2021), both vitamin D and vitamin E may preserve motor neurons and alleviate motor symptoms in people with amyotrophic lateral sclerosis. Vitamin D has been demonstrated to preserve motor neurons in-vitro, and in individuals with amyotrophic lateral sclerosis, plasma levels of the vitamin were directly connected with the disease's severity (Plantone et al., 2022). Genetic research has demonstrated that vitamin D regulates multiple immune components, including toll-like receptors, major histocompatibility complex (MHC) class II molecules, poly (ADP-ribose) polymerase 1 (PARP1), and heme oxygenase-1 (HO-1), which is linked to the pathology of amyotrophic lateral sclerosis (Lương and Nguyễn, 2013). It's interesting to note that, according to the results, individuals who did not take their recommended daily amount of vitamin E died earlier than people who did. Additionally, this study found that vitamin E greatly enhanced the motor performance of ALS patients (Traber and Head, 2021).

Neurodegeneration follows increasing neuroinflammation in multiple sclerosis (Pérez-Cerdá et al., 2016). The myelin sheaths that cover and shield nerve fibers in the CNS are attacked by the body's immune system (Stassart et al., 2018). Common symptoms of multiple sclerosis include discomfort, exhaustion, irregularities of movement, and visual difficulties. The preventive effects of vitamin D in multiple sclerosis patients are often dependent on their developmental stage. Accordingly, in a juvenile/adolescent rat model of multiple sclerosis, vitamin D treatment was more effective than vitamin D in treating multiple sclerosis-like neuroinflammation in rats that were older (Rai et al., 2021). Reduced blood levels of vitamin D have been linked to an increased risk of multiple sclerosis, according to observational research and clinical trials (Pierrot-Deseilligny and Souberbielle, 2017). By promoting the oxidation of white matter, vitamin D has strong anti-inflammatory effects (Anjum et al., 2018). On the other hand, hypercalcemia hazardous accumulation of calcium in the blood—can result from a vitamin D overload. The use of vitamin D for the prevention and/or intervention of multiple sclerosis has received a lot of attention due to numerous observational studies that have linked vitamin D levels to the risk of the disease and to the anti-inflammatory properties of the vitamin. Furthermore, because of their anti-inflammatory and antioxidant properties, additional vitamins like A and E may be useful as adjuvants in a variety of adjuvants in multiple sclerosis treatments as the role of vitamins like vitamin B in multiple sclerosis is still controversial.

5. Fats and neuroinflammation

High-fat diets are strongly associated with systemic and neuroinflammation, driven by increased production of pro-inflammatory cytokines such as TNF-α, IL-6, and IL-1β. These cytokines not only disrupt systemic metabolic processes but also compromise the BBB, allowing inflammatory molecules and immune cells to infiltrate CNS (Hargrave et al., 2016; Więckowska-Gacek et al., 2021Heneka et al., 2015). This infiltration triggers microglial activation, leading to the sustained release of inflammatory mediators and reactive oxygen species (ROS), which exacerbate neuroinflammation (Glass et al., 2010; Perry et al., 2010). This persistent inflammatory state eventually leads to neuronal damage and the advancement of neurodegenerative illnesses like Parkinson's and Alzheimer's (Heneka et al., 2014; Glass et al., 2010; Heidari et al., 2022). Moreover, high-fat diets are often associated with oxidative stress, which creates a toxic environment for neurons and glial cells, further amplifying the pathological processes linked to neurodegeneration (Leyane et al., 2022; Więckowska-Gacek et al., 2021).

Furthermore, High-fat diets disrupt the integrity of the BBB through multiple mechanisms. Elevated inflammatory cytokines degrade tight junction proteins like occludin and claudin-5, which are critical for maintaining BBB selectivity (Tohidpour et al., 2017; Więckowska-Gacek et al., 2021). Additionally, oxidative stress caused by excessive ROS from high-fat diets damages endothelial cells, further weakening the BBB and allowing harmful substances to enter the brain (Więckowska-Gacek et al., 2021). This breach in BBB integrity enables peripheral immune cells and inflammatory molecules to infiltrate the CNS, initiating and sustaining neuroinflammation (Heneka et al., 2014). Rat studies between 10, 40, and 90 days have demonstrated that rodents fed high-fat diets exhibit increased BBB permeability, accumulation of inflammatory markers in the brain, and cognitive deficits on Y-maze task (Hargrave et al., 2016). These findings underscore the critical role of BBB dysfunction as a bridge between systemic metabolic changes and CNS pathologies . This compromised BBB also facilitates a positive feedback loop of systemic and neuroinflammation, amplifying neuronal damage and impairing cognitive functions .

Lipopolysaccharides (LPS), components of the outer membrane of gram-negative bacteria, are endotoxins that play a pivotal role in systemic and neuroinflammation, particularly in the context of high-fat diets. A high-fat diet disrupts gut barrier integrity, a condition known as "leaky gut," which allows LPS to enter the bloodstream (Cani et al., 2007). Once in circulation, LPS acts as a potent activator of the immune system through toll-like receptor 4 (TLR4) signaling, triggering the release of pro-inflammatory cytokines such as TNF-α and IL-6 (Hung et al., 2022; Kong et al., 2020). LPS can also cross the BBB, either by increasing its permeability or by binding to receptors on endothelial cells, thereby infiltrating the CNS and activating microglial cells. These microglia, once activated, release inflammatory mediators and ROS, which exacerbate neuroinflammation and contribute to neuronal damage (Erny et al., 2017). Chronic exposure to LPS has been linked to cognitive deficits and the progression of neurodegenerative diseases like Alzheimer's, as it creates a pro-inflammatory environment in the brain that impairs synaptic plasticity and neurogenesis (Hargrave et al., 2016; Ransohoff and El Khoury, 2016).

In the brain, high-fat diets promote persistent microglial activation, a hallmark of neuroinflammation. Activated microglia produce pro-inflammatory cytokines, ROS, and other toxic molecules that impair synaptic plasticity, disrupt neurotransmitter systems, and reduce neurogenesis (Perry and Teeling, 2013). Similarly, astrocytes undergo astrogliosis in response to high-fat diet-induced inflammation, releasing neurotoxic molecules and glutamate, which can lead to excitotoxicity and further neuronal damage (Sofroniew and Vinters, 2010). These glial responses, coupled with oxidative stress, create an inflammatory environment that accelerates the pathogenesis of neurodegenerative diseases and cognitive decline (Heneka et al., 2015; Ransohoff and El Khoury, 2016). Furthermore, the chronic activation of these glial cells impairs the brain's natural repair mechanisms, contributing to long-term neural dysfunction and increasing susceptibility to disorders like Alzheimer's disease, Parkinson's disease, and multiple sclerosis (Perry et al., 2010).

The link between high-fat diets and obesity further exacerbates these effects, particularly due to the role of visceral fat as an endocrine organ. Visceral fat produces large amounts of inflammatory cytokines and free fatty acids, which can cross the BBB and trigger CNS inflammation (Hotamisligil, 2006; Scherer, 2016). This type of fat is more metabolically active than subcutaneous fat and is strongly associated with systemic inflammation and insulin resistance, both of which contribute to neuroinflammation (Patel and Abate, 2013). Obesity, driven by high fat diets, also impairs hypothalamic function, contributing to leptin resistance and a disrupted energy balance (Thaler et al., 2012; Valdearcos et al., 2014). Chronic inflammation from obesity and high-fat diets disrupts the gut-brain axis, allowing lipopolysaccharides (LPS) from the gut microbiota to enter circulation and activate immune pathways that exacerbate neuroinflammation (Cani et al., 2007; Erny et al., 2017).

Saturated fats, found predominantly in red meat, butter, full-fat dairy products, and tropical oils like coconut and palm oil, are characterized by their chemical structure, which contains no double bonds between carbon atoms, making them solid at room temperature. These fats are strongly associated with pro-inflammatory effects, as they promote the production of cytokines such as TNF-α and IL-6, which play critical roles in systemic inflammation and neuroinflammation. By disrupting the integrity of the

BBB, saturated fats allow harmful substances, including cytokines and immune cells, to enter the brain and activate microglial cells. This activation perpetuates a cycle of inflammation, oxidative stress, and neuronal damage, which are implicated in the progression of neurodegenerative diseases like Alzheimer's and Parkinson's diseases (Glass et al., 2010; Heidari et al., 2022; Heneka et al., 2014; Leyane et al., 2022). Long-term consumption of saturated fats contributes to chronic low-grade inflammation, exacerbating metabolic disorders and increasing the risk of cognitive decline and other neurodegenerative conditions (Heneka et al., 2015).

Unsaturated fats, by contrast, are generally considered healthier and are categorized into monounsaturated fats and polyunsaturated fats (PUFAs). Monounsaturated fats, found in olive oil, avocados, and nuts, contain a single double bond in their chemical structure, making them liquid at room temperature and solid when chilled. Unsaturated fats are known for their anti-inflammatory properties, as they reduce inflammatory markers like C-reactive protein (CRP) and IL-6 and have been associated with lower levels of neuroinflammation and a reduced risk of cognitive decline. Diets rich in monounsaturated fats, such as the Mediterranean diet, are linked to improved brain health and a lower risk of neurodegenerative diseases (McGrattan et al., 2019). Polyunsaturated fats include omega-3 and omega-6 fatty acids, which are found in foods such as fish, flaxseeds, walnuts, and certain vegetable oils. Omega-3 fatty acids, particularly eicosapentaenoic acid (EPA) and docosahexaenoic acid (DHA), are potent anti-inflammatory agents that inhibit the production of pro-inflammatory cytokines like TNF-α and IL-6. They also promote the synthesis of anti-inflammatory mediators, such as resolvins and protectins, which play critical roles in reducing neuroinflammation, protecting BBB integrity, and regulating microglial activation (Calder, 2017). In contrast, omega-6 fatty acids, while essential for immune responses and cellular function, can contribute to pro-inflammatory effects when consumed in excessive amounts or without a balanced intake of omega-3s, as is common in many modern diets (Simopoulos, 2016).

Trans fats, which are artificially created through a process called hydrogenation to solidify liquid oils, are found in many processed foods, including margarine, baked goods, and fried fast food. Chemically similar to saturated fats, trans fats are strongly pro-inflammatory and have been shown to increase the production of cytokines such as TNF-α, IL-6, and IL-1β. These inflammatory responses are further exacerbated by oxidative stress, as trans fats promote the generation of ROS, which damage cellular structures and amplify systemic inflammation. In the brain, trans

Table 1 The ways fats function in mitigating the onset of neuroinflammation and neuroprotection respectively.

Fats (types)	Role in Neuroinflammation	Reference
High-fat diet	Production of pro-inflammatory cytokines. Oxidative stress- toxic environment for neurons and glial cells. Disruption of BBB integrity. Damages endothelial cells. Inflammatory markers; Cognitive deficits. Microglial cells activation. Astrogliosis, excitotoxicity and neuronal damage	Hargrave et al., 2016; Erny et al., 2017; Perry and Teeling, 2013; Sofroniew and Vinters, 2010; Hung et al., 2022
Visceral fat (endocrine gland)	Impairs hypothalamic function Leptin resistance. Disrupted energy balance. Gut-brain axis disruption and lipopolysaccharides (LPS) entry.	Thaler et al., 2012; Valdearcos et al., 2014; Erny et al., 2017; Hung et al., 2022
Saturated fats	Production of cytokines such as TNF-α and IL-6. Chronic low-grade inflammation. Exacerbates metabolic disorders. Increases cognitive decline	Heneka et al., 2015
Unsaturated fats	Anti-inflammatory properties. Reduces inflammatory markers [C-reactive protein (CRP) and IL-6]. Lowers neuroinflammation. Reduces risk of cognitive decline	Estruch et al., 2018; McGrattan et al., 2019
Monounsaturated fats	Improves brain health. Lowers risk of neurodegenerative diseases.	Estruch et al., 2018; McGrattan et al., 2019
Polyunsaturated fats	Inhibits production of pro-inflammatory cytokines like TNF-α and IL-6. Promotes synthesis of anti-inflammatory mediators [resolvins and protectins]. Protects BBB integrity. Regulates microglial activation	Calder, 2017
Trans fats	Pro-inflammatory agents. Increases cytokine production. Contributes to BBB dysfunction Activate microglia and astrocytes. Vascular damage Impaired brain perfusion	Calder, 2017; Morris and Tangney, 2014

fats contribute to BBB dysfunction, allowing inflammatory molecules to infiltrate the CNS and activate microglia and astrocytes. This chronic neuroinflammation is associated with an increased risk of neurodegenerative diseases, including Alzheimer's disease, as the inflammatory environment created by trans fats accelerates neuronal damage and cognitive decline (Morris and Tangney, 2014). Regular consumption of trans fats has also been linked to cardiovascular disease, which further compounds the risk of neuroinflammatory conditions through vascular damage and impaired brain perfusion as illlustrated in Table 1.

Furthermore, reducing intake of pro-inflammatory fats, such as saturated and trans fats, while increasing anti-inflammatory omega-3 polyunsaturated fatty acids (PUFAs), can help mitigate inflammation (Estruch et al., 2018; Calder, 2017). Omega-3 PUFAs, such as docosahexaenoic acid (DHA), have shown significant neuroprotective effects by reducing micro-glial activation and promoting the production of anti-inflammatory mediators (Calder, 2017). Therapeutic approaches targeting oxidative stress and inflammatory pathways, including the NF-κB and JNK signaling cascades, show promise in countering fat-induced neuroinflammation . Additionally, interventions that restore gut health, such as probiotics and prebiotics, may improve gut-brain axis function and reduce systemic inflammation (Lee et al., 2021; Teng et al., 2024; Dinan and Cryan, 2012). These combined strategies emphasize the importance of dietary and lifestyle modifications in preserving brain health and preventing inflammation-driven neurodegenerative diseases.

6. Conclusion

Taken together, the intricate regulation of neuroinflammation by cytokines, chemokines, and other mediators highlights the complexity of immune responses in the CNS. Understanding these mechanisms is critical for developing targeted therapies to mitigate inflammation, restore immune balance, and prevent the progression of neurodegenerative disorders. In addition, understanding the interplay between neuroinflammation and the immune system opens new avenues for therapeutic interventions, including the use of proteins, amino acids, vitamins, and minerals to modulate immune responses. Strategies targeting inflammatory pathways are being actively explored as potential treatments for neurodegenerative diseases. These approaches aim to restore immune balance by either enhancing anti-inflammatory mechanisms or inhibiting pro-inflammatory cytokines. The recognition of immune responses in dementia has gained traction, with immune-related genetic abnormalities increasingly identified as risk factors for neurodegeneration. As key molecular pathways become elucidated, there is robust support for developing therapies to regulate neuroinflammation and prevent CNS disorders of which nutritional avenue could serve to an end.

References

Abd El Salam, A.S.G., Abd Elrazik, N.A., 2024. Cinnamaldehyde/lactulose combination therapy alleviates thioacetamide-induced hepatic encephalopathy via targeting P2X7R-mediated NLRP3 inflammasome signaling. Life Sci. 344, 122559. doi:10.1016/j.lfs.2024.122559.

Alam, J., 2022. Vitamins: a nutritional intervention to modulate the Alzheimer's disease progression. Nutr. Neurosci. 25 (5), 945–962.

Anjum, I., Jaffery, S.S., Fayyaz, M., Samoo, Z., Anjum, S., 2018. The role of vitamin D in brain health: a mini literature review. Cureus 10 (7), e2960.

Borst, K., Dumas, A.A., Prinz, M., 2021. Microglia: immune and non-immune functions. Immunity 54 (10), 2194–2208.

Calou, I.B.F., de Barros Viana, M., Ribeiro, D.A., Estadella, D., de Barros Viana, G.S., 2023. Vitamin D3 (VD3) effects on neuroinflammation and neurodegenerative disorders: focus on Alzheimer's and Parkinson's diseases. Vitamins and Minerals in Neurological Disorders, 023, pp. 417–437.

Castelo Rueda and Maria Paulin. 2021. The role of mitochondria in reduced penetrance of heterozygous Parkin mutations. Doctoral Thesis. https://dx.doi.org/10.15168/11572_319705

Chao, Y.W., Tung, Y.T., Yang, S.C., Shirakawa, H., Su, L.H., Loe, P.Y., Chiu, W.C., 2024. The effects of rice bran on neuroinflammation and gut microbiota in ovariectomized mice fed a drink with fructose. Nutrients 16 (17), 2980. doi:10.3390/nu16172980.

Chatterjee, R., Chatterjee, K., Sen, C., 2017. Reversible parkinsonism due to vitamin D toxicity. J. Neurosci. Rural Pract. 8 (02), 305–306.

Dalangin, R., Kim, A., Campbell, R.E., 2020. The role of amino acids in neurotransmission and fluorescent tools for their detection. Int. J. Mol. Sci. 21 (17), 6197.

Dalangin, R.R., 2021. Engineering genetically encoded indicators for neurotransmission. Doctoral Thesis. ualberta.scholaris.ca.

Didonna, A., 2020. Tau at the interface between neurodegeneration and neuroinflammation. Genes Immun. 21 (5), 288–300.

Dinan, T.G., Cryan, J.F., 2012. Regulation of the stress response by the gut microbiota: implications for psychoneuroendocrinology. Psychoneuroendocrinology 37 (9), 1369–1378.

DiNicolantonio, J.J., O'Keefe, J.H., 2021. Magnesium and vitamin D deficiency as a potential cause of immune dysfunction, cytokine storm and disseminated intravascular coagulation in COVID-19 patients. Mo. Med. 118 (1), 68.

Erny, D., Hrabě de Angelis, Prinz, M., 2017. Communicating systems in the body: how microbiota and microglia cooperate. Immunology 150 (1), 7–15.

Feijó, G.D.S., Jantsch, J., Correia, L.L., Eller, S., Furtado-Filho, O.V., Giovenardi, M., et al., 2022. Neuroinflammatory responses following zinc or branched-chain amino acids supplementation in obese rats. Metab. Brain Dis. 37 (6), 1875–1886.

Gammoh, N.Z., Rink, L., 2017. Zinc in infection and inflammation. Nutrients 9 (6), 624.

Gammoh, N.Z., Rink, L., 2019. Zinc and the immune system. In: *Nutrition and immunity*. Springer International Publishing, Cham, pp. 127–158.

Ghazy, A.A., Soliman, O.A., Elbahnasi, A.I., Alawy, A.Y., Mansour, A.M., Gowayed, M.A., 2022. Role of oxytocin in different neuropsychiatric, neurodegenerative, and neurodevelopmental disorders. Rev. Physiol. Biochem. Pharmacol. 186, 95–134.

Glass, C.K., Saijo, K., Winner, B., Marchetto, M.C., Gage, F.H., 2010. Mechanisms underlying inflammation in neurodegeneration. Cell 140 (6), 918–934. doi:10.1016/j.cell.2010.02.016.

Glenn, J.M., Madero, E.N., Bott, N.T., 2019. Dietary protein and amino acid intake: links to the maintenance of cognitive health. Nutrients 11 (6), 1315.

Gong, F., Ge, T., Liu, J., Xiao, J., Wu, X., Wang, H., et al., 2022. Trehalose inhibits ferroptosis via NRF2/HO-1 pathway and promotes functional recovery in mice with spinal cord injury. Aging (Albany NY) 14 (7), 3216–3232. doi:10.18632/aging.204009.

Gugliandolo, A., Bramanti, P., Mazzon, E., 2017. Role of vitamin E in the treatment of Alzheimer's disease: evidence from animal models. Int. J. Mol. Sci. 18 (12), 2504.

Haase, S., Kuhbandner, K., Mühleck, F., Gisevius, B., Freudenstein, D., Hirschberg, S., et al., 2024. Dietary galactose exacerbates autoimmune neuroinflammation via advanced glycation end product-mediated neurodegeneration. Front. Immunol. 15, 1367819. doi:10.3389/fimmu.2024.1367819.

Hargrave, S.L., Davidson, T.L., Zheng, W., Kinzig, K.P., 2016. Western diets induce blood-brain barrier leakage and alter spatial strategies in rats. Behav. Neurosci. 130 (1), 123.

Heidari, A., Yazdanpanah, N., Rezaei, N., 2022. The role of Toll-like receptors and neuroinflammation in Parkinson's disease. J. Neuroinflam. 19 (1), 135.

Heneka, M.T., Carson, M.J., El Khoury, J., Landreth, G.E., Brosseron, F., Feinstein, D.L., et al., 2014. Neuroinflammation in Alzheimer's disease. Lancet Neurol. 14 (4), 388–405. doi:10.1016/S1474-4422(14)70016-6.

Heneka, M.T., Golenbock, D.T., Latz, E., 2015. Innate immunity in Alzheimer's disease. Nat. Immunol. 16 (3), 229–236. doi:10.1038/ni.3102.

Holton, K.F., 2021. Micronutrients may be a unique weapon against the neurotoxic triad of excitotoxicity, oxidative stress and neuroinflammation: a perspective. Front. Neurosci. 15, 726457.

Hsu, H.W., Bondy, S.C., Kitazawa, M., 2018. Environmental and dietary exposure to copper and its cellular mechanisms linking to Alzheimer's disease. Toxicol. Sci. 163 (2), 338–345.

Huang, Z., Liu, Y., Qi, G., Brand, D., Zheng, S.G., 2018. Role of vitamin A in the immune system. J. Clin. Med. 7 (9), 258.

Hung, H.C., Tsai, S.F., Sie, S.R., Kuo, Y.M., 2022. High glucose enhances lipopolysaccharide-induced inflammation in cultured BV2 microglial cell line. Immun. Inflamm. Dis. 10 (5), e610. doi:10.1002/iid3.610.

Hwang, D., Boehm, A., Rostami, A., Zhang, G.X., Ciric, B., 2022. Oral d-mannose treatment suppresses experimental autoimmune encephalomyelitis via induction of regulatory T cells. J. Neuroimmunol. 362, 577778. doi:10.1016/j.jneuroim.2021.577778.

Jiang, P.P., Peng, S.S., Pankratova, S., Luo, P., Zhou, P., Chen, Y., 2022. Proteins involved in synaptic plasticity are downregulated in the cerebrospinal fluid of infants with clinical sepsis complicated by neuroinflammation. Front. Cell. Neurosci. 16, 887212.

Kandasamy, P., Gyimesi, G., Kanai, Y., Hediger, M.A., 2018. Amino acid transporters revisited: new views in health and disease. Trends Biochem. Sci 43 (10), 752–789.

Kölliker-Frers, R., Udovin, L., Otero-Losada, M., Kobiec, T., Herrera, M.I., Palacios, J., et al., 2021. Neuroinflammation: an integrating overview of reactive-Neuroimmune cell interactions in health and disease. Mediators Inflamm. 2021 (1), 9999146.

Kong, J., Du, Z., Dong, L., 2020. Pinitol prevents lipopolysaccharide (LPS)-induced inflammatory responses in BV2 microglia mediated by TREM2. Neurotox. Res. 38 (1), 96–104. doi:10.1007/s12640-020-00187-z.

Lee, Y.S., Lai, D.M., Huang, H.J., Lee-Chen, G.J., Chang, C.H., Hsieh-Li, H.M., et al., 2021. Prebiotic lactulose ameliorates the cognitive deficit in Alzheimer's disease mouse model through macroautophagy and chaperone-mediated autophagy pathways. J. Agric. Food Chem. 69 (8), 2422–2437. doi:10.1021/acs.jafc.0c07327.
Leyane, T.S., Jere, S.W., Houreld, N.N., 2022. Oxidative stress in ageing and chronic degenerative pathologies: molecular mechanisms involved in counteracting oxidative stress and chronic inflammation. Int. J. Mol. Sci. 23 (13), 7273.
Li, P., He, W., Wu, G., 2021. Composition of amino acids in foodstuffs for humans and animals. In *Amino acids in nutrition and health: amino acids in gene expression, metabolic regulation, and exercising performance*. Springer International Publishing, Cham, pp. 189–210.
Liddelow, S.A., Barres, B.A., 2017. Reactive astrocytes: production, function, and therapeutic potential. Immunity 46 (6), 957–967. doi:10.1016/j.immuni.2017.06.006.
Lucas, S.M., Rothwell, N.J., Gibson, R.M., 2006. The role of inflammation in CNS injury and disease. Br. J. Pharmacol. 147 (S1), S232–S240. doi:10.1038/sj.bjp.0706400.
Lương, K.V.Q., Nguyễn, L.T.H., 2013. Roles of vitamin D in amyotrophic lateral sclerosis: possible genetic and cellular signaling mechanisms. Mol. Brain 6 (1), 16.
Lv, L., Tan, X., Peng, X., Bai, R., Xiao, Q., Zou, T., et al., 2020. The relationships of vitamin D, vitamin D receptor gene polymorphisms, and vitamin D supplementation with Parkinson's disease. Transl. Neurodegener. 9, 1–13.
Maier, J.A., Locatelli, L., Fedele, G., Cazzaniga, A., Mazur, A., 2022. Magnesium and the brain: a focus on neuroinflammation and neurodegeneration. Int. J. Mol. Sci. 24 (1), 223.
Man Anh, H., Linh, D.M., My Dung, V., Thi Phuong Thao, D., 2019. Evaluating dose-and time-dependent effects of vitamin C treatment on a Parkinson's disease fly model. Parkinsons Dis. 2019 (1), 9720546.
Martin, S., Al Khleifat, A., Al-Chalabi, A., 2017. What causes amyotrophic lateral sclerosis? F1000Res. 6, 371.
Maugard, M., Vigneron, P.A., Bolaños, J.P., Bonvento, G., 2021. l-Serine links metabolism with neurotransmission. Prog. Neurobiol. 197, 101896.
Mayne, K., White, J.A., McMurran, C.E., Rivera, F.J., de la Fuente, A.G., 2020. Aging and neurodegenerative disease: is the adaptive immune system a friend or foe? Front. Aging Neurosci. 12, 572090.
McGrattan, A.M., McGuinness, B., McKinley, M.C., et al., 2019. Diet and Inflammation in Cognitive Ageing and Alzheimer's Disease. Curr Nutr Rep 8, 53–65. doi:10.1007/s13668-019-0271-4.
Moreno-García, A., Kun, A., Calero, M., Calero, O., 2021. The neuromelanin paradox and its dual role in oxidative stress and neurodegeneration. Antioxidants 10 (1), 124.
Moretti, R., Caruso, P., 2019. The controversial role of homocysteine in neurology: from labs to clinical practice. Int. J. Mol. Sci. 20 (1), 231.
Murao, A., Aziz, M., Wang, H., Brenner, M., Wang, P., 2021. Release mechanisms of major DAMPs. Apoptosis 26 (3), 152–162.
Nimgampalle, M., Chakravarthy, H., Sharma, S., Shree, S., Bhat, A.R., Pradeepkiran, J.A., Devanathan, V., 2023. Neurotransmitter systems in the etiology of major neurological disorders: emerging insights and therapeutic implications. Ageing Res. Rev., 101994.
Nimmerjahn, A., Kirchhoff, F., Helmchen, F., 2005. Resting microglial cells are highly dynamic surveillants of brain parenchyma in vivo. Science 308 (5726), 1314–1318. doi:10.1126/science.1110647.
O'Day, D.H., Huber, R.J., 2022. Calmodulin binding proteins and neuroinflammation in multiple neurodegenerative diseases. BMC Neurosci. 23 (1), 10.
Papadakis, S., Thompson, J.R., Feczko, E., Miranda-Dominguez, O., Dunn, G.A., Selby, M., et al., 2024. Perinatal Western-style diet exposure associated with decreased microglial counts throughout the arcuate nucleus of the hypothalamus in Japanese macaques. J Neurophysiol 131 (2), 241–260. doi:10.1152/jn.00213.2023.

Patel, S., 2018. Danger-associated molecular patterns (DAMPs): the derivatives and triggers of inflammation. Curr. Allergy Asthma Rep. 18, 1–12.
Pérez-Cerdá, F., Sánchez-Gómez, M.V., Matute, C., 2016. The link of inflammation and neurodegeneration in progressive multiple sclerosis. Mult. Scler. Demyelinating Disord. 1, 1–8.
Perry, V.H., Nicoll, J.A., Holmes, C., 2010. Microglia in neurodegenerative disease. Nat. Rev. Neurol. 6 (4), 193–201. doi:10.1038/nrneurol.2010.17.
Perry, V.H., Teeling, J., 2013. Microglia and macrophages in the central nervous system. Lancet Neurol. 12 (2), 188–199. doi:10.1016/S1474-4422(12)70276-3.
Pierrot-Deseilligny, C., Souberbielle, J.C., 2017. Vitamin D and multiple sclerosis: an update. Mult. Scler. Relat. Disord. 14, 35–45.
Plantone, D., Primiano, G., Manco, C., Locci, S., Servidei, S., De Stefano, N., 2022. Vitamin D in neurological diseases. Int. J. Mol. Sci. 24 (1), 87.
Rafiee, S., Asadollahi, K., Riazi, G., Ahmadian, S., Saboury, A.A., 2017. Vitamin B12 inhibits tau fibrillization via binding to cysteine residues of tau. ACS Chem. Neurosci. 8 (12), 2676–2682.
Rai, S.N., Singh, P., Steinbusch, H.W., Vamanu, E., Ashraf, G., Singh, M.P., 2021. The role of vitamins in neurodegenerative disease: an update. Biomedicines 9 (10), 1284.
Ransohoff, R.M., Brown, M.A., 2012. Innate immunity in the central nervous system. J. Clin. Invest. 122 (4), 1164–1171. doi:10.1172/JCI58644.
Ransohoff, R.M., El Khoury, J., 2016. Microglia in health and disease. Nat. Neurosci. 19 (3), 453–465. doi:10.1038/nn.4222.
Ransohoff, R.M., Schafer, D., Vincent, A., Blachère, N.E., Bar-Or, A., 2015. Neuroinflammation: ways in which the immune system affects the brain. Neurotherapeutics 12 (4), 896–909.
Saadat, P., Ahmadi Ahangar, A., Samaei, S.E., Firozjaie, A., Abbaspour, F., Khafri, S., et al., 2018. Serum homocysteine level in Parkinson's disease and its association with duration, cardinal manifestation, and severity of disease. Parkinsons Dis. 2018 (1), 5813084.
Sanmarco, L.M., Polonio, C.M., Wheeler, M.A., Quintana, F.J., 2021. Functional immune cell–astrocyte interactions. J. Exp. Med. 218 (9), e20202715.
Saso, L., Giuffre, A., Valacchi, G., Maccarrone, M., 2023. Modulation of Oxidative Stress: Biochemical, Physiological and Pharmacological Aspects. Elsevier, Medical, p. 390.
Savonije, K., Weaver, D.F., 2023. The role of tryptophan metabolism in Alzheimer's disease. Brain Sci. 13 (2), 292.
Scammell, T.E., Jackson, A.C., Franks, N.P., Wisden, W., Dauvilliers, Y., 2019. Histamine: neural circuits and new medications. Sleep 42 (1), p.zsy183.
Schulz, D., Morschel, J., Schuster, S., Eulenburg, V., Gomeza, J., 2018. Inactivation of the mouse L-proline transporter PROT alters glutamatergic synapse biochemistry and perturbs behaviors required to respond to environmental changes. Front. Mol. Neurosci. 11, 279.
Sinha, P., Verma, B., Ganesh, S., 2021. Trehalose ameliorates seizure susceptibility in Lafora disease mouse models by suppressing neuroinflammation and endoplasmic reticulum stress. Mol. Neurobiol. 58 (3), 1088–1101. doi:10.1007/s12035-020-02170-3.
Sofroniew, M.V., 2009. Molecular dissection of reactive astrogliosis and glial scar formation. Trends Neurosci. 32 (12), 638–647. doi:10.1016/j.tins.2009.08.002.
Sonninen, T.M., Goldsteins, G., Laham-Karam, N., Koistinaho, J., Lehtonen, Š., 2020. Proteostasis disturbances and inflammation in neurodegenerative diseases. Cells 9 (10), 2183.
Sorgdrager, F.J., Naudé, P.J., Kema, I.P., Nollen, E.A., Deyn, P.P.D., 2019. Tryptophan metabolism in inflammaging: from biomarker to therapeutic target. Front. Immunol. 10, 2565.
Stassart, R.M., Möbius, W., Nave, K.A., Edgar, J.M., 2018. The axon-myelin unit in development and degenerative disease. Front. Neurosci. 12, 467.

Stefanache, A., Lungu, I.I., Butnariu, I.A., Calin, G., Gutu, C., Marcu, C., et al., 2023. Understanding how minerals contribute to optimal immune function. J. Immunol. Res. 2023 (1), 3355733.

Stevens, S.L., 2021. Fat-soluble vitamins. Nurs. Clin. 56 (1), 33–45.

Teng, C.Y., Kao, N.J., Nguyen, N.T.K., Lin, C.I., Cross, T.L., Lin, S.H., 2024. Effects of xylo-oligosaccharide on gut microbiota, brain protein expression, and lipid profile induced by high-fat diet. J. Nutr. Biochem. 129, 109640. doi:10.1016/j.jnutbio.2024.109640.

Tian, X., Dong, W., Zhou, W., Yan, Y., Lu, L., Mi, J., et al., 2024. The polysaccharides from the fruits of Lycium barbarum ameliorate high fat and high-fructose diet-induced cognitive impairment via regulating blood glucose and mediating gut microbiota. Int. J. Biol. Macromol. 258, 129036. doi:10.1016/j.ijbiomac.2023.129036.

Tjalkens, R.B., Popichak, K.A., Kirkley, K.A., 2017. Inflammatory activation of microglia and astrocytes in manganese neurotoxicity. Adv Neurobiol 18, 159–181. doi:10.1007/978-3-319-60189-2_8.

Tohidpour, A., Morgun, A.V., Boitsova, E.B., Malinovskaya, N.A., Martynova, G.P., Khilazheva, E.D., et al., 2017. Neuroinflammation and infection: molecular mechanisms associated with dysfunction of neurovascular unit. Front. Cell. Infect. Microbiol. 7, 276.

Traber, M.G., Head, B., 2021. Vitamin E: how much is enough, too much and why!. Free Radical Biol. Med. 177, 212–225.

Van de Rest, O., van der Zwaluw, N.L., de Groot, L.C., 2013. Literature review on the role of dietary protein and amino acids in cognitive functioning and cognitive decline. Amino Acids 45, 1035–1045.

Vasquez-Vivar, J., Shi, Z., Tan, S., 2022. Tetrahydrobiopterin in cell function and death mechanisms. Antioxid. Redox Signal. 37 (1–3), 171–183.

Wang, J., Jalali Motlagh, N., Wang, C., Wojtkiewicz, G.R., Schmidt, S., Chau, C., et al., 2021. d-Mannose suppresses oxidative response and blocks phagocytosis in experimental neuroinflammation. Proc. Nat. Acad. Sci. U.S.A. 118 (44), e2107663118. doi:10.1073/pnas.2107663118.

Weyh, C., Krüger, K., Peeling, P., Castell, L., 2022. The role of minerals in the optimal functioning of the immune system. Nutrients 14 (3), 644.

Więckowska-Gacek, A., Mietelska-Porowska, A., Wydrych, M., Wojda, U., 2021. Western diet as a trigger of Alzheimer's disease: from metabolic syndrome and systemic inflammation to neuroinflammation and neurodegeneration. Ageing Res. Rev. 70, 101397. doi:10.1016/j.arr.2021.101397.

Yamanaka, R., Shindo, Y., Oka, K., 2019. Magnesium is a key player in neuronal maturation and neuropathology. Int. J. Mol. Sci. 20 (14), 3439.

Yang, K., Lv, Z., Zhao, W., Lai, G., Zheng, C., Qi, F., et al., 2024. The potential of natural products to inhibit abnormal aggregation of α-Synuclein in the treatment of Parkinson's disease. Front. Pharmacol. 15, 1468850.

Yoo, H.S., Shanmugalingam, U., Smith, P.D., 2022. Potential roles of branched-chain amino acids in neurodegeneration. Nutrition 103, 111762.

Zhang, W., Hong, J., Zheng, W., Liu, A., Yang, Y., 2021. High glucose exacerbates neuroinflammation and apoptosis at the intermediate stage after post-traumatic brain injury. Aging (Albany NY) 13 (12), 16088–16104. doi:10.18632/aging.203136.

Zhang, Y., Du, M.R., Zhang, Q.Y., Yang, S.Y., Chen, J.Q., Dan, C.M., et al., 2024. Armillariella tabescens-derived polysaccharides alleviated α-Gal-induced neuroinflammation and cognitive injury through enterocerebral axis and activation of keap-1/Nrf2 pathway. Int. J. Biol. Macromol. 273, 133035. doi:10.1016/j.ijbiomac.2024.133035.

Zhang, Y., Ye, P., Zhu, H., Gu, L., Li, Y., Feng, S., et al., 2024. Neutral polysaccharide from Gastrodia elata alleviates cerebral ischemia-reperfusion injury by inhibiting ferroptosis-mediated neuroinflammation via the NRF2/HO-1 signaling pathway. CNS Neurosci. Ther. 30 (3), e14456. doi:10.1111/cns.14456.

Zozulya, S.A., Akopyan, A.A., Tenditnik, M.V., Ovsyukova, M.V., Korolenko, T.A., Klyushnik, T.P., et al., 2024. Effect of trehalose disaccharide on activation of microglia and indices of systemic inflammation in an experimental model of Alzheimer's disease. Bull. Exp. Biol. Med. 177 (2), 207–211. doi:10.1007/s10517-024-06157-1.
Das, A., 2022. A metabolic approach to investigate the role of amino acids in brain (Doctoral dissertation, UNSW Sydney). https://doi.org/10.26190/unsworks/24407.
Sofroniew, M.V., Vinters, H.V., 2010. Astrocytes: biology and pathology. Acta Neuropathol 119 (1), 7–35. doi:10.1007/s00401-009-0619-8.
Cani, P.D., Amar, J., Iglesias, M.A., Poggi, M., Knauf, C., Bastelica, D., et al. Metabolic Endotoxemia Initiates Obesity and Insulin Resistance. Diabetes. 2007; 56 (7): 1761–1772. 10.2337/db06-1491.
Hotamisligil, G.S., 2006. Inflammation and metabolic disorders. Nature 444 (7121), 860–867. doi:10.1038/nature05485.
Scherer, P.E., 2016. The Multifaceted Roles of Adipose Tissue-Therapeutic Targets for Diabetes and Beyond: The 2015 Banting Lecture. Diabetes 65 (6), 1452–1461. doi:10.2337/db16-0339.
Thaler, J.P., Yi, C.X., Schur, E.A., Guyenet, S.J., Hwang, B.H., Dietrich, M.O., et al., 2012. Obesity is associated with hypothalamic injury in rodents and humans. J Clin Invest 122 (1), 153–162. doi:10.1172/JCI59660.
Valdearcos, M., Robblee, M., Benjamin, D., 2014. Microglia Dictate the Impact of Saturated Fat Consumption on Hypothalamic Inflammation and Neuronal Function. Cell Rep 9, 2124–2138.
Simopoulos, A.P., 2016. An Increase in the Omega-6/Omega-3 Fatty Acid Ratio Increases the Risk for Obesity. Nutrients 8 (3), 128. doi:10.3390/nu8030128.
Calder, P.C., 2017. Omega-3 fatty acids and inflammatory processes: from molecules to man. Biochem Soc Trans 45 (5), 1105–1115. doi:10.1042/BST20160474.
Patel, P., Abate, N., 2013. Body fat distribution and insulin resistance. Nutrients 5 (6), 2019–2027. doi:10.3390/nu5062019.
Estruch, R., Ros, E., Salas-Salvadó, J., et al., 2018. Primary prevention of cardiovascular disease with a mediterranean diet supplemented with extra-virgin olive oil or nuts. N Engl J Med 378 (25), e34. doi:10.1056/NEJMoa1800389.
Morris, M.C., Tangney, C.C., 2014. Dietary fat composition and dementia risk. Neurobiol Aging 35 (2), S59–64. doi:10.1016/j.neurobiolaging.2014.03.038.

CHAPTER THREE

Influence of non-sugar sweeteners on neural health: The current state of knowledge

Adeboye A. Olaseni, Grazielle C.C. Weis, Luana H. Maurer, Isaac A. Adedara, Tatiana Emanuelli*

Department of Food Technology and Science, Center of Rural Sciences, Federal University of Santa Maria, Camobi, RS, Brazil

*Corresponding author. e-mail address: tatiana.emanuelli@ufsm.br.

Contents

Abstract

There is a growing concern about the use of non-sugar sweeteners (NSS) and their influence on the gut system, resulting in gut dysbiosis. Gut dysbiosis has been known to influence the gut-brain axis, which could potentially hamper brain functions. Recent studies in fish and mammalian models have demonstrated that NSS exposure above

Advances in Neurotoxicology, Volume 16
ISSN 2468-7480
https://doi.org/10.1016/bs.ant.2026.02.002

the acceptable daily intake (ADI) results in an increase in reactive oxygen species, endoplasmic reticulum stress, increased insulin levels, and impaired glucose tolerance. Some studies have also shown that at doses below the ADI or recommended doses, NSS are capable of causing dysbiosis in rodent models and in humans, raising regulatory concerns on the use of these sweeteners. Given the established link between intestinal dysbiosis and neurobehavioral changes, the potential neurotoxicological risks associated with prolonged consumption of these sweeteners necessitate further studies. Emerging studies support an association between NSS consumption and cognitive impairment in some animal models and in humans. This review discussed the current literature on the interplay between NSS and neural health, as well as outlined critical areas for future research.

Keywords: Brain function, Gut microbiota, Gut-brain axis, Neurotoxicological risks, Non-sugar sweetener, Reactive oxygen species

1. Introduction

An increasing body of evidence links diet with gut microbiota and brain function in a phenomenon referred to as the diet-microbiota-gut-brain axis (Schneider et al., 2024). High free sugars consumption is among the dietary factors that are associated with gut dysbiosis, obesity (Guo et al., 2021), and other disorders, such as dental caries, autoimmune syndrome (Johnson et al., 2017), cardiovascular diseases, type II diabetes, and hepatic steatosis (Jensen et a., 2018), as well as neuronal degeneration (Pistollato et al., 2018). Accordingly, the World Health Organization (WHO) released a strong recommendation that daily free sugar intake must be lower than 10 percent of total energy intake and a conditional recommendation for intake to be lower than 5 percent of total energy intake (WHO, 2015).

The consumption of sugar has been shown to be increasing in Western society (Pistollato et al., 2018). However, satisfying the people's desire for sweetness in accordance with the WHO's guidelines on low sugar is a rather herculean undertaking. To counter this, several non-sugar sweeteners (NSS), including sucralose, saccharin, aspartame, and acesulfame-K, have become widely used. Although all NSS have little or no caloric content (Fitch and Keim, 2012) and can create a sweet taste, they differ considerably in their chemical structures, absorption, metabolism, and elimination mechanisms and biological effects (Magnuson et al., 2016). Like other food additives, NSS were submitted for toxicological evaluation before a safety dose was approved for food use. Acceptable daily intake (ADI) limits have been set by agencies, including the Joint FAO/WHO Expert Committee on Food Additives (JECFA) of the United Nations, European Food Safety Authority

(EFSA) of the European Union, and Food and Drug Administration (FDA) of the United States, using toxicological studies. However, some studies demonstrated adverse effects of certain NSS (Mossavar-Rahmani et al., 2019).

Evidence now suggests that NSS alter the gut microbiota (Suez et al., 2014) with a consequent putative risk for neuronal functions (Zheng et al., 2019). In recent years, neurodegenerative diseases have become a significant public health concern and have inadvertently contributed to the fragility of the world's mental health (Ayeni et al., 2022). Although regulatory endorsement of NSS has been preceded by extensive toxicological investigations to decipher how each NSS behaves in the body, emerging evidence on their interaction with gut microbiota has disclosed novel health implications that strengthen the risk of neuronal dysfunction and may influence regulatory directions. Thus, this chapter aims to review the existing research on how NSS affect neural health and provide a comprehensive discussion of the current knowledge, including potential effects of these sweeteners on the brain and nervous system, as well as identify gaps in understanding and suggest areas for future research.

2. Historical perspectives of artificial NSS

NSS are additives that mimic sugar's sweetness with fewer calories, offering a zero or low-calorie option. Artificial NSS are synthetic compounds that provide intense sweetness. Due to their high sweetness power, they are used in trace amounts that provide negligible calories at sweetness levels. Until recently, they have been considered safer and more cost-effective alternatives to sugar, especially for overweight and obese people aiming to decrease calorie consumption and improve weight management (Peters et al., 2016; Daher et al., 2025). Synthetic NSS approved for food use include saccharin, aspartame, neotame, advantame, and sucralose, among others. These NSS have variable chemical structures and sweetness intensities ranging from 200 to 20,000 times higher than sucrose, as depicted in Fig. 1.

Innumerable processed foods, including soft drinks, energy drinks, ice cream, nibbles, toffees, and other candies and chocolate products, flavored milk, yogurt and other fermented milk products, jams, among others, contain NSS, and they are also used as Table-top sweeteners. To keep NSS consumption within safety levels, the maximum level of NSS allowed for use in processed foodstuffs is established, taking into account the ADI of each compound (Fig. 1) and the estimated daily consumption of NSS-containing

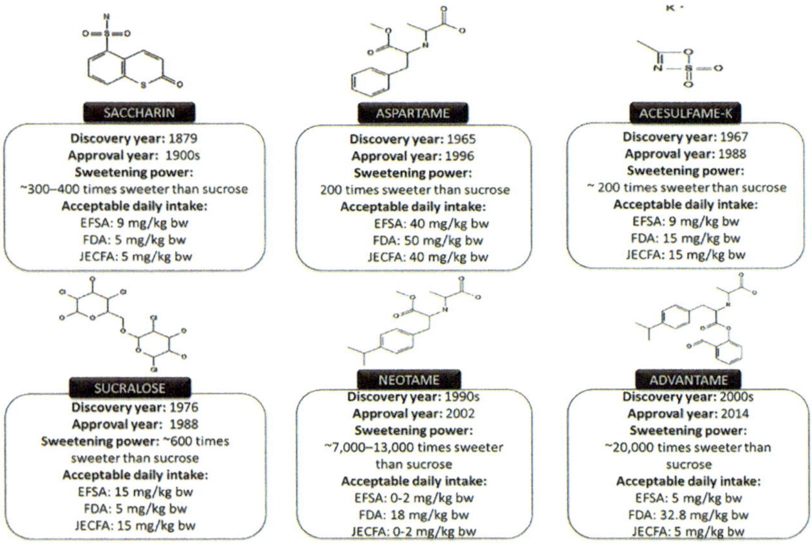

Fig. 1 Historical perspectives of non-sugar sweeteners and their ADIs as approved by the various regulatory bodies.
bw, body weight; FDA, Food and Drug Administration in the United States; EFSA, European Food Safety Authority in the European Union; JECFA, The Joint FAO/WHO Expert Committee on Food Additives.

foods. However, Table-top sweeteners can be used by consumers without restriction.

In the 1800s, NSS were integrated into the food industry and have persistently evolved with some novel discoveries. The historical perspective on the discovery of NSS is illustrated in Fig. 1. Inadvertently, in 1879, saccharin was discovered by Constantin Fahlberg while conducting research on toluene-sulfonamide oxidation. Initially, saccharin was approved by the FDA for use in the production of beverages, gums, cosmetics, among others (De la Peña, 2010; FDA, 2021). Later, studies have proven it to be safe, resulting in the US lifting the restrictions in 2000, and Canada followed suit in 2014 (Weihrauch et al., 2004; CSPI, 2022). James M. Schlatter discovered aspartame while conducting an experiment on anti-ulcer drugs in 1965. Subsequently, it was approved as a Table sweetener in 1981 by the FDA and was primarily accepted as a general-purpose sweetener in 1996 (Czarnecka et al., 2021).

Thereafter, in 1967, Karl Clauss discovered acesulfame-K, which was later approved by FDA for use in beverages in 1988 and obtained generic approval in 2003. It is primarily utilized as an additive in toffee, beverages, dairy, and desserts (Wu et al., 2024). Another member of the artificial NSS is sucralose, which came to limelight in 1976 when Tate and Lyle were working on a chemical intermediate of sugar for use as a sweetener. A derivative of aspartame known as neotame was ratified for use by the FDA in 2002 and, as a result of its heat stability, it is seen as the best-fit for baking (Nofre and Tinti, 2000). Another heat-stable sweetener suitable for baking is advantame which was ratified for use in 2014 (CSPI, 2023).

3. Food use and toxicological aspects of artificial NSS

3.1 Saccharin

Saccharin, derived from 1,2-benzisothiazol-3-(2H), is among the earlier discovered NSS. Despite its appalling bitter and metallic aftertaste, it remains in use as a non-caloric sugar substitute (Riera et al., 2007; Chattopadhyay et al., 2014). In addition, it is used in the manufacture of some products, including chewing gum, beverages, jams, mouthwash, and toothpaste. The ability of saccharin to remain stable even at elevated temperatures makes it an ideal choice for cooking low-calorie and sugar-free products. Its sweetening potency is almost 300 times higher than that of sucrose (Fig. 1). The safety of saccharin has been extensively investigated and some studies are summarized in Table 1 . Based on a literature review of studies published between 1994 and 2024, EFSA has reviewed the ADI for saccharin to 9 mg/kg bw, replacing the previous limit of 5 mg/kg bw (EFSA FAF Panel, 2024), whereas JECFA and U.S. FDA have an ADI set at up to 5 mg/kg bw (JECFA, 1993) (Fig. 1). Also, the ADI suggests that consuming up to 8.5 packets of saccharin daily over a lifetime, which is equivalent to 41.18 mg/packet for an average adult weighing 70 kg, is safe. Given that average consumption is less than 0.028 kg/year, most people are unlikely to exceed the ADI.

A study by Tripathi et al. (2006) evaluated the levels of saccharin in products like crushed ice, pan masala, pan flavorings, and ice candy across different age groups in India. The results showed that the amount of saccharin consumption from products, including the pan masala, containing about 12,750 mg/kg to 24,300 mg/kg, exceeds the acceptable limit set by the Indian Prevention of Food Adulteration (PFA) Act. Also, pan flavorings contained 12.2 percent–20.1 percent, while crushed ice contained 200 mg/kg – 700 mg/kg, and ice candy contained 280 mg/kg – 460 mg/kg saccharin (Tripathi et al., 2006). Further, children aged 6–10 were found to have consumed less than 21 percent of the ADI at

Table 1 Effects of saccharin on the gut-brain health.

Model	Exposure scheme	Gut-brain mechanisms	Health and neurological implications	References
Zebrafish (*Danio rerio*)	0, 1, 10, and 50 mg/L, water exposure, 8 weeks	Lower levels of serotonin and dopamine, and higher levels of GABA in the brain.	Increased anxiety-like behavior (dark side preference in the light/dark test), inhibited exploratory behavior (top arena in the novel tank diving test), social interaction (social-isolation test), and adaptability to the novel object in the NOR test, and interfered in the color-based associative learning and memory ability.	Han et al. (2021)
Adult male mice andHealthy humans (non-diabetic)	C57Bl/6 WT mice: 0.1 mg/mL in drinking water (equivalent to the human FDA ADI 5 mg/kg bw/day) Antibiotic treatment and faecal transplantation to germ-free mice assays were used to investigate the role of gut microbiome. HumansCohort study compared consumers and non-consumers of artificial sweeteners	Increased glycaemia after an oral glucose tolerance test in lean and obese mice. Increased relative abundance of *Bacteroides vulgatus* and decrease of *Akkermansia muciniphila*, and marked increase in SCFAs propionate and acetate in stool. The effect was abolished by antibiotic treatment and transferred by faecal transplantation. Healthy humans that consume artificial sweeteners had glucose intolerance that was not associated with BMI	Glucose intolerance due to gut intestinal dysbiosis.	Suez et al. (2014)

Male Sprague-Dawley rats	3 mg/kg bw/day in drinking water, for 6 weeks	Increased oxidative stress (MDA assay) and reactive hippocampal astrogliosis (GFAP expression).	Impaired memory retention (passive avoidance learning test).	Erbaş et al. (2018)
Healthy humans ($n = 10$–13/group, aged 18–45 y, BMI $\leq$ 25 kg/m^2) and adult mice (wild-type and sweet taste receptors' deficient mice)	Humans: capsules for oral delivery at the maximum acceptable daily intake (ADI), for 2 weeks, in a randomized double-blind, placebo-controlled, parallel arm study; Mice: 250 mg/kg bw/day (equivalent to 4-times the human ADI adjusted for mouse body surface, 62 mg/kg bw), drinking water, 10 weeks.	No changes were observed in the glycaemia or insulinemia response after an oral glucose tolerance test, faecal microbiome (16S rRNA sequencing) or SCFA levels.	No alteration in the gut microbiome, glycaemia control or insulin resistance.	Serrano et al. (2021)

GABA – Gamma-Aminobutyric Acid; SCFA – Short chain fatty acids; rRNA – Ribosomal RNA; BMI – Body Mass Index; GFAP – Glial Fibrillary Artery Protein; MDA – Malondialdehyde.

5 mg/kg for both crushed ice and ice candy. However, the maximum intake for this age group reached 57 percent of the ADI from ice candy and 68 percent from crushed ice. When both items were consumed together, the ADI was exceeded by 54 percent, highlighting the potential health risks for this vulnerable population. For adult pan masala and pan consumers, the risks were even more pronounced. The daily intake of saccharin from pan masala ranged from 1.84 to 13.33 mg/kg bw/day, contributing 37–267 percent of the ADI, with some estimates reaching 810 percent of the ADI. Similarly, the estimated daily intake from pan alone was 6.87 mg/kg bw/day, or 137 percent of the ADI (Tripathi et al., 2006). From this study, it can be deduced that saccharin is capable of having deleterious effects on the health of children and adults if consumed in excess of the ADI. Studies have also shown that there was an increase in the food intake and weight gain in rats that were challenged with saccharin-sweetened liquids in comparison with those that took glucose-sweetened liquids (Swithers et al., 2010). This finding challenges the belief that intense artificial NSS sweeteners aid in weight loss through calorie reduction.

3.2 Aspartame

Aspartame, which has been used in the manufacturing of products such as yogurt, soft drinks, and coffee, among others (Anton et al., 2010), is approximately 200 times sweeter than sucrose. Its ADI has been set at up to 50 mg/kg bw by FDA (1996), whereas JECFA, 1981, 2023a, 2023b) and EFSA (2013) set an ADI up to 40 mg/kg bw (Fig. 1).

Nevertheless, a double-blind randomized crossover clinical trial revealed that aspartame affected certain aspects of neurobehavioral performance in healthy adults (Lindseth et al., 2014). After 8 days of dietary supplementation at levels lower than ADI (10 and 25 mg/kg bw/day of aspartame), there was increased depression, irritability, and impaired spatial orientation, but no changes in working memory (Lindseth et al., 2014) (Table 2). Hence, careful consideration is needed when consuming food products that contain aspartame, as it may affect neurobehavioral function even at low doses. After intake, aspartame breaks down into aspartic acid, phenylalanine, and methanol in the digestive system (EFSA, 2013; FDA, 2013). Due to their metabolic intolerance to phenylalanine, patients with phenylketonuria (PKU) should not consume aspartame (Fitch and Keim, 2012; MacDonald et al., 2020). Therefore, regulatory agencies require that aspartame-containing products clearly indicate its presence due to its potential effects on PKU patients (Whitehouse et al., 2008).

Table 2 Effects of aspartame, advantame and neotame on the gut-brain health.

Model	Exposure scheme	Gut-brain mechanisms	Health and neurological implications	References
Aspartame				
Healthy adults ($n = 28$, average age 21 y, BMI 24 kg/m^2)	High-aspartame diet (25 mg/kg bw/day) and low-aspartame diet (10 mg/kg bw/day) for 8 days, in a randomized crossover clinical trial with 2-week washout period between the treatments.	Not investigated.	Increased depression score (Zung's self-rating depression scale), irritability (Zung's irritability subscale), and impaired spatial orientation (Vandenberg MRT). No significant effect on working memory (Sternberg item recognition test) and rare headaches.	Lindseth et al. (2014)
Healthy male and female humans ($n = 17$, aged 18–45 y, BMI 22.9 kg/m^2)	425 mg/day (14 percent of ADI) for 14 days, in a randomized, double-blind crossover clinical trial with a 4-week washout period and pre–post intervention comparisons.	No significant effect on gut microbiota composition (16S rRNA sequencing) or faecal SCFAs.	No dysbiosis.	Ahmad et al. (2020)

Continued

Table 2 Effects of aspartame, advantame and neotame on the gut-brain health.—cont'd

Model	Exposure scheme	Gut-brain mechanisms	Health and neurological implications	References
Male Wistar rats with folate deficiency (methotrexate-induced)	40 mg/kg bw/day, oral administration, 90 days	Increased plasma corticosterone levels, and brain NO and iNOS levels. Down-regulation of NMDAR-CaMKII-ERK/CREB signaling pathway and the activities of membrane-bound ATPases and AChE in certain brain regions.	Decreased learning and spatial working and spatial recognition memory deficit (Morris water maze and Y-maze tests) in folate deficiency rats. Degeneration of hippocampal pyramidal cells.	Iyaswamy et al. (2018)
Male Sprague–Dawley rats	250 mg/kg bw/day, drinking water, for 120 days	Increased muscarinic receptor density in the frontal cortex, midcortex, hippocampus, hypothalamus, and cerebellum. Increased Na^+, K^+-ATPase activity in the midbrain.	Impaired long-term memory retention (T-maze test).	Christian et al. (2004)
Male Sprague–Dawley rats	3 mg/kg bw/day, drinking water, for 6 weeks	Increased brain oxidative stress (MDA), hippocampal neuronal loss, and reactive astrogliosis (GFAP expression).	Impaired memory retention (passive avoidance learning test) and hippocampal damage.	Erbaş et al. (2018)

Male Wistar rats with folate deficiency (methotrexate-induced)	40 mg/kg bw/day, oral administration, for 90 days	Increased hydrogen peroxide level, and expression of apoptotic genes and proteins, along with neuronal cell death in certain brain regions.	Behavioral changes and neurodegeneration in brain specific regions in folate deficiency rats. Increased immobilization and fecal bolus production with a marked decrease in ambulation, rearing, and grooming (Open field test). Anxiety-like behavior characterized by decreased entry and time in the open arms, and increased number of head dips (elevated plus maze test).	Ashok et al. (2015)
Adult male mice (C57Bl/6 WT)	4 percent commercial aspartame in drinking water for 11 weeks	Increased glycaemia after an oral glucose tolerance test that was abolished by antibiotic treatment.	Glucose intolerance due to gut microbiota composition.	Suez et al. (2014)

Continued

Table 2 Effects of aspartame, advantame and neotame on the gut-brain health.—cont'd

Model	Exposure scheme	Gut-brain mechanisms	Health and neurological implications	References
Male and female (C57BL/6J)	250 mg/L, drinking water (approx. the FDA ADI dose, lifetime exposure (prenatal up to 20-weeks old).	Higher weight gain and fasting glucose levels, and lower insulin in male mice, as well as higher fasting glucose levels in female mice.	Adverse effects on spatial cognitive performance (Morris water maze test) and glucose homeostasis, particularly in males after lifetime exposure.	Collison et al. (2012)
Advantame				
Male and female Crl:CD (SD) IGS BR rats	2000, 10,000, or 50,000 ppm in the diet (equivalent to 164–204, 833–1036, 4410–8447 mg/kg bw/day), ad libitum, for two generations; F0 (16 weeks) and F1 (up to 30 days postpartum).	Not assessed.	No observed effects on mortality, body weights, reproduction, litter observations, or postnatal offspring development.	Otabe et al. (2011)

Neotame				
Male CD-1 mice	0.75 mg/kg bw/day (2.5-times the FDA ADI dose), drinking water, for 4 weeks.	Decreased alpha-diversity and altered beta-diversity in the gut microbiome (16S rRNA sequencing). Reduced Firmicutes (largely corresponding to the current Bacillota phylum) and increased Bacteroidetes (currently Bacteroidota). Increased fecal levels of fatty acids and cholesterol but decreased levels of malic and glyceric acid.	Gut dysbiosis.	Chi et al. (2018)

FDA – Food Drug Administration; NO – Nitric oxide; ADI – Acceptable daily intake; CD – Crohn's disease; GFAP – Glial fibrillary artery protein; MDA – Malondialdehyde; Na+, *K*+-ATPase – Sodium-potassium ATPase; iNOS – Inducible nitric oxide synthase; NMDAR – N-Methyl-D-aspartate receptor; CaMKII – Calcium/Calmodulin-dependent protein kinase II; ERK/CREB – Extracellular signal-regulated kinase/cAMP response element-binding protein; SCFA – Short chain fatty acids; rRNA – Ribosomal RNA; BMI – Body mass index; AChE – Acetylcholinesterase.

The safety of aspartame is debated. While industry-funded research generally supports its safety, many independent studies suggest it might cause adverse health effects. Van den Eeden et al. (1994) linked aspartame to headaches based on the concept of genetic polymorphism (Van den Eeden et al., 1994).

3.3 Acesulfame-K

In 1967, Hoechst pharmaceutical company discovered acesulfame-K, a NSS that is about 200 times sweeter than sucrose (SCF, 1985; Rymon and Lipinski, 1991). Its use was initially restricted to sugar-free baked products, gelatin desserts, and chewing gums, but in 1998, the FDA ratified it for use in beverages. Acesulfame-K is suitable for cooking and baking as it is heat-stable. The ADI for acesulfame-K varies among regulatory agencies. The FDA and JECFA set the ADI at up to 15 mg/kg bw (JECFA, 1991), while the Scientific Committee on Food (SCF) of the European Commission sets it at 9 mg/kg bw (SCF, 2000), as depicted in Fig. 1. In Portugal, 85.7 percent non-alcoholic beverages have been found to contain acesulfame-K at an average concentration of 97.4 mg/L (Basílio et al., 2020). For this population, local soft drinks have been found to be a major contributor to the consumption of acesulfame-K at 0.7 mg/kg bw by both adults and children, which is below the ADI, showing that acesulfame-K poses no dietary risk (Basílio et al., 2020).

Mice studies suggest that acute exposure to acesulfame-K at doses higher than the ADI may reduce intracellular ATP production, which could be harmful to neuronal cell viability and protective functions (Cong et al., 2013). Additionally, chronic acesulfame-K consumption in mice has been shown to impair learning and memory, particularly in the hippocampus, at 12.5 mM and 25 mM, respectively, which is significantly higher than typical human dietary exposure levels and exceeds the ADI established by regulatory authorities (Cong et al., 2013; Ibi et al., 2018). Although the acesulfame-K doses used in this study were quite high (Table 3), when converted to a human equivalent dose based on the body surface area normalization method (Reagan-Shaw et al., 2008), it only exceeds the estimated ADI by 1.2 to 1.9 times, suggesting neurotoxicity concerns for high-level consumer groups.

3.4 Sucralose

Sucralose is a modified form of sucrose in which three hydroxyl groups are replaced by chlorine atoms (AlDeeb et al., 2013), resulting in a

Table 3 Effects of acesulfame-K on the gut-brain health.

Model	Exposure scheme	Gut-brain mechanisms	Health and neurological implications	References
Male mice (ddY strain) under low-carbohydrate diet	25 mM, drinking water, 35 days (equivalent to 1600 mg/kg bw/day).	Decreased glucose level in the frontal cortex. No change in plasma glucose levels. Decreased GLUT1 expression in cortical tissues (Western blotting).	Cognitive impairment in short-term (Y-maze test) and object recognition memory (novel object recognition test), associated with disrupted cortical glucose transport	Ibi et al. (2018)
Male mice (C57BL/6J), 4 months old	12.5 mM, drinking water (equivalent to 231 – 359 mg/kg bw/day), for 40 weeks,	Increased fasting insulin and leptin levels, and plasma cholesterol levels. Metabolic dysregulation (glycolysis inhibition and functional ATP depletion) and neurosynaptic abnormalities (dysregulation of TrkB-mediated BDNF and Akt/Erk-mediated cell growth/survival pathway) in hippocampal neurons. No changes in insulin sensitivity or body weight.	Cognitive impairment in spatial (Morris- water-maze) and object recognition memory (novel object recognition test). No changes in motor coordination (rotarod task), anxiety levels (open field with dark insert, and elevated plus maze tests).	Cong et al. (2013)

Continued

Table 3 Effects of acesulfame-K on the gut-brain health.—cont'd

Model	Exposure scheme	Gut-brain mechanisms	Health and neurological implications	References
Zebrafish (*Danio rerio*)	50, 75, 125 μg/L, water exposure, for 4 months	Increased oxidative stress in brain and liver tissues, and inhibition of AChE activity.	Increased anxiety-like behavior (increased freezing time and reduced swimming activity in the novel tank and dark/light test).	Colín-García et al. (2023)

GLUT1 – Glucose transporter 1; BDNF – Brain derived neurotrophic factor; Akt/Erk- Protein kinase B (PKB)/Extracellular signal-regulated kinase; TrkB – Tropomyosin receptor kinase B; AChE – Acetylcholinesterase; ATP – Adenosine triphosphate.

600 times increase of sweetness compared to sucrose (Goldsmith, 2000; Aguayo-Guerrero et al., 2024) (Fig. 1). The ADI set by the US FDA is up to 5 mg/kg bw, while JECFA and EFSA set the ADI at up to 15 mg/kg bw (Fig. 1). Studies in human subjects have shown that sucralose concentration in breast milk can surpass the threshold for taste detection of sweetness (Rother et al., 2018; Sylvetsky et al., 2019). Additionally, perinatal exposure of neonate mice to sucralose has been shown to alter the gut microbiome (Olivier-Van Stichelen et al., 2019).

Moreover, NSS, including sucralose, saccharin, cyclamate, and acesulfame-K, have been shown to permeate the placenta into the fetal circulation and amniotic fluid (Leth-Møller et al., 2023). Excessive intake of sucralose during pregnancy, has been shown to increase neonates' body and resulted in metabolic alterations and low-grade inflammatory response (Aguayo-Guerrero et al., 2023). Furthermore, chronic intake of sucralose has been linked to gut dysbiosis and changes in insulin and glucose levels in healthy young adults (Méndez-García et al., 2022).

Studies have shown that sucralose causes alterations in the gut microbiome even at the ADI (Schiffman and Nagle, 2019; Ruiz-Ojeda et al., 2019). At low concentration, particularly at 0.1 mM, sucralose causes an increase in intestinal epithelial barrier permeability and reduces claudin-3 expression in Caco-2 cells (Shil et al., 2020). C57BL/6 male mice that received sucralose at 5 mg/kg bw in the drinking water for 6 months showed an increase in pro-inflammatory genes, disruption in fecal metabolites, and in gut microbiota composition (Bian et al., 2017). Similarly, mice exposed to sucralose at 1.1–11 mg/kg bw for 12 weeks exhibited a decrease in the beneficial bacteria, including lactobacilli and bifidobacteria, and a severe intestinal histological damage (Abou-Donia et al., 2008).

A study demonstrated that adult zebrafish exposed to sucralose at concentrations of 50, 75, and 125 μg/L for 4 months exhibited increased anxiety-like behaviors, elevated oxidative stress, and a reduction in neuronal acetylcholinesterase activity, indicating its potential neurotoxic effects (Colín-García et al., 2022, 2023) (Table 4).

3.5 Neotame

Neotame is an amino acid-based low calorie sweetener similar to both aspartame and advantame. Also, neotame is about 800 times sweeter than sucrose and has been found to be easily absorbed and eliminated (Fig. 1). Neotame has been approved for use in beverages and foods in Europe under

Table 4 Effects of sucralose on the gut-brain health.

Model	Exposure scheme	Gut-brain mechanisms	Health and neurological implications	References
Animal Models				
Adult male mice (C57Bl/6 WT)	5 percent commercial sucralose in drinking water for 11 weeks	Increased glycaemia after an oral glucose tolerance test that was abolished by antibiotic treatment.	Glucose intolerance due to gut microbiota composition.	Suez et al. (2014)
Male mice (C57BL/6J)	Low dose (1.5 mg/kg bw/day) and high dose (15 mg/kg bw/day), drinking water, for 8 weeks	Disrupted gut microbiome composition characterized by lower relative amount of *Clostridium cluster* XIVa in feces (16S rRNA sequencing). Increased hepatic cholesterol and cholic acid.	Gut dysbiosis. Disrupted cholesterol and bile acid metabolism.	Uebanso et al. (2017)
Mice (C57BL/6J)	0.1 mg/mL, drinking water, for 6 months (equivalent to the human FDA ADI 5 mg/kg bw/day)	Disrupted gut microbiome composition (16S rRNA sequencing) and functional profiles (fecal amino acid and derivatives profile, increased gene expression of pro-inflammatory mediators like LPS, flagella proteins, and fimbriae).	Gut dysbiosis. Increased liver inflammation and putative implication for low-grade chronic inflammatory response.	Bian et al. (2017)
Zebra fish (*Danio rerio*)	50, 75, 125 μg/L, water exposure, for 4 months	Increased oxidative stress in brain and liver tissues, and inhibition of AChE activity.	Increased anxiety-like behavior (increased freezing time and reduced swimming activity in the novel tank and dark/light test).	Colín-García et al. (2023)

Cyprinus carpio (common carp)	0.05 μg/L and 155 μg/L, water exposure, for 12, 24, 48, 72, and 96 h	Oxidative stress characterized by increased lipid peroxidation, hydroperoxide content, protein carbonyl content, SOD and CAT activity in gill, brain, muscle, and liver.	Not directly assessed. Inferred effects include neurological risks linked to oxidative damage, such as neurodegenerative diseases.	Saucedo-Vence et al. (2017)
Daphnia magna	0.1–5000 μg/L, water exposure, for 24 h	Sucralose concentration had a positive association with lipid peroxidation (TBARS), ROS generation (ORAC), and AChE activity.	Not directly assessed. Inferred effects include neurological dysfunction, behavioral aberrations.	Eriksson Wiklund et al. (2014)
Mouse model of Crohn's Disease (CD)-like ileitis (SAMP1/YitFc strain) and its parent control strain (AKR/J mice)	Low dose (1.08 mg/mL), FDA-approved dose (3.5 mg/mL), and high dose (35 mg/mL), drinking water, for 6 weeks	Overgrowth of Proteobacteria (currently Pseudomonadota) in feces (metagenomic shotgun DNA sequencing, 16S rRNA and enumeration analysis), especially *E. coli*, and increased myeloperoxidase activity in gut tissues of CD-prone mice.	Gut dysbiosis. No morphological exacerbation (histological analysis) of ileitis in CD-prone mice. Increased myeloperoxidase suggests exacerbation of inflammation in CD-prone individuals.	Rodriguez-Palacios et al. (2018)

Continued

Table 4 Effects of sucralose on the gut-brain health.—cont'd

Model	Exposure scheme	Gut-brain mechanisms	Health and neurological implications	References
Mice (C57BL/6J)	0.0003, 0.003, 0.03, and 0.3 mg/mL, drinking water, for 16 weeks (doses lower and equal to the FDA ADI)	Histological damage and distinct lymphocyte aggregation in ileum and colon; reduced abundance of probiotic species (*Lachnoclostridium* and *Lachnospiraceae*) and increased abundance of diabetes-associated genus (*Allobaculum*), and pathogenic genus (*Tenacibaculum, Ruegeria, Staphylococcus, Allobaculum*) in the feces (16S rRNA sequencing).	Gut dysbiosis, disrupted intestinal barrier integrity, acute colitis.	Zheng et al. (2022)
Mice (C57BL/6)	0.1 mg/mL (approx. 5–15 mg/kg bw/day, equivalent to the FDA ADI), drinking water, 12 weeks (maternal exposure during gestation and lactation followed by high-fat diet during adulthood)	Reduced fecal abundance of butyrate-producing bacteria (16S rRNA sequencing), lower butyrate levels, impaired gut barrier function via tight junction disruption, and induction of low-grade intestinal inflammation in the offspring. Impaired gut-liver axis (histology, immunofluorescence), disturbed fatty acid biosynthesis and metabolism (HPLC analysis).	Gut dysbiosis. Maternal exposure exacerbated susceptibility of the offspring to hepatic steatosis induced by high-fat diet in adulthood.	Dai et al. (2020)

Male Sprague–Dawley rats	1.5 mg/kg bw/day, drinking water, for 6 weeks	Increased brain oxidative stress (MDA) and reactive hippocampal astrogliosis (GFAP expression).	Impaired memory retention (passive avoidance learning test).	Erbaş et al. (2018)
Neonatal NMRI mice	5, 25, or 125 mg/kg bw/day, by gavage, during postnatal days 8–12	No biochemical changes in brain protein expression of CaMKII, GAP-43, synaptophysin, and tau (slot-blot) 24 h after exposure.	No behavioral changes in the spontaneous motor activity of adults (locomotion, rearing, total activity) at doses higher than acceptable daily intake.	Viberg and Fredriksson (2011)
Male Wistar rats	15 mg/mL, drinking water, for 16 weeks, with or without HFD	Decreased α-diversity of the gut microbiota (16S rRNA sequencing), increased abundance of *Bacillus fragilis* and bacteria involved in LPS and SCFA synthesis. Decreased protein expression of occludin. Increased circulating levels of proinflammatory cytokines, and insulin resistance. Increased fatty acid oxidation and ketone bodies production.	Gut dysbiosis, metabolic endotoxemia, and glucose intolerance.	Sánchez-Tapia et al. (2020)

Continued

Table 4 Effects of sucralose on the gut-brain health.—cont'd

Model	Exposure scheme	Gut-brain mechanisms	Health and neurological implications	References
Male Sprague-Dawley rats	After obesity induction (8 weeks of HFD), rats received daily gavage (2 mL) of 0.54 or 0.78 mM sucralose (equivalent to human doses of 0.11 and 0.16 mg/kg bw based on a body surface area translation approach), for 4 weeks	Decreased abundance of gut beneficial bacteria (*Lactobacillaceae* and *Akkermansiaceae*), and increased abundance of *Acidaminoccaceae* and *Phascolarctobacterium* (16S rRNA sequencing). Increased predicted profile of carbohydrate (0.54 mM) and amino acid metabolism (0.78 mM).	Altered gut microbiota composition and putative changes in microbiome function (carbohydrate metabolism and amino acid metabolism).	Zhang et al. (2021)
Male and female Wistar rats exposed to early-life stress (PNDs 1–21: dam separation for 180 min/day	Post–weaning exposure to 15 mg/kg bw, drinking water, over PNDs 26–50	No significant effects on body weight gain or blood glucose levels.	Early-life stress caused sex-dependent anxiety-like behavior in the open field test in adolescent female rats that was reversed sucralose. No behavioral changes were observed in male rats.	Rico et al. (2024)

Cell culture and clinical human studies				
HMC3 human microglial cells	0 or 1 mM for 7, 14 or 21 days	Decreased SIRT1, GPx4, GSH, and Fe^{2+} levels. Increased IL-1β, NLRP3, 8-OHdG, MDA, and caspase 3/7 activity. Reduced cell migration rate.	Induced neuroinflammation and ferroptosis.	Hacioglu (2024)
Healthy male adults ($n = 14$–16/group, $18 - 50$ y; BMI 20–30 kg/m^2)	780 mg/day capsules for 7 days in a randomized, double-blind, placebo-controlled, parallel clinical trial with pre–post intervention comparisons.	No changes were observed in the body weight, glycaemia or insulinemia response after an oral glucose tolerance test, or faecal microbiome (16S rRNA sequencing).	No alteration in the gut microbiome, glycaemia control or insulin resistance.	Thomson et al. (2019)
Healthy male and female ($n = 17$, aged 18–45 y, BMI 22.9 kg/m^2)	136 mg/day (20 percent of ADI) for 14 days, in a randomized, double-blind crossover clinical trial with a 4-week washout period and pre–post intervention comparisons.	No significant effect on gut microbiota composition (16S rRNA sequencing) or faecal SCFAs.	No dysbiosis.	Ahmad et al. (2020)

Continued

Table 4 Effects of sucralose on the gut-brain health.—cont'd

Model	Exposure scheme	Gut-brain mechanisms	Health and neurological implications	References
Healthy male and female ($n = 20$/group, aged 18–35 y, BMI 18.5–29.9 kg/m^2)	48 mg/day, oral intake, 10 weeks (approx. 15 percent FDAADI), in a randomized, open-label, placebo- controlled, parallel clinical trial with pre–post intervention comparisons.	Increased abundance of gut bacteroides *Blautia coccoides*, and decreased *L. acidophilus* (species specific quantitative PCR). Increased glycaemia and insulinemia response after an oral glucose tolerance test.	Gut dysbiosis. Alteration in glucose and insulin response.	Méndez-García et al. (2022)

rRNA – ribosomal RNA; FDA – Food and Drug Administration; ADI – Acceptable daily intake; LPS – Lipopolysaccharide; AChE – Acetylcholinesterase; CAT – Catalase; SOD – Superoxide dismutase; TBARS – Thiobarbituric acid reactive substances; ORAC – Oxygen radical absorbance capacity; ROS – Reactive oxygen species; DNA – Deoxyribonucleic acid; HPLC – High-performance liquid chromatography; GFAP – Glial fibrillary artery protein; MDA – Malondialdehyde; CaMKII – Calcium/Calmodulin-dependent protein kinase II; GAP-43 – Growth-associated protein 43; SCFA – Short chain fatty acids; HFD – High fat diet; PNDs – Postnatal days; NLRP3 – NOD-, LRR-, and Pyrin Domain-Containing Protein 3; 8-OHdG – 8-Hydroxy-2′-deoxyguanosine; SIRT1 – Sirtuin 1; GPx4 – Glutathione peroxidase 4; GSH – Reduced glutathione; IL-1β- Interleukin-1 beta; BMI – Body mass index; qPCR – Quantitative polymerase chain reaction.

Annex II of regulation 1333/2008. Also, it has been approved for use in over 35 countries including Brazil, Mexico, China, Nigeria, Argentina, among others. JECFA in 2003 and EFSA in 2007 gave a positive review on neotame and set the ADI at up to 2 mg/kg bw (JECFA, 2003a; EFSA, 2007) (Fig. 1).

3.6 Advantame

Advantame is formed by the catalytic hydrogenation of 3–hydroxy-4-methoxycinnamaldehyde. The ADI set by the regulatory agencies includes EFSA and JECFA at up to 5 mg/kg bw (EFSA, 2013; JECFA, 2013), and FDA at 32.8 mg/kg bw (FDA, 2021) (Fig. 1). Advantame was not considered to be carcinogenic or mutagenic (EFSA, 2013). Although ingested advantame can form phenylalanine, under normal use, advantame level is not relevant for phenylketonuria patients. It also has no adverse effects in type 2 diabetic patients.

4. Estimated daily intake of NSS

Although the maximum limits of NSS allowed for foods are established based on the ADI, the unrestricted use of tabletop NSS, combined with diverse dietary habits in some population groups, poses a risk of exceeding the ADI. Barraj and colleagues recently assessed the intake of six NSS (acesulfame potassium, aspartame, cyclamate, saccharin, steviol glycosides, and sucralose) in Brazil (Barraj et al., 2021a) and Argentina, Chile, and Peru (Barraj et al., 2021b) and compared it to the ADIs established by JECFA. Results showed that the estimated intakes by the total population of the analyzed countries, including children, were well below the JECFA ADIs. This applies to all identified scenarios, including the most conservative ones. These results are in line with the outcomes of other recent studies in these countries, including an analysis performed by the Chilean Food Safety and Quality Agency (ACHIPIA) aiming at assessing the dietary exposure of the Chilean population (including children) to four authorized NSS (acesulfame potassium, aspartame, sucralose, and steviol glycosides). ACHIPIA concluded that the estimated consumption of these four sweeteners remains below the ADI for each sweetener across all exposure scenarios and age groups (ACHIPIA, 2021). However, WHO recently advised against the use of NSS as a means of weight control or for reducing the risk of noncommunicable diseases (conditional recommendation) (WHO, 2023). Studies suggest that

NSS consumption may influence certain health outcomes (Mullee et al., 2019).

5. Biochemical mechanisms of neurotoxicity of NSS

The biochemical processes of neurotoxicity linked to NSS encompassnumerous pathways affecting the neurons and nervous system. Research indicates that these artificial sweeteners can result in glucose intolerance in animal models by causing alterations in gut microbiota, and similar effects have been detected in a cohort study of humans who were short- and long-term NSS consumers (Suez et al., 2014).

5.1 Saccharin

Zebrafish (*Danio rerio*) exposed to saccharin (1 - 50 mg/L in water) for 8 weeks had increased anxiety-like behavior, inhibited exploratory behavior, social interaction, and adaptability to the novel object with impairment in learning and memory functions, accompanied by lower levels of serotonin and dopamine, and higher levels of GABA in the brain (Han et al., 2021; Table 1). However, gut dysbiosis has not been assessed.

Healthy humans who received saccharin at the maximum ADI for 2 weeks and adult mice (wild-type and sweet taste receptors' deficient mice) exposed to 250 mg/kg bw/day of saccharin (equivalent to 4-times the human ADI adjusted for mouse body surface, 62 mg/kg bw) in drinking water for 10 weeks showed no changes in the glycaemia or insulinemia response, faecal composition of microbiome or SCFA levels (Serrano et al., 2021; Table 1). The brief exposure period restricts the ability to extend the conclusion of no gut dysbiosis or carbohydrate metabolism changes to longer exposures, such as those associated with human dietary intake.

5.2 Aspartame

Aspartame hydrolysis during digestion releases aspartic acid, methanol, and phenylalanine (Choudhary and Lee, 2018). Although an ADI dose has been approved for aspartame (Fig. 1), the presence of these metabolites at elevated amounts can potentiate neurotoxic effects (Rycerz and Jaworska-Adamu, 2013; Lindseth et al., 2014). Phenylalanine has been found to potentiate neurotoxic effects in phenylketonuria patients (Muri et al., 2024). Chronic exposure to aspartame at 40 mg/kg bw in rats has been found to alter the brain antioxidant status and induce apoptotic changes in the brain (Ashok and Sheeladevi, 2014).

Furthermore, it has been demonstrated that aspartame at 40 mg/kg bw/day (as recommended by EFSA, 2012) may cause behavioral deficits, including learning and memory impairments, by affecting specific brain regions and altering neurotransmitter function (Choudhary et al., 2016). Aspartame can be safely consumed within the ADI by healthy individuals, but it has long been recognised as toxic to individuals with phenylketonuria, and therefore, it should be avoided (MacDonald et al., 2020).

Aspartame consumption by rats (40 mg/kg bw, 90 days) under folate deficiency (used to mimic formate accumulation that can occur in humans during aspartame metabolization) increased nitric oxide levels as a result of increased gene and protein expression of neuronal nitric oxide synthase (nNOS) and inducible nitric oxide synthase (iNOS) (Iyaswamy et al., 2018; Table 2). There was a parallel decrease in learning and impairment in spatial working and spatial recognition memory on the Morris water maze and Y-maze tests, which could result of reduced hippocampal acetylcholine esterase (AChE) activity observed in the same study. In addition, the brain homogenates from the aspartame-challenged group revealed a decrease in the phosphorylation of NMDAR1-CaMKII-ERK/CREB signaling pathway, which leads to excitotoxicity and memory decline (Iyaswamy et al., 2018). The aspartame metabolite methanol was likely implicated in these effects as it is known to activate NOS, microglia, and astrocytes, inducing neurodegeneration. In fact, abnormal neuronal morphology of pyramidal cell layers was observed in the aspartame-challenged rats (Iyaswamy et al., 2018). Although the mechanisms of neurodegeneration have not been completely elucidated, these findings suggest that aspartame metabolites could be a contributing factor for the development of brain oxidative stress, even during exposure periods shorter than chronic.

Another recently unveiled putative mechanism for aspartame toxicity is its ability to form amyloid-like structures. Under physiological-like conditions, aspartame can undergo spontaneous self-assembly, resulting in cytotoxic nanofibrils (Anand et al., 2019). The amyloid-like structure of aspartame fibrils can trap globular proteins, Aβ peptides, and aromatic metabolites and has been shown to induce hemolysis, causing DNA damage and resulting in both apoptosis and necrosis-mediated cell death (Anand et al., 2019).

Aspartame also induces chemical stress, elevating plasma corticosterone levels, which can negatively impact mood and immune function (Ashok et al., 2015; Table 2). During the open field test, the locomotor activity and emotional behavior of folate-deficient rats challenged with aspartame (40 mg/kg bw, 90 days) exhibited a noticeable increase in immobilization and fecal bolus production with a decrease in ambulation, rearing, and grooming response (Table 2). An anxiety-like behavior was observed for the aspartame-treated rats in the elevated plus maze test, as indicated by the decreased number of head dips, entries, and time spent in the open arms (Ashok et al., 2015). Increased levels of hydrogen peroxide were found in different brain regions of aspartame-treated rats, namely cerebellum, cerebral cortex, hippocampus, midbrain, pons medulla. Also, there was an upregulation of Fas receptor, tumor necrosis factor alpha, c-Jun N-terminal kinase 3, caspase 8, and 9 genes, indicating apoptosis activation, but histological evaluations were not performed.

Chronic lifetime exposure to aspartame at doses higher than the FDA ADI (46 mg/kg bw and 5 mg/kg bw, for 140 days) starting *in utero* resulted in gender-specific toxic effects in C57BL/6J mice (Collison et al., 2012; Table 2). The results obtained from this study showed that at 17 weeks of age, the male mice challenged with aspartame had increased fasting glucose levels, body weight, and diminished insulin sensitivity, while female mice had only elevated fasting glucose levels. Only male mice exposed to aspartame had longer escape latencies in the Morris water maze acquisition trials, indicating impaired learning, along with increased thigmotaxis and floating behavior, and spent less time in the target quadrant, suggesting further deficits in spatial memory (Collison et al., 2012). In the subsequent Morris water maze probe test, both male and female mice exposed to aspartame demonstrated impaired reference memory, as they spent less time searching for the previous platform location (Collison et al., 2012). These results indicate that prolonged exposure to aspartame at doses slightly higher than the ADI, *in utero*, may impair spatial learning and glucose homeostasis, with more pronounced effects in male mice.

Exposure of rats to aspartame at doses well above the ADI (250 mg/kg/day, 3 months) impaired long-term memory retention in the T-maze test. Behavioral changes were accompanied by an overexpression of muscarinic receptors in the frontal cortex, mid-cortex, posterior cortex, hippocampus, hypothalamus, and cerebellum, along with increased Na^+,K^+-ATPase activity in the midbrain (Christian et al., 2004; Table 2).

A randomized, double-blind crossover clinical trial conducted a pre vs. post intervention comparison for 14 days of administration of aspartame at a daily dose equivalent to 14 percent of ADI, found no significant differences in the relative abundance of gut bacterial taxa or faecal SCFA content (0.425 g) (Ahmad et al., 2020; Table 2). Results obtained do not support gut dysbiosis after a short-term aspartame exposure.

A randomized crossover clinical trial that compared the exposure to a high- (25 mg/kg bw/day) or low-aspartame diet (10 mg/kg bw/day) for 8 days found increased depression score and irritability in the Zung's tests (self-rating depression scale and irritability subscale) and impaired spatial orientation score in the Vandenberg MRT after high-aspartame intake period (Lindseth et al., 2014). In contrast, no significant changes were observed in the Sternberg Item Recognition test, which evaluates working memory. Given that the highest dose in this study is well below the ADI and induced neurobehavioral changes after a short-term exposure, proper attention is needed when consuming food products containing aspartame, which has the potential to affect the brain.

5.3 Acesulfame-K

Male mice (ddY strain) under low-carbohydrate diet exposed to 25 mM of acesulfame-K in the drinking water (equivalent to 1600 mg/kg bw/day) for 35 days showed cognitive impairment in short-term and novel object recognition tests, along with decreased glucose level and transport in the frontal cortex, but no change in plasma glucose levels (Ibi et al., 2018; Table 3). Nevertheless, the gut-brain axis relationship was not investigated.

Male mice exposed to 12.5 mM of acesulfame-K in the drinking water (equivalent to 231 – 359 mg/kg bw/day) for 40 weeks had impaired spatial and novel object recognition memory along with increased fasting insulin and leptin levels, metabolic dysregulation evidenced by glycolysis inhibition and functional ATP depletion, neurosynaptic abnormalities in hippocampal neurons but no changes in insulin sensitivity or body weight (Cong et al., 2013; Table 3).

5.4 Sucralose

Sucralose has been found to alter gut microbiota composition, leading to an imbalance between beneficial and detrimental bacteria, which could potentially result in neuronal toxicity (Suez et al., 2014). A randomized, open-label, placebo-controlled, parallel clinical trial with pre–post intervention comparisons showed that sucralose intake at doses higher than ADI increased the abundance of detrimental bacteria, resulting in gut dysbiosis (Méndez-García et al., 2022).

Mice exposed to sucralose (equivalent to 1.5 and 15 mg/kg bw/day for humans) for 8 weeks had decreased relative abundance of *Clostridium cluster XIVa* in feces and increased hepatic cholesterol and cholic acid levels but no changes in food consumption, body weight gain, or fat in the cecum or epididymis (Uebanso et al., 2017). These findings suggest that daily intake of sucralose at the maximum ADI levels (Fig. 1 and Table 4) induces gut dysbiosis and disrupts cholesterol and bile acid metabolism. However, no neurobehavioral assessments were conducted.

Mice exposed to sucralose (equivalent to 5 mg/kg bw/day for humans) for 6 months had disrupted gut bacteria composition, increased expression of pro-inflammatory bacterial genes like lipopolysaccharide, flagella proteins, and fimbriae, along with changes in fecal amino acid and derivatives profile (Bian et al., 2017; Table 4). These changes suggest the chronic sucralose intake at doses within the ADI (Fig. 1) can induce a low-grade inflammatory response by altering gut microbiome composition. Although inflammatory response is known to affect cognitive performance, no neurobehavioral assessments have been performed.

Mice exposure to sucralose at doses lower or equivalent to the FDA ADI for 16 weeks induced changes in the intestinal barrier function in the ileum and colon, while not affecting the body weight (Zheng et al., 2022; Table 4). These changes were accompanied by histological alterations that are characteristic of acute colitis, along with gut dysbiosis. Changes in gut microbiota composition were observed in groups exposed to the lowest and highest sucralose doses, including reduced abundance of probiotic species (*Lachnoclostridium* and *Lachnospiraceae*) and increased abundance of the diabetes-associated genus *Allobaculum,* and other pathogenic genera, namely *Tenacibaculum, Ruegeria,* and *Staphylococcus* (Zheng et al., 2022).

Short-term exposure of common carp (*Cyprinus carpio*) to sucralose in the water (0.05 and 155 μg/L over 12, 24, 48, 72, and 96 h) has been found to induce oxidative stress characterized by increased lipid peroxidation, hydroperoxide, and protein carbonyl content in the muscle, gills, and brain (Saucedo-Vence et al., 2017; Table 4). However, neurobehavioral tests, histology, and gut microbiota composition have not been investigated.

Another study conducted in zebrafish (*Danio rerio*) demonstrated that exposure to sucralose at 50, 75, and 125 μg/L of water for 4 months leads to anxiety-like behavior, as evidenced by the increase in freezing time in the light and dark test and concurrent decrease in swimming activity in the novel tank test (Colín-García et al., 2023; Table 4). These behavioral changes were accompanied by increased brain levels of H_2O_2, lipid peroxidation, and protein carbonyls, along with a concentration-dependent inhibition of AChE activity (Colín-García et al., 2023). This study suggests that even at environmentally relevant concentrations, chronic exposure to sucralose is capable of causing neurotoxic and metabolic impairment in zebrafish, which corroborates the findings that sucralose at environmentally relevant concentrations of 0.05 and 155 μg/L induced oxidative stress in the brain of juveniles *Cyprinus carpio* (Saucedo-Vence et al., 2017). Due to its low digestibility, a large amount of sucralose is excreted into the sewage system, resulting in its presence in various aquatic environments at concentrations ranging from 10^{-5} to 10^{2} mg/L (Wu et al., 2014; Cantwell et al., 2019). The limitations of the study include its focus on oxidative stress indices and anxiety-related behaviors, without considering other parameters, including gut microbiota analysis, histological, and immunohistochemical tests. Also, the reversibility of these effects potentiated by sucralose, and the mechanisms behind the neurotoxic and metabolic disruptions was not explored.

Male and female rats submitted to early life stress during postnatal days 1–21 and subsequently exposed to sucralose at 15 mg/kg bw over post-natal days 26–50 had no changes in body weight gain or blood glucose levels, but had lower anxiety-like behavior in the open field test compared to rats that did not receive sucralose (Rico et al., 2024; Table 4). The mechanism underlying the neurobehavioral changes observed was not explored.

Sucralose exposure caused no exacerbation of ileitis in a mouse model of Crohn's Disease (CD) but increased inflammatory response that was likely caused by the gut dysbiosis associated with the overexpression of Proteobacteria (currently Pseudomonadota), especially *E. coli* (Rodriguez-Palacios et al., 2018; Table 4).

As reviewed by Pepino (2015), some studies suggest that sucralose consumption may affect insulin sensitivity or glucose regulation, which could indirectly influence neural function, as the brain depends on glucose for energy and proper functioning.

Mice exposed to sucralose at 0.1 mg/mL (approximately 5–15 mg/kg bw/day, equivalent to the FDAADI) in the drinking water for 12 weeks during gestation and lactation and later submitted to a high-fat diet (HFD) during adulthood had reduced fecal abundance of butyrate-producing bacteria with down-regulation of G-protein-coupled receptor 43 (GPR43), lowered butyrate levels, impaired gut barrier function via tight junction disruption, and low-grade intestinal inflammation (Dai et al., 2020; Table 4). These changes were associated with disturbed fatty acid biosynthesis and metabolism and exacerbated susceptibility to hepatic steatosis induced by a high-fat diet in adulthood. Although butyrate, GPR43, and inflammatory markers can potentially affect neuronal function, no behavioral outcomes have been assessed.

Neonatal mice exposed to sucralose at 5, 25, or 125 mg/kg bw/day for 8–12 postnatal days had no behavioral changes in spontaneous motor activity during adulthood or in the protein expression of calcium/calmodulin-dependent protein kinase II, growth-associated protein-43, synaptophysin, or tau levels within the cerebral cortex or hippocampus (Viberg and Fredriksson, 2011; Table 4).

Male Wistar rats exposed to sucralose at 15 mg/dL in drinking water for 16 weeks with or without HFD showed gut dysbiosis characterized by decreased α-diversity and concurrent increase in the abundance of *Bacillus fragilis* and bacteria involved in LPS and SCFA synthesis, along with increased circulating levels of proinflammatory cytokines, insulin resistance, fatty acid oxidation, and ketone bodies production (Sánchez-Tapia et al., 2020; Table 4). This finding suggests that sucralose consumption induces metabolic endotoxemia and glucose intolerance, but behavioral changes have not been investigated.

Male Sprague Dawley rats exposed to 0.54 or 0.78 mM sucralose (equivalent to human doses of 0.11 and 0.16 mg/kg bw based on a body surface area translation approach) for 4 weeks after induction of obesity with HFD showed gut dysbiosis characterized by decreased abundance of gut beneficial bacteria (*Lactobacillaceae* and *Akkermansiaceae*) with a concurrent increase in the abundance of *Acidaminoccaceae* and *Phascolarctobacterium,* that suggests potential shifts in carbohydrate and amino acid metabolism (Zhang et al., 2021; Table 4). However, a neurobehavioral assessment was not carried out.

Human microglial clone 3 (HMC3) treated with 1 mM sucralose for 7, 14, or 21 days showed reduced cell migration rate, decreased SIRT1, GPx4, GSH, and Fe^{2+} levels with an increase in IL-1β, NLRP3, 8-OHdG, MDA, and caspase 3/7 activity (Hacioglu, 2024; Table 4). Despite the difficulty of extrapolating cell-based findings to humans, the results suggest that long-term sucralose exposure may induce neuroinflammation and ferroptosis.

Healthy male volunteers who received 780 mg/day of sucralose for 7 days in a randomized, double-blind, placebo-controlled, parallel clinical trial with pre–post intervention comparison showed no changes in body weight, glycaemia, or insulinemia response after an oral glucose tolerance test, or faecal microbiome composition (Thomson et al., 2019; Table 4). Although the study showed no gut dysbiosis or carbohydrate metabolism changes, the short duration of exposure limits the ability to extrapolate the findings to longer periods.

Healthy male and female volunteers who received 136 mg/day of sucralose for 14 days, in a randomized, double-blind crossover clinical trial with pre–post intervention comparisons showed no changes in gut microbiota composition or SCFA production (Ahmad et al., 2020; Table 4). Although no dysbiosis was observed in this short-term study with exposure to a sucralose dose equivalent to 20 percent of the ADI, a longer exposure study revealed gut dysbiosis and changes in glucose metabolism at doses equivalent to 15 percent of the FDA ADI (Méndez-García et al., 2022). In that study, healthy male and female volunteers who received 48 mg/day of sucralose for 10 weeks in a randomized, open-label, placebo-controlled, parallel design with pre–post intervention comparisons had increased abundance of gut bacteroides *Blautia coccoides*, and decreased *Lactobacillus acidophilus* with increased glycaemia and insulinemia response (Méndez-García et al., 2022; Table 4).

5.5 Neotame

Male CD-1 mice exposed to neotame at 0.75 mg/kg bw/day in drinking water (2.5-times FDA ADI of 0.3 mg/kg bw/day) for 4 weeks decreased alpha-diversity and altered beta-diversity of gut microbiome, reduced Firmicutes (largely corresponding to the current Bacillota phylum) and increased Bacteroidetes (currently Bacteroidota), increased fecal levels of fatty acids and cholesterol but decreased levels of malic and glyceric acid (Chi et al., 2018; Table 2). Evidently, this suggests that neotame induces gut dysbiosis. However, neurobehavioral dysfunction has not been investigated.

5.6 Advantame

Male and female Crohn's disease (CD) rats exposed to advantame at 2000, 10,000, or 50,000 ppm in the diet (equivalent to 164–204, 833–1036, 4410–8447 mg/kg bw/day) for two generations F0 (16 weeks) and F1 (up to 30 days postpartum) showed no changes in mortality, body

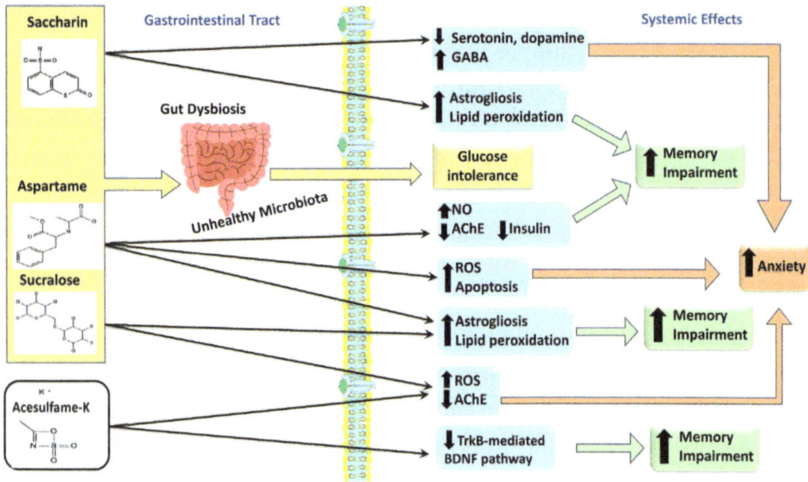

Fig. 2 Effect of NSS on the gut microbiota and the brain.
The intake of saccharin, aspartame, and sucralose has been linked to impaired glucose tolerance mediated by gut dysbiosis. Furthermore, the consumption of saccharin, aspartame, sucralose, and acesulfame-K has been associated with distinct biochemical dysfunctions, memory impairment, and anxiety-like responses, as detailed in Tables 1–4. Gut dysbiosis is a likely cause of the neurobehavioral dysfunctions associated with NSS intake.

weight, reproduction, litter observations, or postnatal offspring development (Otabe et al., 2011; Table 2).

6. Modulatory effect of NSS on gut microbiota and its neural implication

While the health risks linked to excessive sugar intake are well established (Fidler et al., 2017), our understanding of how NSS impacts human physiology either directly on the host or indirectly by altering gut microbiota remains limited. The first evidence on the modulatory effects of NSS on gut microbiota appeared in the last decade when Suez et al. (2014) demonstrated that chronic oral administration of saccharin, aspartame or sucralose to mice leads to reduced glucose tolerance due to changes in gut microbiome composition (Fig. 2). These findings were corroborated by human cohort data showing that metabolic-syndrome-related clinical parameters including impaired glucose tolerance were positively associated with NSS consumption in healthy subjects even after controlling for BMI values (Suez et al., 2014).

Thereafter, most investigations have focused on how NSS affect gut microbiota and their consequent impact on insulin resistance, obesity, and inflammation (Ruiz-Ojeda et al., 2019). Saccharin and sucralose have been shown to cause gut dysbiosis in rodent models, and such an effect was confirmed in human studies of long-term dietary exposure (Tables 1 and 4; Fig. 2). NSS have been found to cause an upsurge in harmful bacteria, especially *Escherichia coli*, and *Clostridium*, as reviewed by Conz et al. (2023).

Gut microbiota has been shown to metabolize xenobiotics, such as pharmaceutical drugs (Calatayud et al., 2021), and there is evidence that gut microbiota can likewise alter NSS (Suez et al., 2015). Although there is a paucity of studies on the mechanisms by which the gut bacteria metabolize NSS, certain strains have been demonstrated to use NSS as a carbon source (Bian et al., 2017). Several studies have found that the gut microbiota mediates certain effects of NSS on human physiology (Bian et al., 2017), raising questions about their safety and the need to understand their interactions with the host and microbiota.

Understanding the impacts of NSS, including sucralose, acesulfame-K, saccharin, and aspartame on the gut system is of utmost relevance because the gut microbiome plays a crucial role in brain development and behavior (Kim and Shim, 2023). Gut microbiota produces SCFAs that have systemic effects, including the modulation of neuroactive compounds and neuronal function that can ultimately affect neuronal health (Ochoa-Repáraz et al., 2020). Additionally, metabolites of the gut microbiota act as neurotransmitter precursors, thereby modulating neural functions. For example, *Lactobacillus* species can produce acetylcholine, while *Bacillus* generates dopamine. Additionally, *Enterococcus, Candida, Streptococcus*, and *Escherichia* species synthesize serotonin; *Bifidobacterium* and *Lactobacillus* generate GABA (Scardaci et al., 2021). Also, some microbial genera, including *Bifidobacterium, Clostridium,* and *Prevotella,* have been allied with dopaminergic signaling (Hamamah et al., 2022). Alterations in the gut composition can also influence the production of proinflammatory metabolites like lipopolysaccharides (Caputi and Giron, 2018). NSS have been shown to disrupt this balance via the alterations of the microbes responsible for maintaining neuronal health (Richardson and Frese, 2022).

Studies have shown that gut bacteria impact the endocrine pathway through the GABAergic neurotransmission, which is linked to neurodegenerative diseases (Strandwitz, 2018). Notably, beneficial gut bacteria, including *Lactobacillus* and *Lactococcus,* among others, have been found to stimulate

GABAergic transmission (Barrett et al., 2012). Gut microbiota also plays an important role in the modulation of the hypothalamic-pituitary-adrenal axis (Kasarello et al., 2023). Although the direct relationship between NSS-induced gut dysbiosis and neurobehavioral dysfunction have not been experimentally assessed, there is consistent evidence that NSS intake is associated with cognitive impairment in animal models and in human studies even at dietary levels lower than ADI (Fig. 2). In fact, an 8-year prospective study with 12,772 participants (average 52 years-old, 92 mg/day self-reported NSS intake) recently demonstrated that consumption of aspartame, saccharin, and acesulfame-K was associated with an accelerated decline in global cognition, particularly in memory and verbal fluency domains in participants that had less 60 years-old (Gonçalves et al., 2025). Impaired memory performance has been previously demonstrated in a rodent model of long-term exposure to sucralose (1.5 mg/kg/day), saccharin (3 mg/kg/day), and aspartame (3 mg/kg/day), which also increased brain lipid peroxidation and hippocampal astrogliosis (Erbas et al., 2018). In addition to the memory impairment, the exposure to saccharin (Table 1), aspartame (Table 2), acesulfame K (Table 3), and sucralose (Table 4) has also been associated with anxiety-like behavior in zebrafish studies and in some rodent models, as illustrated in Fig. 2.

7. Perspectives of NSS on neural health

NSS may impact the neuronal pathways associated with metabolism and appetite (Chakravartti et al., 2025). Notably, acesulfame-K has been shown to cross the blood-brain barrier and potentially result in neurobehavioral deficits in the hippocampal region (Cong et al., 2013; Ibi et al., 2018). Also, studies suggest that NSS can modulate reward responses similarly to caloric sugars, while others find no significant change in reward value with repeated exposure (Griffioen-Roose et al., 2013). This inconsistency highlights the complexity of how NSS interact with neural processes governing food intake.

Future research on NSS and neural health should employ integrative approaches that combine neurobiology, psychology, and nutrition science. The long-term effects of NSS on cognitive functions and their potential associations with obesity and metabolic syndrome, particularly in diverse dietary populations, should be investigated.

8. Conclusion

The impact of NSS on neural health requires attention, as it is a dynamic and multifaceted area of research. There is a paucity of studies addressing the effects of NSS on the gut-brain axis. Most investigations focus on NSS-induced gut dysbiosis while overlooking their potential neurobehavioral consequences. Moreover, studies that have associated NSS intake with neurobehavioral impairment in animal models and in humans have not examined the involvement of the gut-brain axis. Therefore, further research is needed to clarify whether the neurobehavioral dysfunctions observed in animal models exposed to NSS and the cognitive impairments associated with NSS consumption in humans are triggered by gut dysbiosis mechanisms. Such understanding will help guide regulatory policies on the use of NSS.

Declaration of AI and AI-assisted technologies in the writing process

During the preparation of this work, the authors used ChatGPT to assist with content organization and structural flow. Additionally, ChatGPT and Grammarly were used to correct language, spelling, and grammar errors. These tools were not used to generate content. After using these tools, the authors reviewed and edited the content as needed and take full responsibility for the published article.

References

Abou-Donia, M.B., El-Masry, E.M., Abdel-Rahman, A.A., McLendon, R.E., Schiffman, S.S., 2008. Splenda alters gut microflora and increases intestinal p-glycoprotein and cytochrome P-450 in male rats. J. Toxicol. Environ. Health A 71 (21), 1415–1429. doi:10.1080/15287390802328630.

ACHIPIA (Chilean Food Safety and Quality Agency), Miranda, C., Martinez, N., Sotomayor, G., 2021. Chronic Dietary Exposure Assessment on Sweeteners in Food Consumed by the Chilean Population. Available from: https://www.achipia.gob.cl/wp-content/uploads/2021/06/2021_ACHIPIA_Informe-EED-CronicaEdulcorantes-MINSAL-ACHIPIA_Nueva-Version_final-con-abstract-English.pdf (Accessed 19 March 2024).

Aguayo-Guerrero, J.A., Méndez-García, L.A., Manjarrez-Reyna, A.N., Esquivel-Velázquez, M., León-Cabrera, S., Meléndez, G., et al., 2023. Newborns from mothers who intensely consumed sucralose during pregnancy are heavier and exhibit markers of metabolic alteration and low-grade systemic inflammation: a cross-sectional, prospective study. Biomedicines 11 (3), 650. doi:10.3390/biomedicines11030650.

Aguayo-Guerrero, J.A., Méndez-García, L.A., Solleiro-Villavicencio, H., Viurcos-Sanabria, R., Escobedo, G., 2024. Sucralose: from sweet success to metabolic controversies – unraveling the global health implications of a pervasive non-caloric artificial sweetener. Life (Basel) 14 (3), 323. doi:10.3390/life14030323.

Ahmad, S.Y., Friel, J., Mackay, D., 2020. The effects of non-nutritive artificial sweeteners, aspartame and sucralose, on the gut microbiome in healthy adults: secondary outcomes of a randomized double-blinded crossover clinical trial. Nutrients 12 (11), 3408. doi:10.3390/nu12113408.

AlDeeb, O.A., Mahgoub, H., Foda, N.H., 2013. Sucralose. Profiles Drug Subst. Excip. Relat. Methodol. 38, 423–462. doi:10.1016/B978-0-12-407691-4.00010-1.

Anand, B.G., Prajapati, K.P., Dubey, K., Ahamad, N., Shekhawat, D.S., Rath, P.C., et al., 2019. Self-assembly of artificial sweetener aspartame yields amyloid-like cytotoxic nanostructures. ACS Nano 13 (5), 6033–6049. doi:10.1021/acsnano.9b02284.

Anton, S.D., Martin, C.K., Han, H., Coulon, S., Cefalu, W.T., Geiselman, P., et al., 2010. Effects of stevia, aspartame, and sucrose on food intake, satiety, and postprandial glucose and insulin levels. Appetite 55 (1), 37–43. doi:10.1016/j.appet.2010.03.009.

Ashok, I., Sheeladevi, R., 2014. Biochemical responses and mitochondrial mediated activation of apoptosis on long-term effect of aspartame in rat brain. Redox Biol. 2, 820–831. doi:10.1016/j.redox.2014.04.011.

Ashok, I., Wankhar, D., Wankhar, W., Sheeladevi, R., 2015. Neurobehavioral changes and activation of neurodegenerative apoptosis on long-term consumption of aspartame in the rat brain. J. Nutr. Intermed. Metab. 2, 76–85. doi:10.1016/j.jnim.2015.09.001.

Ayeni, E.A., Aldossary, A.M., Ayejoto, D.A., Gbadegesin, L.A., Alshehri, A.A., Alfassam, H.A., et al., 2022. Neurodegenerative diseases: implications of environmental and climatic influences on neurotransmitters and neuronal hormones activities. Int. J. Environ. Res. Public Health 19 (19), 12495. doi:10.3390/ijerph191912495.

Barraj, L., Scrafford, C., Bi, X., Tran, N., 2021a. Intake of low- and no-calorie sweeteners (LNCS) by the Brazilian population. Food Addit. Contam. A Chem. Anal. Control Expo Risk Assess. 38 (2), 181–194. doi:10.1080/19440049.2020.1846083.

Barraj, L., Bi, X., Tran, N., 2021b. Screening level intake estimates of low- and no-calorie sweeteners in Argentina, Chile, and Peru. Food Addit. Contam. Part A Chem. Anal. Control Expo Risk Assess. 38 (12), 1995–2011. doi:10.1080/19440049.2021.1956692.

Barrett, E., Ross, R.P., O'Toole, P.W., Fitzgerald, G.F., Stanton, C., 2012. γ-Aminobutyric acid production by culturable bacteria from the human intestine. J. Appl. Microbiol. 113 (2), 411–417. doi:10.1111/j.1365-2672.2012.05344.x.

Basílio, M., Silva, L.J.G., Pereira, A.M.P.T., Pena, A., Lino, C.M., 2020. Artificial sweeteners in non-alcoholic beverages: occurrence and exposure estimation of the Portuguese population. Food Addit. Contam. A Chem. Anal. Control Expo Risk Assess. 37 (12), 2040–2050. doi:10.1080/19440049.2020.1812734.

Bian, X., Chi, L., Gao, B., Tu, P., Ru, H., Lu, K., 2017. Gut microbiome response to sucralose and its potential role in inducing liver inflammation in mice. Front. Physiol. 8, 487. doi:10.3389/fphys.2017.00487.

Calatayud Arroyo, M., García Barrera, T., Callejón Leblic, B., Arias Borrego, A., Collado, M.C., 2021. A review of the impact of xenobiotics from dietary sources on infant health: early life exposures and the role of the microbiota. Environ. Pollut. 269, 115994. doi:10.1016/j.envpol.2020.115994.

Cantwell, M.G., Katz, D.R., Sullivan, J., Kuhn, A., 2019. Evaluation of the artificial sweetener sucralose as a sanitary wastewater tracer in Narragansett Bay, Rhode Island, USA. Mar. Pollut. Bull. 146, 711–717. doi:10.1016/j.marpolbul.2019.07.036.

Caputi, V., Giron, M.C., 2018. Microbiome-gut-brain axis and Toll-like receptors in Parkinson's disease. Int. J. Mol. Sci. 19 (6), 1689. doi:10.3390/ijms19061689.

Chakravartti, S.P., Jann, K., Veit, R., Liu, H., Yunker, A.G., Angelo, B., et al., 2025. Non-caloric sweetener effects on brain appetite regulation in individuals across varying body weights. Nat. Metab. 7 (3), 574–585. doi:10.1038/s42255-025-01227-8.

Chattopadhyay, S., Raychaudhuri, U., Chakraborty, R., 2014. Artificial sweeteners – a review. J. Food Sci. Technol. 51 (4), 611–621. doi:10.1007/s13197-011-0571-1.

Chi, L., Bian, X., Gao, B., Tu, P., Lai, Y., Ru, H., et al., 2018. Effects of the artificial sweetener neotame on the gut microbiome and fecal metabolites in mice. Molecules 23 (2), 367. doi:10.3390/molecules2302036.

Choudhary, A.K., Lee, Y.Y., 2018. Neurophysiological symptoms and aspartame: what is the connection? Nutr. Neurosci. 21 (5), 306–316. doi:10.1080/1028415X.2017.1288340.

Choudhary, A.K., Sundareswaran, L., Sheela Devi, R., 2016. Effects of aspartame on the evaluation of electrophysiological responses in Wistar albino rats. J. Taibah Univ. Sci. 10 (4), 505–512. doi:10.1016/j.jtusci.2015.07.006.

Christian, B., McConnaughey, K., Bethea, E., Brantley, S., Coffey, A., Hammond, L., et al., 2004. Chronic aspartame affects T-maze performance, brain cholinergic receptors and Na+,K+-ATPase in rats. Pharmacol. Biochem. Behav. 78 (1), 121–127. doi:10.1016/j.pbb.2004.02.017.

Colín-García, K., Elizalde-Velázquez, G.A., Gómez-Oliván, L.M., Islas-Flores, H., García-Medina, S., Galar-Martínez, M., 2022. Acute exposure to environmentally relevant concentrations of sucralose disrupts embryonic development and leads to an oxidative stress response in Danio rerio. Sci. Total Environ. 829, 154689. doi:10.1016/j.scitotenv.2022.154689.

Colín-García, K., Elizalde-Velázquez, G.A., Gómez-Oliván, L.M., García-Medina, S., 2023. Influence of sucralose, acesulfame-K, and their mixture on fish brain: a study of behavior, oxidative damage, and acetylcholinesterase activity in Danio rerio. Chemosphere 340, 139928. doi:10.1016/j.chemosphere.2023.139928.

Collison, K.S., Makhoul, N.J., Zaidi, M.Z., Saleh, S.M., Andres, B., Inglis, A., et al., 2012. Gender dimorphism in aspartame-induced impairment of spatial cognition and insulin sensitivity. PLoS ONE 7 (4), e31570. doi:10.1371/journal.pone.0031570.

Cong, W.N., Wang, R., Cai, H., Daimon, C.M., Scheibye-Knudsen, M., Bohr, V.A., et al., 2013. Long-term artificial sweetener acesulfame potassium treatment alters neurometabolic functions in C57BL/6J mice. PLoS ONE 8 (8), e70257. doi:10.1371/journal.pone.0070257.

Conz, A., Salmona, M., Diomede, L., 2023. Effect of non-nutritive sweeteners on the gut microbiota. Nutrients 15 (8), 1869. doi:10.3390/nu15081869.

CSPI (Center for Science in the Public Interest), 2022. Saccharin. CSPI. Available from https://www.cspinet.org/article/saccharin.

CSPI (Center for Science in the Public Interest), 2023. Advantame. CSPI. Available from https://www.cspinet.org/article/advantame. Accessed 17 December 2023.

Czarnecka, K., Pilarz, A., Rogut, A., Maj, P., Szymańska, J., Olejnik, Ł., et al., 2021. Aspartame – true or false? Narrative review of safety analysis of general use in products. Nutrients 13 (6), 1957. doi:10.3390/nu13061957.

Daher, M., El Darazi, E., Kacim, M., Hobeika, M., Sacre, Y., 2025. The effect of non-nutritive sweeteners' consumption on body weight: a randomized-controlled trial. Food Sci. Nutr. 13 (7), e70691. doi:10.1002/fsn3.70691.

Dai, X., Guo, Z., Chen, D., Li, L., Song, X., Liu, T., et al., 2020. Maternal sucralose intake alters gut microbiota of offspring and exacerbates hepatic steatosis in adulthood. Gut Microbes 11 (4), 1043–1063. doi:10.1080/19490976.2020.1738187.

De la Peña, C.T., 2010. Empty Pleasures: The Story of Artificial Sweeteners from Saccharin to Splenda. University of North Carolina Press, Chapel Hill.

EFSA ANS Panel (EFSA Panel on Food Additives and Nutrient Sources Added to Food), 2013a. Scientific opinion on the safety of advantame for the proposed uses as a food additive. EFSA J. 11 (7), 3301. doi:10.2903/j.efsa.2013.3301.

EFSA ANS Panel (EFSA Panel on Food Additives and Nutrient Sources Added to Food), 2013b. Scientific opinion on the re-evaluation of aspartame (E 951) as a food additive. EFSA J. 11 (12), 3496. doi:10.2903/j.efsa.2013.3496.

EFSA (European Food Safety Authority), 2007. Neotame as a sweetener and flavour enhancer. EFSA J. 5 (4), 581. Available from https://efsa.onlinelibrary.wiley.com/doi/pdf/10.2903/j.efsa.2007.581.

EFSA FAF Panel (EFSA Panel on Food Additives Flavourings), Castle, L., Andreassen, M., Aquilina, G., Bastos, M.L., Boon, P., Fallico, B., et al., 2024. Re-evaluation of saccharin and its sodium, potassium and calcium salts (E 954) as food additives. EFSA J 22 (11), e9044. doi:10.2903/j.efsa.2024.9044.

Erbaş, O., Erdoğan, M.A., Khalilnezhad, A., Solmaz, V., Gürkan, F.T., Yiğittürk, G., et al., 2018. Evaluation of long-term effects of artificial sweeteners on rat brain: a biochemical, behavioral, and histological study. J. Biochem. Mol. Toxicol. 32 (6), e22053. doi:10.1002/jbt.22053.

Eriksson Wiklund, A.K., Adolfsson-Erici, M., Liewenborg, B., Gorokhova, E., 2014. Sucralose induces biochemical responses in Daphnia magna. PLoS One 9 (4), e92771. doi:10.1371/journal.pone.0092771.

FDA (U.S. Food and Drug Administration), 1996. Food additives permitted for direct addition to food for human consumption; aspartame. Fed. Regist. 61 (126), 33654–33656. Available from https://www.govinfo.gov/content/pkg/FR-1996-06-28/pdf/96-16522.pdf. Accessed 15 July 2025.

FDA (U.S. Food and Drug Administration), 2013. Statement on European Aspartame Study. CFSAN/Office of Food Additive Safety. Available from https://downloads.regulations.gov/EPA-HQ-ORD-2009-0398-0021/attachment_8.pdf. Accessed 6 January 2025.

FDA (U.S. Food and Drug Administration), 2021. Food additives & ingredients – Aspartame and other sweeteners in food. Available from: https://www.fda.gov/food/food-additives-petitions/aspartame-and-other-sweeteners-food (Accessed 14 January 2025).

Fidler Mis, N., Braegger, C., Bronsky, J., Campoy, C., Domellöf, M., Embleton, N.D.ESPGHAN Committee on Nutrition, 2017. Sugar in infants, children and adolescents: a position paper of the European Society for Paediatric Gastroenterology, Hepatology and Nutrition Committee on Nutrition. J. Pediatr. Gastroenterol. Nutr. 65 (6), 681–696. doi:10.1097/MPG.0000000000001733.

Fitch, C., Keim, K.S., 2012. Position of the Academy of Nutrition and Dietetics: use of nutritive and nonnutritive sweeteners. J. Acad. Nutr. Diet. 112, 739–758. doi:10.1016/j.jand.2012.03.009.

Goldsmith, L.A., 2000. Acute and subchronic toxicity of sucralose. Food Chem. Toxicol. 38 (Suppl 2), S53–S69. doi:10.1016/S0278-6915(00)00028-4.

Gonçalves, N.G., Martinez-Steele, E., Lotufo, P.A., Bensenor, I., Goulart, A.C., Barreto, S.M., et al., 2025. Association between consumption of low- and no-calorie artificial sweeteners and cognitive decline: an 8-year prospective study. Neurology 105 (7), e214023. doi:10.1212/WNL.0000000000214023.

Griffioen-Roose, S., Smeets, P.A., Weijzen, P.L., van Rijn, I., van den Bosch, I., de Graaf, C., 2013. Effect of replacing sugar with non-caloric sweeteners in beverages on the reward value after repeated exposure. PLoS ONE 8 (11), e81924. doi:10.1371/journal.pone.0081924.

Guo, Y., Zhu, X., Zeng, M., Qi, L., Tang, X., Wang, D., et al., 2021. A diet high in sugar and fat influences neurotransmitter metabolism and then affects brain function by altering the gut microbiota. Transl. Psychiatry 11, 328. doi:10.1038/s41398-021-01443-2.

Hacioglu, C., 2024. Long-term exposure of sucralose induces neuroinflammation and ferroptosis in human microglia cells via SIRT1/NLRP3/IL-1β/GPx4 signaling pathways. Food Sci. Nutr. 12 (11), 9094–9107. doi:10.1002/fsn3.4488.

Hamamah, S., Aghazarian, A., Nazaryan, A., Hajnal, A., Covasa, M., 2022. Role of microbiota-gut-brain axis in regulating dopaminergic signaling. Biomedicines 10 (2), 436. doi:10.3390/biomedicines10020436.

Han, G., Li, X., Dong, G., Zhang, L., Gao, J., Li, M., et al., 2021. Phenotyping aquatic neurotoxicity induced by the artificial sweetener saccharin at sublethal concentration levels. J. Agric. Food Chem. 69 (7), 2041–2050. doi:10.1021/acs.jafc.0c05872.

Ibi, D., Suzuki, F., Hiramatsu, M., 2018. Effect of AceK (acesulfame potassium) on brain function under dietary restriction in mice. Physiol. Behav. 188, 291–297. doi:10.1016/j.physbeh.2018.02.024.

Iyaswamy, A., Kammella, A.K., Thavasimuthu, C., Wankupar, W., Dapkupar, W., Shanmugam, S., et al., 2018. Oxidative stress evoked damages leading to attenuated memory and inhibition of NMDAR-CaMKII-ERK/CREB signalling on consumption of aspartame in rat model. J. Food Drug Anal. 26 (2), 903–916. doi:10.1016/j.jfda.2017.11.001.

JECFA, 2003a. Summary and Conclusions of the Sixty-First Meeting of the Joint FAO/WHO Expert Committee of Food Additives, Rome, 10–19 June 2003. World Health Organization Available from. https://www.who.int/publications/i/item/9241209224 (Accessed 11 January 2025).

Jensen, T., Abdelmalek, M.F., Sullivan, S., Nadeau, K.J., Green, M., Roncal, C., et al., 2018. Fructose and sugar: a major mediator of non-alcoholic fatty liver disease. J. Hepatol. 68 (5), 1063–1075. doi:10.1016/j.jhep.2018.01.019.

Johnson, R.J., Sánchez-Lozada, L.G., Andrews, P., Lanaspa, M.A., 2017. A historical and scientific perspective of sugar and its relation with obesity and diabetes. Adv. Nutr. 8 (3), 412–422. doi:10.3945/an.116.014654.

Joint FAO/WHO Expert Committee on Food Additives (JECFA), 1981. Evaluation of Aspartame. WHO Food Additives Series No. 16. Available from: https://apps.who.int/food-additives-contaminants-jecfa-database/Home/Chemical/62.

Joint FAO/WHO Expert Committee on Food Additives (JECFA), 2023a. Ninety-Sixth Meeting: Safety Evaluation of Certain Food Additives. Available from: https://www.who.int/publications/i/item/9789240092549 (Accessed 8 November 2024).

Joint FAO/WHO Expert Committee on Food Additives (JECFA), 2023b. Evaluation of certain food additives: ninety-sixth report of the Joint FAO/WHO Expert Committee on Food Additives. WHO Technical Report Series No. 1050. Available from https://iris.who.int/handle/10665/376279.

Joint FAO/WHO Expert Committee on Food Additives, 1991. Acesulfame potassium. Toxicological Evaluation of Certain Food Additives, 28. WHO Food Additives Series, Geneva World Health Organization.

.JECFA (Joint FAO/WHO Expert Committee on Food Additives), 2013. Evaluation of certain food additives: seventy-seventh report of the Joint FAO/WHO Expert Committee on Food Additives. WHO Technical Report Series No. 983. Available from https://iris.who.int/handle/10665/98388 (Accessed 7 May 2025).

Joint FAO/WHO Expert Committee on Food Additives, World Health Organization, Food and Agriculture Organization of the United Nations & International Programme on Chemical Safety, 1993. Toxicological Evaluation of Certain Food Additives and Naturally Occurring Toxicants: Thirty-Ninth Meeting of the Joint FAO/WHO Expert Committee on Food Additives, Rome, 3–12 February 1992. World Health Organization, Geneva Available from. https://iris.who.int/handle/10665/40784.

Kasarello, K., Cudnoch-Jedrzejewska, A., Czarzasta, K., 2023. Communication of gut microbiota and brain via immune and neuroendocrine signaling. Front. Microbiol. 14, 1118529. doi:10.3389/fmicb.2023.1118529.

Kim, G.H., Shim, J.O., 2023. Gut microbiota affects brain development and behavior. Clin. Exp. Pediatr. 66 (7), 274–280. doi:10.3345/cep.2021.01550.
Leth-Møller, M., Duvald, C.S., Stampe, S., Greibe, E., Hoffmann-Lücke, E., Pedersen, M., Ovesen, P.G., 2023. Transplacental transport of artificial sweeteners. Nutrients 15 (9), 2063. doi:10.3390/nu15092063.
Lindseth, G.N., Coolahan, S.E., Petros, T.V., Lindseth, P.D., 2014. Neurobehavioral effects of aspartame consumption. Res. Nurs. Health 37 (3), 185–193. doi:10.1002/nur.21595.
MacDonald, A., van Wegberg, A.M.J., Ahring, K., Beblo, S., Bélanger-Quintana, A., Burlina, A., et al., 2020. PKU dietary handbook to accompany PKU guidelines. Orphanet J. Rare Dis. 15 (1), 171. doi:10.1186/s13023-020-01391-y, Erratum in: *Orphanet J. Rare Dis.* 15(1), 230..
Magnuson, B.A., Carakostas, M.C., Moore, N.H., Poulos, S.P., Renwick, A.G., 2016. Biological fate of low-calorie sweeteners. Nutr. Rev. 74, 670–689.
Méndez-García, L.A., Bueno-Hernández, N., Cid-Soto, M.A., De León, K.L., Mendoza-Martínez, V.M., Espinosa-Flores, A.J., et al., 2022. Ten-week sucralose consumption induces gut dysbiosis and altered glucose and insulin levels in healthy young adults. Microorganisms 10 (2), 434. doi:10.3390/microorganisms10020434.
Mossavar-Rahmani, Y., Kamensky, V., Manson, J.E., Silver, B., Rapp, S.R., Haring, B., et al., 2019. Artificially sweetened beverages and stroke, coronary heart disease, and all-cause mortality in the Women's Health Initiative. Stroke 50 (3), 555–562. doi:10.1161/STROKEAHA. 118.023100, Erratum in: *Stroke* 50(6), e176..
Mullee, A., Romaguera, D., Pearson-Stuttard, J., Viallon, V., Stepien, M., Freisling, H., et al., 2019. Association between soft drink consumption and mortality in 10 European countries. JAMA Intern. Med. 179 (11), 1479–1490. doi:10.1001/jamainternmed.2019.2478, Erratum in: *JAMA Intern. Med.* 179(11), 1607..
Muri, R., Rummel, C., McKinley, R., Rebsamen, M., Maissen-Abgottspon, S., Kreis, R., et al., 2024. Transient brain structure changes after high phenylalanine exposure in adults with phenylketonuria. Brain 147 (11), 3863–3873. doi:10.1093/brain/awae139.
Nofre, C., Tinti, J.-M., 2000. Neotame: discovery, properties, utility. Food Chem. 69 (3), 245–257. doi:10.1016/S0308-8146(99)00254-X.
Ochoa-Repáraz, J., Ramelow, C.C., Kasper, L.H., 2020. A gut feeling: the importance of the intestinal microbiota in psychiatric disorders. Front. Immunol. 11, 510113. doi:10.3389/fimmu.2020.510113.
Olivier-Van Stichelen, S., Rother, K.I., Hanover, J.A., 2019. Maternal exposure to non-nutritive sweeteners impacts progeny's metabolism and microbiome. Front Microbiol 10, 1360. doi:10.3389/fmicb.2019.01360.
Otabe, A., Fujieda, T., Masuyama, T., 2011. A two-generation reproductive toxicity study of the high-intensity sweetener advantame in CD rats. Food Chem. Toxicol 49 (Suppl 1), S70–S76. doi:10.1016/j.fct.2011.07.002.
Pepino, M.Y., 2015. Metabolic effects of non-nutritive sweeteners. Physiol. Behav 152 (Pt B), 450–455. doi:10.1016/j.physbeh.2015.06.024.
Peters, J.C., Beck, J., Cardel, M., Wyatt, H.R., Foster, G.D., Pan, Z., et al., 2016. The effects of water and non-nutritive sweetened beverages on weight loss and weight maintenance: a randomized clinical trial. Obesity (Silver Spring) 24 (2), 297–304. doi:10.1002/oby. 21327.
Pistollato, F., Iglesias, R.C., Ruiz, R., Aparicio, S., Crespo, J., Lopez, L.D., et al., 2018. Nutritional patterns associated with the maintenance of neurocognitive functions and the risk of dementia and Alzheimer's disease: a focus on human studies. Pharmacol. Res. 131, 32–43. doi:10.1016/j.phrs.2018.03.012.
Reagan-Shaw, S, Nihal, M, Ahmad, N., 2008. Dose translation from animal to human studies revisited. FASEB J. 22 (3), 659–661. doi:10.1096/fj.07-9574LSF.

Richardson, I.L., Frese, S.A., 2022. Non-nutritive sweeteners and their impacts on the gut microbiome and host physiology. Front. Nutr 9, 988144. doi:10.3389/fnut.2022.988144.

Rico, J.L., Aya-Ramos, L., Dueñas, Z., 2024. Effects of early-life stress followed by access to stevia or sucralose during adolescence on weight gain, glycemia, and anxiety-related behaviors in male and female rats. Physiol. Behav 280, 114529. doi:10.1016/j.physbeh.2024.114529.

Riera, C.E., Vogel, H., Simon, S.A., le Coutre, J., 2007. Artificial sweeteners and salts producing a metallic taste sensation activate TRPV1 receptors. Am. J. Physiol. Regul. Integr. Comp. Physiol. 293 (2), R626–R634. doi:10.1152/ajpregu.00286.2007.

Rodriguez-Palacios, A., Harding, A., Menghini, P., Himmelman, C., Retuerto, M., Nickerson, K.P., et al., 2018. The artificial sweetener Splenda promotes gut Proteobacteria, dysbiosis, and myeloperoxidase reactivity in Crohn's disease-like ileitis. Inflamm. Bowel Dis. 24 (5), 1005–1020. doi:10.1093/ibd/izy060.

Rother, K.I., Sylvetsky, A.C., Walter, P.J., Garraffo, H.M., Fields, D.A., 2018. Pharmacokinetics of sucralose and acesulfame-potassium in breast milk following ingestion of diet soda. J. Pediatr. Gastroenterol. Nutr. 66 (3), 466–470. doi:10.1097/MPG.0000000000001817.

Ruiz-Ojeda, F.J., Plaza-Díaz, J., Sáez-Lara, M.J., Gil, A., 2019. Effects of sweeteners on the gut microbiota: a review of experimental studies and clinical trials. Adv. Nutr 10 (suppl_1), S31–S48. doi:10.1093/advances/nmy037, Erratum in: *Adv. Nutr.* 11(2), 468..

Rycerz, K., Jaworska-Adamu, J.E., 2013. Effects of aspartame metabolites on astrocytes and neurons. Folia Neuropathol 51 (1), 10–17. doi:10.5114/fn.2013.34191.

Rymon, V., Lipinski, G.-W., 1991. Properties and applications of acesulfame-K. In: Mayer, D.G., Kemper, F.H. (Eds.), Acesulfame-K. Marcel Dekker, New York, pp. 209–225.

Sánchez-Tapia, M., Miller, A.W., Granados-Portillo, O., Tovar, A.R., Torres, N., 2020. The development of metabolic endotoxemia is dependent on the type of sweetener and the presence of saturated fat in the diet. Gut Microbes 12 (1), 1801301. doi:10.1080/19490976.2020.1801301.

Saucedo-Vence, K., Elizalde-Velázquez, A., Dublán-García, O., Galar-Martínez, M., Islas-Flores, H., SanJuan-Reyes, N., et al., 2017. Toxicological hazard induced by sucralose to environmentally relevant concentrations in common carp (Cyprinus carpio). Sci. Total Environ. 575, 347–357. doi:10.1016/j.scitotenv.2016.09.230.

Scardaci, R., Manfredi, M., Barberis, E., Scutera, S., Marengo, E., Pessione, E., 2021. Serotonin exposure improves stress resistance, aggregation, and biofilm formation in the probiotic *Enterococcus faecium* NCIMB10415. Microbiol. Res. 12 (3), 606–625. doi:10.3390/microbiolres12030043.

SCF (Scientific Committee on Food), 1985. Reports of the Scientific Committee for Food on Sweeteners. Commission of the European Communities, Luxembourg, Belgium Food – Science and Techniques. 16th series. Available from. https://ec.europa.eu/food/fs/sc/scf/reports/scf_reports_16.pdf (Accessed 7 April 2025).

SCF (Scientific Committee on Food), 2000. Opinion of the European Commission Scientific Committee on Food on "Re-evaluation of acesulfame K". Available from: http://ec.europa.eu/food/fs/sc/scf/out52_en.pdf (Accessed 7 April 2025).

Schiffman, S.S., Nagle, H.T., 2019. Revisited: assessing the in vivo data on low/no-calorie sweeteners and the gut microbiota. Food Chem. Toxicol 132, 110692. doi:10.1016/j.fct.2019.110692.

Schneider, E., O'Riordan, K.J., Clarke, G., Cryan, J.F., 2024. Feeding gut microbes to nourish the brain: unravelling the diet-microbiota-gut-brain axis. Nat. Metab 6 (8), 1454–1478. doi:10.1038/s42255-024-01108-6.

Serrano, J., Smith, K.R., Crouch, A.L., Sharma, V., Yi, F., Vargova, V., et al., 2021. High-dose saccharin supplementation does not induce gut microbiota changes or glucose intolerance in healthy humans and mice. Microbiome 9 (1), 11. doi:10.1186/s40168-020-00976-w.

Shil, A., Olusanya, O., Ghufoor, Z., Forson, B., Marks, J., Chichger, H., 2020. Artificial sweeteners disrupt tight junctions and barrier function in the intestinal epithelium through activation of the sweet taste receptor, T1R3. Nutrients 12 (6), 1862. doi:10.3390/nu12061862.

Strandwitz, P., 2018. Neurotransmitter modulation by the gut microbiota. Brain Res. 1693 (Pt B), 128–133. doi:10.1016/j.brainres.2018.03.015.

Suez, J., Korem, T., Zeevi, D., Zilberman-Schapira, G., Thaiss, C.A., Maza, O., et al., 2014. Artificial sweeteners induce glucose intolerance by altering the gut microbiota. Nature 514 (7521), 181–186. doi:10.1038/nature13793.

Suez, J., Korem, T., Zilberman-Schapira, G., Segal, E., Elinav, E., 2015. Non-caloric artificial sweeteners and the microbiome: findings and challenges. Gut Microbes 6 (2), 149–155. doi:10.1080/19490976.2015.1017700.

Swithers, S.E., Martin, A.A., Clark, K.M., Laboy, A.F., Davidson, T.L., 2010. Body weight gain in rats consuming sweetened liquids: effects of caffeine and diet composition. Appetite 55 (3), 528–533. doi:10.1016/j.appet.2010.08.021.

Sylvetsky, A.C., Figueroa, J., Zimmerman, T., Swithers, S.E., Welsh, J.A., 2019. Consumption of low-calorie sweetened beverages is associated with higher total energy and sugar intake among children, NHANES. Pediatr. Obes. 10, e12535. doi:10.1111/ijpo.12535.

Thomson, P., Santibañez, R., Aguirre, C., Galgani, J.E., Garrido, D., 2019. Short-term impact of sucralose consumption on the metabolic response and gut microbiome of healthy adults. Br. J. Nutr. 122 (8), 856–862. doi:10.1017/S0007114519001570.

Tripathi, M., Khanna, S.K., Das, M., 2006. Usage of saccharin in food products and its intake by the population of Lucknow, India. Food Addit. Contam 23 (12), 1265–1275. doi:10.1080/02652030600944395.

Uebanso, T., Ohnishi, A., Kitayama, R., Yoshimoto, A., Nakahashi, M., Shimohata, T., et al., 2017. Effects of low-dose non-caloric sweetener consumption on gut microbiota in mice. Nutrients 9 (6), 560. doi:10.3390/nu9060560.

Van den Eeden, S.K., Koepsell, T.D., Longstreth Jr, W.T., van Belle, G., Daling, J.R., McKnight, B., 1994. Aspartame ingestion and headaches: a randomized crossover trial. Neurology 44 (10), 1787–1793. doi:10.1212/wnl.44.10.1787.

Viberg, H., Fredriksson, A., 2011. Neonatal exposure to sucralose does not alter biochemical markers of neuronal development or adult behavior. Nutrition 27 (1), 81–85. doi:10.1016/j.nut.2009.10.007.

Weihrauch, M.R., Diehl, V., 2004. Artificial sweeteners—do they bear a carcinogenic risk? Ann. Oncol. 15 (10), 1460–1465. doi:10.1093/annonc/mdh256.

Whitehouse, C.R., Boullata, J., McCauley, L.A., 2008. The potential toxicity of artificial sweeteners. AAOHN J. 56 (6), 251–261. doi:10.1177/216507990805600604.

WHO (World Health Organization), 2015. Guideline: Sugars intake for adults and children. World Health Organization, Geneva Available from. https://www.who.int/publications/i/item/9789241549028 Accessed 8 November 2024.

WHO (World Health Organization), 2023. *Use of non-sugar sweeteners: WHO guideline*. Geneva, World Health Organization. 90 p. Available at: https://www.who.int/publications/i/item/9789240073616

Wu, M., Qian, Y., Boyd, J.M., Hrudey, S.E., Le, X.C., Li, X.-F., 2014. Direct large volume injection ultra-high performance liquid chromatography-tandem mass spectrometry determination of artificial sweeteners sucralose and acesulfame in well water. J. Chromatogr. A 1359, 156–161. doi:10.1016/j.chroma.2014.07.035.

Wu, H.T., Chiang, C.C., Wang, C.T., Chen, Y.H., Hsu, S.Y., Chen, Y.C., 2024. Consumption of the nonnutritive sweetener acesulfame potassium increases central precocious puberty risk. J. Hazard Mater. 461, 132529. doi:10.1016/j.jhazmat.2023.132529.

Zhang, M., Chen, J., Yang, M., Qian, C., Liu, Y., Qi, Y., et al., 2021. Low doses of sucralose alter fecal microbiota in high-fat diet-induced obese rats. Front. Nutr. 8, 787055. doi:10.3389/fnut.2021.787055.
Zheng, P., Zeng, B., Liu, M., Chen, J., Pan, J., Han, Y., et al., 2019. The gut microbiome from patients with schizophrenia modulates the glutamate-glutamine-GABA cycle and schizophrenia-relevant behaviors in mice. Sci. Adv. 5 (2), eaau8317. doi:10.1126/sciadv.aau8317.
Zheng, Z., Xiao, Y., Ma, L., Lyu, W., Peng, H., Wang, X., et al., 2022. Low dose of sucralose alter gut microbiome in mice. Front. Nutr. 9, 848392. doi:10.3389/fnut.2022.848392.

CHAPTER FOUR

Nutraceuticals and obsessive-compulsive disorder therapy

Amir Hossein Mohammadi[a], **Negin Amin**[b], **Sayed Abolfazl Tavousi**[c], **Mahmood Khaksary Mahabady**[d], **Michael Aschner**[e], **Hamed Mirzaei**[a,*]

[a] Research Center for Biochemistry and Nutrition in Metabolic Diseases, Institute for Basic Sciences, Kashan University of Medical Sciences, Kashan, Iran
[b] School of Health Science, College of Health, Massey University, Palmerston North, New Zealand
[c] Student Research Committee, Kashan University of Medical Sciences, Kashan, Iran
[d] Anatomical Sciences Research Center, Institute for Basic Sciences, Kashan University of Medical Sciences, Kashan, Iran
[e] Department of Molecular Pharmacology, Albert Einstein College of Medicine, Bronx, NY, United States
*Corresponding authors. e-mail address: h.mirzaei2002@gmail.com.

Contents

Abstract

Neurological disorders are a major public health concern. Notably, the rate of chronic neurological diseases is affected by lifestyle changes, leading to considerable healthcare and economic burden worldwide. Considerable interest has emerged on functional foods due to their beneficial health efficacy. Thus, efforts have been made to extract different bioactive phytochemicals or compounds from therapeutic natural-based products, which has led to functional foods referred to as "vitafoods," "medifoods," "pharmafoods," or "medicinal foods." To overcome neurological illness such as obsessive-compulsive disorder (OCD), nutraceuticals have been applied as prophylactic options This review addresses nutraceutical intervention in the OCD treatment.

Keywords: Food supplements, Mental disorder, Nutraceuticals, Obsessive-compulsive disorder, Probiotic

Advances in Neurotoxicology, Volume 16
ISSN 2468-7480
https://doi.org/10.1016/bs.ant.2026.03.001

1. Introduction

Obsessive-compulsive disorder (OCD) is a prevalent psychiatric illness characterized by repetitive unwanted thoughts (obsessions) as well as recurrent behaviors (compulsions) (Mohammadi et al., 2023; 2024). OCD is commonly found in childhood or adolescence and often coexists with anxiety, mood, somatoform, and psychotic disorders (Ruscio et al., 2010; Boileau, 2011). The annual incidence rate of OCD morbidity worldwide is about 1.1–1.8 percent (Edition, 2013) with the lifetime prevalence of 2–3 percent (Ruscio et al., 2010). Prevailing theories posit that OCD can be driven by etiology and pathophysiology issues such as neurochemical, structural, immunological and genetic factors (Boileau, 2011). There are several lines of evidence showing that nutraceuticals (natural products) have promising modulatory effects on the main pathways involved in the OCD pathogenesis (Camfield et al., 2011). Because of shortfalls and the adverse side-effects of conventional antidepressant treatment options for OCD, it is critical to develop novel promising approaches to ameliorate OCD symptoms (Camfield et al., 2011). Although there is a growing attention in nutraceuticals, theses bioactive factors have been poorly investigated than pharmaceutical substances. A nutraceutical is regarded as a medicinal or nutritional substance in a form of a food, mineral, vitamin, a part of a food, or a herb that affords health effectiveness (Kalra, 2003). The current synopsis aims at providing a review of experimental studies and clinical trials that evaluated the effects of nutraceuticals on OCD management.

1.1 Obsessive-compulsive disorder pathogenesis

OCD is associated with abnormality in the anterior cingulate cortex (ACC), orbitofrontal cortex (OFC), thalamus, and striatum (Milad and Rauch, 2012; Haber and Heilbronner, 2013). In practice, OFC is divided into 2 parts (Kringelbach and Rolls, 2004): medial OFC is hypo-activated (with inverse correlation to the seriousness of symptom) (Couto and Moreira, 2018) and lateral OFC is hyper-activated (in the course of the symptoms' provocation (Couto and Moreira, 2018) that is proportionate to the severity of the OCD symptoms (Rauch et al., 2007). Researchers have shown the role of the dorsal ACC, but it has been not verified by functional neuro-imaging (Radua et al., 2010; Schlösser et al., 2010) and treatment investigations (Perani et al., 1995), as well as by decreasing the OCD symptoms following anterior cingulotomy (Dougherty et al., 2002; Bourne et al., 2013).

Numerous studies have addressed at the neurocircuit level, combining neuro-psychological, neuro-imaging, and treatment into OCD pathophysiology corroborating that it involves abnormal functions in the cortico–striato–thalamo–cortical (CSTC) pathways (Cannistraro et al., 2004; Couto and Moreira, 2018). Activities in the mentioned circuit or pathways enhanced in the basal state and in the course of the symptoms' provocation that is normal to some extent with substantial treatment (Baxter et al., 1992; Swedo et al., 1992; McGuire et al., 1994; Rauch et al., 1994; Simon et al., 2010). Moreover, beneficial effects of the selective serotonin reuptake inhibitors (SSRI) in OCD treatment suggest that serotonin may factor into OCD pathophysiology (Goddard et al., 2008). Brain imaging has corroborated reduced serotonin availability in specific areas of the OCD brain (Couto and Moreira, 2018). Nearly 50 percent of the patients respond to the treatments with SSRIs (Soomro et al., 2008); therefore, in case of serotonin deficits, the patho-physiology of the disease cannot be fully justified.

In addition to the serotonin transporter system, several studies reported on the role of glutamatergic dysfunction in OCD pathogenesis (Pittenger et al., 2011; Wu et al., 2012). Several studies found excess glutamate in cerebrospinal fluid (CSF), caudate, and OFC of patients with OCD (Couto and Moreira, 2018). OS is verified by changes in the anti-oxidant systems in the specimens of such patients' serum and lipid peroxidation. It is notable that Yale-Brown Obsessive Compulsive Scale (Y-BOCS) score was designed to evaluate the severity of clinical symptoms of the disease, which showed a correlation with greater levels of OS (Couto and Moreira, 2018; Balandeh et al., 2021; Mohammadi et al., 2021). Since OCD has been considered as one of the phenotypically heterogeneous disorders with higher symptomatic expression level (Pauls et al., 2014), higher smooth sub-types with distinctive patterns of comorbidities and outputs have been detected (Leckman et al., 2010).

2. Treatment of obsessive-compulsive disorder

Some studies have shown high selective responses of OCD to the serotonergic medicines. Moreover, clomipramine (Anafranil), which is one of the tricyclic antidepressants with a great serotonergic impact, has been among the first options for pharmacologic treatments of OCD. Nevertheless, some issues have been raised regarding its safety and therefore selective serotonin reuptake inhibitors (SSRIs) have been presently prioritized for earlier treatments (Association, 2006; Koran et al., 2007). Moreover, Cochrane's review verified SSRIs' efficiency in treating OCD (absolute risk decrease of 8–17 percent, the number required for treatment is between 6 and 12) (Soomro et al., 2008). The U.S. Food and Drug Administration (FDA) has approved fluvoxamine, Fluoxetine (Prozac), sertraline (Zoloft), as well as paroxetine (Paxil) for the treatment of OCD. It should be noted that escitalopram (Lexapro) and Citalopram (Celexa) have a also widespread usage; however, the FDA has initiated the recommendation for limited dosage for citalopram in 2011 due to issues with QT prolongation. In addition, inadequate studies have been reported on SSRI superiority, and thus choosing SSRI must be individualized with regard to tolerability and interaction with other drugs (Fineberg et al., 2012). To obtain optimal responses, SSRI must be increased in doses of OCD cases in comparison to other medicines over 4–6 weeks to obtain a high dose (Association, 2006; Bloch et al., 2010; U.S. Food and Drug Administration, 2012). It is possible to use a dose greater than the usual maximum doses, with careful supervision for consequences such as serotonin syndrome. Furthermore, treatment must be trialed for 8–12 weeks, with not less than 4–6 weeks at the greatest tolerable doses (Koran et al., 2007). Prominent improvements in the related symptoms generally are seen after 4–6 weeks and sometimes after 10 weeks or more. In case of a successful medical treatment, this trend must be followed for at least 1–2 years (Fineberg and Gale, 2005; Association, 2006). Another study suggested relapse prevention with the continual SSRI treatment as one of the acceptable therapeutic goals (Fineberg et al., 2007). In case of discontinued pharmacotherapy by the patient, physicians should consider decreasing the dose over several months and restarting the original doses in case of the worsened symptoms. Consequently, responses to the psychological treatment can take longer than the responses to the medicines (Whittal et al., 2008). Periodical "booster" sessions of response and exposure prevention have been advised for lowering the relapse risks in case of discontinued psychological therapy (Koran et al., 2007). In case of lack of efficacy of a SSRI physicians must use a combined approach (Simpson et al., 2006). When patients prefer the continuation with just medical treatment, they must try another SSRI (Koran et al., 2007). Since this treatment may result in anti-cholinergic consequences, and rarely seizure or arrhythmia, the treatment process must be initiated at lower doses (25 mg/day) with gradual titrations for minimizing reverse reactions. Adding an atypical antipsychotic affects a number of cases with insufficient responses to SSRI therapies (Koran et al., 2007; Komossa et al., 2010). However, conflicting evidence suggests that atypical antipsychotics may show the greatest efficacy, although their use may carry higher risks of adverse outcomes compared with SSRI monotherapy, potentially limiting the benefit of the aforementioned treatment (Koran et al., 2007). Experts in the field have provided diverse opinions for cases with treatment-resistant OCD; however, there is insufficient support for the majority of treatments. Several investigations supported the efficacy of mirtazapine (Remeron) or serotonin-norepinephrine reuptake inhibitors for OCD (Koran et al., 2005; Dell'Osso et al., 2008). Moreover, other studied medicines include stimulants, glutamatergic agents, ondansetron (Zofran), pindolol, anti-convulsants, as well as acetylcysteine (Fenske and Schwenk, 2009). In addition, deep brain stimulation has been found beneficial for severe treatment-resistant OCD, though few cases have been tested in this regard. Finally, such a therapy has been employed as one of the last resorts with the limited approval from FDA under the humanitarian device exemptions (Goodman and Alterman, 2012).

Modern pre-clinical investigations exploring nutraceutical for OCD referred to the promising activities in the modulation of main pathways that contribute to OCD pathogenesis (serotonergic pathway dys-regulation & glutamatergic). New clinical evidence has confirmed plant-based interventions and specific nutrients with particular active constituents that involve in regulating the above pathways: myo-inositol, N-acetlycysteine, milk thistle (Silybum marianum), and glycine (Camfield et al., 2011). However, it is not possible for serotonin precursor tryptophan to be used in the treatment of OCD whereas 5-HTP is one of the more efficient precursor strategies. Nonetheless, any clinical documents have been not reported for testing the efficiency of each material. Although, there has been a few studies on this field, additional investigations on nutraceuticals must be accomplished by novel pre-clinical information on the respective mechanism and their acceptable consequences as compared to the new SSRI therapies. Future studies using randomized placebo-controlled designs and larger samples are warranted to better established the effectiveness of nutraceutical in OCD management (Camfield et al., 2011).

3. Nutraceuticals: Potentials and mechanisms of action

For many centuries, spices and natural products have been shown therapeutic effects in various ailments (Bungău and Popa, 2015). The advanced scientific investigations have led to finding therapeutic plant- and non-plant-derived agents, resulting in innovative researches in the field of nutritional value of novel dietary products (Makkar et al., 2020). It should bring to light that combination of nutrition and pharmaceutical words make a novel term known as nutraceuticals created by Dr Stephen De Felice in 1989 (Ramalingum and Mahomoodally, 2014). Based on the American Nutraceutical Association definition, nutraceuticals are considered as a food or its supplement with beneficial healthy properties, including genetically modified foods, dietary nutrient products, beverages, herbal products, soups, fruits, vegetables, and processed foods (cereals), *etc.* (Prakash and van Boekel, 2010). Thus far, it has been determined that nutraceuticals are primarily composed of minerals, vitamins, and amino acids, and more than 1000 other probiotics (Makkar et al., 2020). Nutraceuticals are categorized based on their activities as follows; antifungal, antibacterial, anti-inflammatory (Fig. 1), antioxidants, and anti-obesity, which highlight their role and applications. Nevertheless, food-borne infections pose a threat to the health condition. In detail, many fruits and vegetables-derived bioactive factors have been revealed as effective antibacterial agents in the treatment of infectious ailments, such as quercetin (polyphenols), carsonic acid (terpenoids), to name a few (Gutiérrez-del-Río et al., 2018). Regarding antifungal therapy, medicinal mushrooms, with widespread popularity all over Asia, are recognized as worthy candidate due to the least adverse effects (Giavasis, 2014). Since nutraceuticals possess anti-inflammatory function, they can improve various inflammatory problems, including rheumatoid arthritis (Al-Okbi, 2014). Of note, foods full of carotenoids, antioxidants, and tocopherols have greater anti-inflammatory effectiveness. Nutraceuticals have not only well-known radical scavenging activities, but also antioxidant properties in the control of other health problems, such as obesity (Cornelli, 2009). However, toxicity may occur after excessive consumption of anything. Overall, nutraceuticals based to their mode of action and their therapeutic activity at the monitored recommended dosages may reduce toxicity and improve health condition (Makkar et al., 2020).

4. The safety and bioavailability of nutraceuticals

To assess efficient bioactivities of nutraceuticals substances, different parameters are examined such as bioaccessibility, bioavailability, bioconversion, bioactivity, and bioequivalence (Parada and Aguilera, 2007). Frequently, analysis of bioactive compounds can be problematic and confusing because of inaccurate use and definition of above mentioned concepts. In this

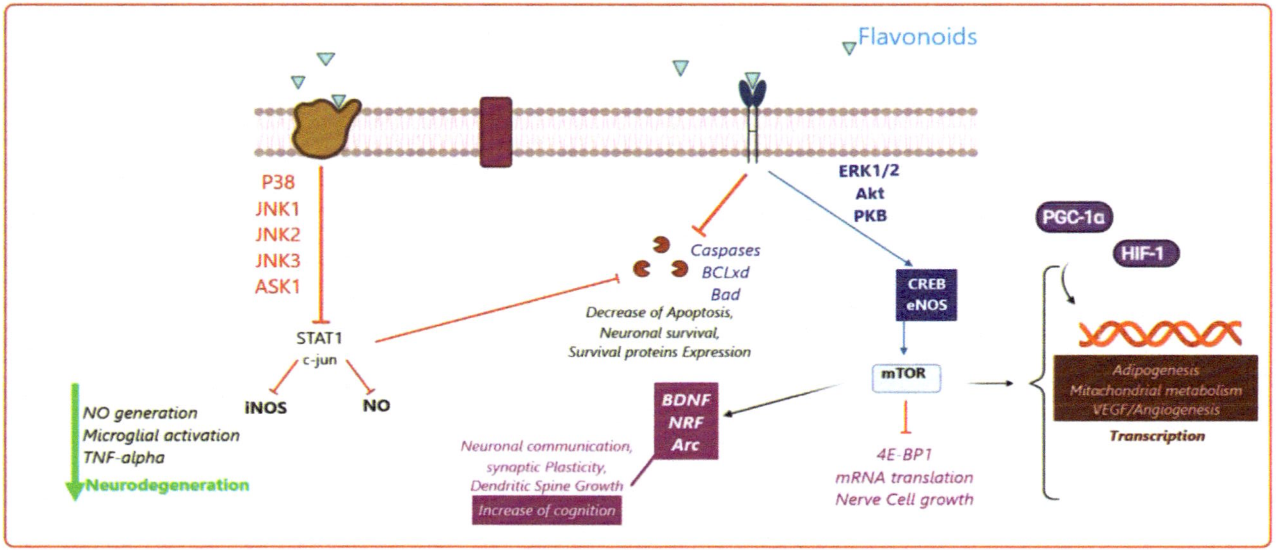

Fig. 1 Scme signaling pathways that are affected by flavonoids in OCD.

context, there is no distinction between the bioavailability of a medicine administrated into the systemic circulation and the biocomponent introduced via the gastrointestinal tract (GIT). Absolute bioavailability is defined as the amount of bioactive substance fundamentally reached into the systemic circulation following non-intravenous (IV) administration compared to IV administration. It should be noted that the relative bioavailability is applied for the bioactive substance cannot be introduced intravenously (Sinko, 2024). It would take in to account that oral bioavailability includes all steps that bioactive compounds go through from oral administration to excretion. Bioavailability of nutraceutical can be defined as the amounts of swallowed substance absorbed at the level of GIT and reached to the tissues and organs (Parada and Aguilera, 2007). Taken together, nutraceuticals bioavailability involves three major stages: bioaccessibility, absorption, and transformation (18). In some studies, bioaccessibility describes uptake of biocomponent via epithelium and bioactivity of bioactives refers to tissue absorption, metabolism, and physiological reaction (Dima et al., 2020)." In the bioaccessibility step, the fraction of swallowed nutraceutical can be absorbed by intestinal epithelium (Jafari and McClements, 2017). Epithelial absorption of biocomponents depends on liberation from food matrix, solubilization in the biological fluid and interaction with other molecules. The bioactive compounds can be affected by the food matrix texture, physicochemical properties of the delivery system, and the enzymatic agents in the digestion environment (Dima et al., 2020). The solubility of bioactive substances can be influenced by their chemical structure as well as the conditions of digestive fluids. In a way that the hydrophilic compounds (bases, acids, and salts) are usually solubilized in the GIT environment based on the ionic strength, pH, and aggregation condition, while the hydrophobic compounds are mostly soluble in the intestinal fluid by lipolytic enzymes (Katouzian et al., 2017; Rafiee et al., 2019). The solubilization of hydrophobic molecules occurs in a micellar form composed of phospholipids, free fatty acids, bile salts, and others. In the second phase of bioavailability, biocomponent absorption takes place at the epithelial cells of GIT (Dima et al., 2020), with high absorption in the small intestine. Considering the modality of the target biocomponent, they transport via different mechanisms (active and passive transport) into the intestinal epithelium. Nonetheless, the bioavailability of biocomponents in the lumen of the intestine can be reversed by efflux transporters (Jafari and McClements, 2017). Another stage of bioavailability inhibition refers to the process of biocomponents transformation particularly liver metabolic process. In addition, GIT fluids can transform some bioactive compounds chemically or biochemically, including degradation of curcumin by alkaline environment (Rafiee et al., 2019), oxidation of poly unsaturated fatty acids by pro-oxidant metals (Dima et al., 2020), *cis*–trans isomerization of carotenoids into the inactive forms (Rostamabadi et al., 2019), hydrolysis of proteins and peptides in presence of acid condition (proteolytic enzymes), to name a few (Dima et al., 2020).

5. Nutraceuticals and obsessive-compulsive disorder

Nutraceuticals have significant impacts on the management of mental disorders, including schizophrenia bipolar complications (Cloutier et al., 2016) and OCD (Camfield et al., 2011; Couto and Moreira, 2018; Karcı and Celik, 2020). It was reported that patients suffering from neurodegenerative and psychotic disorders were managed by nutraceuticals as adjunctive therapy or monotherapy. Mechanistically, nutraceuticals intensify the therapeutic efficacy of the psychiatric medications as monoamines re-uptake inhibitor and neurobiological factor leading to amplified neuroprotection (Makkar et al., 2020), and decreased side effects of psychiatric drugs.

Different parts of the nervous system especially the basal ganglia, prefrontal cortex (anterior cingulate and orbitofrontal cortices), and thalamus have been implicated in the pathology of OCD. Emerging evidence shows that neurotransmitters importantly dopamine, glutamate, serotonin, and γ-aminobutyric acid (GABA) associated with OCD pathophysiology (Huey et al., 2008; Pittenger, 2017). Glutamatergic neurons exist throughout different sections of the brain including cerebellum, basal ganglia, brainstem circuits, and different cortico-subcortical and intracortical connections. Glutamate receptors can be divided into metabotropic and ionotropic groups. To date, the metabotropic receptor, mGLU1, has been categorized to 8 subtypes (mGluR1–mGluR8). Besides, the key ionotropic receptors consist of α-amino-3–hydroxy-5-methyl-4-isoxazole propionic acid, N-methyl-D-aspartate (NMDA), and kainate receptors. Some evidence suggests that glutamate transmission dysfunction in cortico-striato-thalamo-cortical (CSTC) circuits may be responsible for pathogenesis of OCD (Sheshachala and Narayanaswamy, 2019). The common therapeutic anti-obsessional window is selective serotonine (5-HT) reuptake inhibitors (SSRIs) which suppress the re-uptake of 5-HT into the transporter followed by astonishing serotonin reuptake (Baumgarten and Grozdanovic, 1998; Lissemore et al., 2014). Moreover, dopaminergic impairments have been reported in OCD patients. Remarkably, a notable number of OCD patients (about 40–60 percent) have not been treated properly with SRI-monotherapy, suggesting a requisite for a combination of dopaminergic antagonists and SRIs in OCD treatment (Hemmings et al., 2003; Koo et al., 2010). Recently, the use of nutraceuticals in the management of OCD has emerged as an attractive area of research, as the mechanisms of action and effects of a growing number of nutraceutical compounds have begun to be elucidated. Studies highlighted the potential effects of several probiotics on the stimulation of 5-HT synthesis and inhibition of dopamine-β-hydroxylase. In addition, some probiotics induce GABA production that can modulate GABAergic signaling (Yong et al., 2019). One of these nutraceuticals is caffeine, which interferes with adenosine A1, A2 receptor antagonist leading to serotonin reuptake. Additionally, only high amounts of caffeine can inhibitA2 receptor followed by increased dopamine release (Shams et al., 2019). Nutraceuticals may show efficacy in OCD when combined with zinc supplementation and SSRIs, likely via reduction of synaptic glutamate secretion via N-acetyl cysteine (NAC), and regulation of 5-HT transporter function by Mioinozitol.

Probiotics are live yeast sand bacteria with different health benefits when consumed chronically in adequate concentrations (Sherman et al., 2009). Attention has been drawn toward the therapeutic potential of probiotics. Promising findings regarding the efficacy of either single strain or mixture of probiotics in the treatment of several health outcomes including, but not limited to, gastrointestinal (GI) disorders, female reproductive disorders, and cardiovascular disorders (CVDs) have been provided (Khalesi et al., 2019). Probiotic have been also found to be of benefit in the modification of some complications of central nervous system (CNS) as demonstrated in Fig. 2 (Dinan and Cryan, 2013; Cerdó et al., 2017).

Gut microbiome are composed of different genera such as Bacteroidetes and Firmicutes with greater abundance, and Actinobacteria, Proteobacteria, and Fusobacteria with lower frequency (Lankelma et al., 2015). A bidirectional interaction between the gut-brain axis (GBA) and microbiota was detected, and its disruption contributes to several deleterious health outcomes (Dinan and Cryan, 2013). The significance of GBA could be further evidenced by increased risk of co-occurring of psychological disorders (Fig. 3) such as anxiety combined with GI ailments, for instance IBS (Kennedy et al., 2012). The GBA is connected by complicated crosstalk between the immune, endocrine, vagal nerve, and humoral links and combines cognitive and emotional centers of the brain with intestinal activities (Carabotti et al., 2015). There is lines evidence supporting that chronic usage of probiotic Lactobacillus improves health condition in the experimentally

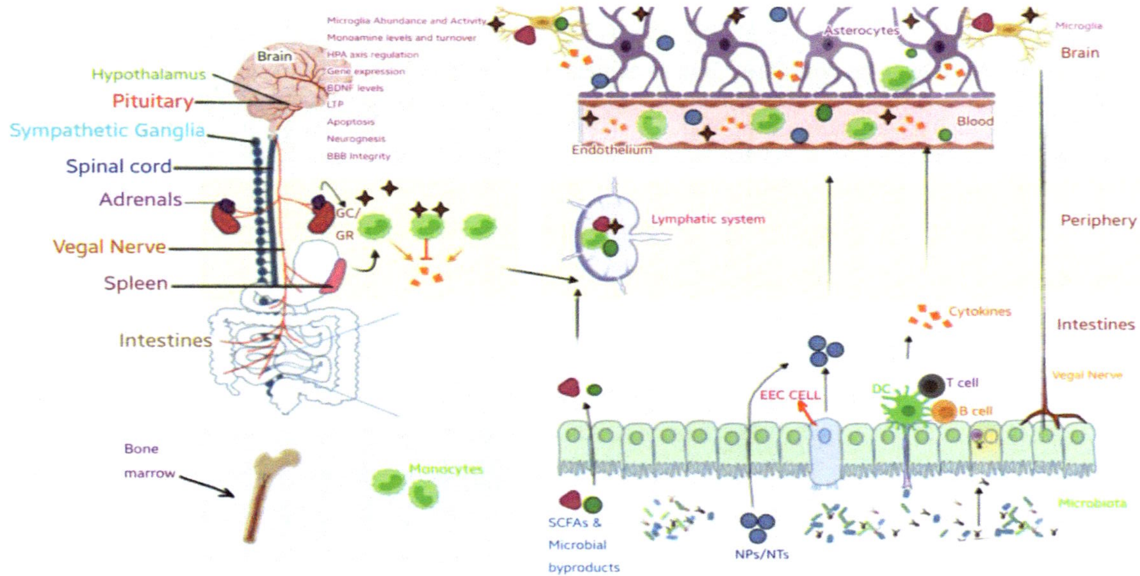

Fig. 2 Schema of micro-biota regulation of the activities of HPA axis and neuro-inflammation. Communicating into the microbiota-gut-brain axis requires a complicated coordination of several systems and factors. In fact, microbiota may dominate events in CNS and periphery nervous system via diverse communication instruments like generation of cytokine, release of neuro-transmitter and neuropeptide, activation of vagal nerve, release of microbial by-products and SCFA, as well as via utilization of systemic and lymphatic circulation. In case of penetration of the signals into blood brain barrier (BBB) and reaching the brain, they may impact activation and influence maturation state of microglia. Upon the activation, microglia contributes importantly to the immune surveillance, debris clearance, as well as synaptic pruning. Moreover, they simplify several daily functions in the brain like modulation of the HPA axis activation state. In this way, releasing glucocorticoids (cortisol) as one of the consequences of HPA axis activation may modulate the activation mood of brain microglia, and impact the release of cytokine and trafficking of the monocytes from the periphery to the brain.

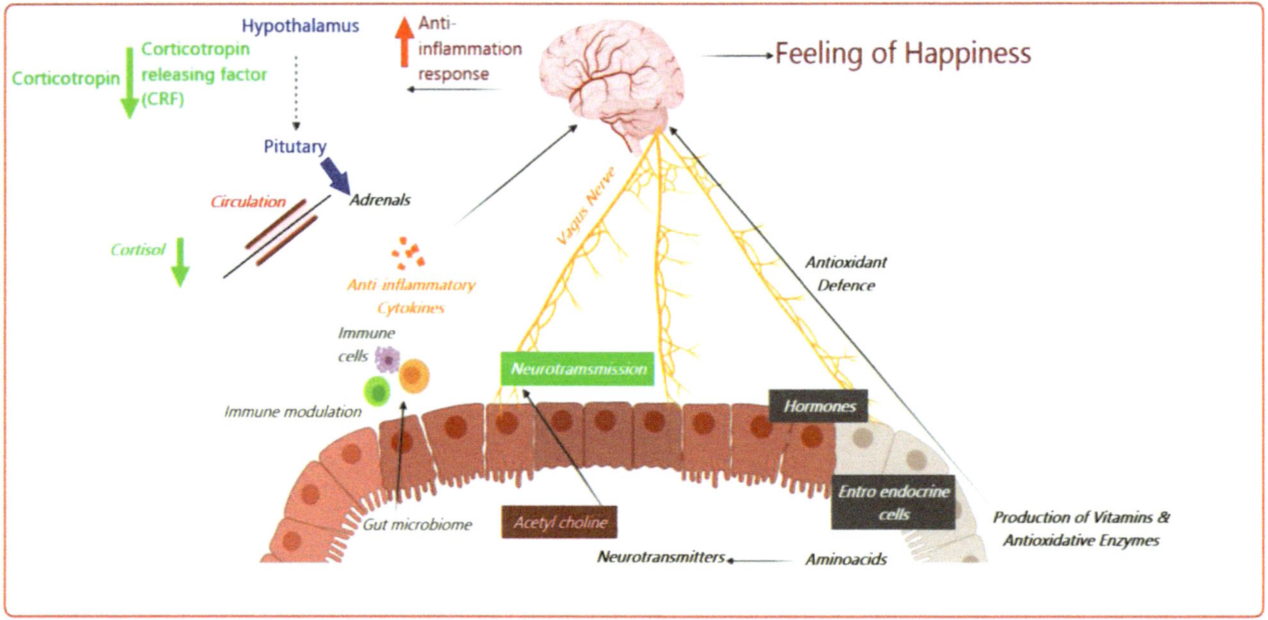

Fig. 3 Probiotics mechanism for overcoming the depressive mood disorders. In fact, probiotics is responsible for regulating immune system, causes resistance to infection, declines the impact of OS and inflammation, and finally creates antioxidant and anti-inflammatory compounds. Moreover, probiotic and other gut commensal bacteria create sufficient level of fatty acids and amino acids necessary to synthesize the neurochemicals crucial to feel well-being and happiness in individuals. Additionally, probiotics create specific bio-active molecules (like hormone & peptide), which have been shown to have the same structural and functional pattern as the ones generated by hosts that modulate the host behavior directly or indirectly.

manipulated mice. Chronic treatment with L. rhamnosus promoted not only the immunity of immunocompromised mice (Wei et al., 2007), but also insulin sensitivity in diabetic mice induced by high-fat regime (Kim et al., 2013). L. rhamnosus intake did not show adverse effects on body weight supporting its safety (Sanikhani et al., 2020). Sanikhani et al. (2020) sough toevaluate the efficacy of probiotic L. casei in the management of OCD symptoms in an animal model (Table 1). Their findings revealed that L. casei, fluoxetine, and a combination of drugs improve OCD symptoms via enhancement of brain-derived neurotrophic factor (*BDNF*) expression and reduction of 5-hydroxytryptamine receptor type 2A (*HTR2A*) expression. *BDNF* is one of a brain-derived neurotrophic factor that activates the tropomyosin receptor kinase B and has propounding role in survival, development, synaptic function (Chao, 2003), and stimulation of experience-dependent plasticity in the brain (Park and Poo, 2013). *HTR2A* belongs to the G protein–coupled serotonin receptor family (Park and Poo, 2013), which is the main excitatory kinds of receptor for serotonin (Martin et al., 1998).

Polyphenols are categorized as secondary natural metabolites, which are broadly found in the plant kingdom. Structurally, polyphenolic compounds consist of more than one hydroxyl substituents in the aromatic ring. Currently, over 8000 phenolic structures have been reported (Chao et al., 2012). Generally, polyphenols include hydrolysable tannins (esters of gallic acids and glucose, as well as other sugars) and phenylpropanoids (flavonoids, lignins, and condensed tannins). Polyphenolic compounds have different ranges of biological effects, such as antioxidant, anti-carcinogenic, anti-inflammatory, antithrombotic and anti-mutagenic effects (Urquiaga and Leighton, 2000). Among polyphenolic flavonoid, curcumin (1,7-bis[4–hydroxy 3–methoxy phenyl]−1,6-heptadiene-3,5–dione), comprises about 4 percent of turmeric (Curcuma Longa). From ancient times curcumin has been applied in traditional Asian regimes and herbal remedies. In addition to curcumin, turmeric-derived curcuminoids include methoxycurcumin and desmethoxycurcumin. Analogous to curcumin, curcuminoids are commercially accessible and are usually studied in experimental models. Since curcumin constitutes about 80 percent of curcuminoids content, the bioactive effects of curcuminoids have frequently been attributed to the curcumin (Chao et al., 2012). In accordance with other polyphenolic compounds (epigallocatechin-3-gallate, caffeic acid, (-)-and resveratrol), curcumin has influential antioxidant activities. Jagatha et al. (2008) Displayed therapeutic value of curcumin treatment both in the But hioninesulfoximine animal model and

Table 1 Clinical studies investigating **probiotics/synbiotics** in the treatment of OCD.

Type of Nutraceutical	Dose and Study Duration	Model	Clinical Findings	Country	Reference
Probiotic (Bifidobacterium lactis UBBLa-70, Bacillus coagulans Unique IS-2, Lactobacillus rhamnosus UBLR-58, Lactobacillus plantarum UBLP-40, Bifidobacterium infantis UBBI-01, Bifidobacterium breve UBBr-01, and glutamine)	8-week	Rats	Rats injected with quinpirole for 5 weeks exhibited increased marble burying and self-grooming behaviors; reduced head dipping; avoided open spaces (anxiety-like). These phenotypes were abolished after 8 weeks of multistrain probiotic treatment. Probiotic normalized elevated mRNA expression of IL-6, TNF-α, and CRP in amygdala, and dysregulated levels of 5-HT, dopamine, and noradrenaline in frontal cortex. BDNF in frontal cortex remained unaffected. Probiotic prevented gut histology changes. 16S analysis showed increased Allobaculum and Bifidobacterium spp. (B. animalis) and decreased Lactobacillus spp.	India	Ghuge et al. (2023)
Probiotic (Lactobacillus casei)	10^9 CF/g daily for 4 weeks	Five groups of 6 male Wistar rats	L. casei, fluoxetine, and combination significantly reduced OCD-related behaviors. L. casei and fluoxetine upregulated BDNF and downregulated HTR2A expression.	Iran	Sanikhani et al. (2020)

Continued

Table 1 Clinical studies investigating **probiotics/synbiotics** in the treatment of OCD.—cont'd

Type of Nutraceutical	Dose and Study Duration	Model	Clinical Findings	Country	Reference
Synbiotic mixture (10^9 CFU Lactobacillus casei, L. acidophilus, L. rhamnosus, L. bulgaricus, Bifidobacterium breve, B. infantis, Streptococcus thermophilus + fructooligosaccharides)	Daily access to 6.25, 12.5, and 25 × 10^6 CFU for 2 weeks	Male Swiss mice	Both doxepin and venlafaxine significantly reduced immobility time after 7 days of synbiotic; immobility lessened after 14 days in fluvoxamine-injected rats. Synbiotic pretreatment affected marble burying test in fluvoxamine-injected rats but not doxepin/venlafaxine groups.	Iran	Mesripour et al. (2020)
Probiotic (Lactobacillus rhamnosus)	10^9 CFU/day for 2 or 4 weeks (10 mg/kg)	12 Male Balb/cJ mice	Compared to saline control, 2 and 4 weeks pretreatment decreased OCD-related behaviors. Two weeks probiotic pretreatment and four weeks fluoxetine yielded comparable OCD-related behavior reductions.	USA	Kantak et al. (2014)
Probiotic (Lactobacillus helveticus R0052 and Bifidobacterium longum R0175)	—	Human	—	France	Messaoudi et al. (2011)

dopaminergic neurons cells (N27) via diminished oxidative stress mediated by induction of glutathione synthesis followed by increased protection of neurons against protein oxidation and mitochondrial complex I function. Moreover, curcumin increases the amounts of different biogenic amines and suppresses their degradation by preventing MAO-A and B in brain (Yu et al., 2002), which is involved predominantly in OCD etiology. Curcumin's efficacy in OCD treatment has been tested in quinpirole-treated animals as an OCD model (Szechtman et al., 1998). Water maze test has shown a protective effect of curcumin on memory task. In addition, curcumin showed anti-OCD properties exerted by increased serotonin levels and decreased dopamine levels (Chimakurthy and Murthy, 2010). However, validation of the authenticity of curcumin as a potential alternative in OCD management needs more experimental and clinical researches.

Caffeine (1,3,7-trimethylxanthine), a natural alkaloid, is the most consumed psychoactive compound worldwide (Fredholm et al., 1999). Caffeine has been shown to improve health conditions in humans, including alertness (Smith, 2002), long-term memory (Mitra et al., 2020), and motor performance (Almosawi et al., 2018). An in vivo study showed the positive impacts of caffeine administration on spatial learning deficit in a rat model of attention deficit hyperactivity disorder (ADHD) (Prediger et al., 2005). In mice, caffeine induces aggressive-like behaviors and exploratory activity (Mitra et al., 2020). In detail, whereas low caffeine doses exert its locomotor-stimulating effect through blocking A2 adenosin receptor, high caffeine doses exert its locomotor-depressant effects via blocking A1 adenosine receptor (Yacoubi et al., 2000), due to the antagonizing effects of caffeine on A1 and A2 adenosin receptors of the brain (Mitra et al., 2020). The potential effects of caffeine on performance have been raised consumption of caffeinated products mainly in young individuals (Claghorn et al., 2017). Importantly, varied caffeinated drinks have triggered heavy consumption of caffeine intake disproportionately. Of concern, deleterious effects of long-term caffeine consumption include sleep disorders, cardiovascular anomalies, and substance abuse (Temple et al., 2017). Furthermore, caffeine usage has various effects in subjects with mental illness, such as reduced manic symptoms, depression, and anxiety (Mitra et al., 2020). On top of this, caffeine treatment has reduced OCD symptoms modestly in the resistant OCD individuals (Koran et al., 2009) following gradual dose increase up to 300 mg/day over several weeks (Shams et al., 2019) (Table 2).

N-acetyl-cysteine (NAC), an amino acid cysteine precursor, has been used as an antioxidant due to its role on glutathione (GSH) metabolism

Table 2 Clinical studies investigating plant-derived compounds and caffeine in the treatment of OCD.

Type of Nutraceutical	Dose and Study Duration	Model	Clinical Findings	Country	Reference
Berberine	7 days, 1–25 mg/kg, intraperitoneally	Adult male Swiss mice	Acute and subchronic berberine reduced marble burying without affecting locomotor activity (dose-dependent). Berberine (1 mg/kg) + fluoxetine (5 mg/kg) also reduced marble burying. Chlorophenylamine blocked fluoxetine but not berberine effects.	India	Dixit et al. (2012)
Curcumin (ethanolic extract of Curcuma longa; EECL)	EECL 10, 15, 25, 40 mg/kg; or fluoxetine 5, 10, 15 mg/kg	Swiss albino male mice (20–25 g)	EECL (40 mg/kg) significantly attenuated marble burying; no critical impact on motor activity.	India	Mishra et al. (2021)
Curcumin	5 and 10 mg/kg for 35 days	Adult Wistar rats	Curcumin improved OCD-related symptoms and memory task; associated with higher serotonin and lower dopamine.	India	Chimakurthy and Murthy (2010)
Genistein	5.0 and 10.0 mg/kg i.p., twice daily, 14 days	6 Swiss albino male mice	Chronic genistein recovered OCD symptoms (marble burying) in normal and diabetic mice; no locomotor impact. Reduced serotonin turnover in diabetic mice to control levels.	India	Phadnis et al. (2018)

Continued

Caffeine	High dose (25 mg/kg)	Mouse strains (HA)	Acute high-dose caffeine reduced nest-building in first hour; later increased at hours 3–5 then decreased over 24 h. Chronic high-dose caffeine increased nest-building. Acute high-dose reduced marbles buried; chronic had little effect. Acute decreased anxiety-like behaviors and motor activity; chronic did not significantly affect open field.	USA	Mitra et al. (2020)
Caffeine	Low (3 mg/kg) and high (25 mg/kg), once daily for 4 weeks	Mouse	Acute (not chronic) 25 mg/kg associated with lower anxiety and higher locomotor activity. Acute reduced nest-building in first hour; increased 3–5 h; decreased 5–24 h. Chronic high-dose increased nest building. Acute reduced marble burying; chronic minimal effect.	USA	Mitra et al., 2020
Caffeine	300 mg/day for 8 weeks	62 patients	Caffeine associated with significantly lower Y-BOCS score vs control.	Iran	Shams et al. (2019)
Caffeine	30 mg D-amphetamine and 300 mg caffeine, once daily for 12 weeks	24 subjects	After week 1, Y-BOCS decreased ≥20 percent in 50 percent (d-amphetamine) and 58 percent (caffeine). At week 5, mean Y-BOCS of responders lower in D-amphetamine (48 percent) vs caffeine (55 percent). OCD recovery independent of depression recovery; no side effects observed.	—	Koran et al. (2009)

as mucolytic therapy in the acetaminophen overdose (Cotgreave, 1996). NAC is almost completely metabolized by the liver and form cysteine, which is incorporated into GSH and the remaining cysteine reach the brain (Couto and Moreira, 2018) and is turned to cystine. The cystine–glutamate antiporter (or x(c)-system) on brain glial cells exchanges extracellular cystine for intracellular glutamate (Pow, 2001) increasing extracellular glutamate levels and intracellular cystine levels (Kau et al., 2008). Intracellular cystine can be converted to cysteine, which is known as the limiting amino acid for GSH formation (Cotgreave, 1996). In OCD, NAC might afford a beneficial effects by two mechanisms. First, NAC exerts antioxidant functions by induction of GSH synthesis mediated by providing an extra availability of cysteine (Couto and Moreira, 2018). Then, increased GSH level scavenges toxic by-products of dopamine and glutamatergic metabolism and is associated with lower levels of oxidative and nitrosative stress, prompting remarkable lower cellular damage (Berk et al., 2013). Second, NAC modifies the transmission of non-vesicular glutamate to extracellular space, stimulates group II metabotropic glutamate receptors (mGluR2/3). These presynaptic receptors negatively regulate the excitotoxicity and glutamatergic neurotransmission (Burdo et al., 2006; Berk et al., 2013).

Sarcosine (N-methylglycine), a natural glycine transporter-1 (GlyT-1) inhibitor, which increases synaptic glycine concentrations near NMDA receptors (NMDAR) (Bergeron et al., 1998; Chen et al., 2003). In addition, Sarcosine directly act as a co-agonist at the glycine-binding site of NMDARs (Wu et al., 2011). Furthermore, sarcosine and glycine are different from each other in the NMDAR desensitization (Zhang et al., 2009). Evidence revealed that enhanced neurotransmission via NMDAR could be a promising option for OCD treatment. Although the function of frontal-subcortical circuits increases in OCD (Saxena and Rauch, 2000), it may not show enhanced glutamatergic concentrations. Rosenberg et al. have demonstrated that glutamatergic activity decreased in the anterior cingulate gyrus in drug-naive children with OCD (Wu et al., 2011). Additionally, OCD is affected by frontal-subcortical circuits activity which triggers anatomical and functional distinctions of NMDA-mediated neurotransmission. For instance, NMDAR-mediated excitatory postsynaptic currents can be increased selectively by D–serine or glycine in dorsolateral compare to ventromedial striatum (Chapman et al., 2003). Potentiation of NMDAR activity may reverse NMDAR antagonistsY stimulated glutamate efflux pathologically in the prefrontal cortex and hippocampus,

(Wu et al., 2011) while NMDAR antagonist MK-801 intensified leaping and repetitive climbing functions in OCD and comorbid Tourette syndrome mouse models (McGrath et al., 2000). G72/G30 is a supposed activator of D-amino acid oxidase (DAAO) (Chumakov et al., 2002). Compulsive behavior also has been identified in G72 transgenic mouse (Otte et al., 2009). To increase NMDA neurotransmission in animal studies, multiple GlyT-1 inhibitors have been proposed because of the effects of GlyT-1 on subsaturating level of glycine at the NMDARs coagonist site (Wu et al., 2011) (Table 3).

Several clinical trials have also examined vitamins, minerals, fatty acids, and combination nutraceutical formulations as adjunctive or monotherapy approaches in OCD (Table 4).

6. Conclusion

Recently, the number of randomized controlled trials regarding the efficacy of nutraceuticals in the management of different psychiatric problems such as OCD has increased gradually especially probiotic supplementation. Nevertheless, controversial and questionable findings cause an incomparable efficacy of nutraceuticals because of differences in selection criteria, sample size, type, concentration, and duration of nutraceuticals. As the usage of current OCD therapies are challenging, finding novel alternatives (nutraceutical and plant-derived compounds) has a great level of importance. Standalone efficacy is another important question that needs to be answered in OCD novel strategy treatments. Another significant issue is a period of 30 day drug wash out before starting the clinical trial. In addition, the efficacy of nutraceuticals should be to evaluated in comparison with cognitive behavioral and pharmacological treatments. Further, combination of behavioral therapies with nutraceutical treatments could provide novel insight into the efficient treatment for clinicians.

Taken together, more research is needed to find alternative option for OCD disorders because of limitations of usual pharmacotherapy for OCD management. Since nutraceutical and plant-derived compounds modulate serotonergic or glutamatergic mechanisms, they could be potential effective therapies for OCD treatment with better tolerance and less adverse effects. To test the efficacy of nutraceutical properly, more research should be performed with randomized placebo-controlled designs and larger sample sizes.

Table 3 Clinical studies investigating amino acids / glutamatergic–NMDA modulators in the treatment of OCD.

Type of Nutraceutical	Dose and Study Duration	Model	Clinical Findings	Country	Reference
Sarcosine (N-methylglycine)	Start 500 mg/day → up to 2000 mg/day (1520 ± 549 mg/day), titrated biweekly; 10 weeks	26 outpatients	Sarcosine significantly reduced mean Y-BOCS. 32 percent responders (≥35 percent Y-BOCS reduction). 62 percent of responders achieved satisfactory response by week 4. One withdrawal due to headache.	China	Wu et al. (2011)
Glycine	60 g/day for 12 weeks	14 adult outpatients	Mean reduction significantly higher than placebo (6.04 vs 1.00). 40 percent responders in glycine group; none in placebo.	USA	Greenberg et al. (2009)
D-cycloserine	35 or 70 mg	Youth ($n = 100$; 7–17 years), randomized DCS + exposure vs placebo + exposure	Significant improvements in all outcomes posttreatment and maintained at 6 months; no arm differences over time (no time-by-medication interaction). Fewer responders/remitters at follow-ups among youth in DCS condition stabilized on SSRIs vs those not taking SSRIs. DCS augmentation did not add benefit over placebo.	Australia	Farrell et al. (2022)

Continued

Table 3 Clinical studies investigating amino acids / glutamatergic–NMDA modulators in the treatment of OCD.—cont'd

Type of Nutraceutical	Dose and Study Duration	Model	Clinical Findings	Country	Reference
D-cycloserine	Bergen 4-day ERP + 100 mg DCS, 250 mg DCS, or placebo	OCD patients relapsed ($n = 100$) or non-responders ($n = 63$) to prior ERP	Posttreatment remission in 91; 70 maintained at 12 months; no difference among groups. Symptoms reduced at 12 months for 250 mg and 100 mg groups, but no significant effect vs placebo on OCD/depression/anxiety/well-being; no serious adverse effects.	Norway	Kvale et al. (2020)
D-cycloserine	DCS or placebo augmented ERP	161 youth (7–17 years)	Irritability declined after treatment; not related to OCD improvement.	USA	Guzick et al. (2021)
D-cycloserine; NVP-AAMO77; Ro25-6981	5 mg/kg	Rats	NVP-AAMO77 and Ro25-6981 inhibited anxiety-like and compulsive behaviors; DCS suppressed anxiety-like and marble-burying. Hippocampal glutamate decreased in NVP-AAMO77 and DCS groups. NVP-AAMO77 decreased NR2A; Ro25-6981 suppressed NR2B; DCS decreased NR2A and NR2B.	China	Zhan et al. (2020)

D-cycloserine	10 sessions CBT + placebo or DCS	139 youth	Lower baseline OCD/depression severity predicted better early response/remission to CBT; supports stepped care.	USA	Rech et al. (2020)
D-cycloserine	ERP trial (two-site)	142 children/adolescents (7–17 years)	No support for OC symptoms mediating ERP→depression; depressive change mediated ERPs impact on later OC symptom change (small effect; non-significant after controlling for prior depression).	USA	Bakhshaie et al. (2020)
D-cycloserine	25 or 50 mg (weight-dependent)	17 youth	Y-BOCS decreased after ERP + DCS and ERP + placebo. At 1-month follow-up, ERP + DCS improved obsessional severity/diagnostic severity/parent severity ratings vs ERP + placebo.	Australia	Farrell et al. (2013)
D-cycloserine	25 or 50 mg	30 youth	Low effects; no significant difference vs placebo; no side effects.	USA	Storch et al. (2010)
D-cycloserine	250 mg for 12 weeks	24 adults	No significant difference between groups; no beneficial effect.	USA	Storch et al. (2007)
N-acetylcysteine (NAC)	20 weeks, 2–4 g/day	98 patients	Modified ITT: no evidence NAC reduced OCD vs placebo; no improvement in anxiety/depression/QoL; well tolerated with mild GI adverse events.	Australia	Sarris et al. (2022)

Continued

Table 3 Clinical studies investigating amino acids / glutamatergic–NMDA modulators in the treatment of OCD.—cont'd

Type of Nutraceutical	Dose and Study Duration	Model	Clinical Findings	Country	Reference
N-acetylcysteine (NAC)	Up to 2700 mg/day for 12 weeks	9 children	At 8 weeks, NAC group had significantly lower Y-BOCS vs placebo. At 12 weeks, mean Y-BOCS 14.4 ± 5.55 vs baseline 21.4 ± 4.65. 20 percent had >35 percent reduction vs none in placebo. One mild side effect each group.	USA	Li et al. (2020)
N-acetylcysteine (NAC)	10-week, titrated to 2400 mg/day	34 outpatients	Y-BOCS reduced from 21.0 ± 8.2 to 11.3 ± 5.7 in NAC; placebo unchanged. QoL improved (except physical activity). No adverse events recorded.	Iran	Ghanizadeh et al. (2017)
N-acetylcysteine (NAC)	3000 mg/day for 16 weeks	30 outpatients	Y-BOCS decreased in both NAC and placebo; difference not significant. NAC improved anxiety symptoms. Abdominal pain more prevalent with NAC	Brazil	Costa et al. (2017)

N-acetylcysteine (NAC)	2000 mg/day for 10 weeks	46 patients	Significant decrease in total Y-BOCS in treated vs placebo. Remission: 12 (NAC) vs 5 (placebo), marginal. Partial/complete response: 11 (NAC) vs 8 (placebo), NS. No significant obsession/compulsion subscale differences over time.	Iran	Paydary et al. (2016)
N-acetylcysteine (NAC)	3 g/day (1.5 g BID), 16 weeks	44 participants	No significant difference NAC vs control at week 16; younger (<34) and shorter illness (<17 years) showed more favorable responses. Baseline depression and illness duration predicted Y-BOCS change.	Australia	Sarris et al. (2016)
N-acetylcysteine (NAC)	3 g/day (1.5 g BID), 16 weeks	—	Y-BOCS change not significant vs placebo. Compulsion subscale decreased at week 12; obsession subscale not significantly reduced. ≥35 percent reduction: 20 percent NAC vs 27 percent placebo. Heartburn higher with NAC	Australia	Sarris et al. (2015)
N-acetylcysteine (NAC)	12-week, up to 2400 mg/day	48 patients	Mean Y-BOCS reduction significantly higher with NAC vs placebo; improvements in CGI-S. More patients achieved ≥35 percent Y-BOCS reduction with NAC	Iran	Afshar et al. (2012)

Table 4 Clinical studies investigating vitamins, minerals, fatty acids, and multi-nutrient formulations in the treatment of OCD.

Type of Nutraceutical	Dose and Study Duration	Model	Clinical Findings	Country	Reference
Inositol	18 mg/day for 6 weeks	10 patients	No significant improvements in most patients.	South Africa	Seedat and Stein (1999)
Inositol	18 g/day for 6 weeks	10 patients	Significant placebo effects; no inositol effects.	Israel	Fux et al. (1999)
Inositol	18 g/day for 6 weeks	13 patients	Inositol group had significantly lower Y-BOCS vs placebo.	Israel	Fux et al. (1996)
Selenium	200 mcg/day for 6 weeks	32 patients	At 6 weeks, Y-BOCS significantly lower with selenium vs control; no significant difference in adverse events.	Iran	Sayyah et al. (2018)
Selenium, l-theanine, zinc, magnesium, pyridoxal-5′ phosphate, and N-acetylcysteine	20 weeks	28 participants with OCD	ITT: mean reduction in Y-BOCS over time; 23 percent achieved ≥35 percent reduction at 20 weeks. Improvements in mood, anxiety, QoL. Better response with lower baseline symptoms.	Australia	Sarris et al. (2022)
Zinc sulfate (adjunct to fluoxetine)	440 mg/day for 8 weeks	23 patients	Both groups decreased Y-BOCS; at weeks 2 and 8, fluoxetine + zinc had significantly lower scores vs fluoxetine + placebo.	Iran	Sayyah et al. (2012)
Omega-3 (EPA)	2 g/day for 6 weeks	11 patients	Adjunctive EPA ineffective against OCD	Israel	Fux et al. (2004)
Vitamin B9 (folic acid)	5 mg/day for 12 weeks	36 patients	No significant differences between folic acid and placebo in Y-BOCS, HAM-A, HAM-D, CGI	Turkey	Tural et al. (2019)
Vitamin B6	100 mg/kg i.p., daily for 9 days	Male mice	Immobility time reduced significantly after 6 and 9 days vs controls.	Iran	Mesripour et al. (2017)

References

Afshar, H., Roohafza, H., Mohammad-Beigi, H., Haghighi, M., Jahangard, L., Shokouh, P., Sadeghi, M., Hafezian, H., 2012. N-acetylcysteine add-on treatment in refractory obsessive-compulsive disorder: a randomized, double-blind, placebo-controlled trial. J. Clin. Psychopharmacol. 32 (6), 797–803.

Almosawi, S., Baksh, H., Qareeballa, A., Falamarzi, F., Alsaleh, B., Alrabaani, M., Alkalbani, A., Mahdi, S., Kamal, A., 2018. Acute administration of caffeine: the effect on motor coordination, higher brain cognitive functions, and the social behavior of BLC57 mice. Behav. Sci. 8 (8), 65.

Al-Okbi, S.Y., 2014. Nutraceuticals of anti-inflammatory activity as complementary therapy for rheumatoid arthritis. Toxicol. Ind. Health 30 (8), 738–749.

Association, C.P., 2006. Clinical practice guidelines. Management of anxiety disorders. Can. J. Psychiatry 51 (8 Suppl 2), 9S–91S.

Bakhshaie, J., Geller, D.A., Wilhelm, S., McGuire, J.F., Small, B.J., Cepeda, S.L., Schneider, S.C., Murphy, T.K., Porth, R., Storch, E.A., 2020. Temporal precedence of the change in obsessive-compulsive symptoms and change in depressive symptoms during exposure and response prevention for pediatric obsessive-compulsive disorders. Behav. Res. Ther. 133, 103697.

Balandeh, E., Karimian, M., Behjati, M., Mohammadi, A.H., 2021. Serum vitamins and homocysteine levels in obsessive-compulsive disorder: a systematic review and meta-analysis. Neuropsychobiology 80 (6), 502–515.

Baumgarten, H., Grozdanovic, Z., 1998. Role of serotonin in obsessive-compulsive disorder. Br. J. Psychiatry 173 (S35), 13–20.

Baxter, L.R., Schwartz, J.M., Bergman, K.S., Szuba, M.P., Guze, B.H., Mazziotta, J.C., Alazraki, A., Selin, C.E., Ferng, H.-K., Munford, P., 1992. Caudate glucose metabolic rate changes with both drug and behavior therapy for obsessive-compulsive disorder. Arch. Gen. Psychiatry 49 (9), 681–689.

Bergeron, R., Meyer, T.M., Coyle, J.T., Greene, R.W., 1998. Modulation of N-methyl-D-aspartate receptor function by glycine transport. Proc. Natl. Acad. Sci. 95 (26), 15730–15734.

Berk, M., Malhi, G.S., Gray, L.J., Dean, O.M., 2013. The promise of N-acetylcysteine in neuropsychiatry. Trends Pharmacol. Sci. 34 (3), 167–177.

Bloch, M.H., McGuire, J., Landeros-Weisenberger, A., Leckman, J.F., Pittenger, C., 2010. Meta-analysis of the dose-response relationship of SSRI in obsessive-compulsive disorder. Mol. Psychiatry 15 (8), 850–855.

Boileau, B., 2011. A review of obsessive-compulsive disorder in children and adolescents. Dialogues Clin. Neurosci. 13 (4), 401.

Boileau, B., 2011. A review of obsessive-compulsive disorder in children and adolescents. Dialogues Clin. Neurosci. 13 (4), 401–411.

Bourne, S.K., Sheth, S.A., Neal, J., Strong, C., Mian, M.K., Cosgrove, G.R., Eskandar, E.N., Dougherty, D.D., 2013. Beneficial effect of subsequent lesion procedures after nonresponse to initial cingulotomy for severe, treatment-refractory obsessive-compulsive disorder. Neurosurgery 72 (2), 196–202 discussion 202.

Bungău, S.G., Popa, V.-C., 2015. Between religion and science some aspects concerning illness and healing in antiquity. Transylv. Rev. 24 (3), 3–18.

Burdo, J., Dargusch, R., Schubert, D., 2006. Distribution of the cystine/glutamate antiporter system x− c in the brain, kidney, and duodenum. J. Histochem. Cytochem. 54 (5), 549–557.

Camfield, D.A., Sarris, J., Berk, M., 2011. Nutraceuticals in the treatment of obsessive compulsive disorder (OCD): a review of mechanistic and clinical evidence. Prog. Neuropsychopharmacol. Biol. Psychiatry 35 (4), 887–895.

Cannistraro, P.A., Wright, C.I., Wedig, M.M., Martis, B., Shin, L.M., Wilhelm, S., Rauch, S.L., 2004. Amygdala responses to human faces in obsessive-compulsive disorder. Biol. Psychiatry 56 (12), 916–920.
Carabotti, M., Scirocco, A., Maselli, M.A., Severi, C., 2015. The gut-brain axis: interactions between enteric microbiota, central and enteric nervous systems. Ann. gastroenterol. Q. Publ. Hell. Soc. Gastroenterol. 28 (2), 203.
Cerdó, T., Ruíz, A., Suárez, A., Campoy, C., 2017. Probiotic, prebiotic, and brain development. Nutrients 9 (11), 1247.
Chao, J., Leung, Y., Wang, M., Chang, R.C.-C., 2012. Nutraceuticals and their preventive or potential therapeutic value in Parkinson's disease. Nutr. Rev. 70 (7), 373–386.
Chao, M.V., 2003. Neurotrophins and their receptors: a convergence point for many signalling pathways. Nat. Rev. Neurosci. 4 (4), 299–309.
Chapman, D.E., Keefe, K.A., Wilcox, K.S., 2003. Evidence for functionally distinct synaptic NMDA receptors in ventromedial versus dorsolateral striatum. J. Neurophysiol. 89 (1), 69–80.
Chen, L., Muhlhauser, M., Yang, C.R., 2003. Glycine tranporter-1 blockade potentiates NMDA-mediated responses in rat prefrontal cortical neurons in vitro and in vivo. J. Neurophysiol. 89 (2), 691–703.
Chimakurthy, J., Murthy, T.E., 2010. Effect of curcumin on quinpirole induced compulsive checking: an approach to determine the predictive and construct validity of the model. N. Am. J. Med. Sci. 2 (2), 81.
Chumakov, I., Blumenfeld, M., Guerassimenko, O., Cavarec, L., Palicio, M., Abderrahim, H., Bougueleret, L., Barry, C., Tanaka, H., La Rosa, P., 2002. Genetic and physiological data implicating the new human gene G72 and the gene for D-amino acid oxidase in schizophrenia. Proc. Natl. Acad. Sci. 99 (21), 13675–13680.
Claghorn, G.C., Thompson, Z., Wi, K., Van, L., Garland Jr., T., 2017. Caffeine stimulates voluntary wheel running in mice without increasing aerobic capacity. Physiol. Behav. 170, 133–140.
Cloutier, M., Aigbogun, M.S., Guerin, A., Nitulescu, R., Ramanakumar, A.V., Kamat, S.A., DeLucia, M., Duffy, R., Legacy, S.N., Henderson, C., 2016. The economic burden of schizophrenia in the United States in 2013. J. Clin. Psychiatry 77 (6), 764–771.
Cornelli, U., 2009. Antioxidant use in nutraceuticals. Clin. Dermatol. 27 (2), 175–194.
Costa, D.L.C., Diniz, J.B., Requena, G., Joaquim, M.A., Pittenger, C., Bloch, M.H., Miguel, E.C., Shavitt, R.G., 2017. Randomized, double-blind, placebo-controlled trial of N-acetylcysteine augmentation for treatment-resistant obsessive-compulsive disorder. J. Clin. Psychiatry 78 (7), e766–e773.
Cotgreave, I.A., 1996. N-acetylcystei ne: pharmacological considerations and experimental and clinical applications. Adv. Pharmacol. 38, 205–227.
Couto, J.P., Moreira, R., 2018. Oral N-acetylcysteine in the treatment of obsessive-compulsive disorder: a systematic review of the clinical evidence. Prog. Neuropsychopharmacol. Biol. Psychiatry 86, 245–254.
Couto, J.P., Moreira, R., 2018. Oral N-acetylcysteine in the treatment of obsessive-compulsive disorder: a systematic review of the clinical evidence. Prog. Neuropsychopharmacol. Biol. Psychiatry 86, 245–254.
Dell'Osso, B., Mundo, E., Marazziti, D., Altamura, A.C., 2008. Switching from serotonin reuptake inhibitors to duloxetine in patients with resistant obsessive compulsive disorder: a case series. J. Psychopharmacol. 22 (2), 210–213.
Dima, C., Assadpour, E., Dima, S., Jafari, S.M., 2020. Bioavailability of nutraceuticals: role of the food matrix, processing conditions, the gastrointestinal tract, and nanodelivery systems. Compr. Rev. Food Sci. Food Saf. 19 (3), 954–994.
Dinan, T.G., Cryan, J.F., 2013. Melancholic microbes: a link between gut microbiota and depression? Neurogastroenterol. Motil. 25 (9), 713–719.

Dixit, P.V., Parihar, G., Jain, D.K., Jaiswal, P., 2012. Increased serotonergic neurotransmission is not responsible for the anticompulsive effect of berberine in a murine model of obsessive–compulsive disorder. Behav. Pharmacol. 23 (7), 716–721.

Dougherty, D.D., Baer, L., Cosgrove, G.R., Cassem, E.H., Price, B.H., Nierenberg, A.A., Jenike, M.A., Rauch, S.L., 2002. Prospective long-term follow-up of 44 patients who received cingulotomy for treatment-refractory obsessive-compulsive disorder. Am. J. Psychiatry 159 (2), 269–275.

Edition, F., 2013. Diagnostic and statistical manual of mental disorders. Depr. Anxi. 30 (8), 723–731. doi:10.1002/da.22132

Farrell, L.J., Waters, A.M., Boschen, M.J., Hattingh, L., McConnell, H., Milliner, E.L., Collings, N., Zimmer-Gembeck, M., Shelton, D., Ollendick, T.H., Testa, C., Storch, E.A., 2013. Difficult-to-treat pediatric obsessive-compulsive disorder: feasibility and preliminary results of a randomized pilot trial of D-cycloserine-augmented behavior therapy. Depress Anxiety 30 (8), 723–731.

Farrell, L.J., Waters, A.M., Tiralongo, E., Mathieu, S., McKenzie, M., Garbharran, V., Ware, R.S., Zimmer-Gembeck, M.J., McConnell, H., Lavell, C., Cadman, J., Ollendick, T.H., Hudson, J.L., Rapee, R.M., McDermott, B., Geller, D., Storch, E.A., 2022. Efficacy of D-cycloserine augmented brief intensive cognitive-behavioural therapy for paediatric obsessive-compulsive disorder: a randomised clinical trial. Depress Anxiety 39 (6), 461–473.

Fenske, J.N., Schwenk, T.L., 2009. Obsessive compulsive disorder: diagnosis and management. Am. Fam. Physician 80 (3), 239–245.

Fineberg, N.A., Brown, A., Reghunandanan, S., Pampaloni, I., 2012. Evidence-based pharmacotherapy of obsessive-compulsive disorder. Int. J. Neuropsychopharmacol. 15 (8), 1173–1191.

Fineberg, N.A., Gale, T.M., 2005. Evidence-based pharmacotherapy of obsessive–compulsive disorder. Int. J. Neuropsychopharmacol. 8 (1), 107–129.

Fineberg, N.A., Pampaloni, I., Pallanti, S., Ipser, J., Stein, D.J., 2007. Sustained response versus relapse: the pharmacotherapeutic goal for obsessive–compulsive disorder. Int. Clin. Psychopharmacol. 22 (6), 313–322.

Fredholm, B.B., Bättig, K., Holmén, J., Nehlig, A., Zvartau, E.E., 1999. Actions of caffeine in the brain with special reference to factors that contribute to its widespread use. Pharmacol. Rev. 51 (1), 83–133.

Fux, M., Benjamin, J., Belmaker, R.H., 1999. Inositol versus placebo augmentation of serotonin reuptake inhibitors in the treatment of obsessive-compulsive disorder: a double-blind cross-over study. Int. J. Neuropsychopharmacol. 2 (3), 193–195.

Fux, M., Benjamin, J., Nemets, B., 2004. A placebo-controlled cross-over trial of adjunctive EPA in OCD. J. Psychiatr. Res. 38 (3), 323–325.

Fux, M., Levine, J., Aviv, A., Belmaker, R.H., 1996. Inositol treatment of obsessive-compulsive disorder. Am. J. Psychiatry 153 (9), 1219–1221.

Ghanizadeh, A., Mohammadi, M.R., Bahraini, S., Keshavarzi, Z., Firoozabadi, A., Alavi Shoshtari, A., 2017. Efficacy of N-acetylcysteine augmentation on obsessive compulsive disorder: a multicenter randomized double blind placebo controlled clinical trial. Iran J. Psychiatry 12 (2), 134–141.

Ghuge, S., Rahman, Z., Bhale, N.A., Dikundwar, A.G., Dandekar, M.P., 2023. Multistrain probiotic rescinds quinpirole-induced obsessive-compulsive disorder phenotypes by reshaping of microbiota gut-brain axis in rats. Pharmacol. Biochem. Behav. 232, 173652.

Giavasis, I., 2014. Bioactive fungal polysaccharides as potential functional ingredients in food and nutraceuticals. Curr. Opin. Biotechnol. 26, 162–173.

Goddard, A.W., Shekhar, A., Whiteman, A.F., McDougle, C.J., 2008. Serotoninergic mechanisms in the treatment of obsessive–compulsive disorder. Drug Discov. Today 13 (7-8), 325–332.

Goodman, W.K., Alterman, R.L., 2012. Deep brain stimulation for intractable psychiatric disorders. Annu. Rev. Med. 63, 511–524.

Greenberg, W.M., Benedict, M.M., Doerfer, J., Perrin, M., Panek, L., Cleveland, W.L., Javitt, D.C., 2009. Adjunctive glycine in the treatment of obsessive-compulsive disorder in adults. J. Psychiatr. Res. 43 (6), 664–670.

Gutiérrez-del-Río, I., Fernández, J., Lombó, F., 2018. Plant nutraceuticals as antimicrobial agents in food preservation: terpenoids, polyphenols and thiols. Int. J. Antimicrob. Agents 52 (3), 309–315.

Guzick, A.G., Geller, D.A., Small, B.J., Murphy, T.K., Wilhelm, S., Storch, E.A., 2021. Irritability in children and adolescents with OCD. Behav. Ther. 52 (4), 883–896.

Haber, S.N., Heilbronner, S.R., 2013. Translational research in OCD: circuitry and mechanisms. Neuropsychopharmacology 38 (1), 252.

Hemmings, S.A.M., Kinnear, C.J., Niehaus, D.J., Moolman-Smook, J.C., Lochner, C., Knowles, J.A., Corfield, V.A., Stein, D.J., 2003. Investigating the role of dopaminergic and serotonergic candidate genes in obsessive-compulsive disorder. Eur. Neuropsychopharmacol. 13 (2), 93–98.

Huey, E.D., Zahn, R., Krueger, F., Moll, J., Kapogiannis, D., Wassermann, E.M., Grafman, J., 2008. A psychological and neuroanatomical model of obsessive-compulsive disorder. J. Neuropsychiatry Clin. Neurosci. 20 (4), 390–408.

Jafari, S.M., McClements, D.J., 2017. Nanotechnology approaches for increasing nutrient bioavailability. Adv. Food. Nutr. Res. 81, 1–30.

Jagatha, B., Mythri, R.B., Vali, S., Bharath, M.S., 2008. Curcumin treatment alleviates the effects of glutathione depletion in vitro and in vivo: therapeutic implications for Parkinson's disease explained via in silico studies. Free Radic. Biol. Med. 44 (5), 907–917.

Kalra, E.K., 2003. Nutraceutical-definition and introduction. Aaps Pharmsci. 5 (3), 27–28.

Kantak, P.A., Bobrow, D.N., Nyby, J.G., 2014. Obsessive–compulsive-like behaviors in house mice are attenuated by a probiotic (Lactobacillus rhamnosus GG). Behav. Pharmacol. 25 (1), 71–79.

Karcı, C.K., Celik, G.G., 2020. Nutritional and herbal supplements in the treatment of obsessive compulsive disorder. Gen. Psychiatry 33 (2), e100159.

Katouzian, I., Esfanjani, A.F., Jafari, S.M., Akhavan, S., 2017. Formulation and application of a new generation of lipid nano-carriers for the food bioactive ingredients. Trends Food Sci. Technol. 68, 14–25.

Kau, K.S., Madayag, A., Mantsch, J.R., Grier, M.D., Abdulhameed, O., Baker, D.A., 2008. Blunted cystine–glutamate antiporter function in the nucleus accumbens promotes cocaine-induced drug seeking. Neuroscience 155 (2), 530–537.

Kennedy, P.J., Clarke, G., Quigley, E.M., Groeger, J.A., Dinan, T.G., Cryan, J.F., 2012. Gut memories: towards a cognitive neurobiology of irritable bowel syndrome. Neurosci. Biobehav. Rev. 36 (1), 310–340.

Khalesi, S., Bellissimo, N., Vandelanotte, C., Williams, S., Stanley, D., Irwin, C., 2019. A review of probiotic supplementation in healthy adults: helpful or hype? Eur. J. Clin. Nutr. 73 (1), 24–37.

Kim, S.-W., Park, K.-Y., Kim, B., Kim, E., Hyun, C.-K., 2013. Lactobacillus rhamnosus GG improves insulin sensitivity and reduces adiposity in high-fat diet-fed mice through enhancement of adiponectin production. Biochem. Biophys. Res. Commun. 431 (2), 258–263.

Komossa, K., Depping, A.M., Meyer, M., Kissling, W., Leucht, S., 2010. Second-generation antipsychotics for obsessive compulsive disorder. Expert Rev. Neurother. 10 (2), 275–290.

Koo, M.-S., Kim, E.-J., Roh, D., Kim, C.-H., 2010. Role of dopamine in the pathophysiology and treatment of obsessive–compulsive disorder. Expert. Rev. Neurother. 10 (2), 275–290.

Koran, L.M., Aboujaoude, E., Gamel, N.N., 2009. Double-blind study of dextroamphetamine versus caffeine augmentation for treatment-resistant obsessive-compulsive disorder. J. Clin. Psychiatry 70 (11), 1530–1535.
Koran, L.M., Gamel, N.N., Choung, H.W., Smith, E.H., Aboujaoude, E.N., 2005. Mirtazapine for obsessive-compulsive disorder: an open trial followed by double-blind discontinuation. J. Clin. Psychiatry 66 (4), 515–520.
Koran, L.M., Hanna, G.L., Hollander, E., Nestadt, G., Simpson, H.B., 2007. Practice guideline for the treatment of patients with obsessive-compulsive disorder. Koran Am. J. Psychiatry. 164 (7), 5–53.
Koran, L.M., Hanna, G.L., Hollander, E., Nestadt, G., Simpson, H.B., 2007. Practice guideline for the treatment of patients with obsessive-compulsive disorder. Am. J. Psychiatry 164 (7 Suppl), 5–53.
Kringelbach, M.L., Rolls, E.T., 2004. The functional neuroanatomy of the human orbitofrontal cortex: evidence from neuroimaging and neuropsychology. Prog. Neurobiol. 72 (5), 341–372.
Kvale, G., Hansen, B., Hagen, K., Abramowitz, J.S., Børtveit, T., Craske, M.G., Franklin, M.E., Haseth, S., Himle, J.A., Hystad, S., Kristensen, U.B., Launes, G., Lund, A., Solem, S., Öst, L.G., 2020. Effect of D-Cycloserine on the effect of concentrated exposure and response prevention in difficult-to-treat obsessive-compulsive disorder: a randomized clinical trial. JAMA Netw. Open 3 (8), e2013249.
Lankelma, J.M., Nieuwdorp, M., de Vos, W.M., Wiersinga, W.J., 2015. The gut microbiota in internal medicine: implications for health and disease. Neth. J. Med. 73 (2), 61–68.
Leckman, J.F., Denys, D., Simpson, H.B., Mataix-Cols, D., Hollander, E., Saxena, S., Miguel, E.C., Rauch, S.L., Goodman, W.K., Phillips, K.A., 2010. Obsessive–compulsive disorder: a review of the diagnostic criteria and possible subtypes and dimensional specifiers for DSM-V. Depress. Anxiety 27 (6), 507–527.
Li, F., Welling, M.C., Johnson, J.A., Coughlin, C., Mulqueen, J., Jakubovski, E., Coury, S., Landeros-Weisenberger, A., Bloch, M.H., 2020. N-acetylcysteine for pediatric obsessive–compulsive disorder: a small pilot study. J. Child. Adolesc. Psychopharmacol. 30 (1), 32–37.
Lissemore, J.I., Leyton, M., Gravel, P., Sookman, D., Nordahl, T.E., Benkelfat, C., 2014. OCD: serotonergic mechanisms. Int. J. Mol. Sci. 21 (12), 4424. doi:10.3390/ijms21124424.
Makkar, R., Behl, T., Bungau, S., Zengin, G., Mehta, V., Kumar, A., Uddin, M., Ashraf, G.M., Abdel-Daim, M.M., Arora, S., 2020. Nutraceuticals in neurological disorders. Int. J. Mol. Sci. 21 (12), 4424.
Martin, P., Waters, N., Schmidt, C., Carlsson, A., Carlsson, M., 1998. Rodent data and general hypothesis: antipsychotic action exerted through 5-HT2A receptor antagonism is dependent on increased serotonergic tone. J. Neural. Transm. 105 (4-5), 365–396.
McGrath, M.J., Campbell, K.M., Parks 3rd, C.R., Burton, F.H., 2000. Glutamatergic drugs exacerbate symptomatic behavior in a transgenic model of comorbid Tourette's syndrome and obsessive–compulsive disorder. Brain Res. 877 (1), 23–30.
McGuire, P., Bench, C., Frith, C., Marks, I., Frackowiak, R., Dolan, R.J., 1994. Functional anatomy of obsessive–compulsive phenomena. Br. J. Psychiatry 164 (4), 459–468.
Mesripour, A., Hajhashemi, V., Kuchak, A., 2017. Effect of concomitant administration of three different antidepressants with vitamin B6 on depression and obsessive compulsive disorder in mice models. Res. Pharm. Sci. 12 (1), 46–52.
Mesripour, A., Meshkati, A., Hajhashemi, V., 2020. A synbiotic mixture augmented the efficacy of doxepin, venlafaxine, and fluvoxamine in a mouse model of depression. Turk. J. Pharm. Sci. 17 (3), 293.
Messaoudi, M., Violle, N., Bisson, J.-F., Desor, D., Javelot, H., Rougeot, C., 2011. Beneficial psychological effects of a probiotic formulation (Lactobacillus helveticus R0052 and Bifidobacterium longum R0175) in healthy human volunteers. Gut Microbes 2 (4), 256–261.

Milad, M.R., Rauch, S.L., 2012. Obsessive-compulsive disorder: beyond segregated cortico-striatal pathways. Trends Cogn. Sci. 16 (1), 43–51.
Mishra, R.K., Mishra, A., Gupta, A., 2021. Magic shotgun nature with scattergun approach of curcumin repurposing in obsessive-compulsive disorder: a novel metaphysician of drug discovery. CNS Neurol. Disord. Drug Targets-CNS Neurol. Disord. 20 (10), 975–981.
Mitra, S., Miranda, V.S., McMillan, C., Dykes, D., Mucha, M., Marth, T.E., Poe, B., Basu, D.G., Bult-Ito, A., 2020. Trait specific modulatory effects of caffeine exposure on compulsive-like behaviors in a spontaneous mouse model of obsessive-compulsive disorder. Behav. Pharmacol. 31 (7), 622–632.
Mitra, S., Miranda, V.S., McMillan, C., Dykes, D., Mucha, M., Marth, T.E., Poe, B., Basu, D.G., Bult-Ito, A., 2020. Trait specific modulatory effects of caffeine exposure on compulsive-like behaviors in a spontaneous mouse model of obsessive-compulsive disorder. Behav. Pharmacol. 31 (7), 622–632.
Mohammadi, A.H., Balandeh, E., Milajerdi, A., 2021. Malondialdehyde concentrations in obsessive-compulsive disorder: a systematic review and meta-analysis. Ann. Gen. Psychiatry 20 (1), 34.
Mohammadi, A. H., Karimian, M., Mirzaei, H., and Milajerdi, A., 2023. Epigenetic modifications and obsessive–compulsive disorder: what do we know? Brain Structure and Function, 228 (5), 1295–1305.
Mohammadi, A. H., Balandeh, E., Hasani, J., Karimian, M., Arabshahi, V., Pourfarzam, M., et al. (2024). The oxidative status and Na+/K+-ATPase activity in obsessive-compulsive disorder: a case-control study. BioMed Res. Int. 2024 (1), 9979582.
Otte, D.M., Bilkei-Gorzó, A., Filiou, M.D., Turck, C.W., Yilmaz, Ö., Holst, M.I., Schilling, K., Abou-Jamra, R., Schumacher, J., Benzel, I., 2009. Behavioral changes in G72/G30 transgenic mice. Eur. Neuropsychopharmacol. 19 (5), 339–348.
Parada, J., Aguilera, J., 2007. Food microstructure affects the bioavailability of several nutrients. J. Food Sci. 72 (2), R21–R32.
Park, H., Poo, M.-m., 2013. Neurotrophin regulation of neural circuit development and function. Nat. Rev. Neurosci. 14 (1), 7–23.
Pauls, D.L., Abramovitch, A., Rauch, S.L., Geller, D.A., 2014. Obsessive–compulsive disorder: an integrative genetic and neurobiological perspective. Nat. Rev. Neurosci. 15 (6), 410–424.
Paydary, K., Akamaloo, A., Ahmadipour, A., Pishgar, F., Emamzadehfard, S., Akhondzadeh, S., 2016. N-acetylcysteine augmentation therapy for moderate-to-severe obsessive-compulsive disorder: randomized, double-blind, placebo-controlled trial. J. Clin. Pharm. Ther. 41 (2), 214–219.
Perani, D., Colombo, C., Bressi, S., Bonfanti, A., Grassi, F., Scarone, S., Bellodi, L., Smeraldi, E., Fazio, F., 1995. [18F] FDG PET study in obsessive–compulsive disorder: a clinical/metabolic correlation study after treatment. Br. J. Psychiatry 166 (2), 244–250.
Phadnis, P., Sarkar, P.D., Rajput, M.S., 2018. Improved serotonergic neurotransmission by genistein pretreatment regulates symptoms of obsessive-compulsive disorder in streptozotocin-induced diabetic mice. J. Basic Clin. Physiol. Pharmacol. 29 (4), 421–425.
Pittenger, C., 2017. Obsessive-Compulsive Disorder: Phenomenology, Pathophysiology, and Treatment. Pharmacol. Ther. 132 (3), 314–332. doi:10.1016/j.pharmthera.2011.09.006.
Pittenger, C., Bloch, M.H., Williams, K., 2011. Glutamate abnormalities in obsessive compulsive disorder: neurobiology, pathophysiology, and treatment. Pharmacol. Ther. 132 (3), 314–332.
Pow, D.V., 2001. Visualising the activity of the cystine-glutamate antiporter in glial cells using antibodies to aminoadipic acid, a selectively transported substrate. Glia 34 (1), 27–38.
Prakash, V., van Boekel, M.A.J.S., 2010. Nutraceuticals: possible future ingredients and food safety aspects. In: Ensuring Global Food Safety: Exploring Global Harmonization. Elsevier, pp. 333–338.

Prediger, R.D., Pamplona, F.A., Fernandes, D., Takahashi, R.N., 2005. Caffeine improves spatial learning deficits in an animal model of attention deficit hyperactivity disorder (ADHD)–the spontaneously hypertensive rat (SHR). Int. J. Neuropsychopharmacol. 8 (4), 583–594.

Radua, J., van den Heuvel, O.A., Surguladze, S., Mataix-Cols, D., 2010. Meta-analytical comparison of voxel-based morphometry studies in obsessive-compulsive disorder vs other anxiety disorders. Arch. Gen. Psychiatry 67 (7), 701–711.

Rafiee, Z., Nejatian, M., Daeihamed, M., Jafari, S.M., 2019. Application of curcumin-loaded nanocarriers for food, drug and cosmetic purposes. Trends Food Sci. Technol. 88, 445–458.

Rafiee, Z., Nejatian, M., Daeihamed, M., Jafari, S.M., 2019. Application of curcumin-loaded nanocarriers for food, drug and cosmetic purposes. Trends Food Sci. Technol. 88, 445–458.

Ramalingum, N., Mahomoodally, M.F., 2014. The therapeutic potential of medicinal foods. Adv. Pharmacol. Sci. 2014, 354264.

Rauch, S.L., Jenike, M.A., Alpert, N.M., Baer, L., Breiter, H.C., Savage, C.R., Fischman, A.J., 1994. Regional cerebral blood flow measured during symptom provocation in obsessive-compulsive disorder using oxygen 15-labeled carbon dioxide and positron emission tomography. Arch. Gen. Psychiatry 51 (1), 62–70.

Rauch, S.L., Wedig, M.M., Wright, C.I., Martis, B., McMullin, K.G., Shin, L.M., Cannistraro, P.A., Wilhelm, S., 2007. Functional magnetic resonance imaging study of regional brain activation during implicit sequence learning in obsessive–compulsive disorder. Biol. Psychiatry 61 (3), 330–336.

Rech, M., Weinzimmer, S., Geller, D., McGuire, J.F., Schneider, S.C., Patyk, K.C., De Nadai, A.S., Cepeda, S.C., Small, B.J., Murphy, T.K., Wilhelm, S., Storch, E.A., 2020. Symptom trajectories of early responders and remitters among youth with OCD. J. Obs. Compuls. Relat. Disord. 27.

Rostamabadi, H., Falsafi, S.R., Jafari, S.M., 2019. Nanoencapsulation of carotenoids within lipid-based nanocarriers. J. Control. Release 298, 38–67.

Ruscio, A.M., Stein, D.J., Chiu, W.T., Kessler, R.C., 2010. The epidemiology of obsessive–compulsive disorder in the National Comorbidity Survey Replication. Mol. Psychiatry 15 (1), 53–63.

Ruscio, A.M., Stein, D.J., Chiu, W.T., Kessler, R.C., 2010. The epidemiology of obsessive–compulsive disorder in the National Comorbidity Survey Replication. Mol. Psychiatry 15 (1), 53–63.

Sanikhani, N.S., Modarressi, M.H., Jafari, P., Vousooghi, N., Shafei, S., Akbariqomi, M., Heidari, R., Lavasani, P.S., Yazarlou, F., Motevaseli, E., 2020. The Effect of Lactobacillus casei consumption in improvement of obsessive–compulsive disorder: an animal study. Probiotics Antimicrob. Proteins 12 (4), 1409–1419.

Sarris, J., Byrne, G., Castle, D., Bousman, C., Oliver, G., Cribb, L., Blair-West, S., Brakoulias, V., Camfield, D., Ee, C., Chamoli, S., Boschen, M., Dean, O.M., Dowling, N., Menon, R., Murphy, J., Metri, N.J., Nguyen, T.P., Wong, A., Jordan, R., Karamacoska, D., Rossell, S.L., Berk, M., Ng, C.H., 2022. N-acetyl cysteine (NAC) augmentation in the treatment of obsessive-compulsive disorder: a phase III, 20-week, double-blind, randomized, placebo-controlled trial. Prog. Neuropsychopharmacol. Biol. Psychiatry 117, 110550.

Sarris, J., Byrne, G.J., Oliver, G., Cribb, L., Blair-West, S., Castle, D., Dean, O.M., Camfield, D.A., Brakoulias, V., Bousman, C., Dowling, N., Ee, C., Murphy, J., Menon, R., Berk, M., Chamoli, S., Boschen, M., Ng, C.H., 2022. Treatment of refractory obsessive-compulsive disorder with nutraceuticals (TRON): a 20-week, open label pilot study. CNS Spectr. 27 (5), 588–597.

Sarris, J., Oliver, G., Camfield, D.A., Dean, O.M., 2016. Participant characteristics as modifiers of response to N-acetyl cysteine (NAc) in obsessive-compulsive disorder. Clin. Psychol. Sci. 4 (6), 1104–1111.

Sarris, J., Oliver, G., Camfield, D.A., Dean, O.M., Dowling, N., Smith, D.J., Murphy, J., Menon, R., Berk, M., Blair-West, S., 2015. N-acetyl cysteine (NAC) in the treatment of obsessive-compulsive disorder: A 16-week, double-blind, randomised, placebo-controlled study. CNS Drugs 29 (9), 801–809.

Saxena, S., Rauch, S.L., 2000. Functional neuroimaging and the neuroanatomy of obsessive–compulsive disorder. Psychiatr. Clin. North Am. 23 (3), 563–586.

Sayyah, M., Andishmand, M., Ganji, R., 2018. Effect of selenium as an adjunctive therapy in patients with treatment-resistant obsessive-compulsive disorder: a pilot randomized double blind placebo-controlled clinical trial. Arch. Psych. Psych. 20 (4), 57–65.

Sayyah, M., Olapour, A., Saeedabad, Y., Yazdan Parast, R., Malayeri, A., 2012. Evaluation of oral zinc sulfate effect on obsessive-compulsive disorder: a randomized placebo-controlled clinical trial. Nutrition 28 (9), 892–895.

Schlösser, R.G., Wagner, G., Schachtzabel, C., Peikert, G., Koch, K., Reichenbach, J.R., Sauer, H., 2010. Fronto-cingulate effective connectivity in obsessive compulsive disorder: a study with fMRI and dynamic causal modeling. Hum. Brain Mapp. 31 (12), 1834–1850.

Seedat, S., Stein, D.J., 1999. Inositol augmentation of serotonin reuptake inhibitors in treatment-refractory obsessive-compulsive disorder: an open trial. Int. Clin. Psychopharmacol. 14 (6), 353–356.

Shams, J., Soufi, E.S., Zahiroddin, A., Shekarriz-Foumani, R., 2019. Using caffeine on the patients as therapeutic option against treatment-resistant obsessive-compulsive disorder. J. Fam. Med. Prim. Care 8 (5), 1741–1747.

Sherman, P.M., Ossa, J.C., Johnson-Henry, K., 2009. Unraveling mechanisms of action of probiotics. Nutr. Clin. Pr. 24 (1), 10–14.

Sheshachala, K., Narayanaswamy, J.C., 2019. Glutamatergic augmentation strategies in obsessive–compulsive disorder. Indian J. Psychiatry 61 (Suppl 1), S58.

Simon, D., Kaufmann, C., Müsch, K., Kischkel, E., Kathmann, N., 2010. Fronto-striato-limbic hyperactivation in obsessive-compulsive disorder during individually tailored symptom provocation. Psychophysiology 47 (4), 728–738.

Simpson, H.B., Huppert, J.D., Petkova, E., Foa, E.B., Liebowitz, M.R., 2006. Response versus remission in obsessive-compulsive disorder. J. Clin. Psychiatry 67 (2), 269–276.

Sinko, P.J. (Ed.), 2024. Martin's physical pharmacy and pharmaceutical sciences: physical chemical and biopharmaceutical principles in the pharmaceutical sciences, 8th ed.. Wolters Kluwer.

Smith, A., 2002. Effects of caffeine on human behavior. Food Chem. Toxicol. 40 (9), 1243–1255.

Soomro, G.M., Altman, D., Rajagopal, S., Oakley-Browne, M., 2008. Selective serotonin re-uptake inhibitors (SSRIs) versus placebo for obsessive compulsive disorder (OCD). Cochrane Database Syst. Rev. 2008 (1), Cd001765.

Storch, E.A., Merlo, L.J., Bengtson, M., Murphy, T.K., Lewis, M.H., Yang, M.C., Jacob, M.L., Larson, M., Hirsh, A., Fernandez, M., 2007. D-cycloserine does not enhance exposure–response prevention therapy in obsessive–compulsive disorder. Int. Clin. Psychopharmacol. 22 (4), 230–237.

Storch, E.A., Murphy, T.K., Goodman, W.K., Geffken, G.R., Lewin, A.B., Henin, A., Micco, J.A., Sprich, S., Wilhelm, S., Bengtson, M., Geller, D.A., 2010. A preliminary study of D-cycloserine augmentation of cognitive-behavioral therapy in pediatric obsessive–compulsive disorder. Biol. Psychiatry 68 (11), 1073–1076.

Swedo, S.E., Pietrini, P., Leonard, H.L., Schapiro, M.B., Rettew, D.C., Goldberger, E.L., Rapoport, S.I., Rapoport, J.L., Grady, C.L., 1992. Cerebral glucose metabolism in childhood-onset obsessive-compulsive disorder: revisualization during pharmacotherapy. Arch. Gen. Psychiatry 49 (9), 690–694.

Szechtman, H., Sulis, W., Eilam, D., 1998. Quinpirole induces compulsive checking behavior in rats: a potential animal model of obsessive-compulsive disorder (OCD). Behav. Neurosci. 112 (6), 1475.

Temple, J.L., Bernard, C., Lipshultz, S.E., Czachor, J.D., Westphal, J.A., Mestre, M.A., 2017. The safety of ingested caffeine: a comprehensive review. Front. Psychiatry 8, 80.

Tural, Ü., Çorapçıoğlu, A., Boşgelmez, Ş., Köroğlu, G., Ünver, H., Duman, C., Önder, E., 2019. Double blind controlled study of adding folic acid to fluoxetine in the treatment of OCD. Psychiatr. Danub. 31 (1), 69–77.

Urquiaga, I., Leighton, F., 2000. Plant polyphenol antioxidants and oxidative stress. Biol. Res. 33 (2), 55–64.

U.S. Food and Drug Administration, 2012. FDA Drug Safety Communication: Revised recommendations for Celexa (citalopram hydrobromide) related to a potential risk of abnormal heart rhythms with high doses. Available at: https://www.fda.gov/drugs/drug-safety-and-availability/fda-drug-safety-communication-revised-recommendations-celexa-citalopram-hydrobromide-related-potential. (accessed 20 June 2026).

Wei, H., Xu, Y., Cheng, B., Xiong, Y., 2007. Synergistic effects of Lactobacillus rhamnosus ZDY114 and bovine colostrums on the immunological function of mouse in vivo and in vitro. Appl. Microbiol. Biotechnol. 75 (2), 427–434.

Whittal, M.L., Robichaud, M., Thordarson, D.S., McLean, P.D., 2008. Group and individual treatment of obsessive-compulsive disorder using cognitive therapy and exposure plus response prevention: a 2-year follow-up of two randomized trials. J. Consult. Clin. Psychol. 76 (6), 1003.

Wu, K., Hanna, G.L., Rosenberg, D.R., Arnold, P.D., 2012. The role of glutamate signaling in the pathogenesis and treatment of obsessive–compulsive disorder. Pharmacol. Biochem. Behav. 100 (4), 726–735.

Wu, P.-L., Tang, H.-S., Lane, H.-Y., Tsai, C.-A., Tsai, G.E., 2011. Sarcosine therapy for obsessive compulsive disorder: a prospective, open-label study. J. Clin. Psychopharmacol. 31 (3), 369–374.

Yacoubi, M.E., Ledent, C., Ménard, J.F., Parmentier, M., Costentin, J., Vaugeois, J.M., 2000. The stimulant effects of caffeine on locomotor behaviour in mice are mediated through its blockade of adenosine A2A receptors. Br. J. Pharmacol. 129 (7), 1465–1473.

Yong, S.J., Tong, T., Chew, J., Lim, W.L., 2019. Antidepressive mechanisms of probiotics and their therapeutic potential. Front. Neurosci. 13, 1361.

Yu, Z., Kong, L., Chen, Y., 2002. Antidepressant activity of aqueous extracts of Curcuma longa in mice. J. Ethnopharmacol. 83 (1-2), 161–165.

Zhan, Y., Xia, J., Wang, X., 2020. Effects of glutamate-related drugs on anxiety and compulsive behavior in rats with obsessive-compulsive disorder. Int. J. Neurosci. 130 (6), 551–560.

Zhang, H.X., Hyrc, K., Thio, L.L., 2009. The glycine transport inhibitor sarcosine is an NMDA receptor co-agonist that differs from glycine. J. Physiol. 587 (13), 3207–3220.

CHAPTER FIVE

The nexus between the catabolism of tryptophan and neurotoxicity

Azubuike P. Ebokaiwe[a,*], **Olusanya Olasehinde**[b]

[a] Toxicology and Immunotherapy Research Unit, Department of Medical Biochemistry, Delta State University Abraka, Nigeria

[b] Department of Biochemistry, University of Benin

*Corresponding author. e-mail address: azubike.ebokaiwe@funai.edu.ng.

Contents

Abstract

L-tryptophan (TRP) is an essential component of the human diet, that is critical in numerous metabolic functions and has been widely used in several research and clinical trials. The metabolism of l-TRP regulates homeostasis, immunity, and neuronal function. Distortion in TRP metabolism has been implicated in the pathophysiology of various neurodegenerative diseases. In this chapter, emphasis is on the nutritional values, neuropathogenic impacts/properties of key metabolites, and furthermore, to summarize preclinical and clinical studies, in recent years, that investigated the metabolic pathway of TRP, focused on changes in biomarkers of these neurologic disorders, their pathogenic implications, and potential therapeutic strategies targeting this metabolic pathway. This chapter, provides up-to-date information to help identify promising directions for future

Advances in Neurotoxicology, Volume 16
ISSN 2468-7480
https://doi.org/10.1016/bs.ant.2026.02.003

preclinical, clinical, and translational research on the modulation of TRP metabolism and its impact on neuro-degenerative disorders.

Keywords: Catabolism, Immunity, Neurotoxicity, Nutrition, Tryptophan

1. History and structure of tryptophan

The discovery of tryptophan (TRP) (Hopkins and Cole, 1901) in the early 1900s after isolating it from casein protein, and the determination of its molecular structure by Ellinger and Flamand (Ellinger and Flamand, 1908) afterwards marked the beginning of research interest in TRP. L-tryptophan is among the eight essential amino acids i.e. the human body cannot synthesize, but only supplied by the diet. It's instructive to note that only the L isomer of TRP is used in protein synthesis (Larkins et al., 2011) and can pass across the blood-brain barrier for all amino acids, as L-tryptophan. There is generally low tissue storage (Reilly et al., 1997) of TRP in humans which makes it overall lowest concentration in the body among all amino acids, even though only minute amounts are necessary for general healthy nutrition. The typical intake for many individuals is approximately 900 to 1000 mg daily, whereas, the recommended daily allowance for adults is estimated to be between 250 mg/day and 425 mg/day, (Food et al., 1973) which translates to a dietary intake of 3.5 to 6.0 mg/kg of body weight per day. Notable among common dietary sources of tryptophan are oats, bananas, dried prunes, milk, tuna fish, cheese, bread, chicken, turkey, peanuts, and chocolate (Ahuja et al., 2013). The structure of L-Tryptophan is depicted in (Fig. 1).

Nutritional aspect of TRP

There are nine essential amino acids and aromatic amino acid tryptophan is one of them, and is the least available of all proteinogenic amino acids. Fundamentally, tryptophan availability is an imperative factor in the control of protein biosynthesis. This may be one significant reason the immune

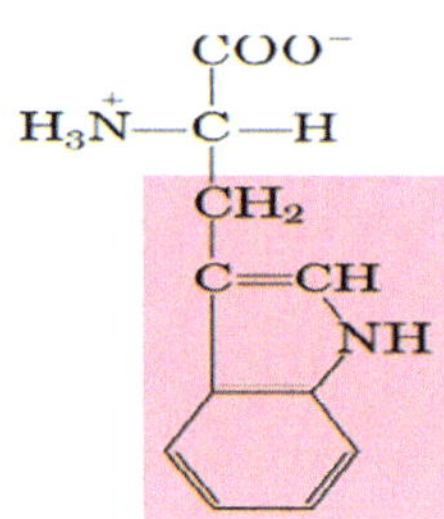

Fig. 1 Structure of TRP.

system employs tryptophan starvation to control the unwanted proliferation of pathogens and malignant cells (Pfefferkorn, 1984). Above and beyond the diverse biological activities of tryptophan catabolites, contributive potentials to a nicotinamide nucleotide pool when the pathway is activated is considered important in situations of dietary shortage, as under normal conditions food is the main source of this metabolite in mammals. Furthermore, kynurenine metabolites are implicated in other metabolic pathways, for example, picolinic acid is associated with the intestinal absorption of zinc and quinolinic acid regulates gluconeogenesis by blocking phosphoenolpyruvate (Gostner et al., 2020). Consequently, the alteration of the tryptophan-kynurenine pathway can be recommended as a pointer for a comprehensible metabolic shift. The consequences of altered tryptophan metabolism as mentioned, indicate that the treatments or conditions which affect these metabolic routes may result in altered Central Nervous System (CNS) function with mood-related, behavioral, and/or cognitive effects. Also, numerous in vitro and in vivo studies involving human primary cells and cell lines show that the TRP pathways are highly inclined to modulation by dietary constituents which include phytochemicals, preservatives, colorants, and probiotics as well as by other lifestyle-related factors (Strasser, Geiger et al., 2016). Both upstream regulatory events and direct interfering activities of TRP metabolism pathway may be relevant. Recently in 2013, tryptophan metabolism was deliberated based on its role as a mediator of environmental factors on cognition and behavior. In line with the roles played by TRP catabolites, the modulation of the TRP metabolism pathway by a variety of nutritional-related factors further underlines the role of Indoleamine 2,3-dioxygenase (IDO)-mediated tryptophan catabolism as a pathway that is particularly sensitive to environmental influences (Gostner et al., 2020).

2. Carbohydrates and protein interference with tryptophan circulation

The transport of tryptophan into the brain is important. However, there is evidence that eating carbohydrates along with protein can increase the tryptophan available to the brain (Fernstrom and Wurtman, 1971). When carbohydrates are consumed, the body produces insulin which directs other amino acids to muscle tissue, but tryptophan remains in the bloodstream. Thus, competing amino acids in the circulation decline and tryptophan more freely enters the brain. Taking pure tryptophan works in a qualitatively different way than when it is obtained from a food source

(Strasser, Gostner et al., 2016). Where peripheral tryptophan concentrations are low following inflammatory conditions, the kynurenine transport becomes even more relevant. L-leucine was shown to competently inhibit the kynurenine transport into the brain of lipopolysaccharide confronted mice, which could efficiently counteract depressive-like behavior (Walker et al., 2019). Another possibility is that the tryptophan concentration in the blood stream may increase when the breakdown of the amino acid is counter acted. This can take place when the intake of compounds suppressed inflammation and impair Indoleamine 2,3-dioxygenase-1 (IDO-1) activity (Gostner et al., 2015). Food rich in antioxidant compounds could improve both serotonin and tryptophan status in the brain. These compounds counteract immune response and tryptophan breakdown by IDO, e.g., compounds like vitamin C and E, and also stilbene resveratrol and coffee flavonoids, were observed to slow down T cell activation and IFN-γ production in vitro in mitogen-stimulated human peripheral blood mononuclear cells (PBMC) from healthy donors, followed by lower activity of the downstream biochemical pathways (like tryptophan breakdown) by IDO-1 production. In neurodegenerative disease, high kynurenine (KYN) concentrations were found to correlate with a decrease of the plasma antioxidants lycopene, lutein, and zeaxanthin, and the vitamins ascorbic acid and α-tocopherol (Murr et al., 2009), so several metabolites could benefit from decreasing oxidative stress. An antioxidative environment stabilizes the oxidation-sensitive molecule BH4, thus affecting the activity of BH4-dependent enzymes. Nonetheless, the estimate of the net effects of dietary components of phytochemicals in humans based on in vitro data only is rather limited due to missing ADME, individual susceptibility, interferences with the microbiome, or interactions with other food components, some of the effects on tryptophan metabolism and related pathways have been reported. In a pilot study with healthy volunteers who received a single dose (5*g*) of resveratrol administered orally after overnight fasting, the Kynurenine/Tryptophan was found to be slightly but significantly elevated, and tryptophan concentrations were decreased up to 5 h after the intervention (Gualdoni et al., 2016). Resveratrol treatment in PBMC in vitro suppressed IDO-1 activity dose-dependently under inflammatory conditions; at low concentrations, a nonmonotonic response provoked an increased activity in otherwise-untreated PBMC (Wirleitner et al., 2005). Similar outcomes were observed when investigating the effect of the intake of regular black tea over a 6-month period: plasma Kyn/Trp and kynurenine concentrations increased in volunteers compared to the placebo group, while tryptophan concentrations were not affected. The tea-drinkers were healthy individuals, who had no major pathological disorders and normal to- mildly elevated systolic blood pressure. In vitro, the stimulating effects of black tea were also shown on other wise-unstimulated Peripheral Blood Mononuclear Cell (PBMC), reflected by the elevation of neopterin concentrations between 2.8-fold induction with cold extract at 1:10 in medium, and 1.9-fold induction with cold extract at 1:1 after 48 h incubation (Zvetkova et al., 2001). Though polyphenol-rich food/beverages and phytochemicals such as resveratrol and other flavonoid compounds are generally considered to act as antioxidants, the data suggest that the net effect depends on an individual's immune status, being immuno-stimulatory in healthy individuals while a dampening of responses may occur on an inflammatory background.

3. Tryptophan metabolism and microbiome/probiotics

The significance of tryptophan metabolism in the cross-talk of gut and microbiome was addressed in a recent study by Laurans et al. (2018), wherein ameliorating influence on obesity, liver steatosis, and insulin sensitivity in the absence of IDO-1 activity in mice, were mediated by microbiota-dependent production of interleukin 22 was reported. In addition, microbiota can be a source of tryptophan and tryptophan-derived metabolites, some of which can be transported across membranes, and of neurotransmitters and precursors such as noradrenalin, γ-aminobutyric acid (GABA), and other neuroactive substances (Cryan and Dinan, 2012). Thus, interferences with the microbiome are likely to influence the availability of tryptophan in gut-brain axis. No major effects regarding exercise-induced Kyn/ Trp concentrations were reported in a study with trained athletes receiving probiotics compared to those receiving placebo. Though there were lower tryptophan degradation rates in subjects supplemented with probiotics, this effect was not significant; however, there was a reduced incidence of upper respiratory tract infections compared to the placebo group. However, patients with liver cirrhosis who suffered from intestinal dysbiosis and pro-inflammatory conditions after probiotic supplementation provoked an increase in KYN/TRP concentrations and a subclinical increase in neutrophil resting burst compared to placebo treated group (Horvath et al., 2016). Treatment with probiotics for 28 days, (aqueous solutions of the probiotic Omnibiotic Stress Repair (Allergosan, Graz, Austria) were given every morning before breakfast). One sachet's contents were suspended in 1/8 L of lukewarm water for 15 min. The suspensions were then once again agitated and consumed. Lactobacillus casei W56, Lactococcus lactis W19, Lactobacillus acidophilus W22, Bifidobacterium lactis W52, Lactobacillus paracasei W20, Lactobacillus plantarum W62, Bifidobacterium lactis W51, Bifidobacterium bifidum W23, and Lactobacillus salivarius W24 made up the composition. Serum kynurenine concentrations were reported to increase slightly but significantly in AD patients (Leblhuber et al., 2018). It is well known that the gut microbiome has significant influence on the outcome of neuropsychiatric conditions like anxiety, depression, cognition, and autism spectrum disorder, likewise that there is a link between gut-derived signals and healthy brain development but also with the progression of neurological disorders (Sharon et al., 2016). The leaky gut syndrome, which represents an increase of the intestinal barrier permeability allowing the entry of microbes and metabolites into the bloodstream, is an indication of AD pathogenesis and provides a mechanistic explanation for this connection (Köhler et al., 2016).

4. Catabolism of tryptophan

The major direction of TRP metabolism is the kynurenine pathway (KP), yielding neuroactive metabolites and nicotinamide adenine dinucleotide (NAD^+). Well above 95 percent of TRP is metabolized through the KP (Wirleitner et al., 2005), while the residual Trp is metabolized by the serotonin pathway. In the first step of the KP, TRP is transformed to N-formyl-L-kynurenine, a compound that is very unstable, by the enzymes tryptophan 2,3-dioxygenase (TDO) and indoleamine 2,3-dioxygenase-1 and 2 (IDO-1, IDO-2). These enzymes represent the main rate limiting enzymes of the KP. Interferon-α, interferon-β (IFN-β) and interferon-γ (IFN-γ) are all potent inducers of IDO but the foremost inducer/activator of IDO is IFN-γ (Heyes et al., 1997). Furthermore, tumor necrosis factor-α and interleukin-6 are also able to activate IDO in an IFN-γ-independent way (Connor et al., 2008). Previous investigations also showed that IDO has an important role in maternal-fetal tolerance, and both enzymes bear immunosuppressive properties (Sedlmayr, 2007). Further degradation of N-formyl-L-kynurenine is by formamidase to L-kynurenine (L-KYN). L-KYN was long thought not to have neuroactive properties. L-KYN participation in oxidative processes has been proposed; pro-oxidative and anti-oxidative properties have been documented (Reyes Ocampo et al., 2014). Recently, evidence emerged that L-KYN is an endogenous ligand for the human aryl-hydrocarbon receptor (AHR). This receptor has a significant role in cellular responses aroused by environmental toxins, such as 2,3,7,8-tetrachlorodibenzo-p-dioxin and polycyclic aromatic hydrocarbons (Kitamura and Kasai, 2007). The AHR is a ligand-activated transcription factor, that is adjourned to be a member of the family of basic helix-loop-helix transcription factors (Bunger et al., 2008). It is an inactive cytosolic protein normally, but after ligand binding, it translocates to the nucleus, where it binds to the regulatory regions of xenobiotic-responsive elements. DNA binding indicates diverse transcriptional responses, such as the activation of enzymes participating in xenobiotic metabolism. AHR also participates in immune response and in tumorigenesis (Murray et al., 2014). Besides environmental pollutants, numerous endogenous AHR ligands have been recognized, including kynurenines L-KYN and kynurenic acid (KYNA), suggesting the involvement of the KP in the diverse transcriptional pathways, regulated by AHR. In the CNS, 40 percent of L-KYN is generated locally, and 60 percent is taken up from the blood (Gál and Sherman, 1978). The KP continues in different branches deriving from L-KYN, the first one leading to the synthesis of KYNA by kynurenine aminotransferases (KATs). The availability of the KYNA in the human and rat brain is in nanomolar concentrations (Moroni et al., 1988), and it is an endogenous broad-spectrum antagonist of N-Methyl-D-aspartate (NMDA) receptors acting at low concentrations on the strychnine-insensitive glycine site (Kessler et al., 1989), while it acts on the glutamate recognition site at higher concentrations (Danysz et al., 1989). Furthermore, it is a feeble antagonist of kainate and AMPA ionotropic glutamate receptors (Perkins and Stone, 1985). At low concentrations (nM–μM), KYNA has a facilitatory effect on AMPA receptors, while at higher concentrations, it acts as a competitive antagonist (Rózsa et al., 2008).

The broad spectrum of KYNAs receptorial action is in favor of its important role in regulation of glutamatergic neurotransmission. On α7-nicotinic acetylcholine receptors (α7nAch), KYNA exerts a non-competitive antagonistic effect (Hilmas et al., 2001), therefore, it participates in both glutamatergic and nicotinergic neurotransmission. Modulation of α7nAch, thus presynaptic glutamate release by KYNA, seems to be an important site of action in protection against glutamate induced excitotoxicity. KYNA has been proposed as an endogenous agonist of the G-protein coupled receptor 35 (GPR35), the function of which is yet poorly elucidated. The gastrointestinal system and immune cells are the expression site of GPR35, but the dorsal root ganglia and hippocampal neurons are the functional sites

(Alkondon et al., 2015). There is a reported demonstration, that KYNA, similarly to L-KYN, is a ligand to AHR (DiNatale et al., 2010). The immunomodulatory role for this compound (KYNA) is predicated on its effects on GPR35 and AHR. Recently, it was shown that KYNA is able to scavenge ROS in an NMDA- and nicotinic receptor-independent way; thus, it can be a potential endogenous antioxidant (Alkondon et al., 2015). Overall, KYNA is involved in possible neurotoxic processes as a protective agent, which exclude its involvement in neurodegenerative mechanisms. The second branch of the KP stemming from L-KYN leads to anthranilic acid with the aid of kynureninase. Anthranilic acid inhibited citric acid cycle and the respiratory chain complexes I–III (Lugo-Huitrón et al., 2011), which ultimately interferes with mitochondrial function. The implication of anti-inflammatory response has been reported via creating a complex with copper and acting as an $^{\bullet}OH$ inactivating ligand (Gaubert et al., 2000). Anthranilate 3-monooxygenase converts anthranilic acid to 3-hydroxyanthranilic acid (3-HA), thus rallying to the third branch of L-KYN metabolism. The first step of the third branch is the conversion of L-KYN by kynurenine monooxygenase (KMO) to 3-hydroxykynurenine (3-HK). KMO has the highest affinity for L-KYN among the enzymes of the three branches, suggesting that under physiological conditions, L-KYN is metabolized by this third branch (Bender and McCreanor, 1982). 3-HK undergoes auto-oxidation, forming highly reactive o-aminoquinone and ROS $O_2^{\bullet-}$ and H_2O_2 and dimerizes to hydroxanthommatin and xanthommatin under physiological conditions. 3-HK is generally considered as a neurotoxic agent in vivo, causing convulsive attacks when administered intraventricularly or leading to tissue damage when administered intrastriatally (Nakagami et al., 1996) 3-HK is also present in eye lens, and it has been connected with cataract formation (Mizdrak et al., 2008). The toxicity of 3-HK can be attributable to its capability to produce free radicals during its auto-oxidation. However, the free radical scavenging effects of this compound have also been observed in vitro in rat cortex and in C6 glioma cells (Leipnitz et al., 2007). There was a thorough review on the potentials of 3-HK to act as pro- or anti-oxidative agent in response to physiological changes by Colín-González et al. (2013) The next step in the metabolism of L-KYN is the generation of 3-HA from either 3-HK by kynureninase or from anthranilic acid by anthranilate 3-monooxygenase. 3-HA is also prone to auto-oxidation, generating superoxide radicals, H_2O_2, and cinnabarinic acid (Dykens et al., 1987). Cinnabarinic acid is a ligand for the type 4 metabotropic glutamate receptor and also for AHR (Lowe et al., 2014).

3-HA can induce apoptosis in monocytes/macrophages (Morita et al., 2001), and it can inhibit the mitochondrial respiratory chain. Furthermore, it has important immunoregulatory functions by interfering with T-cell survival (Fallarino et al., 2002). 3-HK also can be transaminated by KAT to xanthurenic acid, which, similarly to 3-HK and 3-HA, possesses both pro-oxidative and anti-oxidative properties (Murakami et al., 2007). 3-HA is further processed by 3-hydroxyanthranillic acid oxygenase (3-HAO) to 2-amino-3 carboxymuconate semialdehyde. This intermediate can be metabolized by picolinic carboxylase to produce picolinic acid (PIC) or can be transformed by nonenzymatic cyclisation to quinolinic acid (QUIN). PIC is a non-selective metal ion chelating agent (Aggett et al., 1989), induces morphological changes in the rat hippocampus, substantia nigra and striatum when administered intraperitoneally and has a macrophage induction activity (Bosco et al., 2000). PIC is able to prevent QUIN-induced neurotoxicity when injected into the nucleus basalis magnocellularis of rats, at 120 nmol for 7 days and it is able to modulate kainate-induced glutamate release from the striatum (Jhamandas et al., 1990). QUIN under normal conditions is present in the brain in nanomolar concentrations and metabolized for the synthesis of NAD^+. In vitro studies revealed that QUIN is toxic for brain cells from above 150 nM. QUIN is a weak endogenous agonist on NMDA receptors, the action of which is selective, involving the receptor subtypes containing the NR2A and NR2B subunits (de Carvalho et al., 1996). The greatest excitotoxic damage in brain areas rich in NMDA receptors containing NR2A and NR2B subunits is caused by QUIN, mainly in the striatum and in the hippocampus (Tavares et al., 2002). Furthermore, QUIN can increase excitotoxicity via glutamate release by neurons and inhibit glutamate uptake by astrocytes, maintaining a high level to constantly stimulate NMDA receptors (Rios and Santamaria, 1991). Another mechanism that also contribute to QUIN toxicity is lipid peroxidation (Goda et al., 1996). Earlier investigations advocate that QUIN forms a complex with iron, and this complex can contribute to the formation of ROS (Chen et al., 2011). The toxicity of QUIN on brain cells is exerted mainly through NMDA-mediated excitotoxicity (Espey et al., 1997). The metabolic process is completed by quinolinic acid phosphoribosyl transferase which converts QUIN to NAD^+. As an electron transfer molecule, NAD^+ is thereafter utilized by different intracellular processes. The enzymes of the KP are differently distributed among the various cell types in the CNS, providing an important way of controlling the synthesis of different metabolites. Macrophages and microglia express the entire enzymatic machinery of the KP, and neurons are able to

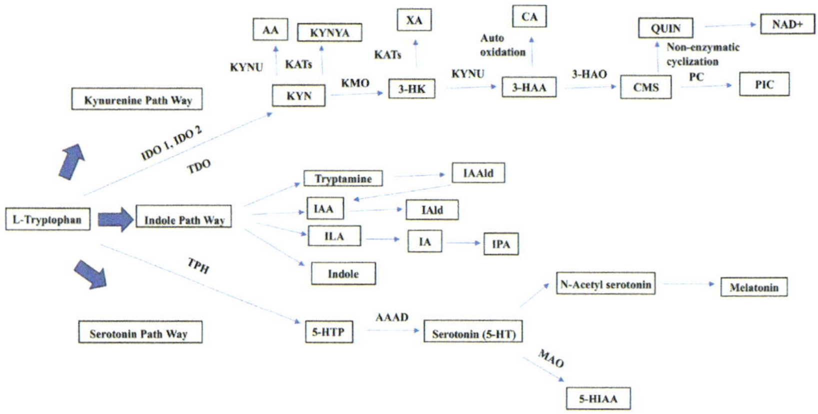

Fig. 2 Tryptophan metabolism by the kynurenine, serotonin and indole pathways. The breakdown of tryptophan (TRP) produces catabolites such as kynurenine (KYN) and kynurenic acid (KYNA) via Kynurenine pathway. Among the catabolites of TRP breakdown, only KYN, and 3-hydroxy kynurenine (3-HK) can enter the brain through the blood brain barrier (BBB), which in turn is involved in the neurotoxic and neuroprotective branches of TRP influence.

synthesize KYNA and PIC (Guillemin et al., 2007). However, astrocytes lack the enzyme KMO and are not able to synthesize 3-HK under physiological conditions. All the enunciated pathways of the L-Trp catabolism are depicted in Fig. 2.

5. Neurotoxic/protective influence of tryptophan catabolites

5.1 Neurotoxic impact

The main synthetase of the QUIN pathway, which converts kynurenine to 3-HK is the outer mitochondrial membrane enzyme KMO. This component crosses the BBB and increases the substrate required for QUIN production, producing neuronal apoptosis, free radical generation, and oxidative stress damage in the brain at nanomolar concentrations, leading to elevated metal toxicity in rat astrocyte cultures and effectively synergizing the neurotoxic effects of QUIN (Behl et al., 2021). Subsequently, 3-HK is converted to 3-hydroxyanthranilic acid (3-HAA) by the action of KYNU which is subsequently converted to QUIN by the action of 3 hydroxyanthranilic acid dioxygenase (3-HAO) before final degradation to nicotinamide adenine dinucleotide (NAD^+) (Myint and Kim, 2014). Additionally, Xanthurenic acid can be produced by 3-HK in the presence of KAT, and

3-HAA can produce picolinic acid (PIC) in the presence of 2-amino3-carboxymuconate-6 semialdehyde decarboxylase (ACMSD). Xanthurenic acid is a regulator of glutamatergic synaptic transmission, which can directly or indirectly regulate metabolic Glu receptors (Fazio et al., 2017). PIC is a non-selective metal ion chelator and neuroprotective agent, which can inhibit QUIN induced neurotoxicity, but the inhibitory effect is weaker than KYNA (Mor et al., 2021). Microglia and neurons are restricted in their catabolism of NAD+, which in turn leads to cumulative neurotoxicity of QUIN, and this occurs when QUIN is in a high expression state, which is manifested in recurrent major depression patients having higher levels of KYN metabolites (Brown et al., 2021). QUIN is an NMDAR activator that enhances Glu release from neurons, leading to excess microenvironmental Glu concentrations and neurotoxicity, and can also induce selective apoptosis of novel glial cells. Meanwhile, prolonged exposure of neurons to elevated QUIN disrupts the structure of dendrites and reduces the immunoreactivity of microtubule associated protein 2, thereby perpetuating neurodegenerative disease (Hunt et al., 2020). Moreover, production of neurotoxicity and oxidative stress may also be fostered by QUIN-induced free radical generation. Its induction of lipid peroxidation is regulated by interaction with Fe^{2+}, forming QUIN-Fe^{2+} complexes that promote ROS production (Fujigaki et al., 2017).

5.2 Neuroprotection influence

KYN generates KYNA in the mammalian CNS in the presence of KAT. Among the four KATs, KAT II is mainly expressed in astrocytes and is responsible for most KYNA synthesis. KYNA exerts a neuroprotective effect against the excitotoxic and apoptotic effects of NMDAR due to its ability to antagonize this receptor (Perkins and Stone, 1982). In addition, KYNA can antagonize a7-nicotinic acetylcholine receptor (a7-nAChR), reducing extracellular Glu and dopamine (DA) levels (Sharaf El Din et al., 2017). And it plays a direct immunomodulatory role through the role of G-protein coupled receptor 35 (GPR35) and Aryl hydrocarbon receptor (Liu et al., 2019). However, high levels of KYNA can be detrimental, leading to low Glu levels that interfere with cognitive function and are associated with psychiatric disorders such as Schizophrenia, including memory deficits and reduced dopaminergic and glutamatergic neurotransmission (Brown et al., 2021). The neuroprotective and neurotoxic influence of the L-Trp catabolites is shown in Fig. 3.

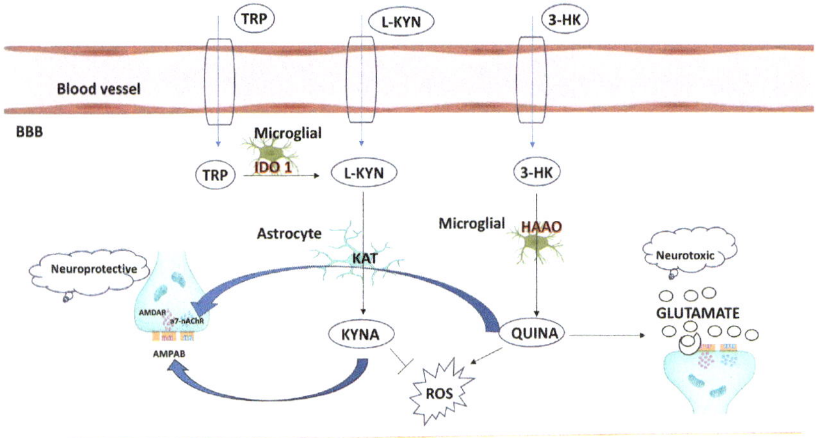

Fig. 3 Neurotoxic/protective influence of TRP catabolites.

6. Neurodegenerative disorders associated with tryptophan catabolites

The involvement of key TRP catabolites and their pathogenic functions in some neurodegenerative disorders such as: Alzheimer's disease, Parkinson's disease, Huntington's disease, Amyotrophic Lateral Sclerosis, Acquired Immunodeficiency Syndrome Dementia Complex is summarized below.

6.1 Alzheimer's disease

Alzheimer's disease (AD) is characterized by progressive cognitive decline and memory loss, mainly in the elderly population. Pathological protein aggregates in the form of amyloid β (Aβ)-formed plaques, and phosphorylated tau (p-tau)-constituted neurofibrillary tangles are the main biomarkers of the disorder. An increased L-KYN/TRP ratio in AD patients suggests an enhanced TRP metabolism and the increased activation of IDO, which could be connected to neuroinflammation being activated by IFN-γ (Widner et al., 2000). The peripheral blood levels of 3-HK are higher in AD patients compared to controls (Schwarz et al., 2013) further confirming a metabolic shift in diseased patients. Furthermore, results showed that QUIN accumulates in the brain of AD patients, and it is co-localized with p-tau and neurofibrillary tangles (Rahman et al., 2009). Levels of KYNA are also elevated in AD patients, mainly in the striatum and in the hippocampus

(Baran et al., 1999). These results suggest that the kynurenine pathway in general is upregulated in AD, but the origin of this phenomenon needs further clarification. Activated microglia may produce an increased amount of KP metabolites in response to Aβ and p-tau, and it is possible that the excess amount of the produced QUIN leads to the invigoration of tau phosphorylation and to the development of a vicious circle (Rahman et al., 2009).

6.2 Parkinson's disease

Parkinson's disease is the second most prevalent neurodegenerative disease, characterized by pathological presence of Lewy bodies and Lewy neurites, comprised mainly of α-synuclein, in the neuromelanin-containing dopaminergic cells. Selective death of dopaminergic neurons is most abundant in the substantia nigra pars compacta; however, expansive neurodegeneration is present in the CNS (Braak et al., 2003). The motor symptoms characteristic of PD is propelled by the selective degradation of the nigrostriatal pathway. The major significant factors in the selective death of dopaminergic cells in PD include oxidative stress, excitotoxicity and neuroinflammation (Zádori et al., 2013). Numerous studies have investigated the involvement of the KP in PD. Widner and colleagues found an increased L-KYN/TRP ratio in serum and cerebrospinal fluid (CSF) of PD patients (Widner et al., 2002), while Ogawa and co-workers demonstrated decreased L-KYN and KYNA and increased 3-HK levels in brain samples of PD patients (Ogawa et al., 1992). These results suggest the upregulation of TRP metabolism in PD, further supported by the finding that 3-HK levels increased in CSF (Lewitt et al., 2013).

6.3 Huntington's disease

Huntington's disease is an autosomal dominantly inherited disorder caused by the CAG expansion in the HD gene on chromosome 4. The protein huntingtin is encoded by this gene, the normal functions of which are still under rigorous investigation. The disease presents with motor symptoms and progressive cognitive decline. Excitotoxicity is considered as one of the main factors in the disease. Administration of QUIN can serve as an animal model of HD, because it causes lesions pathologically similar to that of human samples, as it spares the cholinergic and a spiny neuron of the striatum (Beal et al., 1986). Based on these observations, the quinolinate hypothesis of HD emerged, suggesting that QUIN may have a causative role in this disease.

Several studies measured the levels of the KP metabolites, especially QUIN, in brain and CSF in postmortem samples of HD patients. Brain tissue levels of QUIN were found to be reduced (Heyes et al., 1991) whereas the levels of KYNA decreased both in CSF and brain in HD (Jauch et al., 1995). An increase in 3-HK levels was reported in Huntington's disease (Pearson and Reynolds, 1992). Measuring the same compounds during the early stages of HD revealed that both 3-HK and QUIN are elevated, while at later stages, no change or a decrease can be observed (Pearson and Reynolds, 1992). These findings demonstrate that 3-HK and QUIN may be participants in the degenerative processes early in the course of HD.

6.4 Amyotrophic lateral sclerosis

Amyotrophic lateral sclerosis (ALS) is a progressive neurodegenerative disorder, affecting motor neurons at both spinal, brainstem and cortical levels. ALS is mainly sporadic, with a 5 percent–10 percent occurrence of familial cases, affecting the adult population. About 20 percent of the familial cases is attributable to the mutation of the superoxide dismutase 1 (SOD1) gene (Pearson and Reynolds, 1992), coding an important free radical scavenging enzyme. The symptoms include muscle weakness, atrophy and paralysis, and in most cases, the involvement of breathing muscles results in death within 3–5 years from disease onset. Perturbations of the kynurenines in ALS were studied by several groups, Chen and colleagues found increased L-KYN and QUIN levels in CSF and serum of ALS patients (Chen et al., 2010), while KYNA levels were elevated in CSF of bulbar onset patients, but decreased in the serum of patients with severe clinical status (Monaco et al., 1979). The involvement of the mechanisms of glutamate excitotoxicity, mitochondrial dysfunction, oxidative stress and inflammation marks the exact patho-mechanism of ALS which is not still lucidly understood, but the general view is that it is a multifactorial disease. The kynurenines are involved in all of the above-mentioned processes, thus providing a multitarget option for therapeutic intervention in ALS.

The most well-known characteristic of the multiple sclerosis is demyelination and the forming of sclerotic plaques. The loss of the myelin sheath and the inflammation at different sites of the CNS causes diverse symptoms and distinct disease courses in Multiple sclerosis. Multiple sclerosis can be categorized based on the disease course as clinically-isolated syndrome, relapsing-remitting Multiple sclerosis, primary progressive Multiple sclerosis, secondary progressive Multiple sclerosis and progressive relapsing Multiple sclerosis. Alterations in the KP were noted in Multiple sclerosis; levels of TRP were found to be reduced both in the CSF and plasma of Multiple sclerosis patients compared with control subjects (Ott et al., 1993). Furthermore, KYNA levels were shown to be significantly decreased in CSF of Multiple sclerosis patients during remission, while a significant increase is present both in the plasma and CSF in the course of relapse (Hartai et al., 2005; Rejdak et al., 2002, 2007). These findings suggest that levels of KYNA may be directly involved in the alternation of relapsing-remitting phases of the disease. IFN-β treatment, a first line therapy applied in Multiple sclerosis, has been shown to significantly increase L-KYN levels and the L-KYN/TRP ration, which is indicative of IDO activity. These results suggest that IFN-β treatment leads to the induction of IDO (Durastanti et al., 2011). Thus, the important role of IDO and the KP in immune response and autoimmunity is further confirmed (Mándi and Vécsei, 2012).

6.5 Acquired immunodeficiency syndrome dementia complex

Infection with human immunodeficiency virus type 1 (HIV-1) can lead to the development of acquired immunodeficiency syndrome dementia complex (ADC), also termed HIV-associated dementia. ADC is a subcortical type of dementia characterized by marked memory impairment and psychomotor slowing. It occurs in the severe forms of HIV-1 infection, affecting about 2 percent of the infected population receiving combined antiretroviral therapy (Heaton et al., 2010). Heyes and colleagues proposed that QUIN could play a direct role in the development of ADC, as QUIN levels were elevated in HIV-infected patients and correlated with the severity of neurological deficits (Heaton et al., 2010). A more feasible hypothesis suggests that QUIN is elevated due to the inflammatory process generated by HIV-1, nevertheless emphasizing the importance of the KP in inflammatory processes (Guillemin et al., 2005). The neurotoxic impact of tryptophan catabolites and their associated diseases are presented in Table 1.

7. Potential neurotherapeutic strategies of regulating tryptophan metabolism

Neurodegenerative diseases are a great socio-economic burden, and in most cases, the therapies available provide only symptomatic relief; they do not treat the cause of the disorders. Consequently, there is a constant need for new and effective therapeutic solutions. Manipulating the KP could

Table 1 The neurotoxic impact of tryptophan catabolites and their associated diseases.

S/N	Diseases associated with neurotoxicity of Tryptophan catabolites	Activity of the Catabolites	Organs/Tissues/Biological fluids affected
1	Alzheimer's disease (AD)	L-KYN/TRP	Peripheral blood [88]
		IDO 3-HK	Peripheral blood [88]
		Quin	Brain [89,90]
		KYNA	Striatum; Hippocampus [91]
2	Parkinson's Disease (PD)	L-KYN/TRP	Serum; Cerebrospinal fluid (CSF) [95]
		L-KYN; KYNA	Brain [96]
		3-HK	Brain and CSF [97]
3	Huntington's Disease (HD)	QUIN	CSF [99]
		KYNA	CSF and Brain [100–103]
		3-HK	Brain [101]
		QUIN	Brain [103]
4	Amyotrophic Lateral Sclerosis (ALS)	L-KYN	CSF and Serum [107]
		QUIN	
		KYNA	
5	Acquired Immunodeficiency Syndrome Dementia Complex (AIDS)	QUIN	Serum [117]

result in beneficial effects, affecting each of the contributory mechanisms of neuroinflammation, oxidative stress and excitotoxicity. One of the possible beneficiary therapeutic interventions is the shifting of KP metabolism towards the formation of protective agents, mainly KYNA (Fig. 4). This metabolic shift can be achieved by specific enzyme inhibitors of KMO, kynureninase and 3-HAO. Several compounds have been developed for the inhibition of these enzymes (Chiarugi et al., 1995; Colabroy et al., 2005; Walsh et al., 2003), but the inhibitors of KMO have been the most widely studied and may provide the most beneficial effects, because inhibition of KMO prevents the formation of the most neurotoxic kynurenines, 3-HK, 3-HAA and QUIN.

Inhibition of KMO leads to a decrease in the levels of 3-HK and QUIN in rats (Amori et al., 2009). Blockade with 3,4-dimethoxy-N-[4-(3-nitrophenyl) thiazol-2-yl] benzenesulfonamide (Ro-61–8048), a potent KMO inhibitor, results in increased KYNA content in parkinsonian monkeys, and when combined with levodopa, it reduces the severity of dyskinesias, both acutely and after prolonged administration (Samadi et al., 2005).

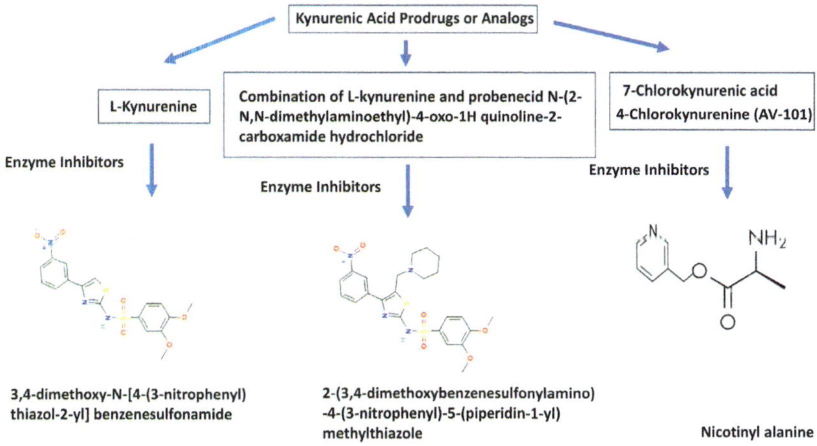

Fig. 4 Therapeutic possibilities for modulating the kynurenine pathway in neurodegenerative diseases, with some of the candidates developed and tested so far.

A prodrug of Ro-61–8048, 2-(3,4-dimethoxybenzenesulfonylamino)-4-(3-nitrophenyl)-5-(piperidin-1-yl) methylthiazole (JM6), was able to ameliorate anxiety-related behavior, spatial memory deficits and synaptic loss in a transgenic mouse model of AD and was able to decrease microglial activation and extend life span of R6/2 mice (a genetic mouse model of HD) (Zwilling et al., 2011). However, the true prodrug properties of JM6 were questioned, and there is no doubt that Ro-61–8048 is an effective KMO inhibitor (Beconi et al., 2012). Another possibility for therapeutic intervention is the pharmacological increase of KYNAs effect, either by administration of its prodrug or by synthetic analogs. Combined administration of L-KYN and probenecid, an inhibitor of organic anion transport, causes elevation in the cortical levels of KYNA (Chauvel et al., 2012). The same combination was shown to reduce the toxic effects of 6-hydroxydopamine, a widely-used model of PD in rats; joint administration was able to mitigate rotation behavior and neurodegeneration (Silva-Adaya et al., 2011). L-KYN administered together with probenecid was able to attenuate histopathological changes and improve spatial memory in the Aβ model of AD (Carrillo-Mora et al., 2010). Furthermore, L-KYN alone mitigated the neuronal cell loss and damage after ischemic insult in rats (Sas et al., 2008). A combination of enzyme inhibition and pharmacological supplementation of KYNA also proved to be protective in a model of PD. L-KYN and probenecid were combined with nicotinyl alanine, an inhibitor of kynureninase and KMO, and were able to modulate QUIN-induced turning behavior

(Miranda et al., 1999). During the past few years, several types of KYNA analogs have been synthesized. A novel KYNA amide, N-(2-N, N-dimethylaminoethyl)-4-oxo-1H-quinoline-2-carboxamide hydrochloride, exerted beneficial effects in numerous paradigms. In micromolar concentrations, it reduced the amplitude of field excitatory postsynaptic potentials in the CA1 region of the hippocampus, while in nanomolar concentrations, it exerted a facilitatory effect, similar to KYNA (Marosi et al., 2010). In a transgenic mouse model of HD, the same analog increased survival time and prevented weight loss and striatal neuron loss of the animals (Marosi et al., 2010). The tests of the anti-inflammatory properties of this compound showed that it has a higher potency to inhibit tumor necrosis factor-α production than KYNA, suggesting a more potent immunoregulatory effect (Tiszlavicz et al., 2011). Similarly to KYNA, it also possesses protective characteristics against ischemia-induced neuronal loss (Gellért et al., 2011). The exact mechanism of action of this new compound is not known, but the lack of cognitive side effects makes it a promising candidate for further investigations (Gellért et al., 2012).

8. Conclusion

A necessary amino acid, tryptophan is involved in a number of vital metabolic processes, such as the production of kynurenine, melatonin, and serotonin. Over 95 percent of dietary tryptophan is broken down by the kynurenine pathway (KP), which is the main pathway of tryptophan catabolism. This pathway serves as a crucial regulator of central nervous system (CNS) homeostasis by producing a number of neuroactive metabolites, some of which are neuroprotective and others of which are neurotoxic.

Tryptophan catabolism via the kynurenine route is essentially a vital molecular connection between immunological activity, metabolism, and neurotoxicity. The brain's susceptibility to illness and injury depends on the ratio of its neuroprotective and neurotoxic metabolites.

References

Aggett, P.J., Fenwick, P.K., Kirk, H., 1989. An in vitro study of the effect of picolinic acid on metal translocation across lipid bilayers. J. Nutr. 119 (10), 1432–1437. doi:10.1093/jn/119.10.1432.

Ahuja, J.K., Moshfegh, A.J., Holden, J.M., Harris, E., 2013. USDA food and nutrient databases provide the infrastructure for food and nutrition research, policy, and practice. J. Nutr. 143 (2), 241S–249S.

Alkondon, M., Pereira, E.F., Todd, S.W., Randall, W.R., Lane, M.V., Albuquerque, E.X., 2015. Functional G-protein-coupled receptor 35 is expressed by neurons in the CA1 field of the hippocampus. Biochem. Pharmacol. 93 (4), 506–518. doi:10.1016/j.bcp.2014.12.009.

Amori, L., Guidetti, P., Pellicciari, R., Kajii, Y., Schwarcz, R., 2009. On the relationship between the two branches of the kynurenine pathway in the rat brain in vivo. J. Neurochem. 109 (2), 316–325. doi:10.1111/j.1471-4159.2009.05893.x.

Baran, H., Jellinger, K., Deecke, L., 1999. Kynurenine metabolism in Alzheimer's disease. J. Neural. Transm. (Vienna) 106 (2), 165–181. doi:10.1007/s007020050149.

Beal, M.F., Kowall, N.W., Ellison, D.W., Mazurek, M.F., Swartz, K.J., Martin, J.B., 1986 Replication of the neurochemical characteristics of Huntington's disease by quinolinic acid. Nature 321 (6066), 168–171. doi:10.1038/321168a0.

Beconi, M.G., Yates, D., Lyons, K., Matthews, K., Clifton, S., Mead, T., et al., 2012. Metabolism and pharmacokinetics of JM6 in mice: JM6 is not a prodrug for Ro-61-8048. Drug Metab. Dispos. 40 (12), 2297–2306. doi:10.1124/dmd.112.046532.

Behl, T., Kaur, I., Sehgal, A., Singh, S., Bhatia, S., Al-Harrasi, A., et al., 2021. The footprint of kynurenine pathway in neurodegeneration: Janus-faced role in Parkinson's disorder and therapeutic implications. Int. J. Mol. Sci. 22 (13), 6737. doi:10.3390/ijms22136737.

Bender, D.A., McCreanor, G.M., 1982. The preferred route of kynurenine metabolism in the rat. Biochim. Biophys. Acta 717 (1), 56–60. doi:10.1016/0304-4165(82)90379-8.

Bosco, M.C., Rapisarda, A., Massazza, S., Melillo, G., Young, H., Varesio, L., 2000. The tryptophan catabolite picolinic acid selectively induces the chemokines macrophage inflammatory protein-1 alpha and -1 beta in macrophages. J. Immunol. 164 (6), 3283–3291. doi:10.4049/jimmunol.164.6.3283.

Braak, H., Del Tredici, K., Rüb, U., de Vos, R.A., Jansen Steur, E.N., Braak, E., 2003. Staging of brain pathology related to sporadic Parkinson's disease. Neurobiol. Aging 24 (2), 197–211. doi:10.1016/s0197-4580(02)00065-9.

Brown, S.J., Huang, X.F., Newell, K.A., 2021. The kynurenine pathway in major depression: what we know and where to next. Neurosci. Biobehav. Rev. 127, 917–927. doi:10.1016/j.neubiorev.2021.05.018.

Bunger, M.K., Glover, E., Moran, S.M., Walisser, J.A., Lahvis, G.P., Hsu, E.L., et al., 2008. Abnormal liver development and resistance to 2,3,7,8-tetrachlorodibenzo-p-dioxin toxicity in mice carrying a mutation in the DNA-binding domain of the aryl hydrocarbon receptor. Toxicol. Sci. 106 (1), 83–92. doi:10.1093/toxsci/kfn149.

Carrillo-Mora, P., Méndez-Cuesta, L.A., Pérez-De La Cruz, V., Fortoul-van Der Goes, T.I., Santamaría, A., 2010. Protective effect of systemic L-kynurenine and probenecid administration on behavioural and morphological alterations induced by toxic soluble amyloid beta (25–35) in rat hippocampus. Behav. Brain Res. 210 (2), 240–250. doi:10.1016/j.bbr.2010.02.041.

Chauvel, V., Vamos, E., Pardutz, A., Vecsei, L., Schoenen, J., Multon, S., 2012. Effect of systemic kynurenine on cortical spreading depression and its modulation by sex hormones in rat. Exp. Neurol. 236 (2), 207–214. doi:10.1016/j.expneurol.2012.05.002.

Chen, Y., Brew, B.J., Guillemin, G.J., 2011. Characterization of the kynurenine pathway in NSC-34 cell line: implications for amyotrophic lateral sclerosis. J. Neurochem. 118 (5), 816–825. doi:10.1111/j.1471-4159.2010.07159.x.

Chen, Y., Stankovic, R., Cullen, K.M., Meininger, V., Garner, B., Coggan, S., et al., 2010. The kynurenine pathway and inflammation in amyotrophic lateral sclerosis. Neurotox. Res. 18 (2), 132–142. doi:10.1007/s12640-009-9129-7.

Chiarugi, A., Carpenedo, R., Molina, M.T., Mattoli, L., Pellicciari, R., Moroni, F., 1995. Comparison of the neurochemical and behavioral effects resulting from the inhibition of kynurenine hydroxylase and/or kynureninase. J. Neurochem. 65 (3), 1176–1183. doi:10.1046/j.1471-4159.1995.65031176.x.

Colabroy, K.L., Zhai, H., Li, T., Ge, Y., Zhang, Y., Liu, A., et al., 2005. The mechanism of inactivation of 3-hydroxyanthranilate-3,4-dioxygenase by 4-chloro-3-hydroxyanthranilate. Biochemistry 44 (21), 7623–7631. doi:10.1021/bi0473455.

Colín-González, A.L., Maldonado, P.D., Santamaría, A., 2013. 3-Hydroxykynurenine: an intriguing molecule exerting dual actions in the central nervous system. Neurotoxicology 34, 189–204. doi:10.1016/j.neuro.2012.11.007.

Connor, T.J., Starr, N., O'Sullivan, J.B., Harkin, A., 2008. Induction of indolamine 2,3-dioxygenase and kynurenine 3-monooxygenase in rat brain following a systemic inflammatory challenge: a role for IFN-gamma? Neurosci. Lett. 441 (1), 29–34. doi:10.1016/j.neulet.2008.06.007.

Cryan, J.F., Dinan, T.G., 2012. Mind-altering microorganisms: the impact of the gut microbiota on brain and behaviour. Nat. Rev. Neurosci. 13 (10), 701–712. doi:10.1038/nrn3346.

Danysz, W., Fadda, E., Wroblewski, J.T., Costa, E., 1989. Kynurenate and 2-amino-5-phosphonovalerate interact with multiple binding sites of the N-methyl-D-aspartate-sensitive glutamate receptor domain. Neurosci. Lett. 96 (3), 340–344. doi:10.1016/0304-3940(89)90402-3.

de Carvalho, L.P., Bochet, P., Rossier, J., 1996. The endogenous agonist quinolinic acid and the non endogenous homoquinolinic acid discriminate between NMDAR2 receptor subunits. Neurochem. Int. 28 (4), 445–452. doi:10.1016/0197-0186(95)00091-7.

DiNatale, B.C., Murray, I.A., Schroeder, J.C., Flaveny, C.A., Lahoti, T.S., Laurenzana, E.M., et al., 2010. Kynurenic acid is a potent endogenous aryl hydrocarbon receptor ligand that synergistically induces interleukin-6 in the presence of inflammatory signaling. Toxicol. Sci. 115 (1), 89–97. doi:10.1093/toxsci/kfq024.

Durastanti, V., Lugaresi, A., Bramanti, P., Amato, M., Bellantonio, P., De Luca, G., et al., 2011. Neopterin production and tryptophan degradation during 24-months therapy with interferon beta-1a in multiple sclerosis patients. J. Transl. Med. 9, 42. doi:10.1186/1479-5876-9-42.

Dykens, J.A., Sullivan, S.G., Stern, A., 1987. Oxidative reactivity of the tryptophan metabolites 3-hydroxyanthranilate, cinnabarinate, quinolinate and picolinate. Biochem. Pharmacol. 36 (2), 211–217. doi:10.1016/0006-2952(87)90691-5.

Ellinger, A., Flamand, C., 1908. Über Synthetisch Gewonnenes Tryptophan und Einige Seiner Derivate. Biol. Chem. 55 (1), 8–24.

Espey, M.G., Chernyshev, O.N., Reinhard Jr., J.F., Namboodiri, M.A., Colton, C.A., 1997. Activated human microglia produce the excitotoxin quinolinic acid. Neuroreport 8 (2), 431–434. doi:10.1097/00001756-199701200-00011.

Fallarino, F., Grohmann, U., Vacca, C., Bianchi, R., Orabona, C., Spreca, A., et al., 2002. T cell apoptosis by tryptophan catabolism. Cell Death Differ. 9 (10), 1069–1077. doi:10.1038/sj.cdd.4401073.

Fazio, F., Lionetto, L., Curto, M., Iacovelli, L., Copeland, C.S., Neale, S.A., et al., 2017. Cinnabarinic acid and xanthurenic acid: two kynurenine metabolites that interact with metabotropic glutamate receptors. Neuropharmacology 112 (Pt B), 365–372. doi:10.1016/j.neuropharm.2016.06.020.

Fernstrom, J.D., Wurtman, R.J., 1971. Brain serotonin content: increase following ingestion of carbohydrate diet. Science 174 (4013), 1023–1025.

Fujigaki, H., Yamamoto, Y., Saito, K., 2017. L-Tryptophan-kynurenine pathway enzymes are therapeutic target for neuropsychiatric diseases: focus on cell type differences. Neuropharmacology 112 (Pt B), 264–274. doi:10.1016/j.neuropharm.2016.01.011.

Gál, E.M., Sherman, A.D., 1978. Synthesis and metabolism of L-kynurenine in rat brain. J. Neurochem. 30 (3), 607–613. doi:10.1111/j.1471-4159.1978.tb07815.x.

Food, J., Organisation, A., & Committee, W. H. O. A. H. E. (1973). Energy and protein requirements. FAO Nutr. Meet. Rep. Ser., WHO Technical Report Series, No. 522.

Gaubert, S., Bouchaut, M., Brumas, V., Berthon, G., 2000. Copper–ligand interactions and the physiological free radical processes. Part 3. Influence of histidine, salicylic acid and anthranilic acid on copper-driven Fenton chemistry in vitro. Free Radic. Res. 32 (5), 451–461. doi:10.1080/10715760000300451.

Gellért, L., Fuzik, J., Göblös, A., Sárközi, K., Marosi, M., Kis, Z., et al., 2011. Neuroprotection with a new kynurenic acid analog in the four-vessel occlusion model of ischemia. Eur. J. Pharmacol. 667 (1–3), 182–187. doi:10.1016/j.ejphar.2011.05.069.

Gellért, L., Varga, D., Ruszka, M., Toldi, J., Farkas, T., Szatmári, I., et al., 2012. Behavioural studies with a newly developed neuroprotective KYNA-amide. J. Neural Transm. (Vienna) 119 (2), 165–172. doi:10.1007/s00702-011-0692-8.

Goda, K., Kishimoto, R., Shimizu, S., Hamane, Y., Ueda, M., 1996. Quinolinic acid and active oxygens. Possible contribution of active Oxygens during cell death in the brain. Adv. Exp. Med. Biol. 398, 247–254.

Gostner, J.M., Becker, K., Ueberall, F., Fuchs, D., 2015. The good and bad of antioxidant foods: an immunological perspective. Food Chem. Toxicol. 80, 72–79.

Gostner, J.M., Geisler, S., Stonig, M., Mair, L., Sperner-Unterweger, B., Fuchs, D., 2020. Tryptophan metabolism and related pathways in psychoneuroimmunology: the impact of nutrition and lifestyle. Neuropsychobiology 79 (1), 89–99.

Gualdoni, G.A., Fuchs, D., Zlabinger, G.J., Gostner, J.M., 2016. Resveratrol intake enhances indoleamine-2, 3-dioxygenase activity in humans. Pharmacol. Rep. 68 (5), 1065–1068.

Guillemin, G.J., Cullen, K.M., Lim, C.K., Smythe, G.A., Garner, B., Kapoor, V., et al., 2007. Characterization of the kynurenine pathway in human neurons. J. Neurosci. 27 (47), 12884–12892. doi:10.1523/jneurosci.4101-07.2007.

Guillemin, G.J., Kerr, S.J., Brew, B.J., 2005. Involvement of quinolinic acid in AIDS dementia complex. Neurotox. Res. 7 (1–2), 103–123. doi:10.1007/bf03033781.

Hartai, Z., Klivenyi, P., Janaky, T., Penke, B., Dux, L., Vecsei, L., 2005. Kynurenine metabolism in multiple sclerosis. Acta Neurol. Scand. 112 (2), 93–96. doi:10.1111/j.1600-0404.2005.00442.x.

Heaton, R.K., Clifford, D.B., Franklin Jr., D.R., Woods, S.P., Ake, C., Vaida, F., et al., 2010. HIV-associated neurocognitive disorders persist in the era of potent antiretroviral therapy: CHARTER study. Neurology 75 (23), 2087–2096. doi:10.1212/WNL.0b013e318200d727.

Heyes, M.P., Chen, C.Y., Major, E.O., Saito, K., 1997. Different kynurenine pathway enzymes limit quinolinic acid formation by various human cell types. Biochem. J. 326 (Pt 2), 351–356. doi:10.1042/bj3260351, Pt 2.

Heyes, M.P., Swartz, K.J., Markey, S.P., Beal, M.F., 1991. Regional brain and cerebrospinal fluid quinolinic acid concentrations in Huntington's disease. Neurosci. Lett. 122 (2), 265–269. doi:10.1016/0304-3940(91)90874-s.

Hilmas, C., Pereira, E.F., Alkondon, M., Rassoulpour, A., Schwarcz, R., Albuquerque, E.X., 2001. The brain metabolite kynurenic acid inhibits alpha7 nicotinic receptor activity and increases non-alpha7 nicotinic receptor expression: physiopathological implications. J. Neurosci. 21 (19), 7463–7473. doi:10.1523/jneurosci.21-19-07463.2001.

Hopkins, F.G., Cole, S.W., 1901. A contribution to the chemistry of proteids: Part I. A preliminary study of a hitherto undescribed product of tryptic digestion. J. Physiol. 27 (4–5), 418.

Horvath, A., Leber, B., Schmerboeck, B., Tawdrous, M., Zettel, G., Hartl, A., et al., 2016. Randomised clinical trial: the effects of a multispecies probiotic vs. placebo on innate immune function, bacterial translocation and gut permeability in patients with cirrhosis. Aliment. Pharmacol. Ther. 44 (9), 926–935. doi:10.1111/apt.13788.

Hunt, C., Macedo, E.C.T., Suchting, R., de Dios, C., Cuellar Leal, V.A., Soares, J.C., et al., 2020. Effect of immune activation on the kynurenine pathway and depression symptoms

– a systematic review and meta-analysis. Neurosci. Biobehav. Rev. 118, 514–523. doi:10.1016/j.neubiorev.2020.08.010.
Jauch, D., Urbańska, E.M., Guidetti, P., Bird, E.D., Vonsattel, J.P., Whetsell Jr., W.O., et al., 1995. Dysfunction of brain kynurenic acid metabolism in Huntington's disease: focus on kynurenine aminotransferases. J. Neurol. Sci. 130 (1), 39–47. doi:10.1016/0022-510x(94)00280-2.
Jhamandas, K., Boegman, R.J., Beninger, R.J., Bialik, M., 1990. Quinolinate-induced cortical cholinergic damage: modulation by tryptophan metabolites. Brain Res. 529 (1–2), 185–191. doi:10.1016/0006-8993(90)90826-w.
Kessler, M., Terramani, T., Lynch, G., Baudry, M., 1989. A glycine site associated with N-methyl-D-aspartic acid receptors: characterization and identification of a new class of antagonists. J. Neurochem. 52 (4), 1319–1328. doi:10.1111/j.1471-4159.1989.tb01881.x.
Kitamura, M., Kasai, A., 2007. Cigarette smoke as a trigger for the dioxin receptor-mediated signaling pathway. Cancer. Lett. 252 (2), 184–194. doi:10.1016/j.canlet.2006.11.015.
Köhler, C.A., Maes, M., Slyepchenko, A., Berk, M., Solmi, M., Lanctôt, K.L., et al., 2016. The gut-brain axis, including the microbiome, leaky gut and bacterial translocation: mechanisms and pathophysiological role in Alzheimer's disease. Curr. Pharm. Des. 22 (40), 6152–6166. doi:10.2174/1381612822666160907093807.
Larkins, J.-A.M., Mosse, J., Chapman, B., 2011. Is the evolution of biochemistry texts decreasing fitness? A case study of pedagogical error in bioenergetics. Proceedings of The Australian Conference on Science and Mathematics Education.
Laurans, L., Venteclef, N., Haddad, Y., Chajadine, M., Alzaid, F., Metghalchi, S., et al., 2018. Genetic deficiency of indoleamine 2,3-dioxygenase promotes gut microbiota-mediated metabolic health. Nat. Med. 24 (8), 1113–1120. doi:10.1038/s41591-018-0060-4.
Leblhuber, F., Steiner, K., Schuetz, B., Fuchs, D., Gostner, J.M., 2018. Probiotic supplementation in patients with Alzheimer's dementia – an explorative intervention study. Curr. Alzheimer Res. 15 (12), 1106–1113. doi:10.2174/1389200219666180813144834.
Leipnitz, G., Schumacher, C., Dalcin, K.B., Scussiato, K., Solano, A., Funchal, C., et al., 2007. In vitro evidence for an antioxidant role of 3-hydroxykynurenine and 3-hydroxyanthranilic acid in the brain. Neurochem. Int. 50 (1), 83–94. doi:10.1016/j.neuint.2006.04.017.
Lewitt, P.A., Li, J., Lu, M., Beach, T.G., Adler, C.H., Guo, L., 2013. 3-hydroxykynurenine and other Parkinson's disease biomarkers discovered by metabolomic analysis. Mov. Disord. 28 (12), 1653–1660. doi:10.1002/mds.25555.
Liu, J.J., Movassat, J., Portha, B., 2019. Emerging role for kynurenines in metabolic pathologies. Curr. Opin. Clin. Nutr. Metab. Care 22 (1), 82–90. doi:10.1097/mco.0000000000000529.
Lowe, M.M., Mold, J.E., Kanwar, B., Huang, Y., Louie, A., Pollastri, M.P., et al., 2014. Identification of cinnabarinic acid as a novel endogenous aryl hydrocarbon receptor ligand that drives IL-22 production. PLoS ONE 9 (2), e87877. doi:10.1371/journal.pone.0087877.
Lugo-Huitrón, R., Blanco-Ayala, T., Ugalde-Muñiz, P., Carrillo-Mora, P., Pedraza-Chaverrí, J., Silva-Adaya, D., et al., 2011. On the antioxidant properties of kynurenic acid: free radical scavenging activity and inhibition of oxidative stress. Neurotoxicol. Teratol. 33 (5), 538–547. doi:10.1016/j.ntt.2011.07.002.
Mándi, Y., Vécsei, L., 2012. The kynurenine system and immunoregulation. J. Neural. Transm. (Vienna) 119 (2), 197–209. doi:10.1007/s00702-011-0681-y.
Marosi, M., Nagy, D., Farkas, T., Kis, Z., Rózsa, E., Robotka, H., et al., 2010. A novel kynurenic acid analogue: a comparison with kynurenic acid. An in vitro electrophysiological study. J. Neural. Transm. (Vienna) 117 (2), 183–188. doi:10.1007/s00702-009-0346-2.
Miranda, A.F., Sutton, M.A., Beninger, R.J., Jhamandas, K., Boegman, R.J., 1999. Quinolinic acid lesion of the nigrostriatal pathway: effect on turning behaviour and protection by elevation of endogenous kynurenic acid in Rattus norvegicus. Neurosci. Lett. 262 (2), 81–84. doi:10.1016/s0304-3940(99)00036-1.

Mizdrak, J., Hains, P.G., Truscott, R.J., Jamie, J.F., Davies, M.J., 2008. Tryptophan-derived ultraviolet filter compounds covalently bound to lens proteins are photosensitizers of oxidative damage. Free Radic. Biol. Med. 44 (6), 1108–1119. doi:10.1016/j.freeradbiomed.2007.12.003.

Monaco, F., Fumero, S., Mondino, A., Mutani, R., 1979. Plasma and cerebrospinal fluid tryptophan in multiple sclerosis and degenerative diseases. J. Neurol. Neurosurg Psychiatry 42 (7), 640–641. doi:10.1136/jnnp.42.7.640.

Mor, A., Tankiewicz-Kwedlo, A., Krupa, A., Pawlak, D., 2021. Role of kynurenine pathway in oxidative stress during neurodegenerative disorders. Cells 10 (7), 1603. doi:10.3390/cells10071603.

Morita, T., Saito, K., Takemura, M., Maekawa, N., Fujigaki, S., Fujii, H., et al., 2001. 3-Hydroxyanthranilic acid, an L-tryptophan metabolite, induces apoptosis in monocyte-derived cells stimulated by interferon-gamma. Ann. Clin. Biochem. 38 (Pt 3), 242–251. doi:10.1258/0004563011900461.

Moroni, F., Russi, P., Lombardi, G., Beni, M., Carlà, V., 1988. Presence of kynurenic acid in the mammalian brain. J. Neurochem. 51 (1), 177–180. doi:10.1111/j.1471-4159.1988.tb04852.x.

Murakami, K., Haneda, M., Qiao, S., Naruse, M., Yoshino, M., 2007. Prooxidant action of rosmarinic acid: transition metal-dependent generation of reactive oxygen species. Toxicol. In Vitro 21 (4), 613–617. doi:10.1016/j.tiv.2006.12.005.

Murr, C., Winklhofer-Roob, B.M., Schroecksnadel, K., Maritschnegg, M., Mangge, H., Böhm, B.O., et al., 2009. Inverse association between serum concentrations of neopterin and antioxidants in patients with and without angiographic coronary artery disease. Atherosclerosis 202 (2), 543–549.

Murray, I.A., Patterson, A.D., Perdew, G.H., 2014. Aryl hydrocarbon receptor ligands in cancer: friend and foe. Nat. Rev. Cancer 14 (12), 801–814. doi:10.1038/nrc3846.

Myint, A.M., Kim, Y.K., 2014. Network beyond IDO in psychiatric disorders: revisiting neurodegeneration hypothesis. Prog. Neuropsychopharmacol. Biol. Psychiatry 48, 304–313. doi:10.1016/j.pnpbp.2013.08.008.

Nakagami, Y., Saito, H., Katsuki, H., 1996. 3-Hydroxykynurenine toxicity on the rat striatum in vivo. Jpn. J. Pharmacol. 71 (2), 183–186. doi:10.1254/jjp.71.183.

Ogawa, T., Matson, W.R., Beal, M.F., Myers, R.H., Bird, E.D., Milbury, P., et al., 1992. Kynurenine pathway abnormalities in Parkinson's disease. Neurology 42 (9), 1702–1706. doi:10.1212/wnl.42.9.1702.

Ott, M., Demisch, L., Engelhardt, W., Fischer, P.A., 1993. Interleukin-2, soluble interleukin-2-receptor, neopterin, L-tryptophan and beta 2-microglobulin levels in CSF and serum of patients with relapsing-remitting or chronic-progressive multiple sclerosis. J. Neurol. 241 (2), 108–114. doi:10.1007/bf00869773.

Pearson, S.J., Reynolds, G.P., 1992. Increased brain concentrations of a neurotoxin, 3-hydroxykynurenine, in Huntington's disease. Neurosci. Lett. 144 (1–2), 199–201. doi:10.1016/0304-3940(92)90749-w.

Perkins, M.N., Stone, T.W., 1982. An iontophoretic investigation of the actions of convulsant kynurenines and their interaction with the endogenous excitant quinolinic acid. Brain Res. 247 (1), 184–187. doi:10.1016/0006-8993(82)91048-4.

Perkins, M.N., Stone, T.W., 1985. Actions of kynurenic acid and quinolinic acid in the rat hippocampus in vivo. Exp. Neurol. 88 (3), 570–579. doi:10.1016/0014-4886(85)90072-x.

Pfefferkorn, E., 1984. Interferon gamma blocks the growth of Toxoplasma gondii in human fibroblasts by inducing the host cells to degrade tryptophan. Proc. Natl. Acad. Sci. 81 (3), 908–912.

Rahman, A., Ting, K., Cullen, K.M., Braidy, N., Brew, B.J., Guillemin, G.J., 2009. The excitotoxin quinolinic acid induces tau phosphorylation in human neurons. PLoS ONE 4 (7), e6344. doi:10.1371/journal.pone.0006344.

Reilly, J., McTavish, S., Young, A., 1997. Rapid depletion of plasma tryptophan: a review of studies and experimental methodology. J. Psychopharmacol. 11 (4), 381–392.
Rejdak, K., Bartosik-Psujek, H., Dobosz, B., Kocki, T., Grieb, P., Giovannoni, G., et al., 2002. Decreased level of kynurenic acid in cerebrospinal fluid of relapsing-onset multiple sclerosis patients. Neurosci. Lett. 331 (1), 63–65. doi:10.1016/s0304-3940(02)00710-3.
Rejdak, K., Petzold, A., Kocki, T., Kurzepa, J., Grieb, P., Turski, W.A., et al., 2007. Astrocytic activation in relation to inflammatory markers during clinical exacerbation of relapsing-remitting multiple sclerosis. J. Neural. Transm. (Vienna) 114 (8), 1011–1015. doi:10.1007/s00702-007-0667-y.
Reyes Ocampo, J., Lugo Huitrón, R., González-Esquivel, D., Ugalde-Muñiz, P., Jiménez-Anguiano, A., Pineda, B., et al., 2014. Kynurenines with neuroactive and redox properties: relevance to aging and brain diseases. Oxid. Med. Cell. Longev. 2014, 646909. doi:10.1155/2014/646909.
Rios, C., Santamaria, A., 1991. Quinolinic acid is a potent lipid peroxidant in rat brain homogenates. Neurochem. Res. 16 (10), 1139–1143. doi:10.1007/bf00966592.
Rózsa, E., Robotka, H., Vécsei, L., Toldi, J., 2008. The Janus-face kynurenic acid. J. Neural. Transm. (Vienna) 115 (8), 1087–1091. doi:10.1007/s00702-008-0052-5.
Samadi, P., Grégoire, L., Rassoulpour, A., Guidetti, P., Izzo, E., Schwarcz, R., et al., 2005. Effect of kynurenine 3-hydroxylase inhibition on the dyskinetic and antiparkinsonian responses to levodopa in Parkinsonian monkeys. Mov. Disord. 20 (7), 792–802. doi:10.1002/mds.20596.
Sas, K., Robotka, H., Rózsa, E., Agoston, M., Szénási, G., Gigler, G., et al., 2008. Kynurenine diminishes the ischemia-induced histological and electrophysiological deficits in the rat hippocampus. Neurobiol. Dis. 32 (2), 302–308. doi:10.1016/j.nbd.2008.07.013.
Schwarz, M.J., Guillemin, G.J., Teipel, S.J., Buerger, K., Hampel, H., 2013. Increased 3-hydroxykynurenine serum concentrations differentiate Alzheimer's disease patients from controls. Eur. Arch. Psychiatry Clin. Neurosci. 263 (4), 345–352. doi:10.1007/s00406-012-0384-x.
Sedlmayr, P., 2007. Indoleamine 2,3-dioxygenase in materno-fetal interaction. Curr. Drug Metab. 8 (3), 205–208. doi:10.2174/138920007780362491.
Sharaf El Din, U.A.A., Salem, M.M., Abdulazim, D.O., 2017. Uric acid in the pathogenesis of metabolic, renal, and cardiovascular diseases: a review. J. Adv. Res. 8 (5), 537–548. doi:10.1016/j.jare.2016.11.004.
Sharon, G., Sampson, T.R., Geschwind, D.H., Mazmanian, S.K., 2016. The central nervous system and the gut microbiome. Cell 167 (4), 915–932. doi:10.1016/j.cell.2016.10.027.
Silva-Adaya, D., Pérez-De La Cruz, V., Villeda-Hernández, J., Carrillo-Mora, P., González-Herrera, I.G., García, E., et al., 2011. Protective effect of L-kynurenine and probenecid on 6-hydroxydopamine-induced striatal toxicity in rats: implications of modulating kynurenate as a protective strategy. Neurotoxicol. Teratol. 33 (2), 303–312. doi:10.1016/j.ntt.2010.10.002.
Strasser, B., Geiger, D., Schauer, M., Gatterer, H., Burtscher, M., Fuchs, D., 2016. Effects of exhaustive aerobic exercise on tryptophan-kynurenine metabolism in trained athletes. PLoS ONE 11 (4), e0153617.
Strasser, B., Gostner, J.M., Fuchs, D., 2016. Mood, food, and cognition: role of tryptophan and serotonin. Curr. Opin. Clin. Nutr. Metab. Care 19 (1), 55–61.
Tavares, R.G., Tasca, C.I., Santos, C.E., Alves, L.B., Porciúncula, L.O., Emanuelli, T., et al., 2002. Quinolinic acid stimulates synaptosomal glutamate release and inhibits glutamate uptake into astrocytes. Neurochem. Int. 40 (7), 621–627. doi:10.1016/s0197-0186(01)00133-4.
Tiszlavicz, Z., Németh, B., Fülöp, F., Vécsei, L., Tápai, K., Ocsovszky, I., et al., 2011. Different inhibitory effects of kynurenic acid and a novel kynurenic acid analogue on tumour necrosis factor-α (TNF-α) production by mononuclear cells, HMGB1 production

by monocytes and HNP1-3 secretion by neutrophils. Naunyn. Schmiedebergs Arch. Pharmacol. 383 (5), 447–455. doi:10.1007/s00210-011-0605-2.
Walker, A.K., Wing, E.E., Banks, W.A., Dantzer, R., 2019. Leucine competes with kynurenine for blood-to-brain transport and prevents lipopolysaccharide-induced depression-like behavior in mice. Mol. Psychiatry 24 (10), 1523–1532.
Walsh, H.A., O'Shea, K.C., Botting, N.P., 2003. Comparative inhibition by substrate analogues 3-methoxy- and 3-hydroxydesaminokynurenine and an improved 3 step purification of recombinant human kynureninase. BMC Biochem. 4, 13. doi:10.1186/1471-2091-4-13.
Widner, B., Leblhuber, F., Fuchs, D., 2002. Increased neopterin production and tryptophan degradation in advanced Parkinson's disease. J. Neural. Transm. (Vienna) 109 (2), 181–189. doi:10.1007/s007020200014.
Widner, B., Leblhuber, F., Walli, J., Tilz, G.P., Demel, U., Fuchs, D., 2000. Tryptophan degradation and immune activation in Alzheimer's disease. J. Neural. Transm. (Vienna) 107 (3), 343–353. doi:10.1007/s007020050029.
Wirleitner, B., Schroecksnadel, K., Winkler, C., Schennach, H., Fuchs, D., 2005. Resveratrol suppresses interferon-γ-induced biochemical pathways in human peripheral blood mononuclear cells in vitro. Immunol. Lett. 100 (2), 159–163.
Zádori, D., Szalárdy, L., Toldi, J., Fülöp, F., Klivényi, P., Vécsei, L., 2013. Some molecular mechanisms of dopaminergic and glutamatergic dysfunctioning in Parkinson's disease. J. Neural. Transm. (Vienna) 120 (4), 673–681. doi:10.1007/s00702-012-0930-8.
Zvetkova, E., Wirleitner, B., Tram, N.T., Schennach, H., Fuchs, D., 2001. Aqueous extracts of Crinum latifolium (L.) and Camellia sinensis show immunomodulatory properties in human peripheral blood mononuclear cells. Int. Immunopharmacol. 1 (12), 2143–2150. doi:10.1016/s1567-5769(01)00140-0.
Zwilling, D., Huang, S.Y., Sathyasaikumar, K.V., Notarangelo, F.M., Guidetti, P., Wu, H.Q., et al., 2011. Kynurenine 3-monooxygenase inhibition in blood ameliorates neurodegeneration. Cell 145 (6), 863–874. doi:10.1016/j.cell.2011.05.020.

CHAPTER SIX

Neuroprotective phytonutrients: Mechanisms of action and evidence in neurotoxic and degenerative conditions

Benjamin Oritsemuelebi[a], Benneth Ben-Azu[a,b,*,#], Ademola Adetokunbo Oyagbemi[c,d], Ifeoluwa Awogbindin[b,e], Emmanuel O. Chidebe[a], Jackson E. Onuelu[a], Prosper Iwhiwhu[a], Josephine Chigbolum Ohuche[e], Daniel T. Esuku[a], Marvellous A. Oghorodi[f]

[a] DELSU Joint Canada-Israel Neuroscience and Biopsychiatry Laboratory, Department of Pharmacology, Faculty of Allied Health Sciences, College of Health Sciences, Delta State University, Abraka, Delta State, Nigeria
[b] School of Medical Sciences, Faculty of Health, University of Victoria, Victoria, British Columbia, Canada
[c] Department of Applied Microbiology, Nnamdi Azikiwe University, Awka, Nigeria
[d] Department of Biochemistry, College of Medicine, University of Ibadan, Ibadan, Nigeria
[e] Department of Veterinary Physiology and Biochemistry, Faculty of Veterinary Medicine, University of Ibadan, Ibadan, Nigeria
[f] State University of Medical and Applied Sciences, Igbo-eno Enugu, Enugu State, Nigeria
*Corresponding author. e-mail addresses: pharmben4ever@yahoo.com, bben-azu@delsu.edu.ng.

Contents

[#] Senior author.

Advances in Neurotoxicology, Volume 16
ISSN 2468-7480
https://doi.org/10.1016/bs.ant.2026.03.002

Abstract

The rise in neurotoxicity and its link to neurodevelopmental and neurodegenerative diseases has been driven by increased exposure to toxic chemicals, nutritional deficiencies, chronic stress, aging, inflammation, environmental stressors, and psychosocial factors. Over 46.8 million people worldwide suffer from disorders like schizophrenia, autism, ADHD, Parkinson's, and Alzheimer's. Disease-modifying natural products remain an unmet alternative need in the field of neurotoxicology of neurodevelopmental and neurodegenerative diseases. Such treatments can be operationally defined as interventions that slow down, prevent, or abrogate the onset of these diseases and their clinical progression to an advanced stage. Treatment with phytochemical agents may achieve this outcome by either inhibiting primary neurodegenerative events ("neuroprotection") or boosting compensatory and regenerative mechanisms in the brain ("neuro-restoration)". These bioactive compounds include plant phytochemicals and those from animals, bacteria, fungi, and algae, which have been used traditionally and are now recognized for their impact on brain health, especially during aging and neurological diseases. Recent studies confirm their biological activities and therapeutic potential by inhibiting reactive oxygen species, boosting antioxidant levels, reducing neuroinflammation, and modulating brain neurochemicals. They also modulate nervous system cells, including microglia, the resident immune cells of the brain. Therefore, these phytochemicals hold promise for delaying, preventing, or treating such diseases and are valuable for drug discovery. Hence, this review examines the pharmacological roles of naturally occurring phytochemicals such as taurine, morin, curcumin, quercetin, turmeric, ginseng, naringin, and naringenin in animal models, aiming to understand their neuroprotective mechanisms and potential in managing neurodevelopmental and neurodegenerative disorders.

Keywords: Antioxidants, Curcumin, Ginseng, Morin, Naringenin, Naringin, Natural products, Neurodegeneration, Neuroinflammation, Neuroprotection, Neurotoxicity, Phytochemicals, Quercetin, Taurine

1. Introduction

The global increase in the prevalence of neurotoxicity has become a significant public health concern, particularly in relation to its role in the onset of neurodevelopmental and neurodegenerative diseases (Nabi and Tabassum, 2022), both of which cause significant increases in the number of people living with disability (Arab and Mostafalou, 2022; Rezaei et al., 2024). Neurotoxicity refers to the adverse effects of certain substances on the central nervous system (CNS), leading to dysfunction and disease

(Asuku et al., 2024). As the world grapples with rising exposure to neurotoxic agents, understanding the implications of these factors on brain health is crucial (Kumar et al., 2023; Fuloria et al., 2025). Neurodevelopmental disorders, including schizophrenia, autism spectrum disorders (ASD), and attention deficit hyperactivity disorder (ADHD), are characterized by atypical brain development and functional disorders (Jeste, 2015; Park et al., 2018; Ben-Azu et al., 2019a). These conditions, which have underlying utero challenges, manifest in young adults and can have lasting effects on cognitive, emotional, and social functioning (Brewer-Smyth, 2022; Komar-Fletcher et al., 2023; Omeiza et al., 2023). Studies suggest that environmental risk factors, especially those encountered during key periods of brain development, play a crucial role in the origins of these disorders (Marangoni et al., 2016; Comer et al., 2020; Okhue et al., 2024). For example, prenatal exposure to environmental neurotoxicants such as heavy metals, pesticides, and other pollutants, as well as infectious agents, has been associated with an increased risk of neurodevelopmental disorders (Awogbindin et al., 2021; Oritsemuelebi et al., 2021). Additionally, the interaction between genetic predispositions and environmental exposures further complicates health outcomes, indicating that some individuals may be more susceptible to the effects of neurotoxins (Asiwe and Oritsemuelebi, 2024).

In parallel, neurodegenerative diseases such as Parkinson's disease (PD) (Tolosa et al., 2021; Oyovwi et al., 2025a), Alzheimer's disease (AD) (Tahami et al., 2022), Huntington's disease (HD) (Shaw et al., 2022), multiple sclerosis (MS), and amyotrophic lateral sclerosis (ALS) present an increasing challenge among aging populations (Wareham et al., 2022). These disorders are marked by the progressive degeneration of neuronal structures, leading to cognitive decline, motor problems, depression, and ultimately, loss of independence (Rauf et al., 2022). Epidemiological studies have increasingly pointed to environmental factors as important contributors to the development of these diseases (Münzel et al., 2022; Deshpande et al., 2023). For example, exposure to neurotoxicants such as lead, mercury, cadmium, arsenic, and various pesticides and herbicides has been linked to an increased risk of developing neurodegenerative diseases (Oritsemuelebi et al., 2021). The mechanisms by which these toxins exert their effects often involve the induction of inflammation and oxidative stress, which is a state of imbalance in free radical production and decreased antioxidant activity (Sule et al., 2022) in the CNS (Renu et al., 2021), depletion of neurotrophic factors (Martinez-Jaramillo et al., 2024), and disruption of cellular homeostasis (Wallace and Djordjevic, 2020). These processes can lead to neurodegeneration and accelerated aging.

Mitochondrial and microglial dysfunction play vital roles in the pathology of neurodevelopmental and neurodegenerative disorders. Mitochondria, as the energy sources of cells, are crucial for maintaining cellular metabolism and homeostasis. When impaired, they cause increased oxidative stress, inflammation, and neuronal cell death. Meanwhile, microglia, the immune cells of the CNS, become dysfunctional due to environmental stressors, adopting a pro-inflammatory state that worsens neuroinflammation (Ben-Azu et al., 2024d). The interaction between mitochondrial impairment and microglial overactivity fosters a cycle of neurodegeneration, ultimately affecting cognitive and emotional health.

The significance of environmental risk factors in contributing to the development of both neurodevelopmental and neurodegenerative diseases cannot be underestimated. Elements such as air pollution, exposure to industrial chemicals, and dietary influences have been linked to the onset and progression of these conditions (Liu et al., 2021; Xu et al., 2022). For instance, early-life exposure to neurotoxic agents may have different effects compared to exposure later in life, highlighting the importance of timing in the development of neurotoxic effects (Dórea, 2021). Remarkably, early-life dysregulation of nutritional heavy metals or exposure to harmful heavy metals has been associated with mechanisms underlying schizophrenia, ASD, and ADHD, common neurodevelopmental disorders characterized by persistent patterns of hyperactivity, inattention, impulsivity, and social interaction difficulties), as well as the acceleration of aging processes that can lead to the onset of neurodegenerative diseases (Chen et al., 2016; Althobaiti, 2025).

Natural nutritional agents, especially those derived from plants, have gained attention for their potential neuroprotective properties, which are vital for maintaining brain health and reducing the risks of neurodegenerative diseases (Rahman et al., 2021; Mursal et al., 2024). For example, curcumin, the active compound in turmeric, is well-studied for its anti-inflammatory (Liu et al., 2021; Elhawary et al., 2024) and antioxidant effects), helping to prevent amyloid-beta (Aβ) plaque formation associated with AD in animal models (Atoki et al., 2023). Other compounds such as taurine, morin, naringin, naringenin, and diosmin also demonstrate neuroprotective effects by reducing oxidative stress, modulating neurotransmitter systems, and promoting neurogenesis (Ben-Azu et al., 2024e). These natural agents can serve as useful dietary additions or adjunct therapies, potentially decreasing the risk of cognitive decline and neurodegenerative conditions (Martínez-Iglesias et al., 2023; Onyekweli et al., 2025). Their tolerability, safety, and diverse mechanisms of action underscore the importance of dietary choices in supporting cognitive health. As studies continue to examine their efficacy, incorporating these compounds into daily diets or as supplements may provide a proactive way to support brain function and enhance quality of life during healthy aging (Vauzour et al., 2017; Puri et al., 2023). The mechanisms of action of these natural agents are diverse and intricate. For example, curcumin has been shown to prevent the formation of amyloid plaques, a hallmark of AD, while also assisting in the removal of these toxic deposits (Ege, 2021). Similarly, taurine has been linked to cardio- and neuroprotection by helping to regulate calcium levels and reduce excitotoxicity (Ramírez-Guerrero et al., 2022; Asiwe et al., 2024a), and influence the release of neurochemicals (Ben-Azu et al., 2023). Understanding these mechanisms is essential for creating effective treatments that utilize the protective effects of these compounds against neurotoxicity-related neurodevelopmental and neurodegenerative disorders.

A growing body of preclinical and clinical studies supports exploring natural products for neuroprotection. These studies offer valuable insights into the effectiveness and safety of various natural agents in managing neurodegenerative diseases. However, challenges persist, including differences in study designs, dosages, CNS penetration, and the complexity of human biology. Additionally, translating findings from animal models to humans remains difficult, requiring rigorous clinical trials to confirm the therapeutic potential of these natural products. By understanding environmental risk factors and investigating natural nutritional agents, we aim to develop therapeutic pathways for innovative prevention and treatment strategies to protect brain health throughout life. Ongoing research into how these agents work, combined with strong clinical trials, will be crucial in overcoming issues related to neurotoxicity and associated neurodevelopmental and neurodegenerative disorders.

2. Neuropathological mechanisms in neurodevelopmental and neurodegenerative diseases

Neurodegenerative diseases pose significant challenges in modern medicine, with complex, multifactorial pathologies that demand a deep understanding of their underlying mechanisms. Key among these are oxidative stress, the depletion of polyunsaturated fatty acids (PUFAs), mitochondrial dysregulation, and microglial dysfunction. These interconnected processes are crucial in the onset and progression of neurological conditions, making their study vital for developing effective interventions and preventive strategies (Ben-Azu et al., 2024a, 2024b, 2024c, 2024d, 2024e). Oxidative stress is a fundamental factor in the development of both neurodevelopmental and neurodegenerative diseases. It results from an imbalance between reactive oxygen species (ROS) production and the brain's capacity to detoxify these harmful compounds through antioxidant systems (Espinós et al., 2020; Curpan et al., 2021; Asiwe and Oritsemuelebi, 2024b). This imbalance causes cellular damage, which is especially harmful to the brain, an organ with naturally low antioxidant capacity and a neuronal network vulnerable to disturbance. The depletion of PUFAs, particularly n-3 PUFAs, worsens oxidative stress by weakening the brain's antioxidant defenses (Shchepinov, 2020). These essential fatty acids are crucial for maintaining cellular integrity and function. Their deficiency can increase ROS production, leading to oxidative damage to proteins, lipids, and nucleic acids hallmarks of neurodegenerative diseases such as AD, PD, MS, and ALS (Shchepinov, 2020; Buccellato et al., 2021). Moreover, oxidative stress is not only a marker of these diseases but also a contributor to their progression, promoting neurotoxicity and cognitive decline (Buccellato et al., 2021). During key periods of brain development, the impacts of oxidative stress are even more significant, potentially causing long-term impairments in brain function and health (Kandlur, 2020; Leyane et al., 2022). This highlights the importance of maintaining adequate dietary intake of PUFAs to support the brain's antioxidant capacity and reduce the risk of neurodevelopmental and neurodegenerative diseases (Leyane et al., 2022).

Mitochondrial dysfunction is a key feature in the pathology of neurodevelopmental and neurodegenerative diseases (Kandlur, 2020). Mitochondria, often referred to as the powerhouses of the cell, are crucial for the brain's bioenergetics, including the production of adenosine triphosphate (ATP) and the maintenance of cellular homeostasis. Damage to mitochondrial structures and functions can increase ROS production, alter brain bioenergetics (Kandlur, 2020), trigger CNS inflammation (Clemente-Suárez et al., 2023), and lead to apoptosis, which is programmed cell death (Ben-Azu et al., 2024e). This further worsens oxidative stress and contributes to neuronal cell death (Sharma et al., 2021). Mitochondrial dysfunction is often associated with disrupted metabolic processes and energy deficits in the brain, resulting in cognitive decline and behavioral impairments (Sharma et al., 2021; Clemente-Suárez et al., 2023). The interaction between oxidative stress and mitochondrial dysfunction creates a vicious cycle in which increased oxidative damage causes more mitochondrial harm, perpetuating neuronal decline (Clemente-Suárez et al., 2023). This cycle is particularly troubling in neurodevelopmental disorders, where early-life mitochondrial issues can have long-lasting and significant effects on brain development and functionn (Ginhoux et al., 2010).

Microglia, the resident immune cells of the CNS, also play a crucial role in the pathology of neurodevelopmental and neurodegenerative diseases (Lukens and Eyo, 2022). Microglia originate from primitive macrophage progenitors in the yolk sac and migrate into the brain during early embryonic day (ED) 9.5 in mice and approximately 4.5 weeks into fetal development in humans, which aligns with the first trimester (Ginhoux et al., 2010; Menassa et al., 2022). At ED9.5, A2 macrophages enter the mouse brain via the pial surface, transforming into microglia as they navigate the brain's circulatory system. They penetrate the cerebral cortex and respond to signals from the CNS environment (Ginhoux et al., 2010; Menassa et al., 2022). Microglia act as homeostatic immune surveillance cells from early development through aging. In a healthy brain, microglia are involved in maintaining homeostasis (Mendes and Majewska, 2021), supporting neuronal development and health, and responding to injuries and infections through a series of sensomes, which are signaling proteins, genes, or transcripts used to communicate with the CNS microenvironments (Ronaldson and Davis, 2020; Awogbindin et al., 2022). However, when microglia become dysfunctional due to environmental toxicants (Nab and Tabassum, 2022), they can contribute to chronic neuroinflammation, which is involved in the progression of various neurodevelopmental and neurodegenerative diseases (Lukens and Eyo, 2022). Overactive or pro-inflammatory microglia, such as disease-associated microglia (DAM) or senescent microglia (Ng et al., 2023), can worsen neuronal damage, including damage to oligodendrocytes, disrupt normal synaptic function, and speed up brain aging, leading to cognitive and behavioral impairments (Saxena et al., 2021; Ng et al., 2023). This dysregulation can be problematic in neurodevelopmental disorders, where excessive inflammation may interfere with critical developmental stages or contribute to unhealthy brain aging with cognitive decline (Zhang et al., 2024). The long-term effects on cognitive and emotional health emphasize the need for a better understading of microglia dynamics in neurodevelopment and neuronal aging. Several environmental factors can trigger early oxidative stress, mitochondrial dysfunction, and microglial problems, further complicating the pathology of neurodevelopmental and neurodegenerative diseases (Olude et al., 2022; Sadeghdoust and Kaushik, 2024). These factors include pollution, stress, alcohol consumption

(Zhang et al., 2024), infections (Jiang et al., 2024), aging (Bondy, 2024), environmental enrichments (Zhang et al., 2024), and diets, notably food additives such as monosodium glutamate (MSG) (Omogbiya et al., 2020). Each of these environmental stressors can trigger a cascade of pathological processes that impair brain health and function, increasing the risk of developing neurodevelopmental and neurodegenerative diseases (Peters et al., 2021). For example, exposure to environmental pollutants like lead (Pb), mercury (Hg), arsenic (As), cadmium (Cd), organophosphate pesticides herbicides such as paraquat and glyphosate, particulate matter (PM2.5, PM10), nitrogen dioxide (NO_2), ozone (O_3), polychlorinated biphenyls (PCBs), bisphenol A (BPA), phthalates, tobacco smoke, and prenatal exposure to infectious agents has been linked to increased oxidative stress and inflammation, which are associated with cognitive deficits and emotional disturbances (Ventriglio et al., 2021). Similarly, MSG has been linked to excitotoxicity and oxidative damage, contributing to neurobehavioral impairments (Asejeje et al., 2024; Omogbiya et al., 2020). The neurotoxic effects of these substances can present as heightened anxiety, depression, and cognitive decline, especially among vulnerable groups like children and the elderly (Adkins and Brunst, 2021). The complex interaction between oxidative stress, mitochondrial dysfunction, and microglial dysregulation forms a tangled web of pathological processes that greatly affect brain health and behavior. As illustrated in Fig. 1 below, understanding these neuropathological mechanisms is essential for developing targeted treatments and preventative measures. Addressing environmental risk factors, ensuring adequate intake of essential phytonutrients, and exploring therapeutic strategies to lessen the effects of oxidative stress, support mitochondrial function, and maintain microglial sensing and neuron stability are all crucial steps in improving outcomes for those affected by neurodevelopmental and neurodegenerative diseases (Heland et al., 2022).

3. Nutritional agents with capacity for neuroprotection

3.1 Taurine

Taurine (2-aminoethanesulfonic acid), named after the Latin word taurus (meaning bull) because of its initial isolation from ox bile (Capsi et al., 2018), is the most abundant free amino acid in the brain. It has the molecular formula $C_2H_7NO_3S$ and a molecular weight of 125.15 g/mol (Ripps and Shen, 2012). Unlike typical amino acids, taurine contains a sulfonic acid

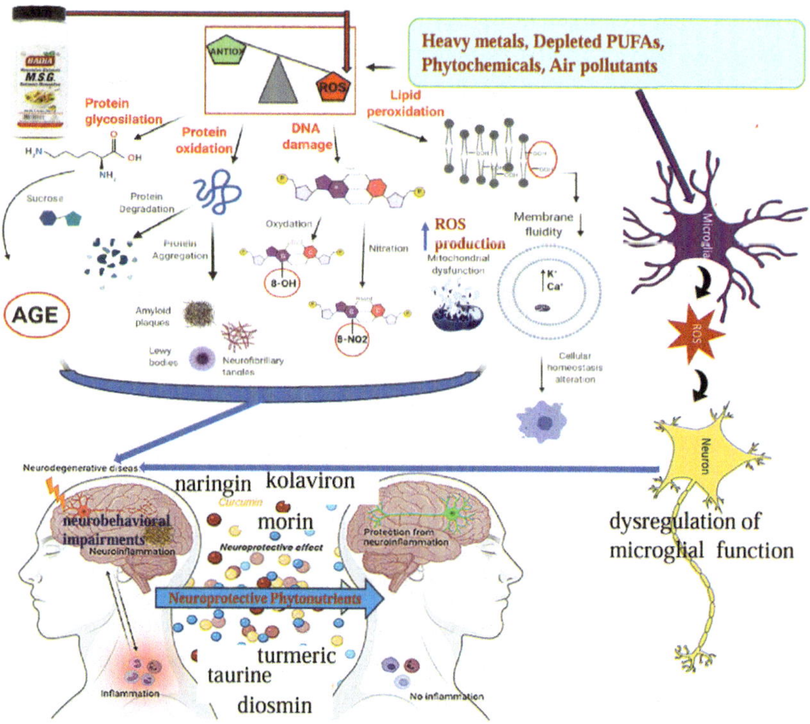

Fig. 1 Neuropathological Processes in Neurodegenerative Diseases: Environmental pollutants such as heavy metals (e.g., lead, mercury), air pollutants, and depleted polyunsaturated fatty acids (PUFAs) increase the production of reactive oxygen species (ROS), unstable molecules that cause oxidative stress. This stress leads to protein oxidation, DNA damage, and lipid peroxidation, resulting in cellular dysfunction. Advanced glycation end-products (AGEs), formed through protein glycosylation, promote the aggregation of misfolded proteins into amyloid plaques, Lewy bodies, and neurofibrillary tangles, which are hallmarks of neurodegenerative diseases. Mitochondrial dysfunction and microglial dysregulation that follow worsen neuronal damage. However, neuroprotective phytonutrients like curcumin, naringin, morin, kolaviron, taurine, and diosmin help counteract these effects by reducing inflammation and oxidative stress, thereby preserving neuronal integrity.

group rather than a carboxylic acid group, which confers distinctive physical and chemical properties (Khodapsand et al., 2015). It is a monobasic acid with a low pKa of about 1.5 and a pKb of 8.82, making it less basic than β-alanine, γ-aminobutyric acid (GABA), or glycine. Taurine also has relatively low solubility in water (10.48 g/100 mL at 25°C). Taurine's unique structure, particularly the location of its sulfonic acid group on the β-carbon, limits its passive diffusion across membranes. Its cyclic conformation, stabilized by intramolecular hydrogen bonding, further restricts its permeability across the blood-brain barrier (BBB). Nonetheless, taurine is abundant in protein-rich foods such as meat, milk, seafood, and energy drinks, and it plays a significant role in cellular function and metabolic processes (Froger et al., 2014; Piccioni et al., 2021; Obukohwo et al., 2024).

Taurine maintains a high transcellular gradient, with higher intracellular than extracellular concentrations. During neuronal depolarization, as shown in Fig. 2 below, taurine is released through volume-sensitive organic anion channels (VSOACs) and transport systems such as TauT1 and TauT2. These transporters operate via reverse transport and calcium-independent mechanisms (Kilb and Fukuda, 2017). Taurine interacts with N-methyl-D-aspartic acid (NMDA) receptors and may activate metabotropic taurine receptors (mTAUR), which are coupled to inhibitory G-proteins. This interaction leads to suppression of phospholipase C (PLC) activity and a reduction in inositol trisphosphate (IP_3) levels. Consequently, calcium (Ca^{2+}) release from intracellular stores is inhibited (Leon et al., 2009; Traynalis et al., 2010; Wu and Prentice, 2010). Additionally, guanosine-5′-triphosphate (GTP) inhibits the taurine receptor in a dose-dependent manner, suggesting the existence of an mTAUR linked to inhibitory G-proteins in the brain (Wu and Prentice, 2010). Taurine also inhibits protein kinase C (PKC), preventing phosphorylation and activation of voltage-gated calcium channels (VGCCs), particularly L-, N-, and P/Q-type channels. This results in reduced calcium influx, contributing to its neuroprotective effects. Taurine provides additional neuroprotection by reducing excitotoxicity. It prevents glutamate-induced damage by modulating postsynaptic responses through the activation of GABA receptors and glycine receptors (GlyRs), as well as by inhibiting chloride ion influx (Paula-Lima et al., 2005). In experimental models of Aβ-induced neurotoxicity, picrotoxin, a $GABA_A$ antagonist, was shown to block taurine's protective effect. This confirms its dependence on GABAergic mechanisms.

Taurine is known to directly affect the NMDA intracellular signaling by an allosteric mechanism without directly binding to glutamate or GlyRs (Ramírez-Guerrero et al., 2022). Glutamate-induced excitotoxicity is a well-established pathological mechanism implicated in various neurodegenerative disorders (Magdaleno Roman and Chapa González, 2024; Nicosia et al., 2024). Under physiological conditions, glutamate functions as the principal excitatory neurotransmitter in the CNS. However, excessive activation of NMDA receptors by glutamate leads to a cascade of deleterious intracellular events. These include mitochondrial membrane depolarization, caspase

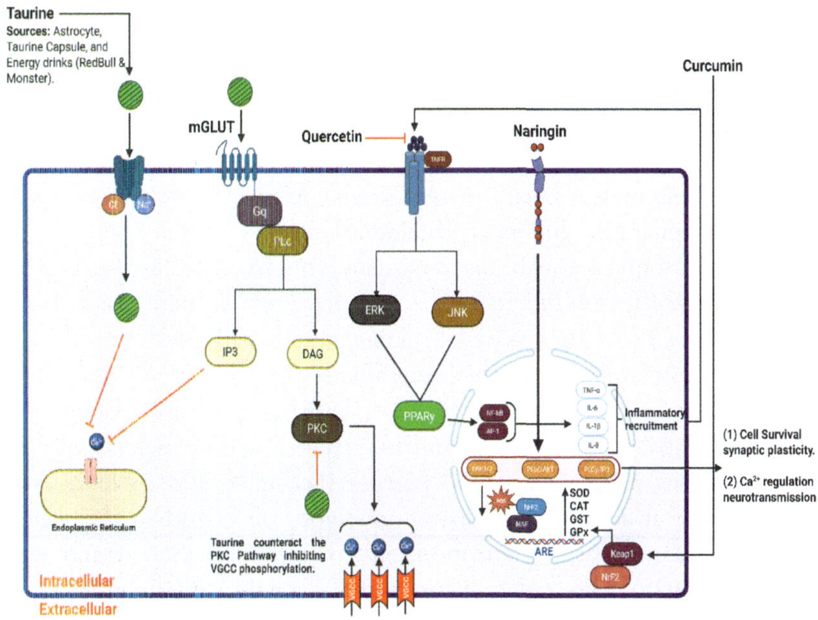

Fig. 2 Mechanisms of action illustrating how Taurine, Quercetin, Naringin, and Curcumin interact within multiple neuroprotective pathways, making them effective interventions for Neurodegenerative, Neuroinflammatory, Neurotoxic, and Neurodevelopmental disorders. The abbreviations used in the image are as follows: Cl^-, Chloride ion; Na^+, Sodium ion; Ca^{2+}, Calcium ion; Gq, G-protein subtype q; PLc, Phospholipase C; IP3, Inositol 1,4,5-triphosphate; DAG, Diacylglycerol; PKC, Protein Kinase C; VGCC, Voltage-Gated Calcium Channel; TNFR, Tumor Necrosis Factor Receptor; ERK, Extracellular Signal-Regulated Kinase; JNK, c-Jun N-terminal Kinase; PPARγ, Peroxisome Proliferator-Activated Receptor Gamma; NF-κB, Nuclear Factor kappa-light-chain-enhancer of activated B cell; AP-1, Activator Protein-1; IL-6, Interleukin-6; IL-1β, Interleukin-1 beta; IL-8, Interleukin-8; TNF-α, Tumor Necrosis Factor-alpha; PI3K/AKT, Phosphoinositide 3-kinase/Protein Kinase B; ERK1/2, Extracellular Signal-Regulated Kinases 1 and 2; PLCγ/PI3K, Phospholipase C gamma/Phosphoinositide 3-kinase; Nrf2, Nuclear Factor Erythroid 2–Related Factor 2; Keap1, Kelch-like ECH-associated protein 1; MAF, Musculoaponeurotic Fibrosarcoma oncogene homolog; ARE, Antioxidant Response Element; SOD, Superoxide Dismutase; CAT, Catalase; GST, Glutathione S-Transferase; GPx, Glutathione Peroxidase.

activation, and the generation of ROS, culminating in neuronal apoptosis and necrosis (Gupta et al., 2021; Kolić and Kovarik, 2024). Concurrently, the sustained influx of calcium ions (Ca^{2+}) enhances the activity of calcium-dependent catabolic enzymes, which further disrupt cellular homeostasis and promote neurodegeneration (Griffioen, 2023). This excitotoxic process is essential to the pathophysiology of AD, PD, AML, HD, and ischemic stroke (Nicosia et al., 2024). In these conditions, sustained NMDA receptor overstimulation leads to intracellular calcium overload, mitochondrial impairment, oxidative stress, and irreversible neuronal injury (Griffioen, 2023; Kolić and Kovarik, 2024).

Experimental studies investigating glutamate excitotoxicity utilize various model organisms. Human studies offer clinically relevant insights, while rodent models such as *Rattus norvegicus* and *Mus musculus* are commonly used for mechanistic and pharmacological research due to their genetic and neuroanatomical similarities to humans (Domínguez-Oliva et al., 2023). Moreover, *Danio rerio* (zebrafish) is a valuable model in neurotoxicological research because of its conserved glutamatergic pathways, optical transparency, and suitability for high-throughput screening (Ahmed et al., 2023). Taurine inhibits programmed cell death, caspase, and calpain activities as part of its anti-apoptotic action and has been shown to increase the Bcl-2:Bax ratio by approximately 60 percent (Sedaghat, 2021; Zhang et al., 2022a; Wang et al., 2024a). Taurine undergoes three distinct degradation processes: transamination, oxidation, and oxygenation (Seneff and Kyriakopoulos, 2025). The transamination process involves the formation of L-alanine and sulfoacetaldehyde from taurine pyruvate (Wei and Zhang, 2021). The oxidation stage involves the formation of isethionate (2-hydroxyethane sulfonate) from taurine (Capsi et al., 2018). The oxygenation process involves taurine being used by sulfoacetaldehyde acetyltransferase (AChE) as a sulfur source, ultimately converting into acetyl-CoA by the phosphatase AChE (Caspi et al., 2018).

3.1.1 Neuroprotective functions of taurine

Taurine has been shown to significantly enhance functional recovery in both traumatic brain injury (TBI) and ischemic stroke models, primarily in rodent species such as R. norvegicus and M. musculus (Sun et al., 2007; Su et al., 2014; Wang et al., 2016). These improvements were associated with reduced neuroinflammation, as evidenced by decreased glial reactivity and modulation of inflammatory cytokines (Dorstyn et al., 2018). Specifically, taurine suppressed microglial and astrocytic reactivity in TBI models, indicated by a downregulation of ionized calcium-binding adaptor molecule 1 (Iba-1), glial fibrillary acidic protein (GFAP), and pro-inflammatory markers such as interleukin-1β (IL-1β) and tumor necrosis factor-alpha (TNF-α) (Nakajima et al., 2010; Su et al., 2014). In experimental epilepsy models, particularly kainate-induced status epilepticus in rodents, taurine reduced seizure frequency and severity by inhibiting hippocampal apoptosis through upregulation of the anti-apoptotic Bcl-2 and suppression of the pro-apoptotic Bax (Junyent et al., 2009). In PD models, taurine conferred neuroprotection to dopaminergic neurons in the substantia nigra pars compacta (SNpc) by reducing microglial reactivity induced by paraquat and maneb. This was evidenced by decreased expression of Iba-1, CD11b, and NADPH oxidase-2 (NOX2), along with suppression of the nuclear factor-kappa B (NF-κB) signaling pathway and reduced transcription of pro-inflammatory cytokines (Che et al., 2018). These changes were observed in murine models, highlighting taurine's potential to modulate innate immune responses in neurodegenerative disease studies.

In chronic stress models, taurine alleviated neurochemical changes in the hippocampus and cortex, improving behavioral outcomes related to memory and mood in rodents (Haider et al., 2020; Jangra et al., 2020). Similarly, in lipopolysaccharide (LPS)-induced neuroinflammation models in mice, taurine decreased microglial reactivity and inflammatory markers, including inducible nitric oxide synthase (iNOS) and cyclooxygenase-2 (COX-2) (Silva et al., 2021; Amenotie et al., 2025). Aged models treated with taurine also showed increased hippocampal neurogenesis, evidenced by elevated expression of doublecortin (DCX) and Ki-67 in M. musculus (Gebara et al., 2015). Regarding oxidative stress, taurine protected neuronal cultures, including primary cortical neurons and SH-SY5Y neuroblastoma cells from Aβ- and glutamate-induced excitotoxicity by enhancing antioxidant defences (Louzada et al., 2004; Paula-Lima et al., 2005). These effects were achieved through higher levels of glutathione (GSH), superoxide dismutase (SOD), and catalase, while malondialdehyde (MDA), a lipid peroxidation marker, was simultaneously reduced. In AD models, especially the 5xFAD transgenic mouse, taurine enhanced spatial memory and synaptic integrity but yielded mixed results on Aβ deposition and Tau fibril formation (Santa et al., 2007; Kim et al., 2014; Oh et al., 2020). As this set of findings pertains specifically to AD pathology, further discussion is omitted here to prevent redundancy.

Taurine also reduced environmental metal-induced neurotoxicity, such as that caused by nickel and manganese in rodent models by restoring redox balance and maintaining neuronal health). Long-term taurine supplementation lessened DNA damage and cognitive impairments driven by oxidative stress, with reductions in 8-OHdG expression and improvements in antioxidant enzyme levels (Caletti et al., 2018; Adedara et al., 2017; Reeta et al., 2017). In schizophrenia models, especially those triggered by ketamine in rats, taurine reversed behavioral issues by decreasing neuroinflammatory mediators (e.g., COX-2, iNOS) in the striatum, prefrontal cortex, and hippocampus (Ben-Azu et al., 2023). Follow-up studies revealed that taurine restored normal dopaminergic activity and boosted antioxidant defenses, increasing GSH, SOD, and catalase, while lowering MDA levels (O'Donnell et al., 2016). These effects demonstrate taurine's ability to modulate both neurotransmitter systems and oxidative stress pathways. Additionally, taurine reduced the extrapyramidal side effects linked to chronic chlorpromazine use in rodent models, by restoring psychoneuroendocrine function and preventing testicular oxidative damage and reproductive toxicity (Oyovwi et al., 2021, 2022, 2025b). Overall, these findings underscore taurine's diverse neuroprotective mechanisms across various disease models and species, highlighting its potential for translational neuropharmacology.

3.2 Morin

Morin (3,5,7,2',4'-pentahydroxyflavone) is a widely available flavonoid derived from plants in the *Moraceae* family, used as a yellow pigment (Jin et al., 2014). Pure morin is a bitter, yellow, stable compound that shows relative neutrality and is insoluble in water. Chemically, it consists of a hydroxyl group at the 3-position and has a flavone structure (2-phenyl-1-benzopyran-4-one). It is classified as a 7–hydroxy flavonol, with three additional hydroxyl groups at positions 2′, 4′, and 5 (Rajupt et al., 2021). It dissolves readily in alcohol, is slightly soluble in ether and acetic acid, and is soluble in aqueous alkaline solutions, producing an intense yellow color that turns brown when exposed to oxygen (Sinha et al., 2016). Morin is sourced from various plants, including Onion (*Allium cepa*), Almond (Prunus dulcis, syn. Prunus amygdalus), Osage orange (*Maclura pomifera*), Guava (*Psidium guajava L.*), Old fustic (*Chlorophora tinctoria or Maclura tinctoria*), *Acridocarpus orientalis*, Apple (*Malus pumila*), Tea (*Camellia sinensis*), and other beverages such as red wine, coffee, and cereals. Phytochemical analyses and their reports have demonstrated that morin varies in quantity and therapeutic potency across sources (Kim et al., 2013a). Notably, morin extracted from onion peel has been shown to protect DNA from toxic exposure, such as lead (Kim et al., 2013b). A high amount of morin is reported in the leaves of guavas, particularly *Psidium guajava*. Along with morin, several morin-like structural components, such as morin-3-O-lyxoside, morin-3-O-arabinoside, quercetin, and quercetin-3-O-arabinoside, are also obtainable from this plant (Rattanachaikunsopon and Phumkhachorn, 2007). The old fustic, or Dyer's mulberry, is a Neotropical tree native to South America that contains a yellow pigment rich in the flavonoid-morin in its leaves, bark, fruit, and other vegetative parts (Sharma et al., 2018), similar to *Acridocarpus orientalis*. Additionally, the apple is the most flavonoid-rich fruit in the North American daily diet and contains a significant quantity of morin in its skin. Morin is also predominantly found in green tea and red wine, contributing to their health benefits (Rajput et al., 2021). Moreover, morin derived from coffee has been reported to serve as a chemopreventive agent in several cancer studies (Safe et al., 2023). Following the ingestion of morin-rich foods, morin sulfate, glycosides, and methylated forms abundantly present in the gut are hydroxylated as they transit from the small to the large intestine, where they are converted into various aglycone forms, reaching a peak plasma concentration of 4 μg/mL (Li et al., 2019). However, morin undergoes prolonged absorption in the stomach, whereas intravenous injection results in a short plasma half-life of 30 min (Li et al., 2019). When multiple doses are administered to rats, morin reduces conjugated metabolites by 78 percent, suggesting saturation of the metabolic process at high doses (Hou et al., 2003). Studies on drug-drug interactions have indicated that morin upregulates the bioavailability of other drugs, such as talinolol, by modulating transport proteins, including P-glycoprotein, following oral administration in rats. This beneficial pharmacokinetic property has attracted the use of morin as a pharmaceutical compound co-formulated with other drugs for improved oral administration. Polyphenols, as highlighted by Kim et al. (2014) , play a crucial role as natural antioxidants, protecting cells from ROS. Morin hydrate is notable for its strong antioxidant properties, largely due to the double bond between C2–C3 atoms and the hydroxyl group (-OH) at the C-3 position (Jomova et al., 2019). Its ability to inhibit lipid peroxidation is linked to the hydroxyl groups at the 2′ and 4′ positions of the B ring (Mendoza-Wilson et al., 2011). The hydroxyl group at the 4′ position is related to anti-radical activity, with the 2′ position being particularly effective. Computational studies (Burda et al., 2001; Morales et al., 2012) indicate that the hydroxyl group at the 2′ position forms a hydrogen bond with the oxygen atom at the 1′ position, enhancing morin's capacity to scavenge free radicals. Lee et al. (2017) demonstrated its antioxidant and cytoprotective effects in V79-4 fibroblasts, showing potential for free radical scavenging, reduced ROS production, and increased HO-1 expression through nuclear factor erythroid 2-related factor 2 (Nrf2) activation and Keap1 downregulation. Morin safeguards healthy cells from chemotherapy damage without reducing the drugs' effectiveness against cancer cells by preventing DNA fragmentation (Sithara et al., 2017).

3.2.1 Neuroprotective effects of Morin

Morin possesses antioxidant and anti-inflammatory properties, helping to restore mitochondrial function in dopaminergic neurons and notably improving PD-like symptoms (Rizvi et al., 2015). While the exact cause of PD remains unclear, oxidative stress, inflammation, and mitochondrial dysfunction are pathological pathways implicated in the disease (Guo et al., 2013; Adebayo et al., 2024; Onuelu et al., 2025). In a study conducted by Anyanwu et al. (2025), dietary morin prevented 1-methyl-4-phenyl-1,2,3,6-tetrahydropyridine (MPTP)-induced motor dysfunction and ameliorated dopaminergic neuronal damage in the striatum and substantia nigra in a mouse model. Several studies have shown that morin improved PD-like symptoms in a mouse model. It enhanced motor function, reduced dopaminergic neuronal loss in the striatum and substantia nigra, and inhibited MPTP-induced glial reactivities, notably primary microglia and astrocytes, by blocking the extracellular signal-regulated kinase (ERK)-p65 pathway. This suggests that morin has the potential to serve as an active food ingredient in the treatment of PD (Anyanwu et al., 2025). Additionally, from an in vitro study, morin protected neurons from MPP+ induced ROS production and disturbances in mitochondrial membrane potential. It also decreased astroglia activation and NF-κB translocation in primary cultured astrocytes (Wang et al., 2023a), thereby normalizing the excitatory/inhibitory balance (Joseph et al., 2025). Besides, morin protected against brain injury induced by middle cerebral artery occlusion (MCAO) by reducing pro-inflammatory mediators such as TNF-α and IL-6, lowering MDA production, and enhancing the levels of antioxidative enzymes like SOD, GSH, and GPx while improving cognitive functions and exploratory behavior (Alla et al., 2024). It also regulated apoptosis via modulation of Bcl-2, Bax, and cysteine aspartate-specific protease-3 (caspase-3) genes (Chen et al., 2017). Morin protected against MCAO-induced cerebral ischemia-associated BBB damage by increasing tight junction proteins (e.g., occludin, claudin, and actin), reducing cerebral infarct size, apoptosis-induced cell death, inhibiting toll-like receptor-4 (TLR4) and NF-κB expressions, and normalizing glial activities by reducing GFAP and Iba1 expressions in the cerebral cortex and striatum (Khamchai et al., 2020). In another recent study, morin suppressed LPS-stimulated RAW264, cell activation and migration, notably inhibiting NO production by reducing mRNA iNOS expression, dose-dependently inhibiting ERK and Jun N-terminal kinase (JNK), and β-catenin expression (Jayakumar et al., 2023). Besides targeting pathogens during bacterial, viral, or fungal infections, excessive production of ROS can also damage surrounding healthy cells. This oxidative stress may trigger chronic inflammation, including neuroinflammation, which is implicated in the progression of neurodegenerative diseases such as PD and AD (Behl et al., 2021; Adamu et al., 2024). In murine models and human studies, elevated ROS levels are linked to microglial activation and neuronal loss, especially in the hippocampus and substantia nigra (Adamu et al., 2024). Morin has demonstrated neuroprotective effects in diabetic neuropathy by reducing ROS-induced neuroinflammation, which involves the immune response within the central or peripheral nervous system, activation of glial cells, and the release of pro-inflammatory cytokines and oxidative damage in a rat model (Bachewal et al., 2018). It also alleviated peripheral neuropathy in chronic constriction injury models by decreasing oxidative stress, neuroinflammation, and poly(ADP-ribose) polymerase (PARP) over-activation (Komirishetty et al., 2016). A related study also demonstrated that morin inhibits oxidative stress and inflammation in the cardiorenal system associated with posttraumatic stress disorder (PTSD) and alcohol use disorder in mice (Ben-Azu et al., 2024a; Ben-Azu et al., 2025). In the APPswe/PS1dE9 mouse model of AD, morin improved cognitive function and reduced brain damage by targeting key pathological mechanisms. It decreased Aβ accumulation, reduced oxidative stress, and suppressed neuroinflammation by limiting glial activation and inflammatory cytokine release. Morin also helped preserve synaptic integrity and may reduce tau hyperphosphorylation, contributing to its overall neuroprotective effect (Du et al., 2016).

Additionally, morin suppressed environmental infectious agents such as LPS-induced neuroinflammation in both in-vivo and in-vitro models by modulating the PKA/Nrf2/HO-1 signaling pathway and the PI3K/Akt and MAPK pathways. This effect ultimately led to reduced production of inflammatory mediators like TNF-α, IL-1β, and nitric oxide, decreased oxidative stress, and protection of neuronal cells from damage. In both cell cultures and animal models, this resulted in improved neuronal health and potentially better cognitive or behavioral outcomes, highlighting morin's therapeutic potential against infection-related neurodegenerative processes. Studies on neurotoxicity caused by environmental toxicants revealed that morin alleviated neurobehavioral deficits, improving anxiogenic tendencies, memory impairment, and motor activity by reversing hippocampal oxidative stress, enhancing the antioxidant system, and reducing inflammation by lowering TNF-α and IL-1β levels, as well as apoptotic markers like Bax and apoptotic factor-1 in rats exposed to bisphenol and diethyl phthalate, which are endocrine-disrupting pollutants. Morin also enhanced the suppression of anti-inflammatory factors such as IL-10 and neurotrophic elements such as brain-derived neurotrophic factor (BDNF), which were induced by bisphenol and diethyl phthalate. It also reduced the hyperactivity of metabolic neurochemical enzymes such as AChE and MAO-A, indicating its potential as a neuroprotective functional food ingredient against neurotoxicity caused by environmental toxicants (Ugwor et al., 2022). Using similar mechanisms, morin inhibited arsenic-induced cell damage in PC12 neurons (Banaeeyeh et al., 2024), as well as neurotoxicity caused by ifosfamide (Celik et al., 2020) and cadmium-induced hepatotoxicity (Sengul et al., 2024) in rats. It achieved this by suppressing the release of enzymes like carbonic anhydrase, AChE, and butyrylcholinesterase (BChE) through pathways involving the upregulation of nuclear factor erythroid 2-releated factor 2 (Nrf-2) and BDNF, and the inhibition of NF-κB, neuronal NOS (nNOS), TNF-α, JNK, and cFos expression, which are molecular markers linked to cellular and metabolic stress. Collectively, these findings highlight morin's ability to exert a chemopreventive effect against neurotoxicity caused by neurotoxicants involved in the development of various neurological disorders.

Regarding neuropsychiatric disorders, morin has been shown to reduce neuropsychiatric disturbances caused by various chemicals and worsened by alcohol consumption in mice (Akinluyi et al., 2022; Ben-Azu et al., 2024c). Morin decreased schizophrenia-like symptoms in mice induced by ketamine, an NMDA receptor antagonist. Morin improved ketamine-induced hyperactivity, social withdrawal, and cognitive deficits by inhibiting oxidative stress (Ben-Azu et al., 2018a, 2018b) and modulating neurochemical changes, especially dopaminergic and serotonergic re eptor dysfunctions (Ben-Azu et al., 2021) in the striatum, prefrontal cortex, and hippocampus of mice (Ben-Azu et al., 2018b). A follow-up study also indicated that the antipsychotic effect of morin is linked to the suppression of nicotinamide adenine dinucleotide oxidase-2 (Nox2), resulting in improvements in GABAergic deficits through the upregulation of glutamic acid decarboxylase-67 (GAD-67) (Ben-Azu et al., 2018c). Additionally, morin reversed LPS-exacerbated ketamine-induced schizophrenia-like behavior by inhibiting the release of pro-inflammatory mediators and inflammatory protein expressions such as iNOS, COX-2, and NF-κB, thereby protecting cortical pyramidal neurons from degeneration (Ben-Azu et al., 2019a). Similarly, morin reduced neuroinflammation in mice exposed to psychosocial and chronic general stress, notably enhancing the expression of cortical-striatal GAD67 and neurotrophic factors, and preventing hippocampal degeneration (Olonode et al., 2019; Ben-Azu et al., 2019a, 2019b, 2020;

Elizabeth et al., 2020). Morin prevented psychiatric behaviors related to chronic stress and PTSD that are worsened by alcohol, reducing depressive symptoms and improving cognitive performance and anxiety by alleviating HPA dysfunction among the mechanisms previously described (Olonode et al., 2019; Emudainohwo et al., 2022; Ben-Azu et al., 2024c). The receptor-interaction mechanistic study showed that morin modulated both monoaminergic and non-monoaminergic receptor systems, including the D2 dopaminergic receptor, 5-hydroxyltryptaminergic-2 (5-HT2), as well as $\alpha 1$ and $\alpha 2$ adrenergic receptors. This modulation occurs following pre-treatment with dopaminergic, serotonergic, and adrenergic antagonists and synthesis inhibitors such as haloperidol, prazosin, yohimbine, metergoline, and para-chlorophenylalanine, which reversed the effects of morin in mice (Ben-Azu et al., 2021).

3.3 Curcumin

Curcumin is the most abundant active compound in the spice turmeric, constituting 2–8 percent of turmeric preparations. It is derived from Curcuma longa, a rhizomatous herbaceous perennial plant belonging to the ginger family *Zingiberaceae*, native to tropical South Asia. Curcumin is a beta-diketone with two hydrogens replaced by feruloyl groups, also known as diferuloylmethane. Around 133 species of *Curcuma* have been identified worldwide. Curcumin is a plant-derived polyphenolic compound found in Curcuma longa, as reported by Stanić (2017). It is noted for its antioxidant, anti-amyloidogenic, neuroprotective, chemopreventive, immunomodulatory, chemotherapeutic, and anti-inflammatory properties (Kotha and Luthria, 2019; Yang et al., 2022; Tripathi et al., 2024). Major bioactive compounds in C. longa include curcumin, demethoxycurcumin, bisdemethoxycurcumin, arturmerone, α-turmerone, and β-turmerone (Burapan et al., 2020). Curcumin is a yellow polyphenolic pigment present in the Curcuma longa L. (turmeric) rhizome. It has been used for centuries for culinary and food coloring purposes and as an ingredient for various medicinal preparations in Ayurveda and Chinese medicine (Sharifi-Rad et al., 2020). Pharmacokinetic evaluations indicated that curcumin demonstrated low bioavailability in healthy human volunteers and cancer patients following oral administration, largely due to a significant first-pass effect and intestinal metabolism that involves the glucuronidation and sulfation of curcumin (Sharma et al., 2007). Peak plasma concentrations of curcumin in glucuronide and sulphate forms (20 and 300 ng/mL) occur after 1.5 h from curcumin liquid droplet micromicellar formulation, while the levels of 95 percent curcumin remain unchanged.

3.3.1 Mechanism of neuroprotective actions of curcumin

As shown in Fig. 2, curcumin has been used in therapeutic preparations across different parts of the world for centuries to effectively manage various disease conditions. Curcumin can cross the BBB in its native form without significant adverse effects (Menon and Sudheer, 2007). A range of pharmacological actions of curcumin, a potent neuroprotective agent, in treating CNS-related diseases such as AD, MS, and PD have been documented (Maurya et al., 2021; Azlan et al., 2023; Balistreri and Monastero, 2023). Consequently, increasing interest considers curcumin as a potential therapeutic option for treating various neurological disorders. Several studies have identified neuroinflammation as a key pathological feature of neurodegenerative diseases (Sundaram et al., 2017). Chronic neuroinflammation plays a critical role in a variety of neurodegenerative conditions, including AD, PD, HD, stroke, ALS, and MS.. The anti-inflammatory action of curcumin has been ascribed to its ability to inhibit the secretion of inflammatory mediators TNFα, NO, and prostaglandin-E2 (PGE2), and expression of iNOS and COX-2 in primary stimulated microglial cell (Keshavarz et al., 2024). During chronic neuroinflammation, NF-κB is a key transcription factor that plays crucial roles in regulating cellular and redox dysfunctions (Feng et al., 2024). NF-kB is regarded as the master regulator of microglial inflammatory responses to neuronal injury, where IκB is known to activate NF-kB dissociation and subsequent translocation to the nucleus, along with the induction of inflammatory mediators (Anilkumar and Wright-Jin, 2024). Interestingly, curcumin has been shown to exert dual inhibition of IκB phosphorylation and degradation, as well as the nuclear translocation of p65, thereby reducing NF-kB activity and microglial reactivity (Wang et al., 2018).

As shown in Fig. 2, depicting the schematic intracellular signaling cascades associated with neuroinflammation and oxidative stress. Curcumin (similar to taurine) modulates PKC and VGCC phosphorylation, downregulates NF-κB, and activates Nrf2-mediated antioxidant pathways to reduce neuroinflammation and neuronal damage. The neuroprotective effects of curcumin and its mechanisms have been demonstrated in mouse-derived BV-2 microglial cells, where it inhibited the production of inflammatory cytokines and prostaglandins in reactive microglia (Cianciulli et al., 2016). Curcumin also decreased the production of TNFα, IL-1β, macrophage inflammatory protein-1β (MIP-1β), monocyte chemoattractant protein-1

(MCP-1), and IL-8 in microglia and astrocytes, as reported by Chen et al. (2014). Curcumin inhibited the production of ROS and reduced inflammatory cytokine levels by suppressing cyclooxygenase and lipoxygenase enzyme activity, which ultimately led to decreased cartilage degradation and joint destruction in osteoarthritic conditions (Koroljević et al., 2023).

The beneficial effects of curcumin have been studied in various research. Curcumin supplementation has also been shown to repair dyskinesia and damage to dopaminergic neurons in mice with MPTP-induced PD by modulating inflammation, oxidative stress, and senescence (Cai et al., 2023; Cai et al., 2024). Curcumin and its analogues enhanced cognitive functions, learning and memory, reduced dopaminergic neuronal damage, and lowered oxidative stress in the substantia nigra pars compacta (SNpc) and the striatum of MPTP-lesioned mice (Wang et al., 2017), including preventing α-synuclein aggregation and increasing its solubility in A53T-mutant α-synuclein expressing SH-SY5Y cells (Ahsan et al., 2015). Curcumin was also reported to inhibit α-synuclein-induced toxicity, mitochondrial depolarization, ROS, caspase-3 activity, caspase-9 activity, and apoptosis in PC12 cells (Pandey et al., 2008; Liu et al., 2011). Furthermore, curcumin improved symptoms of ASD, a group of neurodevelopmental disorders characterized by impaired social interaction, cognitive deficits, restricted repetitive behaviors, altered immune responses, and oxidative stress imbalance (Zhong et al., 2020; Jayaprakash et al., 2021). Treatment with curcumin increases intracellular GSH levels, decreases inflammatory markers (TNF-α, IL-1β, IL-6, COX-2, iNOS, PGE2, MCP-1, ROS, RNS, Iba1, GFAP), alleviates mitochondrial dysfunction, and reduces protein aggregation (Bhandari and Kuhad, 2015; Avola et al., 2024). Curcumin also lowers oxidative stress, mitochondrial defects, TNF-α release, and matrix-degrading metalloproteinases in male Sprague Dawley rats showing autistic traits (Sachdeva et al., 2022). Additionally, curcumin has been recommended for managing AD-like pathology due to its cognitive-enhancing effects observed in experimental animal models and clinical settings (Farooqui, 2016; Voulgaropoulou et al., 2019; ELBini-Dhouib et al., 2021; Fan and Zhang, 2024). For instance, curcumin treatment in mice injected with Aβ1-42 resulted in decreased hippocampal Aβ1-42 deposition, lower levels of TNF-a, IL-6, IL-1β, and MDA, with downregulation of p65 expression and increased AMPK phosphorylation, thereby alleviating AD-like conditions associated with cognitive deficits in C57BL/6 J mice (Mishra and Palanivelu, 2008). Curcumin also reversed the short-term spatial and recognition memory deficits caused by intra-cerebroventricular injection of streptozotocin

(ICV-STZ). This effect is linked to the promotion of new neuron growth, as shown by increased Ki67 and doublecortin-positive cells, which occurred due to the suppression of ICV-STZ-induced neuroinflammation, indicated by decreased immunoreactivity of glial cell markers, such as GFAP and Iba1, in the hippocampus. These findings suggest improvements in neurogenesis within the dentate gyrus of the hippocampus and the subventricular zone (Bassani et al., 2017).

Regarding environmental neurotoxicity, several studies have shown that curcumin prevents heavy metal-induced neurotoxicity, notably inhibiting neuronal lipid peroxidation and reducing harmful adverse effects through antioxidant mechanisms (Gupta et al., 2015). Consistent with this, curcumin has been demonstrated to protect against cadmium-induced neurotoxicity and CA3-hippocampal and cortical degeneration by increasing antioxidant systems, CREB-BDNF, and synapsin-II signaling pathways in mice (Namgyal et al., 2020). Curcumin also inhibited arsenic-induced neurotoxicity by boosting concentrations of biogenic amines such as dopamine, adrenaline, noradrenaline, and serotonin, while decreasing their metabolic markers, including homovanillic acid and dihydroxyphenylacetic acid, in the corpus striatum, frontal cortex, and hippocampus of rats (Yadav et al., 2010). Additionally, lead-induced cerebellar oxidative stress was alleviated through chelation activity and antioxidant mechanisms. Curcumin reversed dichloroacetic acid-induced hippocampal degeneration and cognitive impairment in rats (Wei et al., 2021). It enhanced neuronal synaptic plasticity and cognition by modulating the mRNA expressions of postsynaptic density-95 (PSD-95), synaptophysin, cyclic adenosine monophosphate (cAMP), protein kinase A (PKA), cAMP-response element-binding protein (CREB), and BDNF, notably providing neuroprotection against dichloroacetic acid-induced neuronal damage in rats (Wei et al., 2021).

3.4 Quercetin

Quercetin, a widely studied flavonoid found in many fruits and vegetables, exerts significant neuroprotective effects in various neurodegenerative diseases, including PD (Grewal et al., 2021), AD) (Calderaro et al., 2022), and in models of environmental neurotoxicity (Rarinca et al., 2023). In PD, the pathogenesis involves mitochondrial dysfunction, oxidative stress, protein aggregation, and impaired protein degradation through the ubiquitin-proteasome system (Liang et al., 2023). Quercetin has been shown to protect dopaminergic neurons from MPTP-induced toxicity by activating the nuclear factor Nrf2 signaling pathway. This activation leads to enhanced expression of antioxidant enzymes, better regulation of intracellular iron homeostasis, and inhibition of ferroptosis, thereby reducing oxidative stress and neuronal death (Islam et al., 2025). Iron dysregulation, particularly its accumulation in the substantia nigra, induces lipid peroxidation and GSH depletion, contributing to PD progression. Quercetin's capacity to mitigate these effects underscores its therapeutic potential (Olufunmilayo et al., 2023).

Quercetin also affects the Janus kinase/signal transducer and activator of transcription (JAK/STAT) pathway , which plays a role in cellular survival and inflammation. Disruption of this pathway contributes to demyelination and neuronal degeneration in PD. By restoring JAK/STAT signaling, QUE supports neuroprotection (Lashgari, et al., 2021). Additionally, it activates pro-survival protein kinase D1, increases BDNF levels, and decreases caspase-3 activity, thereby promoting neuronal survival (Lutfi Ismaeel et al., 2023). Furthermore, quercetin reduces the expression of iNOS, nitric oxide (NO), and inflammatory cytokines such as IL-1β, IL-6, COX-2, Bax, and TNF-α (Ashrafizadeh et al., 2021; Aleebrahim-Dehkordi et al., 2022; Alharbi et al., 2025). This anti-inflammatory action is vital for fighting the chronic neuroinflammation seen in PD. Quercetin modulates toll-like receptor 4 (TLR4) signaling pathways (Zhang et al., 2022a), which play a key role in innate immune responses in the CNS. As a TLR4 antagonist, quercetin limits the release of cytokines and chemokines that attract microglia and astrocytes to inflammatory sites, thereby reducing neuronal apoptosis (Rosa et al., 2021). In AD, quercetin inhibits Aβ aggregation, suppresses tau hyper-phosphorylation, and decreases oxidative damage in neuronal tissue (Torres et al., 2021). It activates the PI3K/Akt and MAPK/ERK pathways, which are crucial for neuronal survival, plasticity, and memory formation. Animal studies show that quercetin supplementation improves learning, memory, and cognitive functions (Amidfar et al., 2024). Regarding environmental neurotoxicity, quercetin counteracts neuronal damage induced by agents, such as LPS and particulate matter (PM2.5) (Amidfar et al., 2024). It does so by inhibiting NF-κB activation and blocking the NLRP3 inflammasome, important mediators of inflammation (Chiang et al., 2024). Through these actions, quercetin helps preserve neural structure and function against toxic environmental challenges.

3.4.1 Mechanism of neuroprotective actions of quercetin

Quercetin is recognized for its anti-inflammatory properties, inhibiting enzymes like lipoxygenase and COX, which reduce mediators such as prostaglandins and leukotrienes (Warren et al., 2009). Research from Michigan State University showed that quercetin effectively lowers C-reactive protein (CRP), associated with diseases like lupus and heart disease, and diminishes inflammation and oxidative stress in various neurodegenerative conditions (Lee et al., 2018). Among natural substances, polyphenols like quercetin are the most studied in the context of neuropsychiatric, neurodegenerative, and neurotoxicity conditions (Zhang et al., 2022a). Quercetin is a potent scavenger of ROS and RNS, such as ONOO and NO (Ulusoy and Sanlier, 2020). The molecule's two key pharmacophores contributing to its antioxidant activity are the hydroxyl group at position 3 and the catechol group in the B ring (Ulusoy and Sanlier, 2020). These pharmacophores have an ideal configuration for scavenging free radicals. In vitro, quercetin can directly neutralize ROS at doses ranging from 5 to 50 μM (Bayazid and Lim, 2022). However, quercetin is believed to be present in the brain at a concentration probably in the nanomolar range, much lower than what is needed to produce a significant direct antioxidant effect (Grabska-Kobyłecka et al., 2023). Therefore, even though quercetin has a potent antioxidant capacity in vitro, it is unlikely that its neuroprotective effects in vivo are due to a direct antioxidant activity. Instead, it has been suggested that quercetin and/or its metabolites may act as prooxidants rather than antioxidants (Engwa et al., 2022; Rana et al., 2022) by altering the cell's natural antioxidant defense mechanisms. The cell's inherent antioxidant defences may be strengthened by a small amount of oxidative stress, leading to general cytoprotection.

Quercetin has demonstrated multiple mitochondrial and cytoprotective benefits across various experimental models. In a study by Davis et al. (2009), quercetin supplementation was shown to enhance mitochondrial biosynthesis in mice subjected to endurance training, highlighting its role in boosting mitochondrial capacity under physical stress conditions. Chen et al. (2016) reported that quercetin reduced neuronal cell apoptosis in a rat model of cerebral ischemia, indicating neuroprotective effects under oxidative injury. Aghababaei and Hadidi (2023) further confirmed its antioxidant properties in vitro using cultured hepatocytes under hydrogen peroxide-induced stress. Notably, mice fed quercetin-enriched diets exhibited significantly higher mitochondrial DNA (mtDNA) copy numbers in skeletal muscle, suggesting enhanced mitochondrial replication and biogenesis (Davis et al., 2009). The signaling mechanisms by which quercetin and similar flavonoids exert their neuroprotective effects may involve suppression of PKC, downregulation of NF-κB, and activation of antioxidant response elements via the Nrf2 pathway, as shown in Fig. 2. These findings suggest that quercetin not only supports mitochondrial integrity under stress but may also improve endurance and resilience in disease models characterized by oxidative damage or mitochondrial dysfunction (Aghababaei and Hadidi, 2023). Several studies have reported that quercetin has potent adaptogenic properties against chronic stress and PTSD-induced neurological dysfunction, including the release of oxidative stress and excessive production of inflammatory cytokines in brain regions vulnerable to depression, anxiety, and cognitive impairment (Mehta et al., 2017; Ma et al., 2021; Ugwu et al., 2022; Wang et al., 2024a; Oyovwi et al., 2025b, 2025c). Several studies have investigated the effect of quercetin on neuronal damage caused by apoptosis (Ma et al., 2021; Wang et al., 2024a). Quercetin combined with 2-methoxyestradiol in androgen-dependent LNCaP and androgen-independent PC-3 human prostate cancer cells inhibits apoptosis by decreasing the Bcl-2/Bax ratio alongside upregulation of glucose-regulating protein 78 (GRP78), activating transcription factor 4 (ATF-4), and inositol-requiring enzyme 1 (IRE-1), supporting quercetin's antitumor properties (Wang et al., 2013).

3.5 Turmeric

Turmeric (*Curcuma longa*) is an herbaceous perennial plant that belongs to the ginger family, *Zingiberaceae*, and is widely used as a spice in its native regions of Southeast Asia, India, and China. It serves as an aromatic stimulant, food preservative, and coloring agent, with common names such as turmeric root, yellow root, and saffron root (Dusabumuremy et al., 2022). Turmeric is rich in carbohydrates and fiber and contains a modest amount of protein and fat, and is completely free of cholesterol (Ahmad et al., 2020). Traditionally, turmeric has been valued for its medicinal properties, making it one of the most extensively researched functional foods and supplements for various health conditions, including heart disease, arthritis, AD, gastrointestinal disorders, dyslipidemia, and liver diseases (Ahmad et al., 2020; Rolfe et al., 2020). As a natural and nutritious supplement, turmeric is high in minerals such as potassium, calcium, magnesium, and phosphorus, as well as vital vitamins C, A, D, and E (Anyaoku et al., 2023). Among its bioactive compounds are curcuminoids (curcumin, demethoxycurcumin, bisdemethoxycurcumin), essential oils (turmerone, germacrone, zingiberene, atlantone), and others like diarylheptanoids, turmerones, and sesquiterpenoids (Bulbula, 2021; Fuloria, 2022). Studies have focused on turmeric in recent years due to its wide range of potential therapeutic benefits. Its primary active compound, curcumin, has shown effects that may help in managing various health conditions, including metabolic disorders, neurodegenerative diseases, neuropsychiatric conditions related to oxidative stress, and inflammation (Cox et al., 2020; Shah et al., 2022). This has led to an investigation into its potential to manage neurodegenerative and neurodevelopmental diseases through neuroprotective mechanisms (Bhat et al., 2019). Studies indicate that turmeric possesses extensive pharmacological properties, including anti-inflammatory, antioxidant, anticancer, antimutagenic, antimicrobial, anti-obesity, hypolipidemic, cardioprotective, and neuroprotective effects (Saengprakai and Chudhangkura, 2017; Iweala et al., 2023). The pharmacology of turmeric depends on its active constituents, especially the curcuminoids and turmerones, which have diverse effects. The absorption of turmeric's active components through the gastrointestinal tract is poor due to their hydrophobic nature and low solubility in aqueous environments. However, sesquiterpenoids show favorable absorption characteristics, with high permeability in the intestinal epithelium. Their absorption can be influenced by gastrointestinal pH and efflux transporters. Curcuminoids, however, are poorly absorbed through the intestinal wall. A recent study showed that germacrone, when administered orally to rats, undergoes extensive metabolism, with 60 metabolites identified, mainly distributed in the stomach, intestine, liver, and kidneys (Guo et al., 2024). The active ingredients in turmeric are widely dispersed; however, since curcuminoids are quickly metabolized in the liver and intestines, they can only remain in the bloodstream for limited periods. The combination of piperine, a compound found in black pepper, with curcuminoids, enhances their bioavailability (Balakumar et al., 2022). To improve their absorption, lipid-phospholipid complexes, liposomal formulations, and nanoparticles have all been developed (Liu et al., 2016). Studies conducted in vitro and on animal models suggest turmeric is generally safe for consumption, but excessive or prolonged use may lead to toxicity (Sahoo et al., 2021). In rare instances, oral curcumin has been linked to side effects such as gagging, tongue redness, irregular heartbeat, and digestive issues like gas, diarrhea, nausea, and constipation (Shah et al., 2022). Turmeric may induce pharmacokinetic alterations in various medications, including cardiovascular drugs, antibiotics, antidepressants, chemotherapy agents, immunosuppressants, and corticosteroids by inhibiting cytochrome P450 enzymes and P-glycoprotein (Bahramsolti et al., 2017).

3.5.1 Mechanism of action of curcumin neuroprotection

Turmeric's antioxidant effects are primarly attributed to its active compound, curcumin, which has been shown to scavenge free radicals, neutralize ROS, and reduce oxidative stress in cells (Fuloria et al., 2022). Evidence suggests that turmeric activates Nrf2, which then promotes the production of antioxidant enzymes such as SOD, catalase, and glutathione peroxidase; this activation is vital in protecting against oxidative stress-related neurodegenerative diseases like AD, PD, and MS by reducing neuronal damage and inflammation (Tapia et al., 2012; Méndez-García et al., 2019; Wu et al., 2019b). Curcumin's antioxidative effect is reflected in its ability to lower MDA levels, a marker of lipid peroxidation. Additionally, turmeric extract has demonstrated antioxidative and neuroprotective effects in microglial cells (BV2), dose-dependently reducing LPS-induced NO production (Streyczek et al., 2022). Although these effects are mainly due to curcumin, other compounds such as turmerones (A and B) also contribute (Saji et al., 2023). Additionally, aromatic-turmerone and related compounds, including its s-enantiomer, have shown neuroprotective effects in midbrain slice cultures, preventing dopaminergic degeneration caused by microglial reactivity, likely through Nrf2 activation (Hori et al., 2021). It also lessened dopaminergic neurodegeneration triggered by 1-methyl-4-phenylpyridinium, a neurotoxin. (Hori et al., 2021). Turmeric has also been found to decrease iNOS mRNA expression and iNOS protein production, possibly by inhibiting the phosphorylation and activation of mitogen-activated protein kinases, specifically p42/44 MAPK (Erk1/2) and p38 MAPK, in a concentration-dependent manner. These anti-inflammatory effects are especially beneficial in conditions such as rheumatoid arthritis, inflammatory bowel disease, and neuroinflammation, where the overexpression of iNOS and MAPK signaling leads to tissue damage and chronic inflammation (Streyczek et al., 2022). This mechanism is particularly relevant in inflammatory and neurodegenerative diseases like AD and PD, where excessive iNOS activity contributes to neuronal damage (Streyczek et al., 2022). Given the role of IκB-α in NF-κB activation for conditions associated with chronic neuroinflammation, studies have also demonstrated that turmeric has significant anti-inflammatory effects, particularly at the nuclear level, primarily through the modulation of the NF-κB pathway (Kahkhaie et al., 2019). Turmeric blocks myeloid differentiation primary response 88 and toll-like receptor 4/NF-κB signaling and downregulates NF κB p65 mRNA expression (Zhu et al., 2014). More specifically, other studies have elucidated that turmeric inhibits NF-κB activity by suppressing IκB-α degradation, which supposedly allows NF-κB activation (Kahkhaie et al., 2019).

Furthermore, turmeric has also been studied for its potential to help manage neurodevelopmental disorders such as ASD and ADHD because of its anti-inflammatory, antioxidative, and neuroprotective effects (Abbas et al., 2020; Wei et al., 2024). ASD, marked by deficits in social interaction and repetitive behaviors, is linked to mutations in genes like SH3 (Src Homology 3 domain-containing proteins involved in intracellular signaling), SHANK (SH3 and multiple ankyrin repeat domains proteins that act as scaffolding proteins at synapses), and NRXN1 (Neurexin 1, a synaptic adhesion molecule critical for neuron communication) all of which have vital roles in brain development and neuronal connectivity (Guang et al., 2018). Inflammation is also thought to play a part in ASD, although its exact role in its onset and progression remains unclear. Studies have shown that individuals with ASD tend to have lower levels of anti-inflammatory markers and higher levels of pro-inflammatory markers (Esnafoglu and Subaşi, 2022; Akman and Esnafoglu, 2023; Morimoto et al., 2023). A single dose of curcumin supplementation, a turmeric bioactive compound, increased the expression of antioxidant enzymes in a rat model of VPA-induced oxidative stress. Moreover, pro-inflammatory cytokines, IL-6 and IFNγ, were found at lower levels in rats given curcumin (Dong et al., 2012; Al-Askar et al., 2017). Also, curcumin has been shown to enhance α7 nicotinic acetylcholine receptor (α7nR) activation, reduce social deficits, and alleviate oxidative stress in mouse models of ASD (Jayaprakash et al., 2021), acting as a positive allosteric modulator of α7-nAChRs. ADHD, characterized by persistent behavioral disturbances (Sharma and Couture, 2014), is associated with abnormalities in various brain regions, disruptions in catecholamine pathways, and increased oxidative stress and neuroinflammation (de Sousa Macedo et al., 2021). Recent evidence indicates that curcumin reduces methylphenidate-induced ADHD-related hyperactive and anxious behaviors in rats by improving social interaction, normalizing motor activity, and enhancing spatial cognitive performance (Ashish Jain, 2017; de Sousa Macedo et al., 2021).

Neurodegenerative diseases such as AD, PD, and HD are characterized by progressive neuronal dysfunction and loss (Muddapu et al., 2020; Tran et al., 2022; Mohammed and Akinola, 2024). In the context of AD, curcumin, a key component of turmeric, has been reported to inhibit Aβ plaque formation, disaggregate existing plaques, and regulate tau hyperphosphorylation through the PTEN Akt GSK 3β signaling pathway. Specifically, in Aβ-induced rat models of AD, curcumin improved cognitive impairments and neuroinflammation by affecting the Akt GSK 3β axis. Furthermore, curcumin chelates metal ions that catalyze Aβ aggregation, reduces brain cholesterol levels, and enhances memory performance in both AD patients and animal models (Lakey Beitia et al., 2020). Aromatic turmerone, another active compound in turmeric, has been shown to reverse memory impairment in nontransgenic C57BL/6J mice using aLPS-induced model of neuroinflammation, mediated via TLR4-dependent signaling pathways (Chen et al., 2018). In PD, turmeric-derived curcumin has demonstrated neuroprotective effects in MPTP-induced Parkinsonian mouse models, particularly by reducing dopaminergic neuron loss in the substantia nigra pars compacta, a hallmark of PD pathology (Mythri et al., 2011).

3.6 Naringin and naringenin

Naringenin, along with its analogue naringin, is a well-known dietary flavanone glycoside, commonly found in tomatoes and citrus fruits such as grapefruits and oranges (Shilpa et al., 2023). Naringenin is responsible for the bitter taste (up to 0.97 IBU) often encountered in grapefruits (Drewnowski et al., 1997). Its chemical name is 4,5,7-trihydroxyflavanone, with grapefruit seed extract being its most common supplement. Over the years, this seed extract has been formulated into capsules and liquids and studied for its diverse pharmacological properties with therapeutic potential for treating and managing various diseases (Alam et al., 2014; Ahmad et al., 2015), including neurodegenerative and neurodevelopmental disorders such as AD (Alam et al., 2014), PD (Lou et al., 2014), MS (Wang et al., 2018), schizophrenia, and other neuropsychiatric conditions (George et al., 2020; Ben-Azu et al., 2024g, 2024h, 2024i; Eneni et al., 2023, 2025). The promising therapeutic effects of naringenin and naringin mainly stem from their potent antioxidant , neuroprotective (Ben-Azu et al., 2018d, 2019c) and anti-inflammatory Zhang et al., 2018) properties, which protect against neuronal damage caused by mitochondrial dysfunction, excessive pro-inflammatory cytokines from microglia, environmental neurotoxins (Naraki et al., 2021), and stressful conditions (Ojiakor et al., 2026a, 2026b). These pharmacological traits make naringenin valuable phytonutrients that could provide neuroprotective benefits against neurodevelopmental and neurodegenerative disorders triggered by environmental, biological, or genetic factors.

In an effort to position naringenin as a neuroprotective phytonutrient for managing neurodevelopmental and neurodegenerative disorders, it is important to understand its pharmacokinetic properties and toxicological profiles established through clinical and preclinical studies over the years. Pharmacokinetically, naringenin is hydrolyzed in the intestine by the enzyme naringinase, produced by intestinal bacteria. Both naringenin and naringin are rapidly absorbed from the intestine after oral administration, with rapid bioavailability achieved in the circulation. Naringenin undergoes swift hepatic first-pass metabolism, which reduces its bioavailability in the plasma (Alam et al., 2014). However, naringenin is known to be lipophilic, which leads to the hypothesis that this phytonutrient could accumulate within tissues and membranes in higher amounts than those found in the plasma. Excretion of naringenin occurs via urine and plasma following oral administration of pure naringenin or grapefruit, although detection levels are low due to its poor water solubility (Alam et al., 2014). Long-term toxicological preclinical studies on naringenin and naringin have shown that both compounds are relatively safe, indicated by a low toxicological profile demonstrated through in vitro and in vivo studies. For example, Li et al. (2024) conducted an acute toxicological assessment of naringenin on experimental mice over 13 weeks. The results showed no toxicological changes in clinical biochemistry, hematology, or macroscopic findings. Moreover, no mortality or adverse clinical signs were observed in the test rats, suggesting that naringenin is virtually non-toxic in rats at doses exceeding 250 mg/kg/day.

3.6.1 Mechanisms of the neuroprotective effects of naringenin and its roles in neurotoxicity and neurodegenerative conditions

Neuronal organelles such as mitochondria and microglia cells are known sites of the brain's metabolic activities, performing numerous homeostatic functions that help protect brain immunity. However, in a dysfunctional state due to aging, environmental factors, or injury, these brain organelles and cells are known to generate excessive metabolic by-products that are toxic to neurons, damaging neuronal proteins, lipids, or nucleic acids in the process. Neuronal mitochondria are essential for energy production, metabolism, and cell survival. They serve as a major site of oxidative metabolism and ROS generation due to heightened neuronal activity (Giorgio et al., 2007; S). While ROS are vital for cytotoxicity and cell signaling, excessive ROS resulting from mitochondrial dysfunction can damage nucleic acids, proteins, membranes, and mitochondrial DNA in conditions such as brain injury, neuroinflammation, and environmental toxicity (Adav et al., 2019). In neurotoxicity and neurodegeneration, microglia dysfunction impacts brain function and CNS homeostasis. During these processes, microglia adopt a "primed" or hypertrophic phenotype, characterized by heightened inflammatory responses to immune triggers, leading to the release of TNF-α, IL-1β, and IL-6 (Gao et al., 2023). This occurs following repeated immune challenges, leading to persistent inflammatory cytokine release, prolonged illness, brain injury, and neurotoxicity (Gao et al., 2023). While such changes are common in the natural aging process, they are amplified in neurodegenerative diseases like AD, PD, ALS, and MS (Gao et al., 2023), as well as in chemically induced neurotoxicity (Gao et al., 2023). Another form of microglia under neurotoxic and neurodegenerative conditions is the senescent state, which involves cell arrest along with increased pathological phenotypes (Rim et al., 2024). Senescent microglia can be worsened by neurotoxicity and neurodegenerative states characterized by oxidative stress and iron overload, along with a reduction in neuroprotective microglial phenotypes such as IL-4, IL-10, IL-13, and tissue growth factor-β (TGF-β). In these states, such depletion may drive disease progression (Rim et al., 2024). This senescent state has been observed in conditions including AD, PD, ALS, and MS (Rim et al., 2024). A related subset, disease-associated microglia (DAM), expresses markers such as Iba1 but has lower levels of homeostatic genes, such as *P2ry12*, and increased risk factors for AD, such as Apoe and Trem2. DAMs are found in CNS regions affected by disease and are linked to neurodegenerative disorders and neurotoxicity (Nakaso, 2024). Several studies have shown that the neuroprotective effects of naringenin and naringin against AD pathology mainly involve the inhibition of reactive gliosis. In APP/PS1 mice, characterized by excessive Aβ expression, tau hyperphosphorylation, and impaired cognitive functions, naringenin significantly reduced reactive gliosis, resulting in decreased microglial and astrocyte dysfunctions, as evidenced by lower expression of Iba1 and GFAP in the hippocampus and cortex. It also decreased pro-inflammatory cytokines, such as TNF-α and IL-1β, and lowered Aβ and hyperphosphorylated tau levels. This was also associated with the inhibition of MAPK signaling pathway activation in the APP/PS1 mice. Regarding mitochondrial damage, naringenin and naringin protect against ROS-related mitochondrial dysfunction through their antioxidant and anti-inflammatory mechanisms, activating the Nrf2/ARE pathways to reduce oxidative stress and DNA damage in the brains of Sprague-Dawley rats (Wang et al., 2017; Angelova et al., 2021). Additionally, naringenin and naringin inhibit TNF-α, ROS, iNOS, and NF-κB, offering neuroprotection against neurodevelopmental disorders such as AD, PD, and ADHD.

These protective effects are linked to the modulation of redox-sensitive transcription factors and cytokine signaling cascades such as NF-κB, PKC, and Nrf2/ARE, as illustrated in Fig. 2. AD is notably the most common neurodegenerative disorder, with profound effects on the sufferer's daily activities due to a loss of cognitive and memory functions (Golde et al., 2022). Management of the disease can involve reversing the infamous Aβ aggregation (Zhang et al., 2023a), tau protein hyperphosphorylation, cholinergic dysfunction , and glutamate excitotoxicity (Conway, 2020), which are the pathogenic pathways hypothesized for the disease. In a study involving social defeat stress (SDS)-induced neuroimmune changes in experimental mice, naringin prevented the release of pro-inflammatory cytokines TNF-α and IL-6, thereby mitigating neuroinflammation in mice exposed to psychosocial stress associated with memory impairment (Camargo et al., 2012; Oladapo et al., 2021). Naringin prevented the SDS-related cognitive decline through the inhibition of AChE activity, the neuronal enzyme responsible for the breakdown of ACh, thereby enhancing ACh availability in neurons (Oladapo et al., 2021). Aβ protein cleaving enzyme 1 (BACE1) is the enzyme responsible for breaking down amyloid proteins, which increases the production of Aβ. Meng and his colleagues, in their studies, established that naringenin inhibits the expression of BACE1, leading to a decrease in Aβ production (Oladapo et al., 2021). Naringenin was found to improve impaired cognition and provide neuroprotection against oxidative stress in an AD rat model induced with aluminum by increasing the levels of antioxidants SOD and CAT in the brain and reducing the levels of MDA and AChE activity. Naringenin also alleviated mitochondrial dysfunction in neurons in ischemia-induced rat models by increasing the expression of Nrf2/ARE proteins in the neurons. This resulted in enhanced mitochondrial transport activity and increased mitochondrial phosphate levels (Wang et al., 2017). Furthermore, naringin restores Aβ-induced field excitatory postsynaptic potential (fEPSP) in the CA1 region of the hippocampus in mice, improving spatial and non-spatial cognitive functions by downregulating COX-2, inactivating Bax and Bcl-2, and increasing CREB, BDNF, and TrkB levels. Naringin also dose-dependently enhances field excitatory postsynaptic potentials (fEPSPs) following theta-burst stimulation in organotypic hippocampal slice cultures (Ga-Young et al., 2023). In another study on hydrocortisone-induced AD-like pathology, characterized by elevated Aβ and tau hyperphosphorylation, naringin's neuroprotective effects also involve modulation of p38 phosphorylation, indicating a role for MAPK/p38 pathways (Meng et al., 2021). Together, these findings reinforce the neuroprotective potential of naringenin and naringin against AD-like neurodegenerative hallmarks, primarily through their antioxidant, antiapoptotic, and anti-inflammatory mechanisms targeting brain organelles and cells such as mitochondria and microglia.

Regarding PD, which is also characterized by reactive gliosis, naringenin's antioxidant and anti-inflammatory properties have shown promising phytonutritional potential in the context of PD (Li et al., 2019). A 70 mg/kg p.o. dose of naringenin administered to a PD mouse model induced with 6-hydroxydopamine by Lu and colleagues activated the Nrf2/ARE pathways to prevent oxidative damage in neurons (Lou et al., 2014). In another study by Wang et al., naringenin upregulated the Nrf2 pathway, leading to the release of BDNF and glia cell line-derived neurotrophic factor (GDNF) from dopaminergic neurons in neuron-glia cultures. Besides oxidative stress and neuroinflammation, dopaminergic neuronal loss related to PD has also been linked to the deposition of α-synuclein, a presynaptic neuronal protein. Naringenin was also shown to decrease α-synuclein, pro-inflammatory cytokines IL-1β and TNF-α, and increase SOD, while restoring dopamine metabolite 3,4-dihydroxyphenylacetic acid (DOPAC) in MPTP-induced dopaminergic degeneration. A similar mechanism of naringenin neuroprotection was observed in human neuroblastoma cells (SH-SY5Y) induced with MPP+ (Mani et al., 2018). Considering the results from the studies outlined above, naringenin could be proposed as a phytonutrient that may serve as a neuroprotective agent for neurodegenerative diseases such as PD. Other disorder studies indicate that naringenin has a neuroprotective effect against schizophrenia. In preclinical trials, naringenin showed neuroprotective effects in ketamine-induced schizophrenia models by inhibiting apoptosis and decreasing oxidative stress markers in rats (George et al., 2020). Furthermore, PTSD, which is influenced by neurobiological, genetic, and environmental factors, often results from traumatic events such as violence or brain injuries. PTSD shares symptoms with schizophrenia, including cognitive impairment and anxiety (Bisson et al., 2015). A recent study demonstrated that naringin provided neuroprotective effects in mice with comorbid PTSD and alcohol use disorder by reducing oxidative stress and neuroinflammation, alongside increased GABAergic neurotransmission in the striatum, prefrontal cortex, and hippocampus (Ben-Azu et al., 2024d). These findings suggest that naringin may have potential for treating schizophrenia and related mental health disorders through its antioxidant and anti-inflammatory mechanisms.

Furthermore, multiple sclerosis (MS) is an autoimmune and inflammatory disease that affects the CNS, causing a progressive loss of neuronal function, including myelin. The pathogenesis of MS has been linked to an imbalance between Th1 and Th2 receptors, which are responsible for secreting pro-inflammatory and anti-inflammatory cytokines, respectively, including excessive free radical production (Ahmad et al., 2014). Naringenin, with its multifaceted targeting ability, has been shown by studies to improve the clinical manifestations of MS through its anti-inflammatory and antioxidant effects (Wang et al., 2018). Naringenin was shown to down-regulate the differentiation of CD4(+) T-cells into Th1, Th17, and Th9 cells, which are triggers of inflammatory processes (Wang et al., 2018). Furthermore, naringenin decreased the levels of pro-inflammatory cytokines IL-1 and TNF-α in mice (Wang et al., 2018). In this context, naringenin was recognized as a promising candidate for activating the epidermal growth factor receptor B4 (ErbB4). ErbB4 is known to play a vital role in oligodendrocyte survival, with its up-regulation responsible for remyelination and nerve cell development (Aguirre et al., 2007). Therefore, the results from studies on naringenin's potency against MS suggest that naringenin might be a promising agent for neuronal protection in MS.

3.7 Ginseng

The term "ginseng" refers to a category of herbal remedies mainly derived from the roots of various plant species, including *Panax ginseng* (Korean), *Eleutherococcus senticosus* (Siberian), and *Panax quinquefolius* (American) (Shahrajabian et al., 2019). These plants, which belong to the *Araliaceae* family, are well known for their adaptogenic properties, which help enhance vitality and resilience to stress (Ratan et al., 2021). One notable and widely studied species is *Panax ginseng*, commonly used in traditional Chinese medicine (Shahrajabian et al., 2019). It is believed to boost energy, combat fatigue, and offer therapeutic benefits for conditions such as immune disorders, aging, diabetes, inflammation, depression, and nervous system diseases (Ratan et al., 2021;).

3.7.1 *Neuroprotective benefits of ginseng in neurological disorders and mechanisms*

The neuroprotective effects of ginseng and its active compounds, ginsenosides, have been thoroughly examined in relation to neurodegenerative and neurodevelopmental disorders (Kim et al., 2013. These ginsenosides have shown the capacity to foster neural regeneration, reduce neuronal damage, and could be beneficial in the treatment of various neurological conditions. The disorders most frequently studied include HD, PD, AD, and acute neurodegenerative events like cerebral ischemia and stroke. In AD, the neuroprotective effects of ginsenosides such as Rg1 and Rb1 are significant. Ginsenoside Rg1 has been shown to reduce Aβ plaque formation by inhibiting CDK5-induced phosphorylation of peroxisome proliferator-activated receptor gamma (PPARγ), thereby increasing IDE expression and decreasing BACE1 and APP expression, thereby improving cognitive function (Quan et al., 2020). Ginsenosides also decrease Aβ deposition by inhibiting NADPH oxidase 2 activation in APP/PS1 mice (Zhang et al., 2021). Rg1 protects mitochondria from Aβ-induced toxicity by increasing membrane potential, ATP levels, and cytochrome c oxidase activity, while decreasing cytochrome c release. Another pathway involved in the pathogenesis of AD is mitophagy, a process for mitochondrial breakdown. Rg1 restores mitophagy through the PINK1-Parkin pathway, thereby alleviating memory deficits in AD models (Wang et al., 2022). Additionally, Rg1 improves cognitive function by activating memory-forming pathways such as PKA/CREB, which are essential for synaptic plasticity. Ginsenoside Rb1 combats neuronal death by preventing apoptosis and promoting neural growth, which protects neurons from damage caused by tau phosphorylation, another AD hallmark (Shi et al., 2010; Fang et al., 2012; Wu et al., 2022). These studies collectively demonstrate that Ginsenoside Rg1 exhibits neuroprotective properties and potential for AD treatment through multiple mechanisms targeting Aβ accumulation and mitochondrial dysfunction.

Ginsenosides, especially Rb1 and Rg1, show promise in reducing this neuronal loss in the pathogenesis of PD (Ardah et al., 2015). Ginsenoside Rb1 inhibits α-synuclein fibrillation and toxicity, breaks down preformed fibrils, and stabilizes non-toxic oligomers (Ardah et al., 2015). In a mouse PD model, Rb1 improves motor deficits, prevents dopaminergic neuron loss, and decreases α-synuclein expression by increasing glutamate transporter activity and modulating glutamatergic transmission (Zhang et al., 2018). It also reduces inflammation and oxidative stress, both of which worsen neuronal damage. Rg1 lessens motor impairment and neuroinflammation in a PD mouse model by targeting α-synuclein abnormalities in the substantia nigra, reducing reactive astrocytes and microglia, and lowering pro-inflammatory cytokine levels. These studies suggest that ginsenosides Rb1 and Rg1 may serve as potential therapeutic agents for PD by mitigating neuronal loss and addressing α-synuclein-related pathology.

In HD, ginsenosides such as Rb1, Rc, and Rg5 inhibit apoptosis, especially in neurons affected by glutamate-induced toxicity, which plays a key role in HD pathogenesis. Studies indicate that the striatum of R6/2 HD transgenic mice becomes increasingly susceptible to glutamate toxicity with age, correlating with decreased levels of glutamate transporters GLT-1 and GLAST (Estrada-Sánchez et al., 2009). This susceptibility is related to altered glutamatergic neurotransmission, particularly changes in NMDA receptor conductance and reduced glutamate transporter levels (Sánchez et al., 2008). While early overactivation of glutamate neurotransmission may cause excitotoxicity, later stages of HD are marked by decreased dopamine and glutamate neurotransmission (André et al., 2010). Ginsenoside Rb2 reduced glutamate-mediated oxidative stress and neuronal death by inhibiting MAPK activity and apoptotic induction factor (AIF) translocation (Kim et al., 2019). Rb1, Rc, and Rg5 also lower intracellular calcium levels, thereby protecting neurons from excitotoxicity, a significant factor in cell death in HD (Kim et al., 2019).

Acute neurodegenerative disorders, such as stroke and cerebral ischemia, as well as stress-induced neurodegeneration, also benefit from ginseng's neuroprotective properties (Rastogi et al., 2015; Oladapo et al., 2021). Its mechanisms of action include attenuating inflammation, suppressing oxidative stress, and activating the cytoprotective transcription factor Nrf2 (Rastogi et al., 2015), including upregulation of BDNF and GABAergic neurotransmission, which helps maintain behavioral stability (Oladapo et al., 2021). Ginseng also prevents neuronal death, reducing both anatomical and functional stroke damage (Rastogi et al., 2015; Zhao et al., 2022). Studies demonstrate that Rb1 significantly reduces infarct volume, improves neurological function, and promotes neurogenesis (Zeng et al., 2014; Shi et al., 2020). Its neuroprotective mechanisms involve anti-inflammatory, anti-oxidative, and anti-apoptotic actions (Shi et al., 2020). Rb1 enhances axonal regeneration and motor function recovery through the cAMP/PKA/CREB signaling pathway. Similarly, notoginsenoside R1 alleviates cerebral ischemia/reperfusion injury by promoting neurogenesis through the BDNF/Akt/CREB pathway (Zhu et al., 2021). These findings suggest that ginsenosides, particularly Rb1, have significant therapeutic potential for ischemic stroke, offering both neuroprotection and neuroregeneration benefits. Although research on ginseng's effects on neurodevelopmental disorders is somewhat limited, its antioxidant and neuroprotective properties suggest possible benefits for conditions such as autism and ADHD (Grosso et al., 2023; Kim et al., 2024). The ability of ginsenosides to reduce oxidative stress and enhance neurogenesis could be explored to improve cognitive function in these disorders, though further studies are necessary.

Additionally, research shows potential benefits for ADHD symptoms, particularly inattention, through dopaminergic/noradrenergic modulation and BDNF/TrkB signaling (Kim et al., 2024). Overall, ginseng displays promising potential as a therapeutic agent for various neurological conditions.

Overall, Ginsenosides such as Rb1, Rg1, and compound K have shown effectiveness in preclinical studies, targeting signaling pathways like PI3K/Akt, TLR4/NF-κB, and Nrf2. While ginseng may have potential for treating neurodegenerative diseases, clinical evidence remains limited, and further research is necessary to fully understand its pharmacokinetics and toxicity (Rajabian et al., 2019; Li et al., 2021). Despite the growing interest in herbal medicines, more clinical trials are required to confirm ginseng's therapeutic benefits in neurological disorders (Li et al., 2021).

4. Phytonutrient cocktail: A spotlight on Jobelyn®

Jobelyn® is a nutraceutical derived from the intensely colored leaf sheaths of *Sorghum bicolor (*L.*) Moench*, a plant cultivated from the West African wild variety. The product is recognized for its rich phytochemical profile, including various phenolic flavonoids such as apigeninidin, luteolinidin, luteolin, and naringenin (Rezaei et al., 2024), as well as abundant minerals, including iron, calcium, magnesium, sodium, selenium, zinc, copper, and vitamins, specifically vitamin B12. Notably, the formulation contains 3-deoxyanthocyanidins, which selectively inhibit COX-2 and reduce fluid exudate volume and white blood cell count during inflammatory responses (Umukoro et al., 2015). Other compounds, such as apigeninidin and naringenin, help mitigate mitochondrial damage and activate PPARs (Požgajová et al., 2024), thereby modulating immune responses and promoting the development of precursor cells into functional oligodendrocytes. Additionally, apigenin and luteolin can counteract 4–hydroxy-2-nonenal, a byproduct of lipid peroxidation that activates caspase-3, which is possibly the most executioner caspase responsible for the biochemical and morphological alterations observed in apoptotic cells (Wu et al., 2019a). These compounds reduce brain injury and offer further benefits by inhibiting NF-kB signaling and attenuating neuroinflammation (Umukoro et al., 2019). Moreover, the multi-modal and synergistic actions of these constituents target multiple pathways involved in various neurodegenerative diseases.

Remarkably, Jobelyn® has attracted attention for its effective neuroprotective effects, which have been assessed through a comprehensive approach involving phytochemical analysis, preclinical studies, and clinical trials to confirm its health benefits and establish its antioxidant, adaptogenic (Ben-Azu et al., nd), anti-inflammatory, immunomodulatory, and anti-aging properties (Oke et al., 2022). Preclinical data demonstrate that Jobelyn®'s neuroprotective effects support neuronal integrity and function, especially in conditions with excess free radical production, a key factor in neurodegenerative diseases. Its high oxygen radical absorbance capacity (ORAC) indicates its ability to scavenge chemical radicals, as evidenced by reduced MDA levels in the prefrontal cortex and hippocampus of rats subjected to alcohol-induced brain stress, indicating decreased lipid peroxidation and oxidative damage. Additionally, Jobelyn® can stabilize lysosomal membranes (Umukoro et al., 2015), which helps prevent cellular harm and maintain cellular homeostasis, protecting against neuronal and liver damage by blocking harmful chemicals, such as MSG (Omogbiya et al., 2020), and reinforcing neuronal membrane integrity. Additionally, Jobelyn® treatment significantly reduced neurodegeneration in the cortico-limbic pathway, as assessed by histological and immunohistochemical analyses (Oyinbo et al., 2018). It regulated the expression of key proteins involved in neuronal survival and apoptosis, such as cellular tumor antigen p53 and neuron-specific enolase (NSE), in alcohol-exposed rats, indicating its role in preserving mitochondrial function, supporting energy metabolism, and reducing cell death (Oyinbo et al., 2018). Chronic inflammation in the CNS is a key feature of many neurodegenerative diseases. The effectiveness of Jobelyn® also includes notable anti-inflammatory properties, alleviating the harmful effects of chronic neuroinflammation through COX-2-selective inhibition, reduced production of pro-inflammatory cytokines, and reduced infiltration of WBCs (Umukoro et al., 2015). This helps maintain neuronal viability and reduce neuroinflammation mediated by activated microglia in neurotoxic conditions (Umukoro et al., 2023). The supplement further inhibits inflammasome activation, notably lowering IL-1β and IL-18 secretion and modulating caspase-1 activity in LPS-primed and ATP-activated macrophages (Dia et al., 2019).

Furthermore, Jobelyn® may provide neuroprotection by modulating neurotransmitter function, notably by decreasing AChE activity and thereby increasing acetylcholine levels, thereby enhancing cognitive functions (Umukoro et al., 2019; Afe et al., 2024). Additionally, it influences glutamate and dopamine transmission due to its ability to prevent psychotic-like behavior in mice exposed to apomorphine and amphetamine (Omogbiya et al., 2013). Jobelyn® may indirectly support healthy glutamate transmission and prevent excitotoxicity (Asehinde et al., 2018; Afe et al., 2024). It also promotes dopaminergic motor functions in mice (Umukoro et al., 2023) by supporting neuronal health and reducing stress-related damage and neurodegeneration (Adeleke et al., 2024). The efficacy of Jobelyn® has also been evaluated in clinical settings, particularly in patients with conditions such as anemia and chronic viral infections. These trials assessed outcomes like hemoglobin levels, immune response, and overall health improvements in participants by addressing anemia-related symptoms (Ayuba et al., 2014). From a sub-chronic toxicity assessment, Jobelyn® showed minor behavioral changes, such as reduced motility and piloerection, with histopathological examinations revealing moderate congestion in the liver and lungs at lethal doses, but no significant pathological changes in the heart and kidneys in mice (Eniojukan and Aina, 2009), and it also appears to pose no adverse effects on pregnant rats (Adebayo et al., 2018). Moreover, the US FDA generally considers Jobelyn® safe for consumption. The preservation of the cytoarchitecture of the prefrontal cortex and hippocampus, which are critically involved in cognitive functions, alongside their ability to influence apoptotic and necrotic signaling pathways, modulate key neuroprotective markers, enhance antioxidant properties, improve neurotransmitter function, and stabilize cellular membranes, positions Jobelyn® as a valuable nutraceutical in health promotion and in the management of neurodegenerative disorders such as AD, PD, alcohol-induced neurodegeneration, and other neurodevelopmental disorders.

4.1 Perspectives

The study highlights that advancing clinical applications requires rigorous, well-designed clinical trials that confirm the neuroprotective efficacy and safety of phytonutrients. It underscores the importance of understanding the mechanisms by which these compounds exert their effects, including antioxidant, anti-inflammatory, and neurorestorative pathways. Promoting interdisciplinary collaboration and translational research is essential to bridge the gap between preclinical successes and human applications. Additionally, focusing on standardizing dosing, improving pharmacokinetics, and ensuring CNS penetration are crucial for successful translation. Incorporating patient-centered approaches and real-world evidence will further aid in developing effective, safe, and targeted neuroprotective therapies based on natural phytonutrients.

4.2 Challenges

1. **Variability in Study Designs**: Preclinical studies often differ widely in their methodologies, including dosages, treatment durations, and animal models used, which can lead to inconsistent results when translated to human studies.
2. **Complexity of Human Biology**: Human biological systems are more complex than those in animal models, making it difficult to predict how findings from preclinical studies will translate to human populations.
3. **Regulatory Hurdles**: The pathway for translating findings from preclinical research to clinical applications is often fraught with regulatory challenges, which can delay the development of new treatments.
4. **Limited Understanding of Mechanisms**: While many natural products show promise, the precise mechanisms by which they exert neuroprotective effects are not always fully understood, complicating their use in clinical settings.

Approaches to Improve Applications

1. **Rigorous Clinical Trials**: Conducting well-designed clinical trials based on preclinical findings can help verify the efficacy and safety of phytonutrients in humans. This involves standardizing dosages and treatment protocols to better match preclinical studies.
2. **Translational Research Initiatives**: Establishing translational research programs that focus on the continuum from bench to bedside can facilitate the sharing of knowledge and resources between preclinical and clinical researchers.
3. **Interdisciplinary Collaboration**: Encouraging collaboration among pharmacologists, neuroscientists, clinicians, and regulatory experts can help address the multifaceted challenges of translating preclinical findings into clinical practice.
4. **Focus on Mechanistic Studies**: Further research into the mechanisms of action of phytonutrients can provide insights that enhance their clinical application. Understanding how these compounds interact with biological pathways involved in neurodegeneration can inform treatment strategies 4.
5. **Patient-Centric Approaches**: Incorporating patient feedback and real-world evidence into the research process can help ensure that clinical applications are relevant and effective for the target population.

By addressing these challenges and adopting these strategies, the potential of neuroprotective phytonutrients can be more effectively harnessed in clinical settings, ultimately benefiting patients with neurodevelopmental and neurodegenerative diseases.

5. Conclusion

The comprehensive evidence outlined in this review highlights the significant potential of neuroprotective phytonutrients as promising candidates for preventing and managing neurodevelopmental and neurodegenerative disorders. These natural compounds, such as curcumin, quercetin, morin, and others, exhibit diverse mechanisms of action, including antioxidant, anti-inflammatory, and neurorestorative pathways, which can influence key pathogenic processes such as oxidative stress, mitochondrial dysfunction, and neuroinflammation. However, despite promising preclinical results, translating these benefits into clinical practice remains challenging due to differences in study designs, bioavailability issues, and limited rigorous human trials. To unlock the full therapeutic potential of these phytonutrients, a focused effort involving standardized clinical trials, mechanistic research, interdisciplinary collaboration, and patient-focused strategies is essential. Ultimately, leveraging the power of these bioactive compounds could transform the field of ethno- and neuropharmacology, providing safer, more accessible, and effective options to address the growing burden of neurodegenerative and neurodevelopmental diseases across all stages of life.

Funding

The authors did not receive any funding for this publication

Conflicts of interest

Authors declare that they have no conflict of interest.

Author contributions

Conceptualization BB-A and BO, while BB-A and IA reviewed the first draft of the concept; writing original draft preparation BO, AAO, EOC, JEO, PI, JCO, DTE, MAO, and BB-A, review and editing BB-A, BO, AAO, EOC, JEO, PI, JCO, DTE, MAO, and BB-A, and IA. supervision BB-A. All authors have read and agreed to the publishing of the manuscript.

Acknowledgements

The Authors are grateful to Prof Isaac Adedara for the invitation to contribute to the special issue on Nutrition and Neurotoxicology—Part B, organized by him and his team.

References

Abbas, S., Latif, M., Shafie, N., Ghazali, M.I., Kormin, F., 2020. Neuroprotective expression of turmeric and curcumin. J. Food Sci. 4 (6), 2366–2381. doi:10.26656/fr.2017.4(6).363.

Adamu, A., Li, S., Gao, F., Xue, G., 2024. The role of neuroinflammation in neurodegenerative diseases: current understanding and future therapeutic targets. Front. Aging Neurosci. 16. doi:10.3389/fnagi.2024.1347987.

Adav, S.S., Park, J.E., Sze, S.K., 2019. Quantitative profiling brain proteomes revealed mitochondrial dysfunction in Alzheimer's disease. Mol. Brain 12 (1), 1–2.

Adebayo, O.G., Ben-Azu, B., Akpofure, E., Modo, E.U., Ndidiamaka, I.P., Enya, J.I., Udoeyop, F.A., Oritsemuelebi, B., Chidebe, E.O., Chimezie, J., Omeiza, N.A., Asiwe, J.N., Davies, S.I., Onasanwo, S.A., 2024. Zingiber officinale Roscoe extract improves nigrostriatal dopaminergic activity in rotenone-induced Parkinsonian mice: implication of COX-2/TNF-α/IL-6 and antioxidant enzyme crosstalk in the immunoinflammatory responses. Phytomed. Plus 4, 100610.

Adedara, I.A., Olabiyi, B.F., Ojuade, T.D., Idris, U.F., Onibiyo, E.M., Farombi, E.O., 2017. Taurine reverses sodium fluoride-mediated increase in inflammation, caspase-3 activity, and oxidative damage along the brain-pituitary-gonadal axis in male rats. Can. J. Physiol. Pharmacol. 95 (9), 1019–1029.

Adkins, E.A., Brunst, K.J., 2021. Impacts of fluoride neurotoxicity and mitochondrial dysfunction on cognition and mental health: a literature review. Int. J. Environ. Res. Public Health 18 (24), 12884. doi:10.3390/ijerph182412884.

Aghababaei, F., Hadidi, M., 2023. Recent advances in potential health benefits of quercetin. Pharmaceuticals 16 (7), 1020. doi:10.3390/ph16071020.

Aguirre, A., Dupree, J.L., Mangin, J.M., Gallo, V., 2007. A functional role for EGFR signaling in myelination and remyelination. Nat. Neurosci. 10 (8), 990–1002. doi:10.1038/nn1932.

Ahmad, S.F., Zoheir, K.M., Abdel-Hamied, H.E., Ashour, A.E., Bakheet, S.A., Attia, S.M., Abd-Allah, A.R., 2014. Amelioration of autoimmune arthritis by naringin through modulation of T regulatory cells and Th1/Th2 cytokines. Cell. Immunol. 287 (1), 112–120. doi:10.1016/j.cellimm.2013.12.007.

Ahmad, S.F., Attia, S.M., Bakheet, S.A., Zoheir, K.M., Ansari, M.A., Korashy, H.M., Abdel-Hamied, H.E., Ashour, A.E., Abd-Allah, A.R., 2015. Naringin attenuates the development of carrageenan-induced acute lung inflammation through inhibition of NF-κB, STAT3 and pro-inflammatory mediators and enhancement of IκBα and anti-inflammatory cytokines. Inflammation 38, 846–857. doi:10.1007/s10753-014-0066-3.

Ahmad, R.S., Hussain, M.B., Sultan, M.T., Arshad, M.S., Waheed, M., Shariati, M.A., Plygun, S., Hashempur, M.H., 2020. Biochemistry, safety, pharmacological activities, and clinical applications of turmeric: a mechanistic review. Evid. Based Complement. Altern. Med. 2020, 7656919. doi:10.1155/2020/7656919.

Ahmed, S., Venkatesan, G., Prapul, D.S., Md, A.R., 2023. Zebrafish as a comprehensive model of neurological diseases. Int. J. Sci. Res. Arch. 10 (1), 018–052.

Ahsan, N., Mishra, S., Jain, M.K., Surolia, A., Gupta, S., 2015. Curcumin pyrazole and its derivative (N-(3-nitrophenylpyrazole) curcumin) inhibit aggregation, disrupt fibrils and modulate toxicity of wild-type and mutant α-synuclein. Sci. Rep. 5, 9862. doi:10.1038/srep09862.

Akinluyi, E.T., Edem, E.E., Dakoru, K.M., Nnodim, C.J., Oladipupo, M.A., Adeoluwa, O.A., Obisesan, A.O., Ben-Azu, B., Adebayo, O.G., 2022. Psychobiotic interventions attenuate chronic alcohol use-mediated exacerbation of posttraumatic stress disorder in rats: the role of gut-liver axis response. J. Complement. Integr. Med. 20 (1), 92–105. doi:10.1515/jcim-2022-0353.

Akman, Ö.A., Esnafoğlu, E., 2023. Evaluation of procalcitonin and C-reactive protein levels in children with autism spectrum disorder and attention deficit hyperactivity disorder. Int. J. Dev. Disabil. 71 (1), 159–167. doi:10.1080/20473869.2023.2213923.

Al Askar, M., Bhat, R.S., Selim, M., Al-Ayadhi, L., El-Ansary, A., 2017. Postnatal treatment using curcumin supplements to amend the damage in VPA-induced rodent models of autism. BMC Complement. Altern. Med. 17, Article 1. doi:10.1186/s12906-017-1774-2.

Alam, M.A., Subhan, N., Rahman, M.M., Uddin, S.J., Reza, H.M., Sarker, S.D., 2014. Effect of citrus flavonoids, naringin and naringenin, on metabolic syndrome and their mechanisms of action. Adv. Nutr. 5 (4), 404–417.

Aleebrahim-Dehkordi, E., Soveyzi, F., Arian, A.S., Hamedanchi, N.F., Hasanpour-Dehkordi, A., Rafieian-Kopaei, M., 2022. Quercetin and its role in reducing the expression of pro-inflammatory cytokines in osteoarthritis. Anti-Inflamm. Anti-Allergy Agents Med. Chem. 21 (3), 153–165.

Alharbi, H.O.A., Alshebremi, M., Babiker, A.Y., Rahmani, A.H., 2025. The role of quercetin, a flavonoid in the management of pathogenesis through regulation of oxidative stress, inflammation, and biological activities. Biomolecules 15 (1), 151. doi:10.3390/biom15010151.

Alla, N., Palatheeya, S., Challa, S.R., Kakarla, R., 2024. Morin attenuated the global cerebral ischemia via antioxidant, anti-inflammatory, and antiapoptotic mechanisms in rats. Metab. Brain Dis. 39, 1323–1334. doi:10.1007/s11011-024-01410-y.

Althobaiti, N.A., 2025. Heavy metals exposure and Alzheimer's disease: underlying mechanisms and advancing therapeutic approaches. Behav. Brain Res. 476, 115212. doi:10.1016/j.bbr.2024.115212.

Amenotie, A.J., Ben-Azu, B., Esuku, D.T., Chijioke, B.S., Abo, E., Ozah, E.O., Lawrence, E.O., Efejene, O.I., Onyeukwu, O.B., Alabi, B.A., Ajayi, A.M., 2025. Sabinene inhibits lipopolysaccharide-induced memory decline by enhancing cholinergic function, decreasing molybdenum enzymes, and suppressing oxidative stress and neuroinflammation. Neurotox. Res. 43, 26.

Amidfar, M., Garcez, M.L., Askari, G., Bagherniya, M., Khorvash, F., Golpour-Hamedani, S., de Oliveira, J., 2024. Role of BDNF signaling in the neuroprotective and memory-enhancing effects of flavonoids in Alzheimer's disease. CNS Neurol. Disord. Drug Targets 23 (8), 984–995.

André, V.M., Cepeda, C., Levine, M.S., 2010. Dopamine and glutamate in Huntington's disease: a balancing act. CNS Neurosci. Ther. 16 (3), 163–178. doi:10.1111/j.1755-5949.2010.00144.x.

Angelova, P.R., Esteras, N., Abramov, A.Y., 2021. Mitochondria and lipid peroxidation in the mechanism of neurodegeneration: finding ways for prevention. Med. Res. Rev. 41 (2), 770–784. doi:10.1002/med.21740.

Anilkumar, S., Wright-Jin, E., 2024. NF-κB as an inducible regulator of inflammation in the central nervous system. Cells 13 (6), 485. doi:10.3390/cells13060485.

Anyanwu, G.E., Umeano, A.V., Ojiakor, V.O., Katchy, A.U., Anyanwu, C.N., Fakorede, S., 2025. Morin mitigates methamphetamine-induced neurotoxicity: Effects on motor and cognitive function. J Exp Pharm 307–321.

Anyaoku, I., Igwilo, I.O., Ezekwesili, C.N., 2023. Nutritional profile and health promoting properties of *Curcuma longa* (Linn)(Turmeric). Int. J. Res. Publ. Rev. 4 (11), 2731–2740.

Arab, A., Mostafalou, S., 2022. Neurotoxicity of pesticides in the context of CNS chronic diseases. Int. J. Environ. Health Res. 32 (12), 2718–2755.

Ardah, M.T., Paleologou, K.E., Lv, G., Menon, S.A., Khair, S.B.A., Lu, J.H., Safieh-Garabedian, B., Al-Hayani, A.A., Eliezer, D., Li, M., El-Agnaf, O.M., 2015. Ginsenoside Rb1 inhibits fibrillation and toxicity of alpha-synuclein and disaggregates preformed fibrils. Neurobiol. Dis. 74, 89–101.

Asejeje, F.O., Abiola, M.A., Adeyemo, O.A., Ogunro, O.B., Ajayi, A.M., 2024. Exogenous monosodium glutamate exacerbates lipopolysaccharide-induced neurobehavioral deficits, oxidative damage, neuroinflammation, and cholinergic dysfunction in rat brain. Neurosci. Lett. 825, 137710. doi:10.1016/j.neulet.2024.137710.

Ashish Jain, R., 2017. To study the attention deficit hyperactivity disorder (ADHD) and to evaluate the cognitive, hyperactive and impulsive behavior against 6-OHDA Hbr lesioned Sprague Dawley neonates using Hyperaxe, curcumin & quercetin. Doctoral dissertation. CL Baid Metha College of Pharmacy, Chennai.

Ashrafizadeh, M., Samarghandian, S., Hushmandi, K., Zabolian, A., Shahinozzaman, M., Saleki, H., … Najafi, M., 2021. Quercetin in attenuation of ischemic/reperfusion injury: a review. Curr. Mol. Pharmacol. 14 (4), 537–558.

Asiwe, J.N., Oritsemuelebi, B., 2024a. Environmental toxicant-mediated cardiovascular diseases: an insight into the mechanism and possible preventive strategy. Toxicol. Environ. Health Sci. 16 (1), 1–19.

Asiwe, J.N., Moke, E.G., Ben-Azu, B., Onuelu, J.E., Nwabuoku, U.S., Anachuna, K.K., Demaki, W.E., Chidebe, E.O., Oritsemuelebi, B., 2024b. Hepato-renal oxidative disturbances following acute β-adrenergic stimulation by isoprenaline in male Wistar rat: attenuative role of taurine, a β-amino acid. Nutrire 49, 25.

Asuku, A.O., Ayinla, M.T., Ajibare, A.J., Adeyemo, M.B., Adeyemo, R.O., 2024. Heavy metals and emerging contaminants in foods and food products associated with neurotoxicity. In: Emerging Contaminants in Food and Food Products. CRC Press, pp. 236–250.

Atoki, A.V., Aja, P.M., Ondari, E.N., Shinkafi, T.S., 2023. Advances in Alzheimer's disease therapeutics: biochemistry, exploring bioactive compounds and novel approaches. Int. J. Food Prop. 26 (1), 2091–2127.

Avola, R., Furnari, A.G., Graziano, A.C.E., Russo, A., Cardile, V., 2024. Management of the brain: essential oils as promising neuroinflammation modulators in neurodegenerative diseases. Antioxidants 13 (2), 178.

Awogbindin, I.O., Ben-Azu, B., Olusola, B.A., Akinluyi, E.T., Adeniyi, P.A., Di Paolo, T., Tremblay, M.È., 2021. Microglial implications in SARS-CoV-2 infection and COVID-19: lessons from viral RNA neurotropism and possible relevance to Parkinson's disease. Front. Cell. Neurosci. 15, 670298. doi:10.3389/fncel.2021.670298.

Azlan, U.K., Khairul Annuar, N.A., Mediani, A., Aizat, W.M., Damanhuri, H.A., Tong, X., … Hamezah, H.S., 2023. An insight into the neuroprotective and anti-neuroinflammatory effects and mechanisms of *Moringa oleifera*. Front. Pharmacol. 13, 1035220.

Bachewal, P., Gundu, C., Yerra, V.G., Kalvala, A.K., Areti, A., Kumar, A., 2018. Morin exerts neuroprotection via attenuation of ROS-induced oxidative damage and neuroinflammation in experimental diabetic neuropathy. Biofactors 44, 109–122.

Bahramsoltani, R., Rahimi, R., Farzaei, M.H., 2017. Pharmacokinetic interactions of curcuminoids with conventional drugs: a review. J. Ethnopharmacol. 209, 1–12.

Balakumar, P., Alqahtani, T., Alqahtani, A., Lakshmiraj, R.S., Singh, G., Rupeshkumar, M., Sundram, K., 2022. A unifying perspective in blunting the limited oral bioavailability of curcumin: a succinct look. Curr. Drug Metab. 23 (11), 897–904.

Balistreri, C.R., Monastero, R., 2023. Neuroinflammation and neurodegenerative diseases: how much do we still not know? Brain Sci. 14 (1), 19.

Banaceyeh, S., Razavi, B.M., Hosseinzadeh, H., 2024. Neuroprotective effects of morin against cadmium- and arsenic-induced cell damage in PC12 neurons. Biol. Trace Elem. Res. doi:10.1007/s12011-024-04407-x.

Bassani, T.B., Turnes, J.M., Moura, E.L.R., et al., 2017. Effects of curcumin on short-term spatial and recognition memory, adult neurogenesis and neuroinflammation in a streptozotocin-induced rat model of dementia of Alzheimer's type. Behav. Brain Res. 335, 41–54. doi:10.1016/j.bbr.2017.08.014.

Bayazid, A.B., Lim, B.O., 2022. Quercetin is an active agent in berries against neurodegenerative diseases progression through modulation of Nrf2/HO1. Nutrients 14 (23), 5132.

Behl, T., Makkar, R., Sehgal, A., Singh, S., Sharma, N., Zengin, G., Brisc, C., 2021. Current trends in neurodegeneration: cross talks between oxidative stress, cell death, and inflammation. Int. J. Mol. Sci. 22 (14), 7432.

Ben-Azu, B., Aderibigbe, A.O., Omogbiya, I.A., Ajayi, A.M., Owoeye, O., Olonode, E.T., Iwalewa, E.O., 2018a. Probable mechanisms involved in the antipsychotic-like activity of morin in mice. Biomed. Pharmacother. 105, 1079–1090. doi:10.1016/j.biopha.2018.06.057.

Ben-Azu, B., Aderibigbe, A.O., Eneni, A.O., Ajayi, A.M., Umukoro, S., Iwalewa, E.O., 2018b. Morin attenuates neurochemical changes and increased oxidative/nitrergic stress in brains of mice exposed to ketamine: prevention and reversal of schizophrenia-like symptoms. Neurochem. Res. 43 (9), 1745–1755. doi:10.1007/s11064-018-2590-z.

Ben-Azu, B., Aderibigbe, A.O., Ajayi, A.M., Eneni, A.O., Umukoro, S., Iwalewa, E.O., 2018c. Involvement of GABAergic, BDNF and Nox-2 mechanisms in the prevention and reversal of ketamine-induced schizophrenia-like behavior by morin in mice. Brain Res. Bull. 139, 292–306. doi:10.1016/j.brainresbull.2018.03.006.

Ben-Azu, B., Nwoke, E.E., Umukoro, S., Aderibigbe, A.O., Ajayi, A.M., Iwalewa, E.O., 2018d. Evaluation of the neurobehavioral properties of naringin in Swiss mice. Drug Res. 68 (8), 465–474. doi:10.1055/a-0575-3730.

Ben-Azu, B., Aderibigbe, A.O., Ajayi, A.M., Eneni, A.E.O., Omogbiya, I.A., Owoeye, O., Umukoro, S., Iwalewa, E.O., 2019a. Morin decreases cortical pyramidal neuron degeneration via inhibition of neuroinflammation in mouse model of schizophrenia. Int. Immunopharmacol. 70, 338–353.

Ben-Azu, B., Nwoke, E.E, Aderibigbe, A.O, Omogbiya, I.A, Ajayi, A.M, Olonode, E.T, Umukoro, S., Iwalewa, E.O, 2019b. Possible neuroprotective mechanisms of action involved in the neurobehavioral property of naringin in mice. Biomed. Pharmacother. 109, 536–546.

Ben-Azu, B., Aderibigbe, A.O., Ajayi, A.M., Umukoro, S., Iwalewa, E.O., 2019c. Involvement of l-arginine-nitric oxide pathway in the antidepressant and memory promoting effects of morin in mice. Drug Dev. Res. 80 (8), 1071–1079. doi:10.1002/ddr.21588.

Ben-Azu, B., Emokpae, O., Ajayi, A.M., … Iwalewa, E.O., 2020. Repeated psychosocial stress causes glutamic acid decarboxylase isoform-67, oxidative-Nox-2 changes and neuroinflammation in mice: prevention by treatment with a neuroactive flavonoid, morin. Brain Res. 1744, 146917. doi:10.1016/j.brainres.2020.146917.

Ben-Azu, B., Aderibigbe, A.O, Ajayi, A.M, Omogbiya, A.I, Uruaka, C.I, Umukoro, S., Iwalewa, E.O, 2021. Evaluation of the role of monoaminergic and nonmonoaminergic systems in the psychotropic effects of morin in mice: an interaction study with receptor blockers. Nutrire 46 (8), 1–11.

Ben-Azu, B., Del Re, E.C., VanderZwaag, J., Carrier, M., Keshavan, M., Khakpour, M., Tremblay, M.È., 2023. Emerging epigenetic dynamics in the gut-microglia brain axis: experimental and clinical implications for accelerated brain aging in schizophrenia. Front. Cell. Neurosci. 17, 1139357. doi:10.3389/fncel.2023.1139357.

Ben-Azu, B., Adebesin, A., Moke, G.E., Ojiokor, V.O., Olusegun, A., Jarikre, T.A., Akinluyi, E.T., Olukemi, O.A., Omeiza, N.A., Nkenchor, P., Niemogha, A.R., Ewere, E.D., Igwoku, C., Omamogho, F., 2024a. Alcohol exacerbates psychosocial stress-induced neuropsychiatric symptoms: attenuation by geraniol. Neurochem. Int. 177, 105748. doi:10.1016/j.neuint.2024.105748.

Ben-Azu, B., Fokoua, A.R., Annafi, O.S., Adebayo, O.G., Del Re, E.C., Okuchukwu, N., Aregbesola, G.J., Ejenavi, A.C., Isiwele, D.M., Efezino, A.J., Okpu, I.D., 2024a. Effective action of silymarin against ketamine-induced schizophrenia in male mice: insight into the biochemical and molecular mechanisms of action. J. Psychiatr. Res. 179, 141–155. doi:10.1016/j.jpsychires.2024.09.003.

Ben-Azu, B., Moke, E.G., Chris-Ozoko, L.E., Jaiyeoba-Ojigho, E.J., Adebayo, O.G., Ajayi, A.M., Oyovwi, M.O., Odjugo, G., Omozojie, V.I., Ejomafuwe, G., Onike, N., Eneni, A.O., Ichipi-Ifukor, C.P., Achuba, I.F., 2024b. Diosgenin alleviates alcohol-mediated escalation of social defeat stress and the neurobiological sequalae. Psychopharmacology 241 (4), 785–803. doi:10.1007/s00213-023-06509-1.

Ben-Azu, B., Toloyai, P.Y., Adebesin, A., Ojiokor, V.O., Adebayo, O.G., Fokoua, A.R., Moke, G.E., Ejukolemu, E.J., Akpojevughe, I.-O.O., Abdulkadir, A.M., Okwuchi, E., 2024c. Alcohol exacerbates post-traumatic stress psychiatric behavior and its neuropathological sequelae in experimental mice: preventive effects of morin. Alcohol 15–29. doi:10.1016/j.alcohol.2024.07.009.

Ben-Azu, B., Oritsemuelebi, B., Oghorodi, A.M., Adebesin, A., Isibor, H., Eduviere, A.T., … Uzochukwu-Godfrey, F.C., 2024d. Psychopharmacological interaction of alcohol and posttraumatic stress disorder: effective action of naringin. Eur. J. Pharmacol. 978, 176791. doi:10.1016/j.ejphar.2024.176791.

Ben-Azu, B., Oghorodi, A.M., Oritsemuelebi, B., Chidebe, E.O., Ben-Azu, B., Oghorodi, A.M., Oritsemuelebi, B., Chidebe, E.O. (2024e). A case for the neuroprotective potential of African phytochemicals in the management of Alzheimer's disease. IntechOpen. https://doi.org/10.5772/intechopen.112517.

Ben-Azu, B., Chidebe, E.O, Toloyai, P.Y, Annafic, O.S, Oritsemuelebi, B., Asiwe, J., Moke, G.E, Ajayi, A.M, Chukwuebuka, N.B, Onuelua, J.E, Utomi, C.W, Iregbeyen, U.F, Raheem, D.O, 2024g. Adaptogenic action of diosgenin againsts chronic unpredictable mild stress-induced neuroimmune dysfunction of HPA axis reverses psychiatric behavior in mice. Clin. Tradit. Med. Pharmacol. 5, 200148.

Ben-Azu, B., Adebayo, O.G., Fokoua, A.R., Onuelu, J.E., Asiwe, J.N., Moke, E.G., Omogbiya, A.I., Okpara, O.L., Okoro, J.E., Oghenevwerutevwe, O.M., Uruaka, C.I., 2024h. Containment of neuroimmune challenge by diosgenin confers amelioration of neurochemical and neurotrophic dysfunctions in ketamine-induced schizophrenia in mice. Brain Disord. 13, 100122.

Ben-Azu, B., Adebayo, O.G., Fokoua, A.R., Oritsemuelebi, B., Chidebe, E.O., Nwogueze, C.B., Kumanwee, L., Uyere, G.E., Emuakpeje, M.T., 2024i. Antipsychotic effect of diosgenin in ketamine-induced murine model of schizophrenia: involvement of oxidative stress and cholinergic transmission. IBRO Neurosci. Rep. 16, 86–97.

Ben-Azu, B., Asiwe, J.N, Oritsemuelebi, B., Chidebe, E.O, Onuelua, J.E, Isibor, H., Ogheneoruese, O., Demaki, W.E, Otuachaa, S.O, Omogbiya, A.I, 2025. Morin inhibits oxidative stress and inflammation in the cardiorenal system associated with post-traumatic stress and alcohol use disorders in mice. Clin. Tradit. Med. Pharmacol. 6, 200196.

Bhandari, R., Kuhad, A., 2015. Neuropsychopharmacotherapeutic efficacy of curcumin in experimental paradigm of autism spectrum disorders. Life Sci. 141, 156–169. doi:10.1016/j.lfs.2015.09.012.

Bhat, A., Mahalakshmi, A.M., Ray, B., Tuladhar, S., Hediyal, T.A., Manthiannem, E., Padamati, J., Chandra, R., Chidambaram, S.B., Sakharkar, M.K., 2019. Benefits of curcumin in brain disorders. Biofactors 45 (5), 666–689.

Bisson, J.I., Cosgrove, S., Lewis, C., Roberts, N.P., 2015. Post-traumatic stress disorder. BMJ 351, h6161. doi:10.1136/bmj.h6161.

Bondy, S.C., 2024. Mitochondrial dysfunction as the major basis of brain aging. Biomolecules 14 (4), 402.

Brewer-Smyth, K., 2022. Adverse childhood experiences and neurodevelopment. In: Adverse Childhood Experiences: The Neuroscience of Trauma, Resilience and Healing Throughout the Life Course. Springer International Publishing, pp. 27–53.

Buccellato, F.R., D'Anca, M., Fenoglio, C., Scarpini, E., Galimberti, D., 2021. Role of oxidative damage in Alzheimer's disease and neurodegeneration: from pathogenic mechanisms to biomarker discovery. Antioxidants 10 (9), 1353. doi:10.3390/antiox10091353.

Bulbula, D.D., 2021. Comprehensive review on turmeric (Curcuma longa L.) as a medicinal plant and its nutraceuticals quality to humans. Cancer Ther. Oncol. Int. J. 18 (3) 2473-554X.

Burapan, S., Kim, M., Paisooksantivatana, Y., Eser, B.E., Han, J., 2020. Thai Curcuma species: antioxidant and bioactive compounds. Foods 9 (9), 1219. doi:10.3390/foods9091219.

Burda, S., Oleszek, W., 2001. Antioxidant and antiradical activities of flavonoids. J. Agric. Food Chem. 49 (6), 2774–2779. doi:10.1021/jf0014137.

Cai, B., Zhong, L., Wang, Q., Xu, W., Li, X., Chen, T., 2023. Curcumin alleviates 1-methyl-4-phenyl-1,2,3,6-tetrahydropyridine-induced Parkinson's disease in mice via modulating gut microbiota and short-chain fatty acids. Front. Pharmacol. 14, 1198335. doi:10.3389/fphar.2023.1198335.

Cai, B., Wang, Q., Zhong, L., Liu, F., Wang, X., Chen, T., 2024. Integrating network pharmacology, transcriptomics to reveal the neuroprotective effects of curcumin activating the PI3K/AKT pathway in Parkinson's disease. Drug Des. Dev. Ther. 18, 2869–2881. doi:10.2147/DDDT.S462333.

Calderaro, A., Patanè, G.T., Tellone, E., Barreca, D., Ficarra, S., Misiti, F., Laganà, G., 2022. The neuroprotective potentiality of flavonoids on Alzheimer's disease. Int. J. Mol. Sci. 23 (23), 14835.

Caletti, G., Herrmann, A.P., Pulcinelli, R.R., Steffens, L., Moras, A.M., Vianna, P., Chies, J., Moura, D.J., Barros, H., Gomez, R., 2018. Taurine counteracts the neurotoxic effects of streptozotocin-induced diabetes in rats. Amino Acids 50, 95–104.

Camargo, C.A., Gomes-Marcondes, M.C.C., Wutzki, N.C., Aoyama, H., 2012. Naringin inhibits tumor growth and reduces interleukin-6 and tumor necrosis factor α levels in rats with Walker 256 carcinosarcoma. Anticancer Res. 32 (1), 129–133.

Caspi, R., Billington, R., Fulcher, C.A., Kescler, I.M., Kothari, A., Krummenacker, M., et al., 2018. The MetaCyc database of metabolic pathways and enzymes. Nucleic Acids Res. 46, D639. doi:10.1093/nar/gkx935.

Çelik, H., Kucukler, S., Çomaklı, S., Özdemir, S., Caglayan, C., Yardım, A., Kandemir, F.M., 2020. Morin attenuates ifosfamide-induced neurotoxicity in rats via suppression of oxidative stress, neuroinflammation and neuronal apoptosis. Neurotoxicology 76, 126–137. doi:10.1016/j.neuro.2019.11.002.

Che, Y., Hou, L., Sun, F., Zhang, C., Liu, X., Piao, F., Zhang, D., Li, H., Wang, Q., 2018. Taurine protects dopaminergic neurons in a mouse Parkinson's disease model through inhibition of microglial M1 polarization. Cell Death Dis. 9, 435.

Chen, N., Geng, Q., Zheng, J., He, S., Huo, X., Sun, X., 2014. Suppression of the TGF-β/Smad signaling pathway and inhibition of hepatic stellate cell proliferation play a role in the hepatoprotective effects of curcumin against alcohol-induced hepatic fibrosis. Int. J. Mol. Med. 34 (4), 1110–1116. doi:10.3892/ijmm.2014.1867.

Chen, L., Sun, L., Liu, Z., Wang, H., Xu, C., 2016. Protection afforded by quercetin against H2O2-induced apoptosis on PC12 cells via activating PI3K/Akt signal pathway. J. Recept. Signal Transduct. 36 (1), 98–102. doi:10.3109/10799893.2015.1049516.

Chen, P., Miah, M.R., Aschner, M., 2016. Metals and neurodegeneration. F1000Research 5. doi:10.12688/f1000research.7431.1, F1000 Faculty Rev-366.

Chen, Y., Li, Y., Xu, H., Li, G., Ma, Y., Pang, Y.J., 2017. Morin mitigates oxidative stress, apoptosis, and inflammation in cerebral ischemic rats. Afr. J. Tradit. Complement. Altern. Med. 14, 348–355.

Chen, M., Chang, Y.Y., Huang, S., Xiao, L.H., Zhou, W., Zhang, L.Y., ... Zhang, K., 2018. Aromatic-Turmerone attenuates LPS-induced neuroinflammation and consequent memory impairment by targeting TLR4-dependent signaling pathway. Mol. Nutr. Food Res. 62 (2), 1700281. doi:10.1002/mnfr.201700281.

Chiang, M.C., Tsai, T.Y., Wang, C.J., 2023. The potential benefits of quercetin for brain health: a review of anti-inflammatory and neuroprotective mechanisms. Int. J. Mol. Sci. 24 (7), 6328. doi:10.3390/ijms24076328.

Cianciulli, A., Calvello, R., Porro, C., Trotta, T., Salvatore, R., Panaro, M.A., 2016. PI3k/Akt signalling pathway plays a crucial role in the anti-inflammatory effects of curcumin in LPS-activated microglia. Int. Immunopharmacol. 36, 282–290. doi:10.1016/j.intimp.2016.05.007.

Clemente-Suárez, V.J., Redondo-Flórez, L., Beltrán-Velasco, A.I., Ramos-Campo, D.J., Belinchón-deMiguel, P., Martínez-Guardado, I., … Tornero-Aguilera, J.F., 2023. Mitochondria and brain disease: a comprehensive review of pathological mechanisms and therapeutic opportunities. Biomedicines 11 (9), 2488. doi:10.3390/biomedicines11092488.

Comer, A.L., Carrier, M., Tremblay, M.È., Cruz-Martín, A., 2020. The inflamed brain in schizophrenia: the convergence of genetic and environmental risk factors that lead to uncontrolled neuroinflammation. Front. Cell. Neurosci. 14, 274. doi:10.3389/fncel.2020.00274.

Conway, M.E., 2020. Alzheimer's disease: targeting the glutamatergic system. Biogerontology 21, 257–274. doi:10.1007/s10522-020-09860-4.

Cox, F.F., Misiou, A., Vierkant, A., Ale-Agha, N., Grandoch, M., Haendeler, J., Altschmied, J., 2022. Protective effects of curcumin in cardiovascular diseases—impact on oxidative stress and mitochondria. Cells 11 (3), 342. doi:10.3390/cells11030342.

Curpan, A.S., Luca, A.C., Ciobica, A., 2021. Potential novel therapies for neurodevelopmental diseases targeting oxidative stress. Oxidative Med. Cell. Longev. 2021, 6640206.

Dórea, J.G., 2021. Exposure to environmental neurotoxic substances and neurodevelopment in children from Latin America and the Caribbean. Environ. Res. 192, 110199.

Davis, J.M., Murphy, E.A., Carmichael, M.D., Davis, B., 2009. Quercetin increases brain and muscle mitochondrial biogenesis and exercise tolerance. Am. J. Physiol. Regul. Integr. Comp. Physiol. 296 (4), R1071–R1077.

de Sousa Macedo, L.L.B., Antunes, F.T.T., de Andrade Alvarenga, W., Batista, M.C.C., de Moura, M.S.B., Farias, M.N.L., & de Souza, A.H. (2021). Curcumin for attention deficit hyperactivity disorder: a behavioral preliminary investigation. Naunyn Schmiedebergs Arch Pharmacol. 2022 Jul; 395 (7), 803–813. https://doi.org/10.1007/s00210-022-02236-0.

Deshpande, G., Beetch, J.E., Heller, J.G., Naqvi, O.H., Kuhn, K.G., 2023. Assessing the influence of climate change and environmental factors on the top tick-borne diseases in the United States: a systematic review. Microorganisms 12 (1), 50.

Domínguez-Oliva, A., Hernández-Ávalos, I., Martínez-Burnes, J., Olmos-Hernández, A., Verduzco-Mendoza, A., Mota-Rojas, D., 2023. The importance of animal models in biomedical research: current insights and applications. Animals 13 (7), 1223.

Dong, S., Zeng, Q., Mitchell, E.S., Xiu, J., Duan, Y., Li, C., … Zhao, Z., 2012. Curcumin enhances neurogenesis and cognition in aged rats: implications for transcriptional interactions related to growth and synaptic plasticity. PLoS One 7 (2), e31211.

Dorstyn, L., Akey, C.W., Kumar, S., 2018. New insights into apoptosome structure and function. Cell Death Differ. 25, 1194–1208. doi:10.1038/s41418-017-0025-z.

Drewnowski, A., Henderson, S.A., Shore, A.B., 1997. Taste responses to naringin, a flavonoid, and the acceptance of grapefruit juice are related to genetic sensitivity to 6-n-propylthiouracil. Am. J. Clin. Nutr. 66 (2), 391–397. doi:10.1093/ajcn/66.2.391.

Du, Y., Qu, J., Zhang, W., Bai, M., Zhou, Q., Zhang, Z., Li, Z., Miao, J., 2016. Morin reverses neuropathological and cognitive impairments in APPswe/PS1dE9 mice by targeting multiple pathogenic mechanisms. Neuropharmacology 108, 1–13.

Dusabumuremyi, J.C., Duhoranimana, E., Chen, X., 2022. Health benefits of turmeric and ginger. In: Handbook of Nutraceuticals and Natural Products: Biological, Medicinal, and Nutritional Properties and Applications, 2. pp. 259–268https://doi.org/10.1002/9781119746843.ch24

Ege, D., 2021. Action mechanisms of curcumin in Alzheimer's disease and its brain-targeted delivery. Materials 14 (12), 3332.

Elbini-Dhouib, I., Doghri, R., Ellefi, A., Degrach, I., Srairi-Abid, N., Gati, A., 2021. Curcumin attenuated neurotoxicity in sporadic animal model of Alzheimer's disease. Molecules 26 (10), 3011. doi:10.3390/molecules2610301.

Elhawary, E.A., Moussa, A.Y., Singab, A.N.B., 2024. Genus Curcuma: chemical and ethnopharmacological role in aging process. BMC Complement. Med. Ther. 24 (1), 31.

Elizabeth, A., Adegbuyi, A., Olusegun, A., et al., 2020. Morin hydrate attenuates chronic stress-induced memory impairment and degeneration of hippocampal subfields in mice: the role of oxidative, nitrergic, and neuroinflammatory pathways. Metab. Brain Dis. 35 (7), 1145–1156. doi:10.1007/s11011-020-00595-2.

Emudainohwo, J.OT., Ben-Azu, B., Adebayo, O.G, Aduema, W., Uruaka, C., Ajayi, A.M, Okpakpor, E.E, Ozolua, R.I, 2022. Normalization of HPA axis, cholinergic neurotransmission, and inhibiting brain oxidative and inflammatory dynamics are associated with the adaptogenic-like effect of rutin against psychosocial defeat stress. J. Mol. Neurosci. doi:10.1007/s12031-022-02084-w.

Eneni, A.O, Ben-Azu, B., Ajayi, A.M, Aderibigbe, A.O., Benneth, 2023. Lipopolysaccharide exacerbates ketamine-induced psychotic-like behavior, oxidative stress, and neuroinflammation in mice: ameliorative effect of diosmin. J. Mol. Neurosci. doi:10.1007/s12031-022-02077-9.

Eneni, A.O, Ben-Azu, B., Ajayi, A.M, Aderibigbe, A.O., 2025. Ketamine-induced behavioral perturbations, redox imbalances, and neurotransmitter deficits in mice: the preventive and reversal neuromodulatory potential of diosmin as an antipsychotic. Brain Disord. 18, 100210.

Engwa, G.A., Nweke, F.N., Nkeh-Chungag, B.N., 2022. Free radicals, oxidative stress-related diseases, and antioxidant supplementation. Altern. Ther. Health Med. 28 (1), 26–38.

Esnafoglu, E., Subaşi, B., 2022. Association of low 25-OH-vitamin D levels and peripheral inflammatory markers in patients with autism spectrum disorder: vitamin D and inflammation in autism. Psychiatry Res. 316, 114735. doi:10.1016/j.psychres.2022.114735.

Espinós, C., Galindo, M.I., García-Gimeno, M.A., Ibáñez-Cabellos, J.S., Martínez-Rubio, D., Millán, J.M., … Pallardó, F.V., 2020. Oxidative stress, a crossroad between rare diseases and neurodegeneration. Antioxidants 9 (4), 313. doi:10.3390/antiox9040313.

Estrada-Sánchez, A.M., Montiel, T., Segovia, J., Massieu, L., 2009. Glutamate toxicity in the striatum of the R6/2 Huntington's disease transgenic mice is age-dependent and correlates with decreased levels of glutamate transporters. Neurobiol. Dis. 34 (1), 78–86. doi:10.1016/j.nbd.2009.01.011.

Fan, L., Zhang, Z., 2024. Therapeutic potential of curcumin on the cognitive decline in animal models of Alzheimer's disease: a systematic review and meta-analysis. Naunyn-Schmiedeb. Arch. Pharmacol. 397 (7), 4499–4509. doi:10.1007/s00210-024-02946-7.

Fang, F., Chen, X., Huang, T., Lue, L.F., Luddy, J.S., Yan, S.S., 2012. Multi-faced neuroprotective effects of Ginsenoside Rg1 in an Alzheimer mouse model. Biochim. Biophys. Acta (BBA) Mol. Basis Dis. 1822 (2), 286–292. doi:10.1016/j.bbadis.2011.10.012.

Farooqui, A.A., 2016. Effects of curcumin on oxidative stress in animal models and patients with Alzheimer disease. In: Therapeutic potentials of curcumin for Alzheimer disease. Springer, Cham, pp. 83–112. doi:10.1007/978-3-319-15889-1_6.

Feng, A.C., Thomas, B.J., Purbey, P.K., de Melo, F.M., Liu, X., Daly, A.E., … Smale, S.T., 2024. The transcription factor NF-κB orchestrates nucleosome remodeling during the primary response to Toll-like receptor 4 signaling. Immunity 57 (3), 462–477.e9. doi:10.1016/j.immuni.2024.02.004.

Froger, N., Sahel, J.A., Picaud, S., 2014. Taurine deficiency and the eye. In: Preedy, V.R. (Ed.), Handbook of Nutrition, Diet and the Eye. Academic Press, pp. 505–513. doi:10.1016/B978-0-12-401717-7.00051-4.

Fuloria, S., Mehta, J., Chandel, A., Sekar, M., Rani, N.N.I.M., Begum, M.Y., … Fuloria, N.K., 2022. A comprehensive review on the therapeutic potential of Curcuma longa Linn. in relation to its major active constituent curcumin. Front. Pharmacol. 13, 820806. doi:10.3389/fphar.2022.820806.

Fuloria, N.K., Fuloria, S., Sekar, M., Narain, K., Kumar, S., 2025. Introduction to environmental toxicants. In: Gut Microbiome and Environmental Toxicants. CRC Press, pp. 1–22.

Gao, C., Jiang, J., Tan, Y., Chen, S., 2023. Microglia in neurodegenerative diseases: mechanism and potential therapeutic targets. Signal Transduct Target Ther 8 (1), 359. doi:10.1038/s41392-023-01588-0.

Gebara, E., Udry, F., Sultan, S., Toni, N., 2015. Taurine increases hippocampal neurogenesis in aging mice. Stem Cell Res. 14, 369–379. doi:10.1016/j.scr.2015.04.001.

George, M.Y., Menze, E.T., Esmat, A., Tadros, M.G., El-Demerdash, E., 2020. Potential therapeutic antipsychotic effects of naringin against ketamine-induced deficits in rats: involvement of Akt/GSK-3β and Wnt/β-catenin signaling pathways. Life Sci. 249, 117535. doi:10.1016/j.lfs.2020.117535.

Ginhoux, F., Greter, M., Leboeuf, M., Nandi, S., See, P., Gokhan, S., … Merad, M., 2010. Fate mapping analysis reveals that adult microglia derive from primitive macrophages. Science 330 (6005), 841–845. doi:10.1126/science.1194637.

Giorgio, M., Trinei, M., Migliaccio, E., Pelicci, P.G., 2007. Hydrogen peroxide: a metabolic by-product or a common mediator of ageing signals? Nat. Rev. Mol. Cell Biol. 8 (9), 722–728. doi:10.1038/nrm2240.

Golde, T.E., 2022. Alzheimer's disease—the journey of a healthy brain into organ failure. Mol. Neurodegener. 17, 18. doi:10.1186/s13024-022-00523-1.

Grabska-Kobyłecka, I., Szpakowski, P., Król, A., Książek-Winiarek, D., Kobyłecki, A., Głąbiński, A., Nowak, D., 2023. Polyphenols and their impact on the prevention of neurodegenerative diseases and development. Nutrients 15 (15), 3454. doi:10.3390/nu15153454.

Grewal, A.K., Singh, T.G., Sharma, D., Sharma, V., Singh, M., Rahman, M.H., … Abdel-Daim, M.M., 2021. Mechanistic insights and perspectives involved in neuroprotective action of quercetin. Biomed. Pharmacother. 140, 111729. doi:10.1016/j.biopha.2021.111729.

Griffioen, G., 2023. Calcium dyshomeostasis drives pathophysiology and neuronal demise in age-related neurodegenerative diseases. Int. J. Mol. Sci. 24 (17), 13243. doi:10.3390/ijms241713243.

Grosso, C., Santos, M., Barroso, M.F., 2023. From plants to psycho-neurology: unravelling the therapeutic benefits of bioactive compounds in brain disorders. Antioxidants 12 (8), 1603. doi:10.3390/antiox12081603.

Guang, S., Pang, N., Deng, X., Yang, L., He, F., Wu, L., … Peng, J., 2018. Synaptopathology involved in autism spectrum disorder. Front. Cell. Neurosci. 12, 470. doi:10.3389/fncel.2018.00470.

Guo, C., Sun, L., Chen, X., Zhang, D., 2013. Oxidative stress, mitochondrial damage, and neurodegenerative diseases. Neural Regen. Res. 8 (21), 2003–2014. doi:10.3969/j.issn.1673-5374.2013.21.009.

Guo, F., Xu, F., Yu, J.H., Zou, J.G., Xue, B.J., Shang, M.Y., … Cai, S.Q., 2024. Metabolism and distribution of two major constituents of 'Xing-Nao-Jing Injection'—germacrone and curdione in rats. J. Pharm. Biomed. Anal. 248, 116288. doi:10.1016/j.jpba.2024.116288.

Gupta, V.K., Singh, S., Agrawal, A., Siddiqi, N.J., Sharma, B., 2015. Phytochemicals mediated remediation of neurotoxicity induced by heavy metals. Biochem. Res. Int. 2015, 534769. doi:10.1155/2015/534769.

Gupta, R., Ambasta, R.K., Kumar, P., 2021. Autophagy and apoptosis cascade: which is more prominent in neuronal death? Cell. Mol. Life Sci. 78 (24), 8001–8047. doi:10.1007/s00018-021-03989 5.

Haider, S., Sajid, I., Batool, Z., Madiha, S., Sadir, S., Kamil, N., … Khaliq, S., 2020. Supplementation of taurine insulates against oxidative stress, confers neuroprotective and attenuates memory impairment in noise stress-exposed male Wistar rats. Neurochem. Res. 45, 2762–2774. doi:10.1007/s11064-020-03108-3.

Heland, S., Fields, N., Ellery, S.J., Fahey, M., Palmer, K.R., 2022. The role of nutrients in human neurodevelopment and their potential to prevent neurodevelopmental adversity. Front Nutr 9, 992120. doi:10.3389/fnut.2022.992120.

Hou, Y.C., Chao, P.D., Ho, H.J., Wen, C.C., Hsiu, S.L., 2003. Profound difference in pharmacokinetics between morin and its isomer quercetin in rats. J. Pharm. Pharmacol. 55 (2), 199–203. doi:10.1211/002235702487.

Islam, M.R., Al-Imran, M.I.K., Zehravi, M., Sweilam, S.H., Mortuza, M.R., Gupta, J.K., … Emran, T.B., 2025. Targeting signaling pathways in neurodegenerative diseases: quercetin's cellular and molecular mechanisms for neuroprotection. Anim. Models Exp. Med. 8 (5), 798–818. doi:10.1002/ame2.12551.

Iweala, E.J., Uche, M.E., Dike, E.D., Etumnu, L.R., Dokunmu, T.M., Oluwapelumi, A.E., … Ugbogu, E.A., 2023. Curcuma longa (Turmeric): ethnomedicinal uses, phytochemistry, pharmacological activities and toxicity profiles—a review. Pharmacol. Res. Mod. Chin. Med. 6, 100222.

Jangra, A., Rajput, P., Dwivedi, D., Lahkar, M., 2020. Amelioration of repeated restraint stress-induced behavioural deficits and hippocampal anomalies with taurine treatment in mice. Neurochem. Res. 45, 731–740.

Jayakumar, T., Sheu, J.-R., Yuan, K.-C., Chang, C.-C., Hsieh, C.-Y., Hsu, M.-J., Huang, C.-W., Chiu, Y.-H., Chou, D.-S., 2023. Anti-inflammatory mechanism of morin hydrate by suppressing the NF-κB/MAPKs mediated cell migration and modulating Src/FAK and β-Catenin. Nat. Prod. Commun. 18 (9). https://doi.org/10.1177/1934578X231201.

Jayaprakash, P., Isaev, D., Shabbir, W., Lorke, D.E., Sadek, B., Oz, M., 2021. Curcumin potentiates α7 nicotinic acetylcholine receptors and alleviates autistic-like social deficits and brain oxidative stress status in mice. Int. J. Mol. Sci. 22 (14), 7251. doi:10.3390/ijms22147251.

Jeste, S.S., 2015. Neurodevelopmental behavioral and cognitive disorders. Contin. Lifelong Learn. Neurol. 21 (3), 690–714.

Jiang, T., Zhu, K., Kang, G., Wu, G., Wang, L., Tan, Y., 2024. Infectious viruses and neurodegenerative diseases: the mitochondrial defect hypothesis. Rev. Med. Virol. 34 (4), e2565.

Jin, H., Won, L., So, E., Ji, J., Park, H., Kim, G.-S., Yung, C., Chung, H., Soon, H., Sung, S., Hye, K., 2014. Morin, a flavonoid from Moraceae, suppresses growth and invasion of the highly metastatic breast cancer cell line MDA-MB-231 partly through suppression of the Akt pathway. Int. J. Oncol. 45 (4), 1629–1637.

Jomová, K., Hudecova, L., Lauro, P., Simunkova, M., Alwasel, S.H., Alhazza, I.M., 2019. A switch between antioxidant and prooxidant properties of the phenolic compounds myricetin, morin, 3′,4′-dihydroxyflavone, taxifolin, and 4-hydroxy-coumarin in the presence of copper(II) ions: a spectroscopic, absorption titration and DNA damage study. Molecules 24 (23), 4335. doi:10.3390/molecules24234335.

Joseph, U.G., Oyovwi, M.O, Jeroh, E., Esuku, D.T, Ben-Azu, B., 2025. Dysfunctional astrocyte metabolism: a driver of imbalanced excitatory/inhibitory tone and support for therapeutic intervention targets. J. Mol. Pathol. 6, 12. doi:10.3390/jmp6020012.
Junyent, F., Utrera, J., Romero, R., Pallas, M., Camins, A., Duque, D., 2009. Prevention of epilepsy by taurine treatments in mice experimental model. J. Neurosci. Res. 87, 1508. doi:10.1002/jnr.21950.
Kahkhaie, K.R., Mirhosseini, A., Aliabadi, A., Mohammadi, A., Mousavi, M.J., Haftchesh-meh, S.M., … Sahebkar, A., 2019. Curcumin: a modulator of inflammatory signaling pathways in the immune system. Inflammopharmacology 27, 885–900.
Kandlur, A., Satyamoorthy, K., Gangadharan, G., 2020. Oxidative stress in cognitive and epigenetic aging: a retrospective glance. Front. Mol. Neurosci. 13, 41.
Keshavarz Shahbaz, S., Koushki, K., Keshavarz Hedayati, S., McCloskey, A.P., Kesharwani, P., Naderi, Y., Sahebkar, A., 2024. Polymer nanotherapeutics: a promising approach toward microglial inhibition in neurodegenerative diseases. Med. Res. Rev. doi:10.1002/med.22064.
Khamchai, S., Chumboatrong, W., Hata, J., Tocharus, C., Suksamrarn, A., Tocharus, J., 2020. Morin protects the blood–brain barrier integrity against cerebral ischemia–reperfusion through anti-inflammatory actions in rats. Sci. Rep. 10 (1), 13379.
Khodapsand, E., Jafarzadeh, N., Farrokhi, F., Kamalidehghan, B., Houshmand, M., 2015. Is Bax/Bcl-2 ratio considered as a prognostic marker with age and tumor location in colorectal cancer. Iran. Biomed. J. 19, 69–75. doi:10.6091/ibj.1366.2015.
Kilb, W., Fukuda, A., 2017. Taurine as an essential neuromodulator during perinatal cortical development. Front. Cell. Neurosci. 11, 328. doi:10.3389/fncel.2017.00328.
Kim, H.J., Kim, P., Shin, C.Y., 2013a. A comprehensive review of the therapeutic and pharmacological effects of ginseng and ginsenosides in central nervous system. J. Ginseng Res. 37 (1), 8–29.
Kim, J.S., Kang, O.J., Gweon, O.C., 2013b. Comparison of phenolic acids and flavonoids in black garlic at different thermal processing steps. J. Funct. Foods 5 (1), 80–86.
Kim, H.Y., Kim, H.V., Yoon, J.H., Kang, B.R., Cho, S.M., Lee, S., Kim, J.Y., Kim, J.W., Cho, Y., Woo, J., Kim, Y., 2014. Taurine in drinking water recovers learning and memory in the adult APP/PS1 mouse model of Alzheimer's disease. Sci. Rep. 4, 7467.
Kim, D.H., Kim, D.W., Jung, B.H., Lee, J.H., Lee, H., Hwang, G.S., … Lee, J.W., 2019. Ginsenoside Rb2 suppresses the glutamate-mediated oxidative stress and neuronal cell death in HT22 cells. J. Ginseng Res. 43 (2), 326–334.
Kim, Y., Cho, I.H., Cho, S.H., 2024. Effect of ginseng and ginsenosides on attention deficit hyperactivity disorder: a systematic review. J. Ginseng Res. 48 (5), 437–448. doi:10.1016/j.jgr.2024.05.006.
Kolić, D., Kovarik, Z., 2024. N-methyl-D-aspartate receptors: structure, function, and role in organophosphorus compound poisoning. BioFactors 50 (5), 868–884.
Komar-Fletcher, M., Wojas, J., Rutkowska, M., Raczyńska, G., Nowacka, A., Jurek, J.M., 2023. Negative environmental influences on the developing brain mediated by epigenetic modifications. Explor. Neurosci. 2 (5), 193–211.
Komirishetty, P., Areti, A., Sistla, R., Kumar, A., 2016. Morin mitigates chronic constriction injury (CCI)-induced peripheral neuropathy by inhibiting oxidative stress-induced PARP over-activation and neuroinflammation. Neurochem. Res. 41, 2029–2042.
Koroljević, Z.D., Jordan, K., Ivković, J., Bender, D.V., Perić, P., 2023. Curcuma as an anti-inflammatory component in treating osteoarthritis. Rheumatol. Int. 43 (4), 589–616. doi:10.1007/s00296-022-05244-8.
Kotha, R.R., Luthria, D.L., 2019. Curcumin: biological, pharmaceutical, nutraceutical, and analytical aspects. Molecules 24 (16), 2930. doi:10.3390/molecules24162930.

Lakey-Beitia, J., Burillo, A.M., Penna, G.L., Hegde, M.L., Hegde, M.L., Rao, K.S., 2020. Polyphenols as potential metal chelation compounds against Alzheimer's disease. J. Alzheimers Dis. 75 (3), 895–910.

Lashgari, N.A., Roudsari, N.M., Momtaz, S., Sathyapalan, T., Abdolghaffari, A.H., Sahebkar, A., 2021. The involvement of JAK/STAT signaling pathway in the treatment of Parkinson's disease. J. Neuroimmunol. 361, 577758. doi:10.1016/j.jneuroim.2021.577758.

Lee, M., Cha, H.J., Choi, E., Han, M., Kim, S., Kim, G.Y., Hong, S., Park, C., Moon, S.K., Jeong, S.J., 2017. Antioxidant and cytoprotective effects of morin against hydrogen peroxide-induced oxidative stress are associated with the induction of Nrf-2 mediated HO-1 expression in V79-4 Chinese hamster lung fibroblasts. Int. J. Mol. Med. 39, 672–680.

Lee, H.N., Shin, S.A., Choo, G.S., Kim, H.J., Park, Y.S., Kim, B.S., … Jung, J.Y., 2018. Anti-inflammatory effect of quercetin and galangin in LPS-stimulated RAW264.7 macrophages and DNCB-induced atopic dermatitis animal models. Int. J. Mol. Med. 41 (2), 888–898.

Leon, R., Wu, H., Jin, Y., Wei, J., Buddhala, C., Prentice, H., et al., 2009. Protective function of taurine in glutamate-induced apoptosis in cultured neurons. J. Neurosci. Res. 87 (5), 1185–1194. doi:10.1002/jnr.21926.

Leyane, T.S., Jere, S.W., Houreld, N.N., 2022. Oxidative stress in ageing and chronic degenerative pathologies: molecular mechanisms involved in counteracting oxidative stress and chronic inflammation. Int. J. Mol. Sci. 23 (13), 7273. doi:10.3390/ijms23137273.

Li, J., Yang, Y., Ning, E., Peng, Y., Zhang, J., 2019. Mechanisms of poor oral bioavailability of flavonoid Morin in rats: from physicochemical to biopharmaceutical evaluations. Eur. J. Pharm. Sci. 128, 290–298. doi:10.1016/j.ejps.2018.12.011.

Li, J., Huang, Q., Chen, J., Qi, H., Liu, J., Chen, Z., … Li, X., 2021. Neuroprotective potentials of Panax ginseng against Alzheimer's disease: a review of preclinical and clinical evidences. Front. Pharmacol. 12, 688490. doi:10.3389/fphar.2021.688490.

Li, P., Zhang, M., Chen, M., Liu, G., Meng, L., Zhang, D., 2024. Systematic studies on the kinetic process of 20(S)-protopanaxadiol in rats and dogs: absorption, distribution, metabolism and excretion. Front. Pharmacol. 15, 1430780. doi:10.3389/fphar.2024.1430780.

Liang, Y., Zhong, G., Ren, M., Sun, T., Li, Y., Ye, M., … Liu, C., 2023. The role of ubiquitin–proteasome system and mitophagy in the pathogenesis of Parkinson's disease. Neuromolecular Med. 25 (4), 471–488. doi:10.1007/s12017-023-08784-6.

Liu, Z., Yu, Y., Li, X., Ross, C.A., Smith, W.W., 2011. Curcumin protects against A53T alpha-synuclein-induced toxicity in a PC12 inducible cell model for Parkinsonism. Pharmacol. Res. 63 (5), 439–444. doi:10.1016/j.phrs.2011.01.004.

Liu, W., Zhai, Y., Heng, X., Che, F.Y., Chen, W., Sun, D., Zhai, G., 2016. Oral bioavailability of curcumin: problems and advancements. J. Drug Target. 24 (8), 694–702. doi:10.3109/1061186X.2016.1157883.

Liu, Y., Wang, J., Huang, Z., Liang, J., Xia, Q., Liu, X., 2021. Environmental pollutants exposure: a potential contributor for aging and age-related diseases. Environ. Toxicol. Pharmacol. 83, 103575. doi:10.1016/j.etap.2021.103575.

Lou, H., Jing, X., Wei, X., Shi, H., Ren, D., Zhang, X., 2014. Naringenin protects against 6-OHDA-induced neurotoxicity via activation of the Nrf2/ARE signaling pathway. Neuropharmacology 79, 380–388. doi:10.1016/j.neuropharm.2013.11.026.

Louzada, P.R., Paul Lima, A.C., Mendonca-Silva, D.L., Noel, F., De Mello, F.G., Ferreira, S.T., 2004. Taurine prevents the neurotoxicity of beta-amyloid and glutamate receptor agonists: activation of GABA receptors and possible implications for Alzheimer's disease and other neurological disorders. FASEB J. 18, 511–518.

Lukens, J.R., Eyo, U.B., 2022. Microglia and neurodevelopmental disorders. Annu. Rev. Neurosci. 45, 425–445. doi:10.1146/annurev-neuro-120421-085015.

Lukens, J.R., Eyo, U.B., 2022. Microglia and neurodevelopmental disorders. Annu. Rev. Neurosci. 45 (1), 425–445. doi:10.1146/annurev-neuro-120421-085015.

Lutfi Ismaeel, G., Makki AlHassani, O.J., Alazragi, R.S., Hussein Ahmed, A., Mohamed, A.A.H., Jasim, N.Y., … Almashhadani, H.A., 2023. Genetically engineered neural stem cells (NSCs) therapy for neurological diseases: state-of-the-art. Biotechnol. Prog. 39 (5), e3363. doi:10.1002/btpr.3363.
Méndez-García, L.A., Martínez-Castillo, M., Villegas-Sepúlveda, N., Orozco, L., Córdova, E.J., 2019. Curcumin induces p53-independent inactivation of Nrf2 during oxidative stress–induced apoptosis. Hum. Exp. Toxicol. 38 (8), 951–961.
Münzel, T., Hahad, O., Sørensen, M., Lelieveld, J., Duerr, G.D., Nieuwenhuijsen, M., Daiber, A., 2022. Environmental risk factors and cardiovascular diseases: a comprehensive expert review. Cardiovasc. Res. 118 (14), 2880–2902.
Ma, Z.X., Zhang, R.Y., Rui, W.J., Wang, Z.Q., Feng, X., 2021. Quercetin alleviates chronic unpredictable mild stress-induced depressive-like behaviors by promoting adult hippocampal neurogenesis via FoxG1/CREB/BDNF signaling pathway. Behav. Brain Res. 406, 113245. doi:10.1016/j.bbr.2021.113245.
Magdaleno Roman, J.Y., Chapa González, C., 2024. Glutamate and excitotoxicity in central nervous system disorders: ionotropic glutamate receptors as a target for neuroprotection. Neuroprotection 2 (2), 137–150. doi:10.1016/j.neupr.2024.02.005.
Marangoni, C., Hernandez, M., Faedda, G.L., 2016. The role of environmental exposures as risk factors for bipolar disorder: a systematic review of longitudinal studies. J. Affect. Disord. 193, 165–174. doi:10.1016/j.jad.2015.12.040.
Martínez-Iglesias, O., Naidoo, V., Carrera, I., Corzo, L., Cacabelos, R., 2023. Natural bioactive products as epigenetic modulators for treating neurodegenerative disorders. Pharmaceuticals 16 (2), 216. doi:10.3390/ph16020216.
Martinez-Jaramillo, E., Jamali, F., Abdalbari, F.H., Abdulkarim, B., Jean-Claude, B.J., Telleria, C.M., Sabri, S., 2024. Pro-oxidant Auranofin and glutathione-depleting combination unveils synergistic lethality in glioblastoma cells with aberrant epidermal growth factor receptor expression. Cancers 16 (13), 2319. doi:10.3390/cancers16132319.
Maurya, S.K., Bhattacharya, N., Mishra, S., Bhattacharya, A., Banerjee, P., Senapati, S., Mishra, R., 2021. Microglia-specific drug targeting using natural products for the regulation of redox imbalance in neurodegeneration. Front. Pharmacol. 12, 654489. doi:10.3389/fphar.2021.654489.
Mehta, V., Parashar, A., Udayabanu, M., 2017. Quercetin prevents chronic unpredictable stress-induced behavioral dysfunction in mice by alleviating hippocampal oxidative and inflammatory stress. Physiol. Behav. 171, 69–78. doi:10.1016/j.physbeh.2017.01.006.
Menassa, D.A., Muntslag, T.A.O., Martin-Estebané, M., Barry-Carroll, L., Chapman, M.A., Adorjan, I., Tyler, T., Turnbull, B., Rose-Zerilli, M.J.J., Nicoll, J.A.R., Krsnik, Z., Kostovic, I., Gomez-Nicola, D., 2022. The spatiotemporal dynamics of microglia across the human lifespan. Dev. Cell 57 (15), 2127–2139.e6. doi:10.1016/j.devcel.2022.07.015.
Mendes, M.S., Majewska, A.K., 2021. An overview of microglia ontogeny and maturation in the homeostatic and pathological brain. Eur. J. Neurosci. 53 (11), 3525–3547.
Mendoza-Wilson, A.M., Santacruz-Ortega, H., Balandrán-Quintana, R.R., 2011. Spectroscopic and computational study of the major oxidation products formed during the reaction of two quercetin conformers with a free radical. Spectrochim. Acta A Mol. Biomol. Spectrosc. 81 (1), 481–488.
Meng, X., Fu, M., Wang, S., Chen, W., Wang, J., Zhang, N., 2021. Naringin ameliorates memory deficits and exerts neuroprotective effects in a mouse model of Alzheimer's disease by regulating multiple metabolic pathways. Mol. Med. Rep. 23, 332. doi:10.3892/mmr.2021.11971.
Menon, V.P., Sudheer, A.R., 2007. Antioxidant and anti-inflammatory properties of curcumin. In: Aggarwal, B.B., Surh, Y.J., Shishodia, S. (Eds.), The Molecular Targets and Therapeutic Uses of Curcumin in Health and Disease. Advances in Experimental Medicine and Biology. In: The Molecular Targets and Therapeutic Uses of Curcumin in Health and

Disease. Advances in Experimental Medicine and Biology, 595. Springer, pp. 105–125. doi:10.1007/978-0-387-46401-5_3.
Mishra, S., Palanivelu, K., 2008. The effect of curcumin (turmeric) on Alzheimer's disease: an overview. Ann. Indian Acad. Neurol. 11 (1), 13–19. doi:10.4103/0972-2327.40220.
Mohammed, A.A., Akinola, O., 2024. Curcumin intervention in hippocampal atrophy of diet-induced Alzheimer-like deficits in insulin-resistant rats. Rwanda Med. J. 81 (1), 103–113.
Morales, J., Günther, G., Zanocco, A.L., Lemp, E., 2012. Singlet oxygen reactions with flavonoids: a theoretical–experimental study. PLoS One 7 (7), e40548. doi:10.1371/journal.pone.0040548.
Morimoto, M., Hashimoto, T., Tsuda, Y., Suenaga, M., Nakamura, T., Katoh, S., 2023. Study on oxidative stress and inflammatory/antioxidant substance levels in autism spectrum disorder. J. Chin. Med. Assoc. 86 (5), 489–493. doi:10.1097/JCMA.0000000000000917.
Muddapu, V.R., Dharshini, S.A.P., Chakravarthy, V.S., Gromiha, M.M., 2020. Neurodegenerative diseases—is metabolic deficiency the root cause? Front. Neurosci. 14, 213.
Mursal, M., Kumar, A., Hasan, S.M., Hussain, S., Singh, K., Kushwaha, S.P., … Fatima, S.K., 2024. Role of natural bioactive compounds in the management of neurodegenerative disorders. Intell. Pharm. 2 (1), 102–113.
Mythri, R.B., Veena, J., Harish, G., Shankaranarayana Rao, B.S., Srinivas Bharath, M.M., 2011. Chronic dietary supplementation with turmeric protects against 1-methyl-4-phenyl-1,2,3,6-tetrahydropyridine-mediated neurotoxicity in vivo: implications for Parkinson's disease. Br. J. Nutr. 106 (1), 63–72.
Nabi, M., Tabassum, N., 2022. Role of environmental toxicants on neurodegenerative disorders. Front. Toxicol. 4, 837579.
Nakajima, Y., Osuka, K., Seki, Y., Gupta, R.C., Hara, T., Wakabayashi, T., 2010. Taurine reduces inflammatory responses after spinal cord injury. J. Neurotrauma 27 (3), 403–410.
Nakaso, K., 2024. Roles of microglia in neurodegenerative diseases. Yonago Acta Med. 61 (1), 1–8. doi:10.33160/yam.2024.02.001.
Namgyal, D., Ali, S., Mehta, R., Sarwat, M., 2020. The neuroprotective effect of curcumin against Cd-induced neurotoxicity and hippocampal neurogenesis promotion through CREB-BDNF signaling pathway. Toxicology 442, 152542. doi:10.1016/j.tox.2020.152542.
Naraki, K., Rezaee, R., Karimi, G., 2021. A review on the protective effects of naringenin against natural and chemical toxic agents. Phytother. Res. 35 (8), 4075–4091.
Ng, P.Y., McNeely, T.L., Baker, D.J., 2023. Untangling senescent and damage-associated microglia in the aging and diseased brain. FEBS J. 290 (5), 1326–1339.
Nicosia, N., Giovenzana, M., Misztak, P., Mingardi, J., Musazzi, L., 2024. Glutamate-mediated excitotoxicity in the pathogenesis and treatment of neurodevelopmental and adult mental disorders. Int. J. Mol. Sci. 25 (12), 6521.
O'Donnell, C.P., Allott, K.A., Murphy, B.P., Yuen, H.P., Proffitt, T.-M., Papas, A., Moral, J., Pham, T., O'Regan, M.K., Phassouliotis, C., Simpson, R., McGorry, P.D., 2016. Adjunctive taurine in first-episode psychosis: a phase 2, double-blind, randomized, placebo-controlled study. J. Clin. Psychiatry 77 (12), e1610–e1617. doi:10.4088/JCP.15m10185.
Obukohwo, O.M., Ben-Azu, B., Nwangwa, E.K., Ohwin, E.P., Igweh, J.C., Adeogun Adetomiwa, E., 2024. Adverse hematological profiles associated with chlorpromazine antipsychotic treatment in male rats: preventive and reversal mechanisms of taurine and coenzyme-Q10. Toxicol. Rep. 12, 448–462. doi:10.1016/j.toxrep.2024.04.004.
Oh, S.J., Lee, H.J., Jeong, Y.J., Nam, K.R., Kang, K.J., Han, S.J., Lee, K.C., Lee, Y.J., Choi, J.Y., 2020. Evaluation of the neuroprotective effect of taurine in Alzheimer's disease using functional molecular imaging. Sci. Rep. 10, 1–9.
Ojiakor, V.O., Ben-Azu, B., Ani, P.N., Ozioko, M.O., Oviosun, A., Ozor, I.I., Esom, E.A., Anyanwu, E.G., 2026a. Naringin Reverses Chronic Stress-Induced Cognitive Deficits

and Enhances Hippocampal Neuroplasticity in Mice. J. Neuroimmune. Pharmacol. 21 (1), 19. doi:10.1007/s11481-026-10281-2.

Ojiakor, V.O., Ben-Azu, B., Finbarrs-Bello, E., Esom, E.A., Anibeze, C.P., Anyanwu, E.G., 2026b. Naringin exhibits adaptogen-like and neuroprotective effects on mice under chronic stress, reversing cortical neuroinflammation and neurodegeneration. Eur. J. Med. Chem. Rep. 16, 100323.

Oke, E.O., Adeyi, O., Nnaji, P.C., Okolo, B.I., Abam, F.I., Ude, C.J., Ayanyemi, J., 2022. Techno-economic analysis and neuro-fuzzy production rate prediction of sorghum (Sorghum bicolor) leaf shealth colourant extract production. Agricult. Res. 11 (3), 579–589.

Okhue, E., Kadiri, H.E, Ichipi-Ifukor, P.C, Ben-Azu, B., Asagba, S.O, Achuba, F.I, Oyem, J.C., 2024. Prenatal double-hit with aluminium and cadmium mediate testicular atrophy and hypothalamic hypoplasia: the role of oxido-nitrergic stress and endocrine perturbations. Biometals doi:10.1007/s10534-023-00563-0.

Oladapo, O.M., Ben-Azu, B., Ajayi, A.M., et al., 2021. Naringin confers protection against psychosocial defeat stress-induced neurobehavioral deficits in mice: involvement of glutamic acid decarboxylase isoform-67, oxido-nitrergic stress, and neuroinflammatory mechanisms. J. Mol. Neurosci. 71 (3), 431–445. doi:10.1007/s12031-020-01664-y.

Olonode, E.T., Aderibigbe, A.O., Adeoluwa, O.A., Eduviere, A.T., Ben-Azu, B., 2019. Morin hydrate mitigates rapid eye movement sleep deprivation-induced neurobehavioral impairments and loss of viable neurons in the hippocampus of mice. Behav. Brain Res. 356, 518–525. doi:10.1016/j.bbr.2017.12.024.

Olude, M.A., Mouihate, A., Mustapha, O.A., Farina, C., Quintana, F.J., Olopade, J.O., 2022. Astrocytes and microglia in stress-induced neuroinflammation: the African perspective. Front. Immunol. 13, 795089.

Olufunmilayo, E.O., Gerke-Duncan, M.B., Holsinger, R.D., 2023. Oxidative stress and antioxidants in neurodegenerative disorders. Antioxidants 12 (2), 517.

Omeiza, N.A, Bakre, A., Ben-Azu, B., Sowunmi, A.A, Abdulrahim, H.A, Chimezie, J., Lawal, S.O, Adebayo, O.G, Alagbonsi, A.I, Akinola, O., Abolaji, A.O, Aderibigbe, A.O., 2023. Mechanisms underpinning *Carpolobia lutea* G. Don ethanol extract's neurorestorative and antipsychotic-like activities in an NMDA receptor antagonist model of schizophrenia. J. Ethnopharmacol. 301, 115767.

Omogbiya, I.A., Umukoro, S., Aderibigbe, A.O., Bakre, A.G., 2013. Jobelyn® pretreatment ameliorates symptoms of psychosis in experimental models. J. Basic Clin. Physiol. Pharmacol. 24 (4), 331–336. doi:10.1515/jbcpp-2012-0073.

Omogbiya, A.I., Ben-Azu, B., Eduviere, A.T., Eneni, A.O., Nwokoye, P.O., Ajayi, A.M., Umukoro, S., 2020. Monosodium glutamate induces memory and hepatic dysfunctions in mice: ameliorative role of Jobelyn® through the augmentation of cellular antioxidant defense machineries. Toxicol. Res. 37 (3), 323–335. doi:10.1007/s43188-020-00068-9.

Onuelu, J.E., Ben-Azu, B., Adebayo, O.G., Fokoua, A.R., Nekabari, M.K., Ozah, E.O., Iwhiwhu, P., Ajayi, A.M., Oyovwi, O.M., Omogbiy, I.A., Eduviere, A.T., Ojezele, M.O., 2025. Taurine, an essential amino acid, attenuates rotenone-induced Parkinson's disease in rats by inhibiting alpha-synuclein aggregation and augmenting dopamine release. Behav. Brain Res. 480, 115397.

Onyekweli, C.C, BenAzu, B., Nwangwa, K.E, Oyovwi, M.O, Nwangwa, J.E, AgbonifoChijiokwu, E., Ovuakporaye, S.I, Onome, O.B, Nwabuoku, S.U, Rotu, R.A, Uchechukwu, J.G, 2025. Autophagydependent pathways mediate memory and testicular impairments induced by a highfat diet in rats, which can be reversed by epigallocatechin3gallate. Nutrire 50, 35.

Oritsemuelebi, B., Frazzoli, C., Eze, E.C., Ilo, C.E., Nwaogazie, I.L., Orisakwe, O.E., 2021. Levels of toxic and essential metals in maternal cord blood and anthropometry at birth: a pilot study. J. Glob. Health Rep. 5, e2021104.

Oyinbo, C.A., Robert, F.O., Avwioro, O.G., Igbigbi, P.S., 2018. Jobelyn suppresses hippocampal neuronal apoptosis and necrosis in experimental alcohol-induced brain stress. Pathophysiology 25 (4), 317–325.

Oyovwi, M.O., Nwangwa, E.K., Ben-Azu, B., Edesiri, T.P., Emojevwe, V., Igweh, J.C., 2021. Taurine and coenzyme Q10 synergistically prevent and reverse chlorpromazine-induced psycho-neuroendocrine changes and cataleptic behavior in rats. Naunyn-Schmiedeb. Arch. Pharmacol. 394, 717–734. doi:10.1007/s00210-020-02003-z.

Oyovwi, M.O., Ben-Azu, B., Agbonifo-Chijiokwu, E., et al., 2022. Possible mechanisms involved in the prevention and reversal of chlorpromazine-induced testicular damage by taurine and coenzyme-Q10 in rats. Nutrire 47, 31. doi:10.1186/s41110-022-00180-w.

Oyovwi, M.O, Babawale, K.H, Jeroh, E., Ben-Azu, B., 2025a. Exploring the role of neuromodulation in neurodegenerative disorders: insights from Alzheimer's and Parkinson's diseases. Brain Disord. 17, 100187.

Oyovwi, M.O, Chijiokwu, E.A, Ben-Azu, B., Atere, A.D, Joseph, U.G, Ogbutor, U.G, Udi, O.A, 2025b. Potential roles of natural antioxidants in modulating neurodegenerative disease pathways. Mol. Neurobiol. doi:10.1007/s12035-025-04874-w.

Oyovwi, M.O, Chijiokwu, A.E, Rotu, A.R, Jeroh, E., Chinwuba, P., Ben-Azu, B., Osirim, E., Nwangwa, E.K., 2025c. Quercetin prevents single prolonged stress-induced posttraumatic stress disorder and neurochemical changes via antioxidant, antiinflammatory and anti-apoptotic mechanisms in rats. Neuroscience 576, 199–212.

Pandey, N., Strider, J., Nolan, W.C., Yan, S.X., Galvin, J.E., 2008. Curcumin inhibits aggregation of alpha-synuclein. Acta Neuropathol. 115 (4), 479–489. doi:10.1007/s00401-007-0332-4.

Park, M.T.M., Raznahan, A., Shaw, P., Gogtay, N., Lerch, J.P., Chakravarty, M.M., 2018. Neuroanatomical phenotypes in mental illness: identifying convergent and divergent cortical phenotypes across autism, ADHD and schizophrenia. J. Psychiatry Neurosci. 43 (3), 201–212.

Paula-Lima, A.C., De Felice, F.G., Brito-Moreira, J., Ferreira, S.T., 2005. Activation of GABA(A) receptors by taurine and muscimol blocks the neurotoxicity of beta-amyloid in rat hippocampal and cortical neurons. Neuropharmacology 49, 1140–1148.

Peters, A., Nawrot, T.S., Baccarelli, A.A., 2021. Hallmarks of environmental insults. Cell 184 (6), 1455–1468.

Piccioni, A., Covino, M., Zanza, C., Longhitano, Y., Tullo, G., Bonadia, N., et al., 2021. Energy drinks: a narrative review of their physiological and pathological effects. Intern. Med. J. 51, 636–646. doi:10.1111/imj.14881.

Požgajová, M., Klongová, L., Kovár, M., Navrátilová, A., 2024. Cell protection by oxidative stress mitigation using substances with bioactive properties. In The Power of Antioxidants-Unleashing Nature's Defense Against Oxidative Stress. IntechOpen.

Puri, S., Shaheen, M., Grover, B., 2023. Nutrition and cognitive health: a life course approach. Front. Public Health 11, 1023907.

Quan, Q., Li, X., Feng, J., Hou, J., Li, M., Zhang, B., 2020. Ginsenoside Rg1 reduces β amyloid levels by inhibiting CDK5-induced PPARγ phosphorylation in a neuron model of Alzheimer's disease. Mol. Med. Rep. 22 (4), 3277–3288.

Rahman, M.H., Bajgai, J., Fadriquela, A., Sharma, S., Trinh, T.T., Akter, R., … Lee, K.J., 2021. Therapeutic potential of natural products in treating neurodegenerative disorders and their future prospects and challenges. Molecules 26 (17), 5327.

Rajabian, A., Rameshrad, M., Hosseinzadeh, H., 2019. Therapeutic potential of Panax ginseng and its constituents, ginsenosides and gintonin, in neurological and neurodegenerative disorders: a patent review. Expert Opin. Ther. Pat. 29 (1), 55–72.

Rajput, S.A., Wang, X.Q., Yan, H.C., 2021. Morin hydrate: a comprehensive review on novel natural dietary bioactive compounds with versatile biological and pharmacological potential. Biomed. Pharmacother. 138, 111511. doi:10.1016/j.biopha.2021.111511.

Ramírez-Guerrero, S., Guardo-Maya, S., Medina-Rincón, G.J., Orrego-González, E.E., Cabezas-Pérez, R., González-Reyes, R.E., 2022. Taurine and astrocytes: a homeostatic and neuroprotective relationship. Front. Mol. Neurosci. 15, 937789.
Rana, A., Samtiya, M., Dhewa, T., Mishra, V., Aluko, R.E., 2022. Health benefits of polyphenols: a concise review. J. Food Biochem. 46 (10), e14264.
Rarinca, V., Nicoara, M.N., Ureche, D., Ciobica, A., 2023. Exploitation of quercetin's antioxidative properties in potential alternative therapeutic options for neurodegenerative diseases. Antioxidants 12 (7), 1418.
Rastogi, V., Santiago-Moreno, J., Doré, S., 2015. Ginseng: a promising neuroprotective strategy in stroke. Front. Cell. Neurosci. 8, 117168.
Ratan, Z.A., Haidere, M.F., Hong, Y.H., Park, S.H., Lee, J.O., Lee, J., Cho, J.Y., 2021. Pharmacological potential of ginseng and its major component ginsenosides. J. Ginseng Res. 45 (2), 199–210.
Rattanachaikunsopon, P., Phumkhachorn, P., 2007. Bacteriostatic effect of flavonoids isolated from leaves of Psidium guajava on fish pathogens. Fitoterapia 78 (6), 434–436. doi:10.1016/j.fitote.2007.03.015.
Rauf, A., Badoni, H., Abu-Izneid, T., Olatunde, A., Rahman, M.M., Painuli, S., … Mubarak, M.S., 2022. Neuroinflammatory markers: key indicators in the pathology of neurodegenerative diseases. Molecules 27 (10), 3194.
Reeta, K.H., Singh, D., Gupta, Y.K., 2017. Chronic treatment with taurine after intracerebroventricular streptozotocin injection improves cognitive dysfunction in rats by modulating oxidative stress, cholinergic functions and neuroinflammation. Neurochem. Int. 108, 146–156.
Renu, K., Chakraborty, R., Myakala, H., Koti, R., Famurewa, A.C., Madhyastha, H., … Gopalakrishnan, A.V., 2021. Molecular mechanism of heavy metals (Lead, Chromium, Arsenic, Mercury, Nickel and Cadmium)-induced hepatotoxicity—a review. Chemosphere 271, 129735.
Rezaei, K., Mastali, G., Abbasgholinejad, E., Bafrani, M.A., Shahmohammadi, A., Sadri, Z., Zahed, M.A., 2024. Cadmium neurotoxicity: insights into behavioral effect and neurodegenerative diseases. Chemosphere 364, 143180. doi:10.1016/j.chemosphere.2024.143180.
Rim, C., You, M.J., Nahm, M., Kwon, M.S., 2024. Emerging role of senescent microglia in brain aging-related neurodegenerative diseases. Transl. Neurodegener. 13 (1), 10. doi:10.1186/s40035-024-00402-3.
Ripps, H., Shen, W., 2012. Review: taurine: a "very essential" amino acid. Mol Vis 18, 2673–2686.
Rizvi, F., Mathur, A., Krishna, S., Siddiqi, M.I., Kakkar, P., 2015. Suppression in PHLPP2 induction by morin promotes Nrf2-regulated cellular defenses against oxidative injury to primary rat hepatocytes. Redox Biol. 6, 587–598. doi:10.1016/j.redox.2015.07.001.
Rolfe, V., Mackonochie, M., Mills, S., MacLennan, E., 2020. Turmeric/curcumin and health outcomes: a meta-review of systematic reviews. Eur. J. Integr. Med. 40, 101252.
Ronaldson, P.T., Davis, T.P., 2020. Regulation of blood–brain barrier integrity by microglia in health and disease: a therapeutic opportunity. J. Cereb. Blood Flow Metab. 40 (S1), S6–S24.
Rosa, J.M., Farré-Alins, V., Ortega, M.C., Navarrete, M., Lopez-Rodriguez, A.B., Palomino-Antolín, A., … Egea, J., 2021. TLR4 pathway impairs synaptic number and cerebrovascular functions through astrocyte activation following traumatic brain injury. Br. J. Pharmacol. 178 (17), 3395–3413.
Sánchez, A.M.E., Mejía-Toiber, J., Massieu, L., 2008. Excitotoxic neuronal death and the pathogenesis of Huntington's disease. Arch. Med. Res. 39 (3), 265–276. doi:10.1016/j.arcmed.2007.11.011.

Sachdeva, P., Mehdi, I., Kaith, R., Ahmad, F., Anwar, M.S., 2022. Potential natural products for the management of autism spectrum disorder. iBrain 8 (3), 365–376. doi:10.1002/ibra.12050.

Sadeghdoust, M., Das, A., Kaushik, D.K., 2024. Fueling neurodegeneration: metabolic insights into microglia functions. J. Neuroinflammation 21 (1), 300. doi:10.1186/s12974-024-03383-w.

Saengprakai, J., Chudhangkura, A., 2017. Antimutagenic activity of turmeric extract beverages. Pak. J. Nutr. 16 (7), 514–520.

Safe, S., Kothari, J., Hailemariam, A., Upadhyay, S., Davidson, L.A., Chapkin, R.S., 2023. Health benefits of coffee consumption for cancer and other diseases and mechanisms of action. Int. J. Mol. Sci. 24 (3), 2706. doi:10.3390/ijms24032706.

Sahoo, J.P., Behera, L., Praveena, J., Sawant, S., Mishra, A., Sharma, S.S., … Samal, K.C., 2021. The golden spice turmeric (Curcuma longa) and its feasible benefits in prospering human health—a review. Am. J. Plant Sci. 12 (3), 455–475. doi:10.4236/ajps.2021.123033.

Saji, R., Uchio, R., Fuwa, A., Okuda-Hanafusa, C., Kawasaki, K., Muroyama, K., … Hirose, Y., 2023. Turmeronols (A and B) from Curcuma longa have anti-inflammatory effects in lipopolysaccharide-stimulated BV-2 microglial cells by reducing NF-κB signaling. Biosci. Microbiota Food Health 42 (3), 172–179. doi:10.12938/bmfh.2022-054.

Santa-Miria, I., Hernandez, F., Moreno, F.J., Avila, J., 2007. Taurine, an inducer for tau polymerization and weak inhibitor for amyloid-beta-peptide aggregation. Neurosci. Lett. 429 (2–3), 91–94. doi:10.1016/j.neulet.2007.09.025.

Saxena, S., Kruys, V., Vamecq, J., Maze, M., 2021. The role of microglia in perioperative neuroinflammation and neurocognitive disorders. Front. Aging Neurosci. 13, 671499. doi:10.3389/fnagi.2021.671499.

Sedaghat, M., 2021. Cardiac remodeling, apoptosis-related process (Bax, Bcl-2), and their ratio (Bax/Bcl-2) in cardiomyocytes of diabetic rats after combined exercise training and taurine supplementation. Comp. Clin. Pathol. 30 (5), 801–810.

Seneff, S., Kyriakopoulos, A.M., 2025. Taurine prevents mitochondrial dysfunction and protects mitochondria from reactive oxygen species and deuterium toxicity. Amino Acids 57 (1), 1–15. doi:10.1007/s00726-024-03060-2.

Sengul, E., Yildirim, S., Cinar, İ., Gelen, V., Uslu, H., 2024. Mitigation of acute hepatotoxicity induced by cadmium through morin: modulation of oxidative and pro-apoptotic endoplasmic reticulum stress and inflammatory responses in rats. Biol. Trace Elem. Res. 202, 5106–5117. doi:10.1007/s12011-024-04064-0.

Shah, M., Murad, W., Mubin, S., Ullah, O., Rehman, N.U., Rahman, M.H., 2022. Multiple health benefits of curcumin and its therapeutic potential. Environ. Sci. Pollut. Res. 29 (29), 43732–43744.

Shahrajabian, M.H., Sun, W., Cheng, Q., 2019. A review of ginseng species in different regions as a multipurpose herb in traditional Chinese medicine, modern herbology, and pharmacological science. J. Med. Plants Res. 13 (10), 213–226.

Sharifi-Rad, J., Rayess, Y.E., Rizk, A.A., Sadaka, C., Zgheib, R., Zam, W., Sestito, S., Rapposelli, S., Neffe-Skocińska, K., Zielińska, D., Salehi, B., Setzer, W.N., Dosoky, N.S., Taheri, Y., El Beyrouthy, M., Martorell, M., Ostrander, E.A., Suleria, H.A.R., Cho, W.C., Maroyi, A., Martins, N., 2020. Turmeric and its major compound curcumin on health: bioactive effects and safety profiles for food, pharmaceutical, biotechnological, and medicinal applications. Front. Pharmacol. 11, 1021. doi:10.3389/fphar.2020.01021.

Sharma, A., Couture, J., 2014. A review of the pathophysiology, etiology, and treatment of attention deficit hyperactivity disorder (ADHD). Ann. Pharmacother. 48 (2), 209–225.

Sharma, R.A., Steward, W.P., Gescher, A.J., 2007. Pharmacokinetics and pharmacodynamics of curcumin. Adv. Exp. Med. Biol. 595, 453–470. doi:10.1007/978-0-387-46401-5_20.

Sharma, S.H., Kumar, J.S.S., Chellappan, D.R., Nagarajan, S., 2018. Molecular chemoprevention by morin—a plant flavonoid that targets nuclear factor kappa B in experimental colon cancer. Biomed. Pharmacother. 100, 367–373. doi:10.1016/j.biopha.2018.02.035.

Sharma, C., Kim, S., Nam, Y., Jung, U.J., Kim, S.R., 2021. Mitochondrial dysfunction as a driver of cognitive impairment in Alzheimer's disease. Int. J. Mol. Sci. 22 (9), 4850. doi:10.3390/ijms22094850.

Shaw, E., Mayer, M., Ekwaru, P., McMullen, S., Graves, E., Wu, J.W., … Mestre, T.A., 2022. Epidemiology and economic burden of Huntington's disease: a Canadian provincial public health system perspective. J. Med. Econ. 25 (1), 212–219.

Shchepinov, M.S., 2020. Polyunsaturated fatty acid deuteration against neurodegeneration. Trends Pharmacol. Sci. 41 (4), 236–248.

Shi, Y.Q., Huang, T.W., Chen, L.M., Pan, X.D., Zhang, J., Zhu, Y.G., Chen, X.C., 2010. Ginsenoside Rg1 attenuates amyloid-β content, regulates PKA/CREB activity, and improves cognitive performance in SAMP8 mice. J. Alzheimers Dis. 19 (3), 977–989.

Shi, Y.H., Li, Y., Wang, Y., Xu, Z., Fu, H., Zheng, G.Q., 2020. Ginsenoside-Rb1 for ischemic stroke: a systematic review and meta-analysis of preclinical evidence and possible mechanisms. Front. Pharmacol. 11, 285.

Shilpa, V.S., Shams, R., Dash, K.K., Pandey, V.K., Dar, A.H., Ayaz Mukarram, S., Harsanyi, E., Kovacs, B., 2023. Phytochemical properties, extraction, and pharmacological benefits of naringin: a review. Molecules 28 (15), 5623. doi:10.3390/molecules28155623.

Silva, S.P., Zago, A.M., Carvalho, F.B., Germann, L., Colombo, G.D.M., Rahmeier, F.L., Gutierres, J.M., Reschke, C.R., Bagatini, M.D., Assmann, C.E., et al., 2021. Neuroprotective effects of taurine against cell death, glial changes, and neuronal loss in the cerebellum of rats exposed to chronic recurrent neuroinflammation induced by LPS. J. Immunol. Res. 2021, 1–10.

Sinha, K., Ghosh, J., Sil, P.C., 2016. Morin and its role in chronic diseases. Adv. Exp. Med. Biol. 928, 453–471.

Sithara, T., Arun, K.B., Syama, H.P., Reshmi, T.R., Nisha, P., 2017. Morin inhibits proliferation of SW480 colorectal cancer cells by inducing apoptosis mediated by reactive oxygen species formation and uncoupling of Warburg effect. Front. Pharmacol. 8, 640. doi:10.3389/fphar.2017.00640.

Stanić, Z., 2017. Curcumin, a compound from natural sources, a true scientific challenge: a review. Plant Foods Hum. Nutr. 72 (1), 1–12. doi:10.1007/s11130-016-0590-1.

Streyczek, J., Apweiler, M., Sun, L., Fiebich, B.L., 2022. Turmeric extract (Curcuma longa) mediates anti-oxidative effects by reduction of nitric oxide, iNOS protein-, and mRNA-synthesis in BV2 microglial cells. Molecules 27 (3), 784.

Su, Y., Fan, W., Ma, Z., Wen, X., Wang, W., Wu, Q., Huang, H., 2014. Taurine improves functional and histological outcomes and reduces inflammation in traumatic brain injury. Neuroscience 266, 56–65.

Sule, R.O., Condon, L., Gomes, A.V., 2022. A common feature of pesticides: oxidative stress—the role of oxidative stress in pesticide-induced toxicity. Oxidative Med. Cell. Longev. 2022, 5563759.

Sun, M., Xu, C., 2007. Neuroprotective mechanism of taurine due to up-regulating calpastatin and down-regulating calpain and caspase-3 during focal cerebral ischemia. Cell. Mol. Neurobiol. 28, 593–611.

Sundaram, J.R., Poore, C.P., Sulaimee, N.H.B., Pareek, T., Cheong, W.F., Wenk, M.R., Pant, H.C., Frautschy, S.A., Low, C.M., Kesavapany, S., 2017. Curcumin ameliorates neuroinflammation, neurodegeneration, and memory deficits in p25 transgenic mouse model that bears hallmarks of Alzheimer's disease. J. Alzheimers Dis. 60 (4), 1429–1442. doi:10.3233/JAD-170093.

Tahami Monfared, A.A., Byrnes, M.J., White, L.A., Zhang, Q., 2022. Alzheimer's disease: epidemiology and clinical progression. Neurol. Ther. 11 (2), 553–569.

Tapia, E., Soto, V., Ortiz-Vega, K.M., Santamaría, J., García-Niño, W.R., Correa, F., … Pedraza-Chaverri, J., 2012. Curcumin induces Nrf2 nuclear translocation and prevents glomerular hypertension, hyperfiltration, oxidant stress, and the decrease in antioxidant enzymes in 5/6 nephrectomized rats. Free Radic. Biol. Med. 53 (1), 52–61.

Tolosa, E., Garrido, A., Scholz, S.W., Poewe, W., 2021. Challenges in the diagnosis of Parkinson's disease. Lancet Neurol. 20 (5), 385–397.

Torres, A.K., Jara, C., Park-Kang, H.S., Polanco, C.M., Tapia, D., Alarcón, F., … Tapia-Rojas, C., 2021. Synaptic mitochondria: an early target of amyloid-β and tau in Alzheimer's disease. J. Alzheimers Dis. 84 (4), 1391–1414.

Tran, D., DiGiacomo, P., Born, D.E., Georgiadis, M., Zeineh, M., 2022. Iron and Alzheimer's disease: from pathology to imaging. Front. Hum. Neurosci. 16, 838692.

Traynelis, S.F., Wollmuth, L.P., Mcbain, C.J., Menniti, F.S., Vance, K.M., Ogden, K.K., et al., 2010. Glutamate receptor ion channels: structure, regulation, and function. Pharmacol. Rev. 62, 405–496. doi:10.1124/pr.109.002451.

Tripathi, P.N., Lodhi, A., Rai, S.N., Nandi, N.K., Dumoga, S., Yadav, P., Tiwari, A.K., Singh, S.K., El-Shorbagi, A.A., Chaudhary, S., 2024. Review of pharmacotherapeutic targets in Alzheimer's disease and its management using traditional medicinal plants. Degener. Neurol. Neuromuscul. Dis. 14, 47–74. doi:10.2147/DNND.S452009.

Ugwor, E.I., Dosumu, O.A., Eteng, O.E., James, A.S., Moses, C.A., Ogbonna, C.U., Adeleye, O.A., Omilo, B.C., Fabunmi, R.F., Olakitan, A.M., Ugbaja, R.N., 2022. Morin attenuates neurobehavioral deficits, hippocampal oxidative stress, inflammation, and apoptosis in rats co-exposed to bisphenol S and diethyl phthalate. Brain Res. 1794, 148068. doi:10.1016/j.brainres.2022.148068.

Ugwu, P.I., Ben-Azu, B., Ugwu, S.U., et al., 2022. Preventive putative mechanisms involved in the psychopathologies of mice passively coping with psychosocial defeat stress by quercetin. Brain Res. Bull. 183, 127–141. doi:10.1016/j.brainresbull.2022.03.004.

Ulusoy, H.G., Sanlier, N., 2020. A minireview of quercetin: from its metabolism to possible mechanisms of its biological activities. Crit. Rev. Food Sci. Nutr. 60 (19), 3290–3303.

Umukoro, S., Oluwole, O.G., Eduviere, A.T., Adrian, O.I., Ajayi, A.M., 2015. Jobelyn® exhibited anti-inflammatory, antioxidant, and membrane-stabilizing activities in experimental models. J. Basic Clin. Physiol. Pharmacol. 26 (5), 501–508.

Umukoro, S., Oghwere, E.E., Ben-Azu, B., Owoeye, O., Ajayi, A.M., Omorogbe, O., Okubena, O., 2019. Jobelyn® ameliorates neurological deficits in rats with ischemic stroke through inhibition of release of pro-inflammatory cytokines and NF-κB signaling pathway. Pathophysiology 26 (1), 77–88.

Vauzour, D., Camprubi-Robles, M., Miquel-Kergoat, S., Andres-Lacueva, C., Bánáti, D., Barberger-Gateau, P., Bowman, G.L., Caberlotto, L., Clarke, R., Hogervorst, E., Kiliaan, A.J., Lucca, U., Manach, C., Minihane, A.M., Mitchell, E.S., Perneczky, R., Perry, H., Roussel, A.M., Schuermans, J., Sijben, J., … Ramirez, M., 2017. Nutrition for the ageing brain: towards evidence for an optimal diet. Ageing Res. Rev. 35, 222–240. doi:10.1016/j.arr.2016.09.010.

Ventriglio, A., Bellomo, A., di Gioia, I., Di Sabatino, D., Favale, D., De Berardis, D., Cianconi, P., 2021. Environmental pollution and mental health: a narrative review of literature. CNS Spectr. 26 (1), 51–61.

Voulgaropoulou, S.D., van Amelsvoort, T.A.M.J., Prickaerts, J., Vingerhoets, C., 2019. The effect of curcumin on cognition in Alzheimer's disease and healthy aging: a systematic review of pre-clinical and clinical studies. Brain Res. 1725, 146476. doi:10.1016/j.brainres.2019.146476.

Wallace, D.R., Djordjevic, A.B., 2020. Heavy metal and pesticide exposure: a mixture of potential toxicity and carcinogenicity. Curr. Opin. Toxicol. 19, 72–79.

Wang, G., Song, L., Wang, H., Xing, N., 2013. Quercetin synergizes with 2-methoxyestradiol inhibiting cell growth and inducing apoptosis in human prostate cancer cells. Oncol. Rep. 30 (1), 357–363.

Wang, Q., Fan, W., Cai, Y., Wu, Q., Mo, L., Huang, Z., Huang, H., 2016. Protective effects of taurine in traumatic brain injury via mitochondria and cerebral blood flow. Amino Acids 48, 2169–2177.

Wang, X.S., Zhang, Z.R., Zhang, M.M., Sun, M.X., Wang, W.W., Xie, C.L., 2017. Neuroprotective properties of curcumin in toxin-base animal models of Parkinson's disease: a systematic experiment literatures review. BMC Complement. Altern. Med. 17 (1), 412. doi:10.1186/s12906-017-1922-x.

Wang, Y., Tang, Q., Duan, P., Yang, L., 2018. Curcumin as a therapeutic agent for blocking NF-κB activation in ulcerative colitis. Immunopharmacol. Immunotoxicol. 40 (6), 476–482. doi:10.1080/08923973.2018.1469145.

Wang, Y., Liu, F., Liu, M., Zhou, X., Wang, M., Cao, K., Jin, S., Shan, A., Feng, X., 2022. Curcumin mitigates aflatoxin B1-induced liver injury via regulating the NLRP3 inflammasome and Nrf2 signaling pathway. Food Chem. Toxicol. 161, 112823. doi:10.1016/j.fct.2022.112823.

Wang, N., Yang, J., Chen, R., Liu, Y., Liu, S., Pan, Y., … Li, Z., 2023a. Ginsenoside Rg1 ameliorates Alzheimer's disease pathology via restoring mitophagy. J. Ginseng Res. 47 (3), 448–457.

Wang, L., Xie, Z., Wu, M., Chen, Y., Wang, X., Li, X., Liu, F., 2024a. The role of taurine through endoplasmic reticulum in physiology and pathology. Biochem. Pharmacol., 116386 doi:10.1016/j.bcp.2024.116386.

Wareham, L.K., Liddelow, S.A., Temple, S., Benowitz, L.I., Di Polo, A., Wellington, C., … Calkins, D.J., 2022. Solving neurodegeneration: common mechanisms and strategies for new treatments. Mol. Neurodegener. 17 (1), 23.

Warren, C.A., Paulhill, K.J., Davidson, L.A., Lupton, J.R., Taddeo, S.S., Hong, M.Y., … Turner, N.D., 2009. Quercetin may suppress rat aberrant crypt foci formation by suppressing inflammatory mediators that influence proliferation and apoptosis. J. Nutr. 139 (1), 101–105.

Wei, Y., Zhang, Y., 2021. Glycyl radical enzymes and sulfonate metabolism in the microbiome. Annu. Rev. Biochem. 90 (1), 817–846.

Wei, W., Dong, Q., Jiang, W., Wang, Y., Chen, Y., Han, T., Sun, C., 2021. Dichloroacetic acid-induced dysfunction in rat hippocampus and the protective effect of curcumin. Metab. Brain Dis. 36 (4), 545–556. doi:10.1007/s11011-020-00657-5.

Wei, Y., Li, H., Li, Y., Zeng, Y., Quan, T., Leng, Y., Chang, E., Bai, Y., Bian, Y., Hou, Y., 2024. Advances of curcumin in nervous system diseases: the effect of regulating oxidative stress and clinical studies. Front. Pharmacol. 15, 1496661. doi:10.3389/fphar.2024.1496661.

Wu, J.Y., Prentice, H., 2010. Role of taurine in the central nervous system. J. Biomed. Sci. 17 (Suppl. i), SI. doi:10.1186/1423-0127-17-S1-S1.

Wu, J., Ibtisham, F., Niu, Y.F., Wang, Z., Li, G.H., Zhao, Y., … An, L., 2019a. Curcumin inhibits heat-induced oxidative stress by activating the MAPK-Nrf2/ARE signaling pathway in chicken fibroblasts cells. J. Therm. Biol. 79, 112–119. doi:10.1016/j.jtherbio.2018.12.008.

Wu, Y., Wang, B., Zeng, Z., Liu, R., Tang, L., Gong, L., Li, W., 2019b. Effects of probiotics Lactobacillus plantarum 16 and Paenibacillus polymyxa 10 on intestinal barrier function, antioxidative capacity, apoptosis, immune response, and biochemical parameters in broilers. Poultry Sci. 98 (10), 5028–5039.

Wu, J.J., Yang, Y., Wan, Y., Xia, J., Xu, J.F., Zhang, L., … Peng, C., 2022. New insights into the role and mechanisms of ginsenoside Rg1 in the management of Alzheimer's disease. Biomed. Pharmacother. 152, 113207. doi:10.1016/j.biopha.2022.113207.

Xu, H., Jia, Y., Sun, Z., Su, J., Liu, Q.S., Zhou, Q., Jiang, G., 2022. Environmental pollution, a hidden culprit for health issues. Eco-Environ. Health 1 (1), 31–45.

Yadav, R.S., Shukla, R.K., Sankhwar, M.L., et al., 2010. Neuroprotective effect of curcumin in arsenic-induced neurotoxicity in rats. Neurotoxicology 31 (5), 533–539. doi:10.1016/j.neuro.2010.05.001.

Yang, H., Zeng, F., Luo, Y., Zheng, C., Ran, C., Yang, J., 2022. Curcumin scaffold as a multifunctional tool for Alzheimer's disease research. Molecules 27 (12), 3879. doi:10.3390/molecules27123879.

Zeng, X.S., Zhou, X.S., Luo, F.C., Jia, J.J., Qi, L., Yang, Z.X., … Bai, J., 2014. Comparative analysis of the neuroprotective effects of ginsenosides Rg1 and Rb1 extracted from *Panax notoginseng* against cerebral ischemia. Can. J. Physiol. Pharmacol. 92 (2), 102–108.

Zhang, Y.L., Liu, Y., Kang, X.P., Dou, C.Y., Zhuo, R.G., Huang, S.Q., … Wen, L., 2018. Ginsenoside Rb1 confers neuroprotection via promotion of glutamate transporters in a mouse model of Parkinson's disease. Neuropharmacology 131, 223–237. doi:10.1016/j.neuropharm.2017.09.029.

Zhang, H., Su, Y., Sun, Z., Chen, M., Han, Y., Li, Y., … Li, W., 2021. Ginsenoside Rg1 alleviates Aβ deposition by inhibiting NADPH oxidase 2 activation in APP/PS1 mice. J. Ginseng Res. 45 (6), 665–675. doi:10.1016/j.jgr.2021.11.004.

Zhang, H.X., Li, Y.Y., Liu, Z.J., Wang, J.F., 2022a. Quercetin effectively improves LPS-induced intestinal inflammation, pyroptosis, and disruption of the barrier function through the TLR4/NF-κB/NLRP3 signaling pathway in vivo and in vitro. Food Nutr. Res. 66, 10–29219. doi:10.29219/fnr.v66.29219.

Zhang, L., Han, Y., Wu, X., Chen, B., Liu, S., Huang, J., Kong, L., Wang, G., Ye, Z., 2023a. Research progress on the mechanism of curcumin in cerebral ischemia/reperfusion injury: a narrative review. Apoptosis 28 (9-10), 1285–1303. doi:10.1007/s10495-023-01869-7.

Zhang, M., Liang, C., Chen, X., Cai, Y., Cui, L., 2024. Interplay between microglia and environmental risk factors in Alzheimer's disease. Neural Regen. Res. 19 (8), 1718–1727. doi:10.4103/1673-5374.370659.

Zhao, A., Liu, N., Yao, M., Zhang, Y., Yao, Z., Feng, Y., … Zhou, G., 2022. A review of neuroprotective effects and mechanisms of ginsenosides from *Panax ginseng* in treating ischemic stroke. Front. Pharmacol. 13, 946752. doi:10.3389/fphar.2022.946752.

Zhong, H., Xiao, R., Ruan, R., Liu, H., Li, X., Cai, Y., Zhao, J., Fan, X., 2020. Neonatal curcumin treatment restores hippocampal neurogenesis and improves autism-related behaviors in a mouse model of autism. Psychopharmacology 237 (12), 3539–3552. doi:10.1007/s00213-020-05634-5.

Zhu, H.T., Bian, C., Yuan, J.C., Chu, W.H., Xiang, X., Chen, F., … Lin, J.K., 2014. Curcumin attenuates acute inflammatory injury by inhibiting the TLR4/MyD88/NF-κB signaling pathway in experimental traumatic brain injury. J. Neuroinflammation 11, 1–17. doi:10.1186/s12974-014-0099-4.

Zhu, T., Wang, L., Xie, W., Meng, X., Feng, Y., Sun, G., Sun, X., 2021. Notoginsenoside R1 improves cerebral ischemia/reperfusion injury by promoting neurogenesis via the BDNF/Akt/CREB pathway. Front. Pharmacol. 12, 615998. doi:10.3389/fphar.2021.615998.

CHAPTER SEVEN

Neuroprotective influence of nutritive agents on neurodegenerative diseases

Omowumi FEMI-AKINLOSOTU[a,*], Olumayowa IGADO[b], Funmilayo OLOPADE[a], Ademola Adetokunbo OYAGBEMI[c]

[a]Developmental Neurobiology Unit, Department of Anatomy, College of Medicine, University of Ibadan. Nigeria
[b]Neuroscience Unit, Department of Veterinary Anatomy, Faculty of Veterinary Medicine, University of Ibadan. Nigeria
[c]Department of Veterinary Physiology and Biochemistry, Faculty of Veterinary Medicine, University of Ibadan. Nigeria
*Corresponding author. e-mail address: omowumiakinlosotu@gmail.com.

Contents

Abstract

Quercetin, Moringa oleifera (MO) leaves, and Caffeine have all been identified as potential therapeutic and nutritive agents in neurological maladies. The essence of neuroprotection in neurodegeneration is to stop or delay the disease progression, reverse the effects or completely prevent neurodegeneration from occurring. This can be achieved through different means.

Quercetin is a flavonoid that can be supplied through exogenous route, Moringa oleifera is a shrub native to Africa, India, Arabia, south America, Southeast Asia, including the Pacific and Caribbean Islands while caffeine is the most frequently ingested psychoactive stimulant by humans and a non-selective adenosine receptor antagonist found naturally in the leaves and seeds of many plants such as tea and coffee. Quercetin administration has been shown to decrease microglial-mediated neuronal cell death, attenuate production of reactive oxygen species (ROS), mitigate H_2O_2 and tertiary–butyl hydroperoxide toxicity, with concomitant reduction of cell death. MIMO: an active ingredient extracted from MO leaves reduced neuronal degeneration, Purkinje cell destruction, astrocytic

Advances in Neurotoxicology, Volume 16
ISSN 2468-7480
https://doi.org/10.1016/bs.ant.2026.02.004

and microglia activation, and also significantly ameliorated the severe demyelination observed across all brain regions. Chronic caffeine consumption has revealed enhanced dendritic lengths and arborization observed in the pyramidal neurons of the cerebral cortex, promoting locomotion, causing improved developmental motor activities and reflexes, and increasing dendritic arborization of the neurons in the sensorimotor cortex and striatum of hydrocephalic mice models and improved memory deficits and reduced the expression of reactive oxygen species and proinflammatory cytokines.
These phytochemicals could open a novel therapeutic window for effective management of neurodegenerative diseases that impact negatively on global mental health.

Keywords: Caffeine, Moringa oleifera, Neurodegenerative diseases, Parkinson's disease, Quercetin

1. Introduction

Neurodegenerative diseases are a number of group of diseases as presented by moderate neuronal cell death (Pedersen et al., 2016). The etiology of neurodegenerative diseases has not yet been fully elucidated. Further, array of predisposing factors could also play some critical roles in exacerbating the progression of the disease condition. Elsewhere, exaggerated oxidative stress, mitochondria dysfunction, and neuroinflammation have been documented in the pathophysiology of neurodegenerative diseases (Lomeli-Lepe et al., 2024). Neurodegenerative diseases such as Alzheimer's disease (AD) and Parkinson's disease (PD) have been positively correlated with neuroinflammatory processes (Mi et al., 2024; Emborg et al., 2024). Microglia activation is known to play a vital role in neurodegenerative and neuropsychiatry disorders through the enhancement of neuroinflammation and oxidative stress (Yao et al., 2024).

An increase in human population (leading to overpopulation) and industrial activities have been known to cause an increase in air pollution, due to ozone formation and suspension of particulate matter. The repeated exposure to these elements can lead to chronic oxidative stress, neurodegeneration and a host of other neuropathologies, directly impacting human health and resulting in an increase in the incidence of neurodegenerative diseases (Bello-Medina et al., 2022). The central nervous system (CNS) is particularly susceptible to the particulate matters as these can gain direct access to the brain through the nasal olfactory mucosa (Costa, 2017). Larger particulate matter are filtered out by the nose and upper airways, while smaller ones ($PM_{2.5}$) enter through the olfactory epithelium into the olfactory bulb, and go further into the olfactory cortex and other brain regions

(Costa et al., 2020). The spongy nature of the cribriform plate also contributes to this permeability.

Ageing still remains the main risk factor for Neurodegenerative diseases (ND). It is estimated that by 2060, the European Union population will be greater than 510 million, with a third of this population being 65 years and above (Costa, 2017). By the year 2050, it is estimated that about 7.6 percent of the Sub-Saharan African population (~2.074 billion) will be aged ≥60 years (Lekoubou et al., 2014). These statistics show that the percentage of the population living with age-related ND will increase. These risk factors such as sex, poor lifestyle habits, vascular conditions, and newly identified risk factors such as hearing loss and others not mentioned, make it imperative to prevent ND, effectively manage or slow down the course of progression in affected subjects, and where possible, completely reverse the condition. In the last two decades, a body of growing interests have reported the potential benefits of flavonoids modulating neuronal function and mitigating neurodegenerative diseases (Bayazid and Lim, 2022). Pathological hallmarks of Parkinson's disease (PD), Alzheimer's disease (AD) and Huntington's disease (HD) have been linked to accumulations of mutant proteins such as α-synuclein, amyloid-β (Aβ) and mutant huntingtin (Htt) respectively, in the affected brain regions (Sharma et al., 2021).

2. Quercetin and neurodegenerative disorders

The anti-inflammatory property of quercetin has been established by various *in vitro* studies. Globally, Parkinson's disease (PD) is the second most prevalent neurodegenerative disease (Paulėkas et al., 2024). Parkinson's disease is characterized by movement difficulty (e.g., tremor, stiffness, slowness), and psychological or cognitive problems (e.g., cognitive decline, depression, anxiety) as documented by Armstrong and Okun (2020). The histopathological hallmark of PD is the loss of dopaminergic neurons from the substantia nigra associated with the classical feature of Lewy bodies. Cell loss in the substantia nigra also occurs in a region-specific manner, with the lateral ventral tier of the pars compacta being most affected.

Quercetin (3,3′,4′,5,7-pentahydroxyflavone) is derived from the Latin word "Quercetum" which means Oak Forest. It is a flavonoid that belongs to the class called flavonol, hence, can only be supplied through exogenous route (Lakhanpal and Rai, 2007). It has a characteristic physico-chemical property of yellow color, poorly soluble in hot water, quite soluble in alcohol and lipids, and insoluble in cold water (Anand David et al., 2016). Quercetin is one of the most abundant dietary flavonoids found in fruits and many green leafy vegetables. The highest concentrations of flavonols have been found in fruits and vegetables (Anand David et al., 2016).

Quercetin administration was shown to decrease microglial mediated neuronal cell death in co-culture of microglia PC12 (167), lipopolysaccharide (LPS) at 1μg/mL for 24 h induced mRNA expression of TNF-α and IL-1β in glial cells and astrocytes (Sharma et al., 2017), inhibited TNF-α pathway in PC12 cells and zebrafish models (Bureau et al., 2008). Research documentation has shown that quercetin in astrocytes attenuated production of reactive oxygen species (ROS), mitigated hydrogen peroxide (H_2O_2) and tertiary–butyl hydroperoxide toxicity, with concomitant reduction of cell death via reduction in inducible nitric oxide synthase protein expression as well as mitochondrial superoxide radical (Bournival et al., 2012). The survival of rat oligodendrocytes (OLN-93) by quercetin has also been reported (Fan et al., 2019). Quercetin (50 mg/kg) by gavage for 4 weeks was recently found to offer neuroprotection in HIV-associated neurocognitive disorders via reductio in oxidative and cognitive disorders (Huang et al., 2025).

There are reports of substantial biological and pharmacological properties of quercetin such as the antioxidant, anti-inflammatory, and free radical scavenging activities (Carrillo-Martinez et al., 2024). The anti-inflammatory actions of quercetin have been reported to significantly reduce pro-inflammatory cytokines such as interleukin-1β (IL-1β) (IL-1β), interleukin-6 (IL-6), interleukin-6 (IL-8), and interleukin-6 (TNF-α) expression via downregulation of microglia activation (Das et al., 2024). Antioxidative and reactive oxygen species (ROS) scavenging activities of quercetin in neurodegenerative disorders have been extensively documented elsewhere (Rarinca et al., 2023). In Fig. 1 below, the anti-oxidative, anti-inflammatory and remyelination potential properties and mechanisms of action of quercetin as a possible complementary therapy in neurodegenerative diseases such as multiple sclerosis. The possible mechanisms of action include reduction of pro-inflammatory cytokines and oxidative stress, remyelination, modulation of microglia and differentiation of oligodendrocyte progenitor cells.

Previous investigation revealed the therapeutic effects of quercetin against PD in preclinical models as a treatment for PD, was found to act through the inhibition of oxidative stress, neuroinflammatory response and apoptotic pathways (de Oliveira Vian et al., 2024). Recently, bee-collected pollen (BP), found as an excellent source of quercetin and its derivatives was

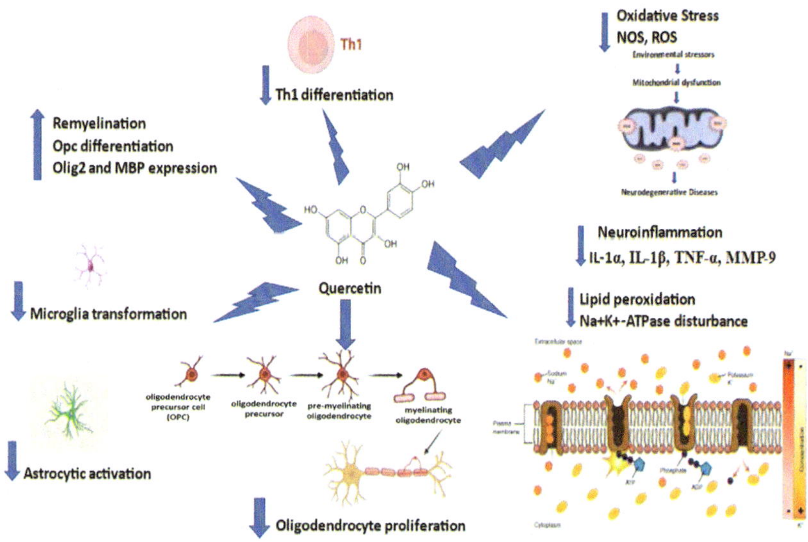

Fig. 1 Anti-oxidative, anti-inflammatory and remyelination potential properties of quercetin.

documented to offer protection against neurodegenerative disorders (Kostić et al., 2024). Notably, flavonoids such as quercetin, kaempferol, hesperitin, anthocyanin, and many more have been reported for their anti- Parkinson's effect (Varshney and Siddique, 2024).

Lakshmi et al. (2023) reported that quercetin nanocrystals (QNC) hold future promise as it effectively reduced toxicity of rotenone-induced oxidative stress in animal model of Parkinson's disease. Further, consumption of foods rich in phytochemicals such as quercetin could be an effective strategy in preventing and treating PD as reported by Sharma et al. (2023) and Bhusal et al. (2023).

3. Neuroprotective actions of quercetin in Alzheimer's disease

Alzheimer's disease (AD) is the most common cause of dementia which negatively impacts families with health and financial complications. Alzheimer's disease is a progressive cognitive impairment with pathological deposition of the neurofibrillary tangles (NFTs) and the senile amyloid plaques of amyloid proteins (Aβ). The Aβ plaques are deposited in the regions of the brain that regulate memory, learning, memory consolidation, retrieval, and cognition, such as the hippocampus, parietal cortex, and frontal cortex, etc. (Sharma et al., 2021). The pathogenesis of AD is associated with the accumulation of amyloid-β (Aβ) aggregates and the hyperphosphorylation of tau proteins, leading to formation of neurofibrillary tangles (NFTs) and synaptic dysfunction (Kommaddi et al., 2018). The therapeutic action of quercetin has been attributed to the inhibition of NFTs formation of amyloid-β proteins, thereby neuronal apoptosis and inflammatory cascade pathways (Khan et al., 2019).

It has been reported that quercetin prevented cell deaths in mouse-derived hippocampal neuronal cell line HT22 by improvement in cell viability and mitigation of oxidative stress biomarkers (Yang et al., 2024). Therefore, quercetin could be used for the prevention and/or treatment of neurodegenerative diseases such as AD. It has been documented that quercetin prevented (Aβ3(pE)-42) uptake in astrocytes thereby enhancing the abrogation of astrogliosis and synaptic dysfunction (Arndt et al., 2024). Zhu and his colleagues reported that *Ginkgo biloba* leaf extract which is ladened with quercetin improved cognitive function in AD (Zhu et al., 2024). Recently, also, quercetin fullerene conjugate (QFC) preparation for nose-to-brain delivery has been shown to be effective for treatment in AD (Palai et al., 2024). Recent scientific evidence has shown that glycogen synthase kinase 3β (GSK-3β) mediates AD progression through several signaling pathways, whereas flavonoids including quercetin are known potent inhibitors of GSK-3β for AD treatment (Zhao et al., 2024). Furthermore, Cheng et al. (2024) reported that intragastric administration of quercetin (100 mg/kg) once every day for 6 months activated Keap1/Nrf2/HO-1 pathway, reduced apoptosis, improved cognitive impairment, and AD-like pathology.

3.1 Neuroprotective actions of quercetin in multiple sclerosis

Multiple sclerosis is a complex autoimmune disorder of the central nervous system (CNS) associated with fatigue, dyscoordination weakness and visual weakness (Javanbakht et al., 2023). Globally, 1.89 million people live with Multiple sclerosis, with over 62,000 new cases diagnosed and documented in 2021. The global prevalence of MS is 23.9 cases per 100,000 population, with a continuous increase over the past 30 years. North America and Western Europe had the highest prevalence, incidence, disability-adjusted life-years (DALYs), and mortality rates. However, countries with the highest prevalence include Sweden (219 /100,000 population), Canada (182),

Norway (176), Ireland (163), and the UK (158) as reported by Khan et al. (2025).

Oxidative stress-induced inflammatory responses have been found to play major roles in the pathophysiology of multiple sclerosis by initiating damage to the myelin sheaths surrounding the axons, together with neuronal and oligodendrocyte cell death (Zha et al., 2022). There has been a strong positive correlation between multiple sclerosis and microglial activation, blood-brain barrier destruction, cytokine production, and oxidative stress, with the pathological hallmark of demyelination (Mojaverrostami et al., 2022). In multiple sclerosis disease, neuroinflammation has been shown to generate ROS/RNS, oxidative stress, and ultimately cause activation of microglia and astrocytes (Gil-Sánchez et al., 2024).

It has also been reported that quercetin prevented demyelination, increased remyelination, improved locomotor activity, inhibited lipid peroxidation and prevented the inhibition of acetylcholinesterase (AChE) activity in beam walking test (Woodfin et al., 2024). Some *in-vitro* studies have reported that quercetin decreased apoptosis, increased proliferation and differentiation of oligodendrocyte progenitor cells (OPCc) after oxygen/glucose deprivation induced injury by activating the PI3K/Akt signalling pathway (Wu et al., 2014). In another *in-vivo* study, quercetin treatment at 25, 50 and 100 mg/kg/day, by oral gavage in Cuprizone induced demyelination model of C57 BL/6 mice, was found to decrease demyelination in the *corpus callosum* (Javanbakht et al., 2023). Again, the remyelination capacity of quercetin was demonstrated as higher MBP and Olig2 protein expressions in the *corpus callosum* (Javanbakht et al., 2023).

3.2 Bioactive compounds from nutritive plants as neuroprotectants against neurodegeneration

The essence of neuroprotection in neurodegeneration is to stop or delay the disease progression, reverse the effects or completely prevent neurodegeneration from occurring. This can be achieved through different means. Neuroprotection can occur through the endogenous defence mechanism or through the use of exogenous substances like antioxidants, which may intervene in the disease progression and slow down the course (Tanaka and Vécsei, 2022).

Since ancient times, medicinal plants have been used to treat a wide array of conditions. Due to the fact that they contain a high concentration of antioxidants, including polyphenols, they have offered neuroprotective effects against various conditions of the nervous system, including ND. This is because they can significantly reduce the concentration of reactive oxygen species (ROS), chelate metal ions and cross the blood-brain barrier. Polyphenols are present in many natural sources (like plant foods, cereals, fruits, legumes, spices, beverages and herbs), and so far, no serious adverse side effect has been reported from their consumption (Arias-Sanchez et al., 2023).

Quercetin as a flavonoid-rich polyphenolic compound is a good candidate for the management of neurodegenerative diseases such Alzheimer's disease, Multiple sclerosis, Parkinson's disease, and Huntington's disease due to antioxidant, free radical scavenging, and anti-inflammatory activities. The use of quercetin and quercetin-rich medicinal plants and phytochemicals could open a novel therapeutic window for effective management of neurodegenerative diseases that impact negatively on global mental health.

3.3 Moringa oleifera

Moringa oleifera (MO) is a shrub, belonging to the family *Moringaceae*; it is one of the 14 species native to Africa, India, Arabia, South America, Southeast Asia, including the Pacific and Caribbean Islands. It is commonly referred to as the horseradish tree or drumstick. The leaves are commonly consumed as vegetables and are reportedly a rich source of essential amino acids. The leaves contain a number of phytochemicals like phenols, flavonoids, glycosides, amino acids, carotenoids, sterols, vitamins and alkaloids (Anwar et al., 2007; Igado et al., 2020; Iqbal and Bhanger, 2006; Verma et al., 2009).

The aqueous and alcoholic extract of the leaves have been used to treat models of neurodegenerative conditions (ageing, metal toxicity etc.) *in vitro* and *in vivo* models, few of which can only be cited here (Adefegha et al., 2021; Ekong et al., 2017; Hindawy et al., 2024; Igado et al., 2018; Kirisattayakul et al., 2013, 2012; Nair et al., 2021). Results obtained showed that the extract gave a protective effect against the induced toxicities. *In vitro* tests showed increased cell viability, reduction in the generation of ROS, modulation of enzyme activity and preservation of the DNA strand. Using hippocampal neurons in cell culture, Hannan and colleagues (Hannan et al., 2014) demonstrated that the addition of the ethanolic extract of MO leaves resulted in a significant increase in the length and number of neurites, and increased their branching, in a dose-dependent manner. They reported improved neuronal viability, decreased cellular injury and accelerated neuronal differentiation. The optimal dose reported was 30 μg/mL. In another study

(Adefegha et al., 2021), the incubation of aqueous extract of MO leaves (0.1–100 μg/mL) to BV-2 microglial cells modulated purinergic and cholinergic enzymes activities and also maintained the viability of the cells. In animal models, there was significantly improved neurobehavioural and cognitive scores, reduced neuro-inflammation, neurodegeneration and demyelination and alleviation of other neuropathologies. Regions mostly examined were the hippocampal regions, cerebral cortices, cerebellum, corpus callosum, cingulum and striatum. Albino mice exposed to ethanolic extract of MO leaves (300 mg/kg) and aluminum chloride (100 mg/kg) orally, simultaneously for 28 days, showed that MO leaves decreased the reduction in weight gain, ameliorated disruption of cellular organisation and neuronal death, and alleviated neuroinflammation in the temporal cortex (Ekong et al., 2017). In another study (Ademosun et al., 2022), MO leaves was incorporated into plain ice cream, and administered at 0.5, 1.0 and 2.0 g of powdered MO leaves in 85 g plain ice cream formulations (0.6 percent, 1.2 percent, 2.4 percent MO respectively) to adult male rats for 30 days. Groups administered MO had reduced brain butyrylcholinesterase and acetylcholinesterase enzymes activities, glycemic index, total cholesterol, triglycerides, while also increasing brain antioxidant activities (catalase and superoxide dismutase). According to Arozal and colleagues (Arozal et al., 2022), administration of aqueous MO leaves extract (500 mg/kg) or MO seed oil administered orally to mice for 28 days, with scopolamine administered on experimental day 22 – 28 to induce cognitive impairment, resulted in a reversal of some of the cognitive decline observed. Using the novel object recognition test, the seed oil (1 mL/kg) effectively restored the scopolamine-induced recognition memory deficit unlike the MO leaves. Both the seed oil and leaves extract however showed an increase in the percentage of the novel object investigation time compared to the scopolamine only group.

However, the use of the extracts has limitations and drawbacks. It is difficult to be specific about the amount, concentration or proportion of the compounds/phytochemicals which are ingested or used; perfect reproducibility may be difficult especially when the specific plants used are from different areas, as soil and water content can affect phytochemical concentration and activity; plants are a plethora of chemical compounds, making it difficult to know which is most responsible for the actions observed (Coppin et al., 2015; Igado et al., 2018, 2020); the use of alcoholic extracts may give false results, as the alcohol used may not fully evaporate; the use of the rotary evaporator mostly used by some authors may give different levels of consistency due to human perception, thereby affecting concentration and dosage . In addition, the process of extraction is a long, tedious, expensive, time-consuming and wasteful method that can result in deforestation if not curbed. Getting 10 millilitres of the extract would have used up to 2 kg of the leaves in dry weight. This process can make the environment prone to deforestation, and the plants not as accessible, especially considering the volume of work being published on MO.

In spite of the neuroprotective effects reported from the use of MO methanol extracts, there have been reports of teratology when used in pregnant dams. Pups showed congenital musculoskeletal anomalies, reduced weight gain with histopathology revealing demyelination, neurodegeneration and neuronal layer distortion observed in the cerebral layers (Atiba et al., 2017). Toxicity reports with the drawbacks and limitations have made it imperative to isolate, characterise and identify the most active antioxidative compound in the MO leaves.

3.4 MIMO2 (p-methoxyphenyl acetic acid butyl ester)

The MO leaves were collected from a single location during the rainy season in the Southwestern region of Nigeria. The methanol extract of the leaves was subjected to a bioassay-guided fractionation, to identify the most active antioxidative component, which was subjected to nuclear magnetic resonance to determine the chemical structure. This compound synthesized and code-named MIMO2 is a phenolic compound (Igado et al., 2018). For ease, MIMO was coined from the name of the scientist and MO (***M**ayowa **I**gado **M**oringa **O**leifera*). Two pure compounds were initially identified and code-named MIMO1 and MIMO2. Carrying out the ferric reducing antioxidative power (FRAP) assay and comparing the MO extract, butanol fraction of the extract (from which MIMO2 was isolated), MIMO1, MIMO2 and tempol (a known antioxidant), showed that the MO extract was slightly lower than tempol, the butanol fraction had the lowest value and was unstable when stored, while MIMO2 had the highest value and was stable at room temperature (Igado et al., 2018). The process of synthesising MIMO2 achieved a number of things – elimination of the long and tedious process of collection and drying of leaves, expensive process of extraction and isolation comprising of different levels of fractionation and testing of each fraction, narrowing down with the use of LC-MS; and conservation of the environment.

Cytological assay with MIMO2 (0.25 and 0.5 μM) in immortalised mouse hippocampal cells (HT22) exposed to 100 μM vanadium (an environmental neurotoxicant) incubated for 30 min, using dihydroethidium, Comet assays and fluorescein diacetate liveability test showed that MIMO2 reduced cell death, improved cell membrane protection, and protected against cellular DNA damage. When compared with results obtained from using MO leaves extract (0.063 mg), MIMO2 was seen to give a better protective effect (Igado et al., 2018).

In vivo tests using 14-day old mice (after establishing the LD_{50}) dosed intraperitoneally for 14 day and exposed to vanadium for 14 days (sacrificed on day 15), revealed that MIMO2 protected against neurobehavioural deficits. Open field tests showed that the mice treated with MIMO2 had more exploratory habits common to rodents relative to those only administered vanadium. Hanging wire test revealed that MIMO2 protected against forelimb weakness, as the mice stayed longer on the wires. Histopathology revealed a significant prevention of demyelination with the co-administration of vanadium and MIMO2. Vanadium causes demyelination. Using basic histological stains, histochemistry, immunohistochemistry and fluorescent immunohistochemistry (confocal imaging), MIMO2 reduced neuronal degeneration, Purkinje cell destruction, astrocytic and microglia activation, and also significantly ameliorated the severe demyelination observed across all brain regions assessed, most especially the corpus callosum, with the retrosplenial and somatosensory cortices. The oligodendrocyte destruction associated with vanadium exposure was significantly reduced with the co-administration of MIMO2, most especially at the corpus callosum, cingulum and the infragranular layers of the M2 cerebral cortex. Administration of MIMO2 alone showed no side effects with the neurobehavioural tests and histopathology. An interesting finding was the fact that MIMO2 appeared to potentiate and improve myelination even when used alone (Igado et al., 2020). *Moringa oleifera* is found in Africa, India and many arid and tropical countries, and is used a food and medicine. The neuroprotective effects exhibited by MO and its derivatives is considered to be due to the amount of antioxidative compounds it contains.

3.4.1 Caffeine

Caffeine, a non-selective adenosine receptor antagonist, is the most widely used psychoactive agent in the world. It is widely consumed and naturally produced from the leaves and seeds of many plants such as tea and coffee. Benefits of caffeine consumption include improvements in physical strength and endurance, cognitive function, particularly alertness and vigilance, perception of fatigue and mood. Caffeine stimulates the central nervous system in both humans and animal models; it easily crosses the blood–brain barrier and is an agonist of adenosine receptors in the brain, acting by competitive inhibition. Caffeine exerts a stimulating effect at low-to-moderate doses, but weaker stimulatory and even depressive effects at higher doses in acute administration.

Researches have demonstrated a significant deterioration in locomotor function in both moderate and high-dose caffeine-treated animal models after chronic administration resulting in the enhancement of motor coordination and functions. Chronic caffeine consumption has been shown to selectively bring about enhanced performance of locomotion and motor coordination. The enhanced dendritic lengths and arborization observed in the pyramidal neurons of the cerebral cortex of mice models might have a strong relationship to the improvement in their motor function, as well as in their hippocampal and prefrontal neurons which was associated with attenuation of previous cognitive decline. Furthermore, in a study in which caffeine was administered for eight weeks in mice models, locomotion and equilibrium of the rodents were promoted but without increase muscular strength; this was associated with increase in the neuronal density or dendritic length and arborization of the neurons in the cerebellar cortex, striatum and cerebral cortex respectively.

Caffeine intake might promote weight, BMI and body fat reduction (Tabrizi et al., 2018). It has been documented that herbal ephedra/caffeine (90/192 mg/day) promoted body weight and body fat reduction (Boozer et al., 2002). It also seems that chronic caffeine consumption in a dose-dependent manner significantly reduced rate of weight gain. This might be attributed to the lipolytic effect of caffeine, especially in chronic consumption of high doses. Some researchers opined that epinephrine, a neurotransmitter known for the fight-or-flight response, through its effect on heart rate and breathing, can burn calories by increasing the former or through its regulation of adipose tissue blood flow, modulating fatty acid delivery from adipose tissue into the systemic circulation. A concurrent increase in lipolysis and adipose tissue blood flow may be advantageous, as it enhances the removal of free fatty acids released during lipolysis from the interstitial space and prevents local accumulation of these acids (Horowitz and Klein, 2000).

The cognitive effects of caffeine are mostly due to antagonism of adenosine A1 receptors in the hippocampus and cerebral cortex (brain areas mostly involved in cognition); however, caffeine also aids information processing and performance, thus improving behavioural routines, arousal enhancement and sensorimotor gating (Olopade et al., 2021a). Report has it that long-term excessive caffeine intake can result in caffeine-induced anxiety disorders, which mimics organic mental disorders, such as panic disorder, generalized anxiety disorder, bipolar disorder, or even schizophrenia (Ribeiro et al., 2002) with experimental animals exhibiting some degree of anxiety following caffeine administration. In a human experiment performed by Lanini and colleagues, acute caffeine dose of low to moderate consumption (25–300mg), surfeited simple attention, improved arousal and selectively enhanced performance in executive functions. In addition, elderly women (but not men) who consumed large amounts of caffeine throughout their lifetimes performed better on memory and other cognitive tasks than non-caffeine drinkers. These positive observations can be attributed to increased cell numbers and increased length of the dendrites in the region of the brain concerned with memory (Olopade et al., 2021b).

4. Conclusion

Combining all, quercetin as a flavonoid-rich polyphenolic compound is a good candidate for the management of neurodegenerative diseases such Alzheimer's disease, Multiple sclerosis, Parkinson's disease, Huntington's disease due to antioxidant, free radical scavenging, and anti-inflammatory activities. The use of quercetin and quercetin-rich medicinal plants and phytochemicals could open a novel window for the management of neurodegenerative diseases that impact negatively on global mental health.

Results obtained showed MIMO2 to be a potential nutritive candidate to combat neurodegeneration, due to its ability to mop up ROS, probably chelate metal ions and cross the blood brain barrier. Further tests are ongoing.

References

Adefegha, S.A., Assmann, C.E., Schetinger, M.R.C., de Andrade, Emanuelli, T., 2021. Moringa oleifera modulates cholinergic and purinergic enzymes activity in BV-2 microglial cells. Metab. Brain Dis. 36 (4), 627–638. doi:10.1002/ptr.2023.

Ademosun, A.O., Oboh, G., Ajeigbe, O.F., 2022. Influence of Moringa (Moringa oleifera) enriched ice creams on rats' brain: Exploring the redox and cholinergic systems. Curr. Res. Food Sci. 5, 366–373. doi:10.1016/j.crfs.2022.01.021.

Anand David, A.V., Arulmoli, R., Parasuraman, S., 2016. Overviews of biological importance of quercetin: a bioactive flavonoid. Pharmacogn. Rev 10 (20), 84–89. doi:10.4103/0973-7847.194044.

Anwar, F., Latif, S., Ashraf, M., Gilani, A.H., 2007. Moringa oleifera: a food plant with multiple medicinal uses. Phyt. Res. Int. J. Dev. Phar. Toxicol. Eval. Nat. Prod. Derivatives 21 (1), 17–25. doi:10.1002/ptr.2023.

Arias-Sanchez, R.A., Torner, L., Fenton Navarro, B., 2023. Polyphenols and neurodegenerative diseases: potential effects and mechanisms of neuroprotection. Molecules 28 (14), 5415. doi:10.3390/molecules28145415.

Armstrong, M.J., Okun, M.S., 2020. Diagnosis and treatment of Parkinson disease: a review. JAMA 323 (6), 548–560. doi:10.1001/jama.2019.22360.

Arndt, H., Bachurski, M., Yuanxiang, P., Franke, K., Wessjohann, L.A., Kreutz, M.R., et al., 2024. A screen of plant-based natural products revealed that quercetin prevents pyroglutamylated amyloid-β (Aβ3(pE)-42) uptake in astrocytes as well as resulting astrogliosis and synaptic dysfunction. Mol. Neurobiol 62, 3730–3745. doi:10.1007/s12035-024-04509-6.

Arozal, W., Purwoningsih, E., Lee, H.J., Barinda, A.J., Munim, A., 2022. Effects of Moringa oleifera in two independents formulation and as neuroprotective agent against scopolamine-induced memory impairment in mice. Front. Nutr. 9, 799127. doi:10.3389/fnut.2022.799127.

Atiba, F.A., Imosemi, I.O., Malomo, A.O., 2017. Noxious effect of Moringa oleifera leave extract on the developing brain, morphology and behaviour of Wistar rat. Italian J. Anat. Embryol. 8–16. doi:10.13128/IJAE-20922.

Bayazid, A.B., Lim, B.O., 2022. Quercetin is an active agent in berries against neurodegenerative diseases progression through modulation of Nrf2/HO1. Nutrients 14 (23), 5132. doi:10.3390/nu14235132.

Bello-Medina, P.C., Rodríguez-Martínez, E., Prado-Alcalá, R.A., and Rivas-Arancibia, S. 2022. Ozone Pollution, Oxidative Stress, Synaptic Plasticity, and Neurodegeneration. Neurología (English Edition) 37 (4): 277–286. https://doi.org/10.1016/j.nrleng.2018.10.025.

Bhusal, C.K., Uti, D.E., Mukherjee, D., Alqahtani, T., Alqahtani, S., Bhattacharya, A., et al., 2023. Unveiling nature's potential: promising natural compounds in Parkinson's disease management. Parkinsonism. Relat. Disord 115, 105799. doi:10.1016/j.parkreldis.2023.105799.

Boozer, C., Daly, P., Homel, P., Solomon, J.L., Blanchard, D., Nasser, J.A., et al., 2002. Herbal ephedra/caffeine for weight loss: a 6-month randomized safety and efficacy trial. Int. J. Obes. 26, 593–604. doi:10.1038/sj.ijo.0802023.

Bournival, J., Plouffe, M., Renaud, J., Provencher, C., Martinoli, M. G., 2012. Quercetin and sesamin protect dopaminergic cells from MPP+-induced neuroinflammation in a microglial (N9)-neuronal (PC12) coculture system. Oxid. Med. Cell. Long. 2012 (1), 921941. doi:10.1155/2012/921941.

Bureau, G., Longpré, F., Martinoli, M.G., 2008. Resveratrol and quercetin, two natural polyphenols, reduce apoptotic neuronal cell death induced by neuroinflammation. J. Neurosci. Res 86 (2), 403–410. doi:10.1002/jnr.21503.

Carrillo-Martinez, E.J., Flores-Hernández, F.Y., Salazar-Montes, A.M., Nario-Chaidez, H.F., Hernández-Ortega, L.D., 2024. Quercetin, a flavonoid with great pharmacological capacity. Molecules 29 (5), 1000. doi:10.3390/molecules29051000.

Cheng, M., Yuan, C., Ju, Y., Liu, Y., Shi, B., Yang, Y., et al., 2024. Quercetin attenuates oxidative stress and apoptosis in brain tissue of APP/PS1 double transgenic AD mice by regulating Keap1/Nrf2/HO-1 pathway to improve cognitive impairment. Behav. Neurol 2024, 5698119. doi:10.1155/2024/5698119.

Coppin, J.P., Juliani, H.R., Wu, Q., Simon, J.E., 2015. Variations in polyphenols and lipid soluble vitamins in Moringa oleifera. In: *Processing and impact on active components in food.* Academic Press, pp. 655–663.

Costa, L.G., 2017. Traffic-Related Air Pollution and Neurodegenerative Diseases: Epidemiological and Experimental Evidence, and Potential Underlying Mechanisms. Advances in Neurotoxicology, 1, 1st ed. Elsevier Inc. https://doi.org/10.1016/bs.ant.2017.07.001.

Costa, L.G., Cole, T.B., Dao, K., Chang, Y.C., Coburn, J., Garrick, J.M., 2020. Effects of Air Pollution on the Nervous System and Its Possible Role in Neurodevelopmental and Neurodegenerative Disorders. Pharm. Ther. 210, 107523. doi:10.1016/j.pharmthera.2020.107523.

Das, D., Banerjee, A., Mukherjee, S., Maji, B.K., 2024. Quercetin inhibits NF-kB and JAK/STAT signaling via modulating TLR in thymocytes and splenocytes during MSG-induced immunotoxicity: an in vitro approach. Mol. Biol. Rep 51 (1), 277. doi:10.1007/s11033-024-09245-7.

de Oliveira Vian, C., Marinho, M.A.G., da Silva Marques, M., Hort, M.A., Cordeiro, M.F., Horn, A.P., 2024. Effects of quercetin in preclinical models of Parkinson's disease: a systematic review. Basic Clin. Pharmacol. Toxicol 135 (1), 3–22. doi:10.1111/bcpt.14011.

Ekong, M.B., Ekpo, M.M., Akpanyung, E.O., Nwaokonko, D.U., 2017. Neuroprotective effect of Moringa oleifera leaf extract on aluminium-induced temporal cortical degeneration. Metab. brain Dis. 32 (5), 1437–1447. doi:10.1007/s11011-017-0011-7.

Emborg, M.E., Gambardella, J.C., Zhang, A., Federoff, H.J., 2024. Autologous vs heterologous cell replacement strategies for Parkinson disease and other neurologic diseases. Handb. Clin. Neurol 205, 41–56. doi:10.1016/B978-0-323-90120-8.00010-1.

Fan, H., Tang, H.B., Shan, L.Q., Liu, S.C., Huang, D.G., Chen, X., et al., 2019. Quercetin prevents necroptosis of oligodendrocytes by inhibiting macrophages/microglia polarization to M1 phenotype after spinal cord injury in rats. J. Neuroinflammation 16 (1), 206. doi:10.1186/s12974-019-1613-2.

Gil-Sánchez, A., Gonzalo, H., Canudes, M., Nogueras, L., González-Mingot, C., Valcheva, P., et al., 2024. Can glatiramer acetate prevent cognitive impairment by modulating oxidative stress in patients with multiple sclerosis? Pharmaceuticals (Basel) 17 (4), 459. doi:10.3390/ph17040459.

Hannan, M.A., Kang, J.Y., Mohibbullah, M.D., Hong, Y.K., Lee, H., Choi, J.S., Choi, I.S., Moon, I.S., 2014. Moringa oleifera with promising neuronal survival and neurite outgrowth promoting potentials. J. Ethnopharm. 152 (1), 142–150. doi:10.1016/j.jep.2013.12.036.

Hindawy, R.F., Manawy, S.M., Nafea, O.E., Abdelhameed, A.A., Hendawi, F.F., 2024. Moringa oleifera leaves ethanolic extract counteracts cortical neurodegeneration induced by aluminum chloride in rats. Toxicol. Res. 13 (2). doi:10.1093/toxres/tfae028.

Horowitz, J.F., Klein, S., 2000. Whole body and abdominal lipolytic sensitivity to epinephrine is suppressed in upper body obese women. Am. J. Physiol. Endocrinol. Metab 278 (6), E1144–E1152. doi:10.1152/ajpendo.2000.278.6.E1144.

Huang, P., Chen, J., Zou, J., Gao, X., Cao, H., 2025. [Quercetin mitigates HIV-1 gp120-induced rat astrocyte neurotoxicity via promoting G3BP1 disassembly in stress granules]. Nan Fang Yi Ke Da Xue Xue Bao 45 (2), 304–312. doi:10.12122/j.issn.1673-4254.2025.02.11.

Igado, O.O., Andrioli, A., Azeez, I.A., Girolamo, F., Errede, M., Aina, O.O., Glaser, J., Holzgrabe, U., Bentivoglio, M., Olopade, J.O., 2020. The ameliorative effects of a phenolic derivative of Moringa oleifera leave against vanadium-induced neurotoxicity in mice. IBRO Rep. 9, 164–182. doi:10.1016/j.ibror.2020.07.004.

Igado, O.O., Glaser, J., Ramos-Tirado, M., Bankoğlu, E.E., Atiba, F.A., Holzgrabe, U., Stopper, H., Olopade, J.O., 2018. Isolation of a novel compound (MIMO2) from the methanolic extract of Moringa oleifera leaves: protective effects against vanadium-induced cytotoxity. Drug Chem. Toxicol. 41 (3), 249–258. doi:10.1080/01480545.2017.1366504.

Iqbal, S., Bhanger, M.I., 2006. Effect of season and production location on antioxidant activity of Moringa oleifera leaves grown in Pakistan. J. Food Comp. Analys. 19 (6–7), 544–551.doi. doi:10.1016/j.jfca.2005.05.001.

Javanbakht, P., Yazdi, F.R., Taghizadeh, F., Khadivi, F., Hamidabadi, H.G., Kashani, I.R., et al., 2023. Quercetin as a possible complementary therapy in multiple sclerosis: anti-oxidative, anti-inflammatory and remyelination potential properties. Heliyon 9 (11), e21741. doi:10.1016/j.heliyon.2023.e21741.

Khan, G., Hashim, M.J., 2025. Epidemiology of multiple sclerosis: global, regional, national and sub-national-level estimates and future projections. J. Epidemiol. Glob. Health 15 (1), 21. doi:10.1007/s44197-025-00353-6.

Khan, H., Ullah, H., Aschner, M., Cheang, W.S., Akkol, E.K, 2019. Neuroprotective effects of quercetin in Alzheimer's disease. Biomolecules 10 (1), 59. doi:10.3390/biom10010059.

Kirisattayakul, W., Wattanathorn, J., Tong-Un, T., Muchimapura, S., Wannanon, P., Jittiwat, J., 2013. Cerebroprotective effect of Moringa oleifera against focal ischemic stroke induced by middle cerebral artery occlusion. Oxid. Med. Cell. Longev. 2013 (1), 951415. doi:10.1155/2013/951415.

Kommaddi, R.P., Das, D., Karunakaran, S., Nanguneri, S., Bapat, D., Ray, A., et al., 2018. Aβ mediates F-actin disassembly in dendritic spines leading to cognitive deficits in Alzheimer's disease. J. Neurosci 38 (5), 1085–1099. doi:10.1523/JNEUROSCI.2127-17.2017.

Kostić, A.Ž., Arserim-Uçar, D.K., Materska, M., Sawicka, B., Skiba, D., Milinčić, D.D., et al., 2024. Unlocking quercetin's neuroprotective potential: a focus on bee-collected pollen. Chem. Biodivers 21 (6), e202400114. doi:10.1002/cbdv.202400114.

Lakhanpal, P., Rai, D.K., 2007. Quercetin: A versatile flavonoid. Int. J. Med. Update 2, 22–37.

Lakshmi, Y.S., Prasanth, D.S.N.B.K., Kumar, K.T.S., Ahmad, S.F., Ramanjaneyulu, S., Rahul, N., et al., 2023. Unravelling the molecular mechanisms of a quercetin nanocrystal for treating potential Parkinson's disease in a rotenone model: supporting evidence of network pharmacology and in silico data analysis. Biomedicines 11 (10), 2756. doi:10.3390/biomedicines11102756.

Lekoubou, A., Echouffo-Tcheugui, J.B., Kengne, A.P., 2014. Epidemiology of neurodegenerative diseases in sub-Saharan Africa: a systematic review. BMC Public Health 14 (1), 653. doi:10.1186/1471-2458-14-653.

Lomeli-Lepe, A.K., Castañeda-Cabral, J.L., Ureña-Guerrero, M.E., Cabrera, G.G., López-Pérez, S.J., 2024. Bee venom reduces early inflammation and oxidative stress associated with lipopolysaccharide-induced alpha-synuclein in the substantia nigra-striatum axis. Cell. Biochem. Biophys 83 (1), 1185–1196. doi:10.1007/s12013-024-01552-x.

Mi, J., Liu, C., Chen, H., Qian, Y., Zhu, J., Zhang, Y., et al., 2024. Light on Alzheimer's disease: from basic insights to preclinical studies. Front. Aging Neurosci 16, 1363458. doi:10.3389/fnagi.2024.1363458.

Mojaverrostami, S., Khadivi, F., Zarini, D., Mohammadi, A., 2022. Combination effects of mesenchymal stem cells transplantation and anodal transcranial direct current stimulation on a cuprizone-induced mouse model of multiple sclerosis. J. Mol. Histol 53 (5), 817–831. doi:10.1007/s10735-022-10092-8.

Nair, D.A., James, T.J., Sreelatha, S.L., Kariyil, B.J., Nair, S.N., 2021. Moringa oleifera (Lam.): a natural remedy for ageing? Nat. Prod. Res. 35 (24), 6216–6222. doi:10.1080/14786419.2020.1837815.

Olopade, F.E., Femi-Akinlosotu, O.M., Adekanmbi, A.J., Fejiro, O., Shokunbi, M.T., 2021. Chronic caffeine ingestion improves motor function and increases dendritic length and arborization in the motor cortex, striatum and cerebellum. J. Caffeine Adenosine Res. 11 (1). doi:10.1089/caff.2020.0017.

Olopade, F.E., Femi-Akinlosotu, O.M., Adekanmbi, A.J., Ighogboja, O.O., Shokunbi, M.T., 2021. Chronic caffeine ingestion improves memory and learning and increases neuronal population and dendritic length in the hippocampus of adult mice. Niger. J. Physiol. Sci. 36, 165–172. www.njps.physiologicalsociety.com.

Palai, A.K., Kumar, A., Mazahir, F., Sharma, A., Yadav, A.K., 2024. Synthesis and characterization of fullerene-based nanocarrier for targeted delivery of quercetin to the brain. Ther. Deliv 15 (7), 545–559. doi:10.1080/20415990.2024.2365620.

Paulėkas, E., Vanagas, T., Lagunavičius, S., Pajėdienė, E., Petrikonis, K., Rastenytė, D., 2024. Navigating the neurobiology of Parkinson's: the impact and potential of α-synuclein. Biomedicines 12 (9), 2121. doi:10.3390/biomedicines12092121.

Pedersen, J.T., Chen, S.W., Borg, C.B., Ness, S., Bahl, J.M., Heegaard, N.H., et al., 2016. Amyloid-β and α-synuclein decrease the level of metal-catalyzed reactive oxygen species by radical scavenging and redox silencing. J. Am. Chem. Soc 138 (12), 3966–3969. doi:10.1021/jacs.5b13577.

Rarinca, V., Nicoara, M.N., Ureche, D., Ciobica, A., 2023. Exploitation of quercetin's antioxidative properties in potential alternative therapeutic options for neurodegenerative diseases. Antioxidants (Basel) 12 (7), 1418. doi:10.3390/antiox12071418.

Ribeiro, J.A., Sebastiao, A.M., de Mendonca, A., 2002. Adenosine receptors in the nervous system: pathophysiological implications. Prog. Neurobiol. 68 (6), 377–392.

Sharma, B.R., Gautam, L.N., Adhikari, D., Karki, R., 2017. A comprehensive review on chemical profiling of Nelumbo nucifera: potential for drug development. Phytother. Res 31 (1), 3–26. doi:10.1002/ptr.5732.

Sharma, P., Kishore, A., De, I., Negi, S., Kumar, G., Bhardwaj, S., et al., 2023. Mitigating neuroinflammation in Parkinson's disease: exploring the role of proinflammatory cytokines and the potential of phytochemicals as natural therapeutics. Neurochem. Int 170, 105604. doi:10.1016/j.neuint.2023.105604.

Sharma, V.K., Singh, T.G., Mehta, V., 2021. Stressed mitochondria: a target to intrude Alzheimer's disease. Mitochondrion 59, 48–57. doi:10.1016/j.mito.2021.04.004.

Tabrizi, R., Saneei, P., Lankarani, K.B., Akbari, M., Kolahdooz, F., Esmaillzadeh, A., et al., 2018. The effects of caffeine intake on weight loss: a systematic review and dose-response meta-analysis of randomized controlled trials. Crit. Rev. Food Sci. Nutr. 59 (16), 2688–2696. doi:10.1080/10408398.2018.1507996.

Tanaka, M., Vécsei, L., 2022. Editorial of special issue 'dissecting neurological and neuropsychiatric diseases. Neurodegeneration and neuroprotection'. Int. J. Mol. Sci. 23 (13), 6991. doi:10.3390/ijms23136991.

Varshney, H., Siddique, Y.H., 2024. Effect of flavonoids against Parkinson's disease. Cent. Nerv. Syst. Agents Med. Chem 24 (2), 145–165. doi:10.2174/0118715249264078231214074107.

Verma, A.R., Vijayakumar, M., Mathela, C.S., Rao, C.V., 2009. In vitro and in vivo antioxidant properties of different fractions of Moringa oleifera leaves. Food Chem. Toxicol. 47 (9), 2196–2201. doi:10.1016/j.fct.2009.06.005.

Woodfin, S., Hall, S., Ramerth, A., Chapple, B., Fausnacht, D., Moore, W., et al., 2024. Potential application of plant-derived compounds in multiple sclerosis management. Nutrients 16 (17), 2996. doi:10.3390/nu16172996.

Wu, X., Qu, X., Zhang, Q., Dong, F., Yu, H., Yan, C., et al., 2014. Quercetin promotes proliferation and differentiation of oligodendrocyte precursor cells after oxygen/glucose deprivation-induced injury. Cell. Mol. Neurobiol 34 (3), 463–471. doi:10.1007/s10571-014-0030-4.

Yang, E.J., Kim, J.C., Na, D.H., 2024. Neuroprotective effects of cerebroprotein hydrolysate and its combination with antioxidants against oxidative stress-induced HT22 cell death. Toxicol. Res 40 (4), 541–550. doi:10.1007/s43188-024-00248-x.

Yao, H., Tong, W., Song, Y., Li, R., Xiang, X., Cheng, W., et al., 2024. Exercise training upregulates CD55 to suppress complement-mediated synaptic phagocytosis in Parkinson's disease. J. Neuroinflammation 21 (1), 246. doi:10.1186/s12974-024-03234-0.

Zha, Z., Liu, S., Liu, Y., Li, C., Wang, L., 2022. Potential utility of natural products against oxidative stress in animal models of multiple sclerosis. Antioxidants (Basel) 11 (8), 1495. doi:10.3390/antiox11081495.

Zhao, Z., Yuan, Y., Li, S., Wang, X., Yang, X., 2024. Natural compounds from herbs and nutraceuticals as glycogen synthase kinase-3β inhibitors in Alzheimer's disease treatment. CNS Neurosci. Ther 30 (8), e14885. doi:10.1111/cns.14885.

Zhu, C., Liu, J., Lin, J., Xu, J., Yu, E., 2024. Investigating the effects of Ginkgo biloba leaf extract on cognitive function in Alzheimer's disease. CNS Neurosci. Ther 30 (9), e14914. doi:10.1111/cns.14914.

CHAPTER EIGHT

Nutraceuticals and neurotoxicity

Samuel Adetunji Onasanwo[a,*], Busayo Oladun[b], Lawrence Dayo Adedayo[c], Ademola Adetokunbo Oyagbemi[d], Olusegun Gafar Adebayo[a], Ismaheel Akinwale Adeniyi[a,e], Kafilat Temidayo Aransi-ola[a]

[a] Department of Physiology, Faculty of Basic Medical Sciences, College of Medicine, University of Ibadan, Ibadan, Nigeria
[b] Exercise and Nutritional Sciences, College of Health Solutions, Arizona State University, Phoenix, United States of America
[c] Department of Human Physiology, Faculty of Basic Medical Sciences, College of Medicine, Federal University Wukari, Taraba, Nigeria
[d] Department of Veterinary Physiology and Biochemistry, Faculty of Veterinary Medicine, University of Ibadan, Ibadan, Oyo State, Nigeria
[e] Department of Neuroscience and Cell Biology, Robert Wood Johnson Medical School, Rutgers University, Piscataway, NJ, United States of America
*Corresponding author. e-mail address: samphil2002@yahoo.com.

Contents

Summary

Many toxic substances have been implicated in neurotoxicity in human ranging from chemicals to environmental hazardous compounds. The nutraceuticals as dietary supplements, utilized to ameliorate health, delay senescence, prevent diseases, and support the proper functioning of the human body. Most of the nutraceuticals used various mechanisms to exert their therapeutic potentials. This chapter enumerated and examined the past and the current trends of nutraceuticals in mitigating neurodegeneration, neurodevelopmental, and neurobehavioral changes. It also discussed the works that were already done with few natural nutraceuticals on neurotoxicity that are promising agent for its management. Therefore, this opens a novel long-term therapeutic window for effective and efficient neuroprotective regimen on a global scale.

Advances in Neurotoxicology, Volume 16
ISSN 2468-7480
https://doi.org/10.1016/bs.ant.2026.02.005

Keywords: Anacardium occidentalis, Cnidoscolus aconitifolius, Garcinia kola, Neurotoxicity, Nutraceuticals, Sorghum bicolor, Vernonia amygdalina

1. Introduction

Nutraceuticals are dietary components which are naturally found in our daily foods, andare believed to have a medical or health benefit on individuals or persons (Chauhan et al., 2013). Nutraceuticals are actually coined from the words, "nutrition" and "pharmaceutical" (Makkar et al., 2020).Nutrition is defined as the biochemical and physiological process by which an organism or human uses food to support its life while a pharmaceutical is any kind of drug used for medicinal purposes. Hence nutraceuticals can be referred to as food substances with medicinal potentials. Nutraceuticals are compounds that are derived from natural substances with substantiated scientific evidence for prevention and/or treatment of a particular disease (Williams et al., 2021; Sachdeva et al., 2020). They also describe nutraceuticals as dietary supplements, utilized to ameliorate health, delay senescence, prevent diseases, and support the proper functioning of the human body. The use of nutraceuticals has become increasingly popular in the last two decades in both humans and animals due to their easy accessibility, affordability, and tolerability with a wide margin of safety (Gupta et al., 2018a). Nutraceuticals may be used to improve health, delay the aging process, prevent chronic diseases, increase life expectancy, or support the structure or function of the body (Puri et al., 2022).

Puri et al. (2022) defined nutraceuticals as "specially designed preparations" that are made to meet nutritional needs and/or provide preventive healthcare; such nutraceuticals are said to be formulated nutrients that, when added to a supplement diet, aid in the prevention and treatment of specific diseases. They have been further described as foods, or portions of foods, that can help with a number of health issues, such as the prevention or treatment of an illness (Puri et al., 2022). Also, studies have reflected the constant mix of nutraceuticals and food supplements as the same entity however, nutraceuticals comprise single or mixed meals for pro- and prebiotic purposes as well as foods for specific medical applications whereas food supplements comprise single or mixed forms of vitamins, minerals, protein supplements, functional foods, and herbal items (Keservani et al., 2017; Puri et al., 2022). A daily diet that includes nutraceuticals may help prevent pathological conditions by extending or completely eliminating the requirement for pharmaceuticals in persons who are suitable for an alternate non-pharmacological treatment to a pathological condition.

The concern for improvement in quality of life has increased in the modern world as rates of disease have also steadily increased. Fruits and vegetables have been systematically linked with the reduction of risk of chronic diseases such as cardiovascular diseases and neurodegenerative diseases. The demand for health-promoting foods has increased among consumers. There is a need for healthy and nutrient-rich foods that can improve the overall health of society, which has further increased consumers' demand for plant-based foods (Gul et al., 2016). There are thousands of bioactive compounds in nature, these metabolites compound (Phytochemicals) produced by plants have the function of biochemical activity that can influence biological processes. In recent years scientific advancements have been demonstrated with the chemical and pharmacological studies of medicinal plants' therapeutic applications (Pérez-González and Jiménez-Arellanes, 2021).

Moser and Li (2020) further described neurotoxicity as unintended detrimental effects on the nervous system's structure and function. It has been stated that in comparison to other organ systems, the nervous system's restricted capacity for regeneration, makes it more vulnerable to certain lesions, even, though it possesses compensatory and adaptive mechanisms. Furthermore, the absence of a completely developed blood–brain barrier and the significant impact of disrupting vital processes like neurogenesis, migration, and programmed cell death on nervous system development make the growing nervous system particularly prone to harm (Moser and Li, 2020). In addition, the possibility of neurotoxicity has been shown to be influenced by the mode of exposure, pattern, length of exposure, rate of metabolism and the substance's capacity to cross the blood–brain barrier, hence the incorporation of concepts from several academic fields such as biochemistry, psychology, physiology, and neurosciences for explicit study of neurotoxicity (Moser and Li, 2020).

Research has shown that many toxic substances are employed as experimental models for investigating the molecular and cellular mechanisms of physiology and pathophysiology in the nervous system of animal tissues *in vivo* and *in vitro,* as well as in human neural progenitor cell lines (Spencer and Lein, 2023). Numerous toxic substances have been used to induce neurotoxic disorders in humans with some directly affecting brain cells, and others obstructing metabolic processes that are particularly important to the nervous system; while certain neurotoxic substances cause harm to the adult nervous system, others change the brain's normal developmental pathways or impair neuronal function. Also, recent research has further depicted those detrimental impacts on other body systems, such as the immune system, endocrine system, and gut microbiota, may also have a harmful influence on the nervous system (Spencer and Lein, 2023).

The National Institute of Health (NIH) had also stated that chemicals used in radiation therapy, chemotherapy, drug therapies, organ transplants, certain foods and food additives, pesticides, industrial and cleaning solvents, cosmetics, and some naturally occurring substances can all cause neurotoxicity (Rishi Paliwal et al., 2022). Damage to the neuronal tissues can result from such neurotoxicants that target a particular area or organ of the nervous system, either permanently or temporarily. Neurotoxicity has also been suggested as a contributing factor to the emergence of neurological disorders via a variety of mechanisms, including oxidative stress, apoptosis, neuroinflammation, neurodegeneration, necrosis, hemorrhage, infiltration of macrophages, dysfunction of the mitochondria, cell proliferation and inhibition of migration, cognitive dysfunction and impairment, altered synaptic transmission, altered nerve conduction, altered synaptic plasticity, genotoxicity, and neurobehavioral toxicity (Rishi Paliwal et al., 2022).

2. Mechanism of neurotoxicity

Studies have shown that some of the most crippling illnesses are neurodegenerative disorders; they are typically brought on by aberrant protein buildup, altered genes, elevated reactive oxygen species (ROS), or the death of neurons in a particular area of the brain (Fu et al., 2014; Keber et al., 2015; Kumar et al., 2015; Zindo et al., 2015). Modern neuroscience research, enhanced experimental models, and research methodologies have contributed to the growing understanding of the distinct mechanisms underlying the initiation and progression of neurodegeneration and toxicant-induced neurotoxicity. Quite a number of established processes such as injury to the mitochondria of neurons, intracellular calcium excess, unchecked production of reactive oxygen species, persistent inflammation, or any combination of these have been linked to neurotoxicity (Dadhania et al., 2016).

a. **Neuronal mitochondria damage:** Studies have attributed the etiology of various toxicant-induced acute neurotoxicity and neurodegenerative diseases to mitochondrial damage (Dadhania et al., 2016). In contrast to other organelles, nuclear DNA (nDNA) and mitochondrial DNA (mtDNA) have been shown to regulate the functions of mitochondria; mutations in either of these genomes can cause mitotoxicity-mediated neurodegeneration (Beal, 2005). However, mitochondrial dysfunction can impact different compartments within a single neuron or a particular area of the brain (Dadhania et al., 2016).

b. Excess intracellular calcium: Intracellular [Ca^{2+}] increase may be caused by cellular processes that impair the integrity of plasma membranes or the actions of ATP-dependent Ca2+-ATPase pumps/uniporter system. However, apart from these methods, there exist various other pathways via which Ca^{2+} can accumulate in neuronal cells; such pathways include the activation of voltage-sensitive Ca^{2+}channels, receptor-operated Ca^{2+} channels, N-methyl-D-Aspartate (NMDA), ATP-dependent Ca^{2+} channels (P2j receptor), and cyclic nucleotide-gated Ca^{2+}channels (Dadhania et al., 2016). The mitochondria, endoplasmic reticulum, and plasma membrane are notable for their limited ability to withstand increases in intracellular Ca^{2+}; hence the persistent increases in intracellular Ca^{2+} disrupt the endoplasmic reticulum and mitochondrial Ca^{2+} homeostasis. This has been linked to the pathophysiology of several neurodegenerative illnesses (Dadhania et al., 2016).

c. Overproduction of reactive oxygen species: Oxidative stress is defined as the overproduction of reactive oxygen species (ROS) due to an imbalance in cellular biochemical processes (Dadhania et al., 2016). ROS and RNS-mediated damage to cellular macromolecules is involved in the pathogenesis of several neurodegenerative diseases (Dadhania et al., 2016). This is due to the fact that the nervous system is particularly vulnerable to ROS due to a number of factors, including high oxygen consumption due to higher energy demand (Adenosine Triphosphate), relatively lower level of the endogenous antioxidant reserve, abundance of polyunsaturated lipids in the plasma membrane of large neuronal cells, and the presence of excitatory amino acids and neurotransmitters whose metabolism can produce ROS (Dadhania et al., 2016).

d. Neuroinflammation: Recent data emphasizes that neuroinflammation has a proactive role in the etiology of neurodegenerative illnesses (Dadhania et al., 2016). It has been shown that the relationship between glia, immune cells, and neurons appears to be crucial for the start and subsequent spread of neuroinflammation; macrophages which are a key component of inflammation-mediated neurodegenerative diseases have been shown to be found in the brain close to glia and microglia. Such microglia are activated in the diseased state thus, producing pro-inflammatory substances such as chemokines and cytokines, which mediate neuronal damage (Gendelman, 2002; Dadhaniaet al., 2016).

Developing therapies that may be able to deter the onset and/or stop the spread of unregulated neurodegeneration has become feasible due to advances in the knowledge of the mechanisms behind neuronal damage. Then, based on the understanding of the causes of neurodegenerative illnesses, target-based treatments including neurotransmitter modulators, agonists/antagonists of direct receptors, second messenger modulators, stem cell-based treatments, hormone replacement therapy, and others have been developed. Meanwhile, these approaches are effective in slowing down the progression of neurodegenerative diseases; they are frequently accompanied by harmful effects and long-term unclear consequences (Dadhania et al., 2016).

Therefore, in order to combat these debilitating neuronal conditions, a safer alternative treatment option must be found and implemented over an extended period of time. Nutraceuticals, with their ability to act on multiple targets simultaneously to improve overall neuronal health, according to existing studies seem to be a promising option (Essa et al., 2015; Trivedi and Jena, 2013, 2014). Moreover, the use of medicinal plants in addressing neurodegenerative diseases has gained increasing attention in both traditional and modern healthcare systems. With the global rise in mental health disorders such as anxiety, depression, Alzheimer's disease, and schizophrenia, medicinal plants offer a promising, natural, and accessible approach to complement conventional treatments.

In recent years, there has been a growing interest in exploring the potential therapeutic properties of medicinal plants for nervous disorders. Scientific research has begun to investigate the bioactive compounds present in these plants and their mechanisms of action. Many of these compounds have been shown to have neuroprotective, anti-inflammatory, and antioxidant effects, which may be useful in treating neurodegenerative illnesses and neurotoxicological conditions. Therefore, scientific validation of plants through useful ethno-medicinal information is necessary for the development of alternative remedies to synthetic pharmaceuticals (Nazeruddin et al., 2011).

3. Neuroprotective mechanisms of nutraceuticals

Mechanisms of nutraceuticals as neuroprotective agents include antioxidant, iron chelation, cell signaling modulation, anti-inflammation, anti-oxidation, anti-apoptosis, and mitochondrial homeostasis (Flieger et al., 2024). Research documentations have shown that dietary phytochemicals suppress neuroinflammation and prevent toxic and degenerative effects in the brain (Almeida et al., 2016). Biological and pharmacological properties of nutraceuticals have proven effective in the management of neurodegenerative disorders. Flavonoids are natural compounds capacity to alleviate neuroinflammation by inhibiting the production of pro-inflammatory mediators, elevating the secretion of anti-inflammatory factors, and modulating the polarization of microglia and astrocyte as reported by (Goyal et al., 2023 Goyal et al., 2022). Flavonoids exert its neuroprotection via suppression of neurodegeneration, neuroinflammation, and oxidative stress by enhancing memory, learning, and cognitive function (Chen et al., 2022).

Neurological or neuropsychiatry disorders are one of the healthcare challenges in the world today as reported. Some of these disorders include Parkinson's disease, epilepsy, Alzheimer disease and other dementias, Huntington's disease (HD), Multiple sclerosis (MS) and Amyotrophic lateral sclerosis (ALS) (Feigin et al., 2020; Dadhania et al., 2016). They are the leading cause of disability and the second leading cause of death globally with particularly in low-income and middle-income countries exacerbated by population growth and ageing (Behl et al., 2021; Feigin et al., 2020).

Some of the nutraceuticals that hold promise against neurological and neuropsychiatry disorders include curcumin, epigallocatechin-3-gallate (EGCG), luteolin, naringin, naringenin, rutin, apigenin, hesperidin, and quercetin (Singh et al., 2024, Goyal et al., 2023, Goyal et al., 2022). Almost all nutraceutical-based flavonoid Ergothioneine (EGT) is one of the naturally occurring amino acids found in mushrooms and has been documented to protect mitochondria from oxidative stress. Similarly, epigallocatechin-3-gallate (EGCG), is a nutraceutical that has been reported to significantly offer neuroprotection against Parkinson's disease via mechanisms mitigation of mitochondrial oxidative stress, iron chelation, ROS scavenging, and anti-apoptotic properties (Pan et al., 2003).

This book chapter discussed the works that were already done with few natural nutraceuticals on neurotoxicity and that are promising agent for its management. These include: *Anacardium occidentalis, Cnidoscolus aconitifolius, Garcinia kola, Sorghum bicolor and Vernonia Amygdalina.*

4. Selected plants with neuroprotective nutraceuticals

4.1 Buchholzia coriacea

The use of medicinal plants in combating neurodegenerative diseases is increasingly recognised for their natural, accessible, and therapeutic potential. With the global rise in mental health conditions such as Alzheimer's disease, depression, and schizophrenia, bioactive compounds derived from these plants, known for their neuroprotective, anti-inflammatory, and antioxidant effects, present promising alternatives to conventional pharmaceuticals. One notable example is *Buchholzia coriacea,* commonly known as Wonder Kola, a plant native to West Africa that has been widely utilised in traditional medicine to treat ailments such as fever, hypertension, diabetes, infertility, and cognitive decline. Its seeds, encased in a distinctive purple aril, are consumed either raw (Chinedu et al., 2012) or fried and are known for their pungent, peppery flavour reminiscent of Capsicum. Often referred to as "brain food," Wonder Kola is believed to support memory enhancement and is traditionally used to manage mental disorders (Erhirhie et al., 2015), gastrointestinal conditions, and even snakebites (Anowi et al., 2012). The fruit of *Buchholzia coriacea* is large, resembling an avocado pear, and turns yellowish when ripe, while the tree flourishes in lowland rainforest regions (Mojab et al., 2003). As scientific exploration continues to validate its medicinal properties, *Buchholzia coriacea* remains a vital resource bridging traditional healing practices and modern therapeutic strategies for neurodegenerative diseases. Researchers have explored and determined the bioactive constituents of Buchholzia coriacea with pharmacological potentials to confirm the ethnopharmacological relevance of the plants' leaves and seeds. These are based on cardinals such as antioxidant, antibacterial, and anti-inflammatory potentials, the antidepressant, and anxiolytic properties. According to Duru et al. (2012), the constituents of *Buchholzia coriacea* seeds comprise the following macronutrients in varied concentrations: carbohydrates, proteins, fats, and oils. (Duru et al., 2012) Alkaloids, tannins, flavonoids, saponins, renin, cyanogenic and cardiac glycosides, steroids, and terpenoids in various amounts were available in the preliminary phytochemical investigation of B. coriacea seeds (Akujobi et al., 2004). The presence of saponins (38.9 percent), steroids (10.6 percent), alkaloids (11.3 percent), and cyanogenic glycosides (in trace amounts) in the seed of B. coriacea has been confirmed (Oladimeji et al., 2009). Wonderkola is extremely significant to many people in Sub-Saharan Africa. Nuts have always been a significant component of their life, from birth to death, as well as in traditional medicine. Furthermore, Erhirhie and co-workers (2015) reported the phytochemical analysis of phlobatannins, carbohydrates, proteins, tannins, saponins, alkaloids, and flavonoids in wonderful kola leaves. The ethanol extract (10–14 mm) and methanol extracts (13–15 mm) showed varying zones of inhibition against the Staphylococcus aureus strains. There are few studies that have shown the nutraceutical potential on the nervous system; these current studies hopefully will be a pathfinder for advanced investigation into the neuropharmacological potentials of Buchholzia coriacea in the management of neurological diseases.

4.2 Cnidoscolus aconitifolius

Leafy green vegetables are a rich source of nutrients, phytochemicals, and other health-promoting components. *Cnidoscolus aconitifolius* (Chaya plant, tree spinach) is a prime example of a green leafy vegetable that has a strong health-promoting connotation, an association that is supported by scientific literature. From folkloric usage, this plant has been associated with several health benefits such as maintaining healthy blood sugar levels and acting as an anti-inflammatory, anti-anemic, anti-microbial (Nurjanah et al., 2024).

Different researchers have explored and evaluated the bioactive components of *Cnidoscolus aconitifolius* with pharmacological potentials to confirm the ethnopharmacological relevance of the plants' leaves. These are based on cardinals such as antioxidants, and anti-inflammatory potentials.

Using high-performance liquid chromatography, *Cnidoscolus aconitifolius* is reported that the phenol, flavonoids, saponin, and alkaloid compositions of the plant were 33.02±0.32, 2.955±0.09, 6.305±0.13 and 1.96±0.26 mg/100*g*, respectively, while the proximate contents of the plants were 41.895, 7.68, 1.145, 13.53, 4.585 and 31.165 (100 percent), respectively, for carbohydrates, protein, crude fat, moisture, ash, and crude fiber. The vitamin composition is also reported to be the vitamin compositions of the plant in mg/100*g* as follows; vitamin C (892.025), vitamin B1 (1.36), vitamin B2 (5.84), vitamin B9 (9.425), vitamin A (59.6445), vitamin D (88.992) and vitamin E (69.785) while there is no presence of vitamin K. Mineral contents of the leaves is also reported to be phosphorus (24.68), copper (7.1), zinc (15.64), iron (14.83), calcium (29.16), potassium (33.88) and sodium (41.47) in mg/100*g* of the leaves (Orji et al., 2016).

The ethanolic extract of the plants was reported to contain phytochemical constituents similar to what was noted above. However, the aqueous extract of the plant contains fewer amounts of the phytochemicals reported above (Chikezie et al., 2016).

Proximate analysis and phytochemical screening of the methanolic extract of the leaves showed that the leaves contained moisture (5.35 percent), ash (13.69 percent), crude protein (18.74 percent), fat (11.95 percent), fiber (9.81 percent) and carbohydrate (40.48 percent), alkaloids, tannins, saponin, flavonoids, and cardiac glycoside while steroids, phlobatannin, and terpenoids were not detected. The mineral analysis in mg/100*g* indicated that leaves contained Sodium (77.32), Potassium (58.45), Calcium (44.82), Magnesium (23.46), Zinc (0.02), Copper (0.004), Iron (0.06) while Lead was absent (Fagbohun et al., 2012).

The presence of these phytochemicals compounds is the basis for recent reports on the antidiabetic, hypotensive, antidiuretic, antioxidants, anti-inflammatory, anticancer, hepatoprotective, hematopoietic, and antimicrobial effects of the plant leaves reported in several preclinical studies (Oyagbemi and Odetola, 2010; Oyagbemi et al., 2008; Samuel et al., 2014; Valenzuela Soto et al., 2015). These have prompted neuroscience natural products researchers to investigate the potential of *Cnidoscolus aconitifolius* as a nutraceutical in the prevention and management of neurological disorders which will be discussed below.

Plants like *Cnidoscolus aconitifolius* present a unique opportunity and potential as a nutraceutical due to its wide geographical distribution and resistance to severe climatic and environmental stresses. Previous studies that documented the phytochemicals in the plant and their effects on preclinical disease models have sparked interest in investigating the potential of the plant on some neurological conditions (Moura et al., 2019). A study in 2012 on the effect of *Cnidoscolus aconitifolius* on the central nervous system, using a hydro-alcoholic extract of the leaves showed that the extract has sedative properties in mice by increasing the onset and duration of sleep, demonstrating the potential of the plant in sleep therapy (Adebiyi et al., 2012). In addition to the sedative property, the extract also demonstrated anti-depressant potential and insignificant anti-convulsant, anxiolytic, and muscle relaxing potentials at doses of 200 and 400 mg/kg, respectively (Iyare et al., 2024).

Onasanwo et al. (2021) also demonstrated the antinociceptive property of the ethanol extract of *Cnidoscolus aconitifolius* leaf on inflammatory pain in laboratory rodents. They reported a reduction in the paw-licking behavior at 100 and 200 mg/kg while the writhing behavior was reduced when treated with at 50, 100 and 200 mg/kg. However, the resulting effects of the treatment were similar to that observed for a known anti-inflammatory drug, indomethacin at 10 mg/kg. Further, at 100 and 200 mg/kg it potential efficacy as a non-steroidal anti-inflammatory drug candidate against carrageenan-induced paw edema was reported (Onasanwo et al., 2011).

One of the few studies involving the potentials of the plant against Alzheimer's disease demonstrated that methanol extract of *Cnidoscolus aconitifolius* prevents lipopolysaccharide-induced impairment in spatial learning and memory by preventing neuroinflammatory and stress response to lipopolysaccharide insult in Wistar rat's brain. This study also showed that acetylcholinesterase activity, an enzyme targeted in the management of Alzheimer's disease was suppressed (Babalola et al., 2023). This shows the potential of *Cnidoscolus aconitifolius* in modulating neurochemicals and neuro-signaling.

The nutraceutical potential was reinforced in another study, where the diet of Wistar rats was supplemented with *Cnidoscolus aconitifolius* for 12 weeks. They observed an increase in brain antioxidant capacity and preserved brain cyto-architecture when fed with 2.5 percent, 5 percent and 10 percent of the supplemented diet leading to improved spatial and non-spatial learning and memory (Adebayo et al., 2022). These studies demonstrated that the plant has a role to play in the management of Alzheimer's disease, which may apply to other neurodegenerative disease such as Parkinson's disease. Although few studies have demonstrated the nutraceutical potential on the brain, these early studies hopefully will chart a pathway for further investigation into the neuropharmacological potentials of *Cnidoscolus aconitifolius* in the management of neurological disorders.

4.3 Garcinia kola

Garcinia kola (*G. kola*), commonly known as bitter kola, which is cultivated in the tropical regions of West Africa (Adam, 2019). The plant is often referred to as a "wonder plant" because every part of it has medicinal importance. Locally known as "Orogbo" in the Yoruba language, "Aku ilu" in the Igbo language, and "Namijingoro" in the Hausa language, *G. kola* is also called "*Male Kola",* a name highlighting its aphrodisiac properties (G and Dalziel, 1938). In African ethnopharmacology, the root of the plant is used as bitter chew-sticks in West Africa. The stem bark is used in folklore remedies as a purgative among the natives of Eastern Nigeria. The seeds have been widely used to treat various ailments such as coughs, bronchitis, liver disorders, and infectious diseases (Adesina, 1982; Farombi., 2003; Iwu, 1985). The seeds, bark, and roots of the plant are frequently used for their potent anti-inflammatory, antimicrobial, antioxidant, and hepatoprotective properties (Farombi and Owoeye, 2011; Iwu et al., 2011; Onasanwo and Olaleye, 2011; Oyagbemi et al., 2016, 2018). These uses illustrate the critical role of the plant in traditional medicine, highlighting its potential for modern therapeutic applications.

Beyond its traditional uses, *G. kola* has gained increasing attention for its pharmacological properties identified in modern research. It is widely consumed in West African countries, particularly in Nigeria, where it plays a crucial role in public health by providing accessible treatment options, particularly in rural areas with limited access to conventional healthcare (Daramola and Adegoke, 2011). *Garcinia kola*'s epidemiological significance includes its socioeconomic value, as it is traded both locally and internationally. Its potential in treating a wide range of diseases has broadened interest in its commercial applications, and it continues to be a subject of pharmacological research worldwide. *G. kola* has been characterized to contain phytochemical compounds which include tannins, saponins, alkaloids, cardiac glycosides, xanthones kolaviron, garcioic, garcinal and tocotrienol (Adegboye et al., 2008; Ijomone et al., 2012).

The pharmacological potential of *G. kola* extends to several neurotoxicological conditions, primarily due to its bioactive compounds, including kolaviron, bioflavonoids, alkaloids, and xanthones. Studies showed that 20mg/mL *Garcinia kola* extracts showed potent antimicrobial capacity (Adegboye et al., 2008; Ijomone et al., 2012). These compounds also exhibit protective effects against oxidative stress, inflammation, and neurotoxicity (Adeoye A. David et al., 2022; Farombi et al., 2002; John et al., 2015; Olatunji et al., 2020). Let us consider some of its actions.

Neuroprotective Effects against Oxidative Stress: Oxidative stress is a key factor in neurodegenerative disorders such as Alzheimer's disease, Parkinson's disease, and stroke. (Farombi et al., 2000) *Garcinia kola*, particularly its biflavonoid component kolaviron has been shown to possess antioxidant properties that scavenge free radicals, reducing neuronal damage in laboratory rodents. For example, 200mg/kg body weight of kolaviron has been observed to delay the onset of stereotypic behavior induced by the administration of methamphetamine (Ijomone et al., 2012; Olanrewaju et al., 2022). Additionally, the pre-treatment with kolaviron protected the hippocampal neurons against the degradative effects of methamphetamine in rats (Ijomone et al., 2012). Likewise, aqueous extract *G. kola* (200mg/kg body weight) protects the neurons within the cerebellar cortex against neuronal degradation induced by monosodium glutamate (John et al., 2015). In another study, ethanol extract of *G. kola* seed, in a dose-dependent manner, reduced lipid peroxidation induced by haloperidol in the brain of mice (Adeoye A. David et al., 2022). These findings highlight the neuroprotective and antioxidative properties of *G. kola* and its bioflavonoid complexes like kolaviron, suggesting its potential in managing neurodegenerative diseases.

Anti-Inflammatory Action: Neuroinflammation is a hallmark of several neurotoxicological conditions including multiple sclerosis and amyotrophic lateral sclerosis (ALS). *G. kola*'s extracts and its bioflavonoids especially kolaviron and garcinoic acid, have been identified as having potent anti-inflammatory properties. For example, in a study that examined the anti-inflammatory potentials of the crude extract of *G. kola* seed and its bioflavonoids- kolaviron and garcinoic acid, using carrageenan-induce paw edema as a model, observed that the potent anti-inflammatory potentials were after 6 h of administration and at the dose of 50mg/kg (Michel et al., 2016). Another study found a dose-dependent relationship in the anti-inflammatory potentials of kolaviron with 150mg/kg being the most effective and 50mg/kg being the least effective compared to aspirin. Additionally, in another study, a dose of 200 mg/kg kolaviron proved very effective in preventing inflammation (Onasanwo and Rotu, 2016). These findings indicate that the *G. kola* seed extract and its bioflavonoids are effective anti-inflammatory alternatives.

Cognitive and Memory Enhancing Potential: *G. kola* has also been investigated for its potential to improve cognitive function and memory. Animal studies have shown that *G. kola* extracts and its bioflavonoids can improve spatial memory and learning abilities in models of cognitive impairment, possibly by modulating acetylcholine levels and reducing neuroinflammation. For example, 100mg/kg ethyl acetate extract of *G. kola* seeds mitigated against cognitive and motor impairment associated with gamma radiation in rats (Ahidjo et al., 2021). Additionally, the oral administration of 50, 100 and 200 mg/kg kolaviron ameliorated the spontaneous alternations impairment caused by scopolamine. In another study, oral administration of 50, 100 and 200 mg/kg kolaviron improved cognitive and learning processes. The researchers found that pre-treatment with 100 mg/kg, 50 mg/kg post-treatment and 200 mg/kg pre- and post-treatment were all effective against lipopolysaccharide induced cognitive decline in Wistar rats (Onasanwo et al., 2021). These findings indicate that *G. kola* extracts as well as its bioflavonoids has potential to prevent cognitive decline associated with many neurodegenerative diseases like Alzheimer's disease.

Neurotoxicological Implications in Drug-Induced Neurotoxicity: Drug-induced neurotoxicity, which can occur with substances like methamphetamine and antipsychotic medications, represents another condition where *G. kola* shows promising intervention. Studies suggest that kolaviron, a compound found in the plant, can help protect brain cells, particularly dopaminergic neurons, from the damaging effects of these drugs. This is important for conditions like Parkinsonism and schizophrenia, where such neurotoxicity plays a key role. For example, 100 and 200 mg/kg ethanolic extract of *G. kola* showed potentials of preventing and reducing ataxia associated with 3,4-methylenedioxymethamphetamine (MDMA) neurotoxicity in Wistar rats (Joshua et al., 2020). Also, the toxicity associated with vanadium intake was ameliorated with the administration of *G. kola* and kolaviron in the cerebrum, cerebellum, hippocampus and olfactory bulb (Igado et al., 2012). Another study identified ethanol extract of *G. kola* to inhibit cataplexy associated with haloperidol administration in mice (Adeoye David et al., 2022). These findings highlight the potential of *G. kola* in reducing the harmful side effects of certain medications.

Hence, *G. kola* represents a promising therapeutic agent in managing various neurotoxicological conditions, with significant antioxidant, anti-inflammatory, and neuroprotective effects. Its traditional use in African ethnomedicine is supported by growing scientific evidence, showing potential in treating conditions such as neurodegenerative diseases, cognitive impairments, and drug-induced neurotoxicity. As research continues, *G. kola* could offer new avenues for developing treatments for complex neurotoxicological disorders.

4.4 Sorghum bicolor

Sorghum bicolor is a dietary supplement obtained from the polyphenol-rich leaf sheath of Sorghum bicolor (L.) Moench *(Poaceae)*. Sorghum bicolor, commonly known as millet, sweet sorghum, broom, or guinea corn has been used in African traditional medicine for the treatment of several diseases for over a 100 years (Okochi et al., 2003; Benson et al., 2013). Jobelyn® has gained a high reputation as a remedy for the treatment of moderate to severe anemia (Okochi et al., 2003).

Jobelyn®® (JB) is a nutraceutical derived from Sorghum bicolor grown in West Africa. Jobelyn® as an herbal antioxidant has been extensively studied for its beneficial effects on chronic inflammatory and oxidative stress conditions (Benson et al., 2013). It is GRAS-certified by The Food and Drug Agency (FDA) USA and utilized globally (Eniojukan and Aina, 2009). Jobelyn®'s potent antioxidant properties, attributed to compounds like apigenin, luteolin, and naringenin, make it effective in managing various diseases, including anemia and arthritis (Benson et al., 2013; Okochi et al., 2003). Jobelyn® has also been recognized for its role in prolonging the latency to convulsions and its potential efficacy in controlling convulsive episodes (Umukoro et al., 2013). Its strong anti-inflammatory and antioxidative properties have been utilized in the management of numerous diseases, ranging from cancer to diabetes (Benson et al., 2013).

4.5 Sorghum bicolor in neurotoxicological conditions

Jobelyn® is known to boost the immune system, thereby enhancing the body's defensive mechanisms in response to stress, infections, or debilitating diseases (Benson et al., 2013; Okochi et al., 2003). Numerous physiologically active substances, including luteolin, naringenin, apigeninidin, luteolin, and apigenin (Khan et al., 2012) have been shown to be present in Jobelyn® and are known to exhibit a variety of neuropharmacological actions (Qiaoa et al., 2012; Yi et al., 2012; Gupta et al., 2018a, 2018b). It has been demonstrated that naringenin at graded doses of 5, 10 and 20 mg/kg markedly decreased depression via the central serotonergic and noradrenergic systems in the tail suspension test (TST) induced mice. Naringenin administration for 14 days (20 mg/kg) elevated hippocampal serotonin (5-HT), norepinephrine (NE) and glucocorticoid receptor (GR) levels, and reduced serum corticosterone levels in mice exposed to the repeated TST (Yi et al., 2012). By lowering pro-inflammatory cytokine levels in the brain and increasing the expression of NF-kB immunopositive cells, Jobelyn® alleviated neurological impairments in rats who had suffered an ischemic stroke (Umukoro et al., 2019; John et al., 2022). Given the roles that dopamine, serotonin, noradrenaline, and BDNF play in the regulation of neuronal activities, it is possible that naringenin's capacity to increase brain concentrations of these chemicals will benefit animals suffering from stroke (Umukoro et al., 2019).

Luteolin is also one of the most studied phytochemicals, reported to protect the rat brain against focal ischemia via several mechanisms, including inhibition of the NF-κB signaling pathway (Manzanero et al., 2013). Previous studies in cultured cells showed that inhibited the infiltration of inflammatory cells, the release of inflammatory mediators, and the formation of free radicals (Oyinbo et al., 2016). Recently, Jobelyn® was reported to attenuate inflammatory responses and neurobehavioral deficits associated with complete Freund's adjuvant-induced arthritis via inhibition of pro-inflammatory cytokines in the mouse brain. Earlier, Zhu et al. reported that 20 μm of luteolin showed significant positive effect of neuroinflammatory event in neurological illnesses induced by lipopolysaccharide (LPS) by inhibition against the production of MAPK, NF-B, and Akt pathways (Zhu et al., 2014; Gupta et al., 2018b). Several other proteins implicating neurological illnesses were ameliorated following luteolin administration. Qiaoa et al. also demonstrated that luteolin intraperitoneally (5, 10, and 25 mg/kg) downregulates TLR4, TLR5, NF-κB and p-p38MAPK expression, and upregulates the p-ERK expression that in turns protects rat brains against focal ischemia after 24 and 72 hr of MCAO surgery (Qiaoa et al., 2012). Although, no further study have ascertain its potential usefulness in stroke disorder but studies were conducted to evaluate its effect on ischemic stroke using a bilateral common carotid artery occlusion model in rats, which closely mimics the clinical and biochemical conditions relevant to the pathology of ischemic stroke in humans (Qiaoa et al., 2012; Umukoro et al., 2019).

In addition, Jobelyn® inhibits apoptosis signaling in neurons of the prefrontal cortex (PFC) by increasing p53 protein expression. The overexpression (up-regulation) and underexpression (down-regulation) of p53 can either protect or harm stressed neurons (Morrison et al., 2003; Mendrysa et al., 2011). This dual role of p53 (both in destroying and repairing damaged DNA) may contribute to these effects. The cellular tumor antigen p53 protects cells by activating DNA repair proteins in cases of damaged or compromised DNA, halting the cell cycle at the G1/S checkpoint, which allows time for repair before growth resumes (Kudo et al., 2012). However, if DNA damage is deemed irreparable, p53 will initiate apoptosis (Efeyan and Serrano, 2007; Kreiner et al., 2013). While the loss of p53 function is linked to CNS tumors, its overexpression or hyperactivity is associated with neural tube defects, neuronal degeneration, and embryonic lethality. Consequently, p53 activity is tightly regulated (Brady and Attardi, 2010; Mendrysa et al., 2011). Over all, Charles et al. reported that 4 mg/kg Jobelyn® supplementation (4 mg/kg) significantly reduced the levels of histologic and biochemical indices of neurodegeneration via increasing the expression of p53 protein and a decreasing expression of NSE immunoreactivity in the prefrontal cortex of rats exposed to ethanol intoxication. Thus, affirming the potential effect of Jobelyn® pharmacotherapy in reducing CNS tumors and other neurodegenerative diseases (Charles et al., 2016). The study by Ca et al. (2015) also demonstrated that Jobelyn® has a neuroprotective effect in the hippocampus of alcoholic rats, mediated in part by alterations in glial cell fibrillary acidic protein (GFAP) and neurofilament protein (NF) protein expressions. Ca et al. showed that following 4 days of binge treatment, JB (4 mg/kg) lowered the histological index of neurodegeneration compared with the alcohol intoxicated rats. JB inhibited GFAP immunoreactivity and prevented astrocyte death in the CA3 and CA1 regions of the hippocampus. Further, JB (4 mg/kg) reduce neuroinflammation in the DG, CA3 and CA1 regions of the hippocampus by elevating the NF expression. In the CA3 region, JB prevented NF protein reduction in alcohol exposed rats, while, JB supplementations caused NF protein overexpression in the CA1 region (Ca et al., 2015).

In addition, Jobelyn® has been reported to gradually alleviate pain and inflammation without having the unfavorable side effects associated with NSAIDs. Additionally, it is said to boosts the body's defenses by fortifying the immune system (Eniojukan and Aina, 2009; Benson et al., 2013). Umukoro et al. reported that JB (10–200 mg/kg) oral administration inhibited inflammatory pain elicited by acetic acid induction as evidenced by reduced abdominal constrictions. The authors established that JB showed higher potency in suppressing the inflammatory pain associated with the second phase of the formalin test in mice model (Umukoro et al., 2014). To substantiate the anti-nociceptive effect of JB, Onasanwo and co-workers experimented that atropine significantly reversed the anti-nociceptive effects of JB (100mg/kg) when compared with JB in acetic acid-induced abdominal writhing test in mice. Pre-treatment with the blockers showed no significance when compared with JB group in the formalin-induced paw licking test. However, JB (10, 50, 100, and 200mg/ kg) when administered orally, inhibited acute inflammation induced by carrageenan in rat paw in which 100 mg/kg JB shows higher activity (Onasanwo et al., 2019).

5. *Vernonia amygdalina*

Vernonia amygdalina has been described as an angiosperm belonging to the Asterales order and *Asteraceae* family; this genus of the plant has been shown to have multiple commercial uses and is primarily grown in tropical climates (Toyang and Verpoorte, 2013). It has also been stated that this plant due to its unpleasant taste is known by its common name in Africa as *Vernonia amygdalina* (bitter leaf) (Toyang and Verpoorte, 2013). The majority of the plant's cultivation occurs in Africa's tropical regions, particularly in West Africa (Tekou et al., 2018) hence its description as "Olugbu," "Ewuro," and "Fetefete" in the Igbo, Yoruba, and Hausa tribes of Nigeria. This soft woody shrub can reach a height of one foot and grows continuously (IfedibaluChukwu *et al.*, 2020).

Vernonia amygdalina leaves have been proven to be an essential component of the human diet because they are rich in vitamins, fiber, carbohydrates, and minerals (Oyeyemi et al., 2018). According to Alara et al. (2018), the plant is said to contain quite a variety of phytochemicals, such as anthraquinone, edotides, xanthones, coumarins, flavonoids, polyphenols, tannins, and sesquiterpenes; these bioactive chemicals have been shown to be extracted and evaluated utilizing a variety of methods, including the soxhlet extraction, microwave-assisted extraction (MAE) (Alara et al., 2018), and liquid chromatography with tandem mass spectrometry (LC-MS/MS) analysis.

Research has shown that due to the *V. amygdalina*'s aforementioned constituting bioactive compounds, its numerous pharmacological properties have been identified; these include antimalarial, antidiabetic, antimicrobial, anti-inflammatory, neuroprotective and hepatoprotective properties (Asante et al., 2023).

Flavonoids found in bitter leaf, specifically luteolin, luteolin 7-O-βglucuronoside, and luteolin 7-O-β-glucoside, have been shown to possess antioxidant properties that can effectively combat radical damage (Farombi and Owoeye, 2011; Edo et al., 2023). Another study on *Vernonia amygdalina*'s pharmacological properties revealed that the aqueous extracts of the leaf were used to reduce oxidative damage in diabetic rats induced with alloxan; this is was seen in the reduced levels of serum malondialdehyde (Edo et al., 2023).

In a review report by Mohammed (2023), several studies have shown that diabetes have been traditionally treated using a decoction made from *Vernonia amygdalina* leaves (Mohammed, 2023); this notable hypoglycemic and hypolipidemic impact of the *Vernonia amygdalina* extracts has been corroborated by a number of investigations using experimental models (Pham et al., 2021; Asante et al., 2023). Pham et al. demonstrated that the leaf methanolic extract of V. amygdalina at the concentration of 500 μg/mL had inhibitory activities on both α-amylase and α-glucosidase. Besides, 400 mg/kg concentration of the extract with 7 days decreased blood glucose level in induced hyperglycemic mice. Total cholesterol, LDL-cholesterol and triglyceride levels in induced hyperlipidemic mice were also reduced following treatment with the extract at the same dose (Pham et al., 2021).

Onasanwo et al. (2017) also conducted research on the anti-nociceptive and anti-inflammatory properties of the methanol extract of *Vernonia amygdalina* leaf at 50, 100 and 200 mg/kg dose, as well as the potential mechanisms of action by examining the roles played by the adrenergic, cholinergic, and opioidergic systems. The findings showed that the number of inflammatory cells infiltrating the dermis of the skin tissues of the animals who had received a prior MEVA treatment (50, 100 mg/kg) had decreased, although in dose-dependent inhibition effect; an implication of the inhibitory effect of MEVA against carrageenan-induced air pouch inflammation. In the carrageenan air pouch model, MEVA (200 mg/kg) inhibited infiltrating leukocytes, protein concentration and MDA levels with unaffected GSH and SOD levels.

The pathophysiology of most neurodegenerative diseases has been linked to neuro-inflammation, microglial activation, oxidative stress, and apoptotic processes, according to scientific research (Pringsheim et al., 2014). Although, it is not yet quite clear, whether the pathology was brought on by one or more of these mediators. Medicinal herbs, their bioactive constituents, and extracts with various pharmacological mechanisms such as anti-inflammatory, anti-apoptotic, and antioxidant—have therefore been highly suggested as possible neuroprotective drugs (Oladele et al., 2020). In order to cause neurodegeneration in rats, Oladele et al., 2020 used nitrobenzene; nitrobenzene, which was originally created in 1834 by processing benzene with fuming nitric acid, has been recognized as a hazardous industrial and environmental chemical. It was first produced for commercial use in England in 1856.

The outcomes of the aforementioned research enhance the therapeutic impacts of *Vernonia amygdalina* leaf extract in neurodegenerative conditions. According to the results, *Vernonia amygdalina* methanol leaf extract significantly reduced the behavioral deficits in rats that were linked to the injection of nitrobenzene (Oladele et al., 2020).

Oladele et al. reported that the treatment with nitrobenzene induced neurodegeneration in rats but was attenuated after 200 and 400 mg/kg of *Vernonia amygdalina* methanol leaf extract administration evidenced by an increase in dopamine, GSH, antioxidant enzymes levels; and a decrease in acetylcholinesterase activity. Also, the authors showed that *Vernonia amygdalina* methanol leaf extract enhanced neurobehavioural and locomotor activities in the nitrobenzene-induced neurodegeneration in rats. In sum, Oladele et al. demonstrated that the extract effectively reduced the molecular processes and pathological characteristics linked to neurodegenerative disorders in the cerebellum, midbrain, and cerebrum through mechanisms involving its antioxidant, acetylcholinesterase suppression, lipid peroxidation inhibition, anti-inflammatory, and neurobehavioral restoring effects (Oladele et al., 2020). This is a reflection of the neuroprotective properties of *Vernonia amygdalina* against neurotoxicological conditions.

The potential neuroprotective effects of *Vernonia amygdalina* methanolic extract against aluminum chloride-induced neurotoxicity damage in Wistar

Table 1 Pharmacological and neuroprotective properties of the nutraceuticals and natural products.

Model	Experiment	Biochemical / Behavioral Assay	Main Result	References
In vitro / microbial assays	Minimum inhibitory concentrations (MICs) of ethanol extract (10–14 mm) and methanol extracts (13–15 mm) of *Buchholzia coriacea* against some strains of Staphylococcus aureus	Microbial inhibition assays	Strong antimicrobial activity of both extracts at 50 mg/mL and 100 mg/mL	Chika et al., 2012
Wistar rats	Kolaviron (200 mg/kg) given before methamphetamine	Behavioral stereotypy assessment; histology (hippocampal neurons)	Delayed methamphetamine-induced stereotypic behavior; protected hippocampal neurons from degeneration	Ijomone et al., 2012; Olanrewaju et al., 2022
Wistar rats	Aqueous extract of *G. kola* (200 mg/kg) before MSG administration	Histology of cerebellar cortex	Prevented MSG-induced neuronal degradation in cerebellum	John et al., 2015
Mice	Ethanol extract of *G. kola* seed against haloperidol toxicity	Lipid peroxidation (MDA), oxidative stress markers	Dose-dependent reduction in haloperidol-induced lipid peroxidation	Adeoye A. David et al., 2022
Wistar rats	Kolaviron & garcinoic acid tested in carrageenan paw edema	Paw edema volume; inflammation markers	Strong anti-inflammatory effect at 6h; 50 mg/kg most potent early effect	Michel et al., 2016
Wistar rats	Kolaviron (200 mg/kg) for inflammation	Paw edema; inflammatory cytokines	Strong prevention of inflammation at 200 mg/kg	Onasanwo and Rotu, 2016
Wistar rats	Ethyl acetate extract of *G. kola* (100 mg/kg) against γ-radiation	Behavioral cognitive tests; motor coordination tests	Improved cognition and motor performance post-radiation	Ahidjo et al., 2021

Wistar rats	Kolaviron (50–200 mg/kg) to reverse scopolamine-induced deficits	Y-maze spontaneous alternation; memory tests	Ameliorated scopolamine-induced memory deficits; enhanced learning	Ishola et al., 2017
Wistar rats	Kolaviron (50, 100, 200 mg/kg) against LPS-induced cognitive decline	Spatial memory tests; learning tasks	Pre- and post-treatment improved memory; 100 mg/kg pre-treatment effective	Onasanwo et al., 2021
Wistar rats	Ethanolic extract of *G. kola* (100 & 200 mg/kg) against MDMA	Motor coordination (ataxia), behavioral tests	Reduced/Prevented MDMA-induced ataxia	Joshua et al., 2020
Wistar rats	*G. kola* and kolaviron treatment in vanadium toxicity	Histology of cerebrum, cerebellum, hippocampus, olfactory bulb	Ameliorated vanadium-induced neurotoxicity in multiple brain regions	Igado et al., 2012
Mice	Ethanol extract of *G. kola* against haloperidol-induced cataplexy	Behavioral scoring of cataplexy	Inhibited haloperidol-induced cataplexy	Adeoye David et al., 2022
In vitro / microbial assays	*G. kola* extract (20 mg/mL) antimicrobial activity	Microbial inhibition assays	Strong antimicrobial activity at 20 mg/mL	Adegboye et al., 2008; Ijomone et al., 2012
Mice	Hydroalcoholic extract *C. aconitifolius* leaves	Sodium pentobarbitone sleeping time test	Elevated the onset and duration of sleep	Adebiyi et al., 2012

Continued

Table 1 Pharmacological and neuroprotective properties of the nutraceuticals and natural products.—cont'd

Model	Experiment	Biochemical / Behavioral Assay	Main Result	References
Mice	Methanol leaf extract of *C. aconitifolius* (200 and 400 mg/kg)	Sedative-hypnotic, anxiolytic, anti-depressant, anti-convulsant and muscle relaxant test	Anti-depressant, anxiolytic, and muscle relaxing potentials	Iyare et al., 2024
Mice and Wistar rats	Ethanol extract of *C. aconitifolius* (200 and 400 mg/kg) against inflammatory and nociception	Hotplate, formalin-induced paw-licking, acetic acid-induced abdominal writhing, and carrageenan-induced paw edema test	Anti-inflammatory and analgesic properties	Onasanwo et al., 2011
Mice	Methanol leaf extract of *C. aconitifolius* against LPS-induced cortico-hippocampal neuroinflammation	Y-maze spontaneous alternation and NOR memory tests. Oxidative stress and pro-inflammatory mediator and acetylcholinesterase assay	Ameliorated LPS-induced memory deficits; oxidative stress and pro-inflammatory cytokines	Babalola et al., 2023
Wistar rats	Food supplementation with *C. aconitifolius* (2.5 percent, 5 percent and 10 percent) against learning and memory and oxidative stress	Spatial memory tests; learning tasks and Oxidative stress and histology of the hippocampus	Elevated antioxidant capacity and preservation of hippocampal cytoarchitecture	Adebayo et al., 2022
Mice	Antidepressant-like activity of 20 mg/kg Jobelyn® (naringenin)	Behavioral scoring; Tail suspension test	Improves central serotonergic and noradrenergic system release and glucocorticoid receptor levels	Yi et al., 2012

Mice	Jobelyn® Pharmacological evaluation of the analgesic and anxiolytic activities	Anxiety scoring, formalin-induced paw-licking, acetic acid-induced abdominal writhing, and carrageenan-induced paw edema test	Anxiolytic and anti-nociceptive properties	Umukoro et al., 2014; Onasanwo et al., 2019
Wistar rats	Jobelyn® ameliorates neurological deficits in rats with ischemic stroke through inhibition of release of pro-inflammatory cytokines and NF-κB signaling pathway	Estimation of pro-inflammatory cytokines and NF-κB signaling pathway	Ameliorates pro-inflammatory cytokines and NF-κB signaling pathway	Umukoro et al., 2019
Mice	Evaluation of the effect of Jobelyn® on chemoconvulsants-induced seizure in mice	Behavioral scoring: seizure score, seizure frequency, latency to seizure	Exhibited anti-convulsants properties	Umukoro et al., 2013
Drosophila melanogaster	Jobelyn® dietary supplementation improves motor function in *Drosophila melanogaster* exposed to lipopolysaccharide via augmentation of antioxidant status	Motor performance test; oxidative stress markers and acetylcholinesterase activity assay	Strong improvement of motor function, antioxidant enzymes and cholinergic neurotransmitter release	John et al., 2022
Co-cultured BV2 cells	*Luteolin* inhibits SH-SY5Y cell apoptosis in LPS-stimulated co-cultured BV2 cells	Estimation of MAPK, NF-B, and Ak signaling	Inhibition of MAPK, NF-B, and Akt in LPS-induced neuroinflammation	Zhu et al., 2014

Continued

Table 1 Pharmacological and neuroprotective properties of the nutraceuticals and natural products.—cont'd

Model	Experiment	Biochemical / Behavioral Assay	Main Result	References
Wistar rats	Anti-neuroinflammatory effect of *luteolin* (5, 10, and 25 mg/kg) against focal ischemia perturbation in rat brains	Estimation of TLR4, TLR5, NF-κB, p-p38MAPK and p-ERK signaling	Downregulation TLR4, TLR5, NF-κB and p-p38MAPK expression, and upregulation of p-ERK expression in focal ischemia after 24 and 72 hr of MCAO surgery	Qiaoa et al., 2012
Wistar rats	Jobelyn® supplementation (4 mg/kg) inhibits apoptosis signaling in neurons of the prefrontal cortex (PFC)	Histologic and biochemical indices	Elevation of p53 protein expression and suppression of NSE immunoreactivity, and repair of DNA proteins in the PFC	Charles et al., 2016
Wistar rats	Jobelyn® supplementation (4 mg/kg) confer neuroprotection in the hippocampus of alcoholic rat	Histologic and immunohistochemistry	Inhibition of GFAP and NF protein expressions	Ca et al., 2015
Mice	Hypo-lipidemic and antidiabetic properties effect of methanol extract of *Vernonia amygdalina* leaf	Biochemical indices: Cholesterol, LDL-cholesterol, triglyceride and fasting blood sugar level	Inhibitory activities on both α-amylase and α-glucosidase	Pham et al., 2021
Mice	Methanol extract of *Vernonia amygdalina* leaf at 50, 100 and 200 mg/kg dose	Acetic acid-induced abdominal writhing, and carrageenan-induced air-pouch and paw edema test	Anti-nociceptive and anti-inflammatory properties via adrenergic, cholinergic, and opioidergic systems	Onasanwo et al., 2017

Wistar rats	Neuroprotective mechanism of *Vernonia amygdalina* (200 and 400 mg/kg) in a rat model of neurodegenerative diseases	Behavioral assessment and Biochemical assay (Oxidant, antioxidant and neurotransmitter parameters)	Reduced behavioral deficit. Increase in dopamine and antioxidant enzymes levels and a decrease in acetylcholinesterase activity in the cerebellum, midbrain, and cerebrum	Oladele et al., 2020
Wistar rats	Neuro-protective and neuro-curative effect of *Vernonia amygdalina* in rat model of neurodegenerative disease induced by aluminum chloride	Biochemical and behavioral assessment; histology (hippocampal neurons)	Reduced depression, neuroinflammation and loss of pyramidal neurons in the hippocampus	Ajeleti et al., 2023

rats was also evaluated in a study conducted by Ajeleti et al. (2023). Nonetheless, it was determined that oral aluminum chloride administration could cause brain damage that could impair memory and learning, as seen in Alzheimer disease, and that this damage could be mitigated with *Vernonia amygdalina* leaf extract; this was seen in the behavioral investigations where the AlCl3 technique was used to simulate the symptoms of depression in rats (Ajeleti et al., 2023). Ajeleti et al. through histological examinations revealed clumpy of cell neurons, condensation of pyramidal cells, necrotic and vascular edema, and very few neurofibrillary tangle which are indicators of neurodegeneration in the aluminum chloride-induced neurotoxicity in rats. Moreover, the authors showed that pre-treatment and post-treatment with 1.31 mL of *Vernonia amygdalina* (VA) extract for 14 days prevented all the pathological features observed in the aluminum chloride-induced neurotoxicity in rats (Ajeleti et al., 2023).

Ultimately, research demonstrating the neuroprotective effects of *Vernonia amygdalina*'s antioxidant and anti-inflammatory qualities has showed the plant compound's promises as a pharmacological treatment in a variety of neurotoxicological conditions. It provides advantageous benefits in illnesses including Alzheimer's, Parkinson's, and metal toxicity since it has been successful in lowering oxidative stress, neuroinflammation, mitochondria dysfunction, and calcium overload in animal models. Also, limited scientific evidence has been shown to support its traditional use in detoxifying neurotoxic chemicals. However, a lack of thorough human clinical trials raises concerns regarding the plant's safety and efficacy in larger populations, and the plant's effectiveness has also been found to vary across literature.

Find the details of the pharmacological and neuroprotective properties of the nutraceuticals and natural products discussed in Table 1.

References

Adam, I.A.A.F., 2019. Wild fruits: composition, nutritional value and products. Wild Fruits: Composition, Nutritional Value and Products. Springer International Publishing, Switzerland AG doi:10.1007/978-3-030-31885-7.

Adebayo, O.G., Onasanwo, S.A., Ajayi, A.M., Aduema, W., Oyebanjo, O.T., Nicodemus, O.U., 2022. Cnidoscolus aconitifolius-supplemented diet enhanced neurocognition, endogenous antioxidants and cholinergic system and maintains hippocampal neuronal integrity in male Wistar rats. Drug Metab. Pers. Ther. 37 (1), 81–93. doi:10.1515/dmpt-2021-0138.

Adebiyi, O., Adebiyi, O., Ilesanmi, O., Raji, Y., 2012. Sedative effect of hydroalcholic leaf extracts of *Cnidoscolous acontifolius*. Int. J. Appl. Res. Nat. Prod. 5 (1), 1–6.

Adegboye, M.F., Akinpelu, D.A., Okoh, A.I., 2008. The bioactive and phytochemical properties of Garcinia kola (Heckel) seed extract on some pathogens. Afr. J. Biotechnol. 7, 3934–3938.

Adesina, S.K., 1982. Studies on some plants used as anti convulsants in amerindian and african traditional medicine. Fitoterapia 53, 147–162.

Ahidjo, N., Ngarka, L., Seke Etet, P.F., Njamnshi, W.Y., Nfor, L.N., Mengnjo, M.K., et al., 2021a. Garcinia kola improves cognitive and motor function of a rat model of acute radiation syndrome in the elevated plus maze. Brain Commun. 3, fcab170. doi:10.1093/braincomms/fcab170.

Ajeleti, A.O., Akunna, G.G., Ojo, J.H., Irozuoke, A., 2023. Neuro-protective and Neuro-curative effect of *Vernonia amygdalina* in rat model of neurodegenerative disease. J. Trace Elem. Miner. 4, 100062. doi:10.1016/j.jtemin.2023.100062.

Akujobi, C., Anyanwu, B., Onyeze, C., Ibekwe, V., 2004. Antibacterial activities and preliminary phytochemical screening of four medicinal plants. J. Appl. Sci. 7 (3), 4328–4338.

Alara, O.R., Abdurahman, N.H., Abdul Mudalip, S.K., Olalere, O.A., 2018. Microwave-assisted extraction of *Vernonia amygdalina* leaf for optimal recovery of total phenolic content. J. Appl. Res. Med. Aromat. Plants 10, 16–24. doi:10.1016/j.jarmap.2018.04.004.

Almeida, S., Alves, M.G., Sousa, M., Oliveira, P.F., Silva, B.M., 2016. Are polyphenols strong dietary agents against neurotoxicity and neurodegeneration? Neurotox. Res 30 (3), 345–366.

Anowi, F., Ike, C., Ezeokafor, E., Ebere, C., 2012. The phytochemical, antispamodic and Anti-diarrhoea properties of the methanol extract of the leaves of *Buchholzia coriacea* family Capparaceae. Int. J. Curr. Pharm. Res 4 (3), 52–55.

Asante, D.B., Wiafe, G.A., Tsegah, K.M., Domey, N.K., 2023. Therapeutic benefits of *Vernonia amygdalina* in the treatment of inflammation and its associated diseases. Clin. Complement. Med. Pharmacol. 4 (1), 100122. doi:10.1016/j.ccmp.2023.100122.

Babalola, K.T., Oyebanjo, O., Adekoya, V.A., Adeniyi, I.A., Ajayi, A.M., Onasanwo, S.A., 2023. Protective effect of methanol leaf extract of Cnidoscolus aconitifolius against lipopolysaccharides-induced cortico-hippocampal neuroinflammation, oxidative stress and memory impairment. Adv. Tradit. Med. 23 (1), 145–155.

Beal, M.F., 2005. Mitochondria take center stage in aging and neurodegeneration. Ann.Neurol. 58, 495–505.

Behl, T., Kaur, G., Sehgal, A., Singh, S., Bhatia, S., Al-Harrasi, A., et al., 2021. Elucidating the multi-targeted role of nutraceuticals: a complementary therapy to starve neurodegenerative diseases. Int. J. Mol. Sci 22 (8), 4045.

Benson, K.F., Beaman, J.L., Ou, B., Okubena, A., Okubena, O., Jensen, G.S., 2013. West African *Sorghum bicolor* leaf sheaths have anti-inflammatory and immune-modulating properties in vitro. J. Med. Food 16 (3), 230–238.

Brady, C.A., Attardi, L.D., 2010. p53 at a glance. J. Cell Sci. 123 (15), 2527–2532.

Ca, O., Dare, W.N., Avwioro, O.G., Igbigbi, P.S., 2015. Neuroprotective effect of Jobelyn® in the hippocampus of alcoholic rat is mediated in part by alterations in GFAP and NF protein expressions. Adv. Biol. Res. 9 (5), 305–317.

Charles, O.A., Patrick, I.S., Godwin, A.O., 2016. Jobelyn®® supplement lowered neuronal degeneration: significance of altered p53 and ɤ-Enolase protein expressions in prefrontal cortex of rat exposed to ethanol. Ann. Neurosci. 23 (3), 139–148.

Chauhan, B., Kumar, G., Kalam, N., Ansari, S.H., 2013. Current concepts and prospects of herbal nutraceutical: a review. J. Adv. Pharm. Technol. Res 4, 4–8.

Chen, Y., Peng, F., Xing, Z., Chen, J., Peng, C., Li, D., 2022. Beneficial effects of natural flavonoids on neuroinflammation. Front. Immunol 13, 1006434. doi:10.3389/fimmu.2022.1006434, PMID: 36353622; PMCID: PMC9638012.

Chika, E., Moses, I., Chigozie, U., Olisa, A., Ifeanyichukwu, I., Michael, A., Charles, E., 2012. Evaluation of antibacterial activity of the leave extracts of Buchholzia coriacea. Asian Journal of Pharmaceutical & Biological Research 2 (4), 204.

Chikezie, U.N., Chijioke, A.N., Adjeroh, L.A., Ogbulie, T.E., Udensi, J.U., Oyirioha, K.C., 2016. An evaluation of the phytochemical and nutritional compositions of fresh leaves of *Cnidoscolus aconitifolius* [Miller] IM Johnston. Int. J. Res. Stud. Biosci. 4 (6), 21–28.

Chinedu, F.A., Chibeze, E.I., Chukwuenweiwe, E.E., 2012. The phytochemical, antispasmodic and antidiarrheal properties of the methanol extract of the leaves of *Buchholzia coriacea* family Capparaceae. Int. J. Pharm. Res. 4 (3), 52–55.
Dadhania, V.P., Trivedi, P.P., Vikram, A., Tripathi, D.N., 2016. Nutraceuticals against neurodegeneration: a mechanistic insight. Curr. Neuropharmacol 14 (6), 627–640. doi:10.2174/1570159x14666160104142223.
Daramola, B., Adegoke, G.O., 2011. Bitter kola (*Garcinia kola*) seeds and health management potential. In: Nuts and Seeds in Health and Disease Prevention. Elsevier, pp. 213–220.
David, A.A., Oladele, A.A., Philip, A.O., Zabdiel, A.A., Olufunso, A.B., Tolulope, O.O., et al., 2022. Neuroprotective effects of *Garcinia kola* ethanol seed extract on haloperidol-induced catalepsy in mice. Trop. J. Nat. Prod. Res. 6, 281–286.
Duru, M.K.C., Agomuo, E.N., Amadi, B.A, 2012. Biochemical studies on –udu || an antimalarial concoction used in Umunchi village, Isiala Mbano L.G.A of Imo State, Nigeria. Contin. J. Pharmacol. Toxicol. Res. 5 (2), 28–34.
Edo, G.I., Samuel, P.O., Jikah, A.N., Onoharigho, F.O., Idu, L.I., Obasohan, P., et al., 2023. Biological and bioactive components of bitter leaf (*vernonia amygdalina* leaf): insight on health and nutritional benefits. A review. Food Chem. Adv. 3, 100488. doi:10.1016/j.focha.2023.100488.
Efeyan, A., Serrano, M., 2007. p53: guardian of the genome and policeman of the oncogenes. Cell Cycle 6 (9), 1006–1010.
Eniojukan, J., Aina, B., 2009. Toxicological profiles of commercial herbal preparation, Jobelyn®. Int. J. Health Res. 2 (4).
Erhirhie, E.O., Ben-Azu, B., Moke, G.E., Chinwuba, P., Omonjiahio, I.A., 2015. Ethno-pharmacological review of *Buchhoizia coriacea* (Wonderful Kola). Int. J. Adv. Pharm. Biol. Chem 4 (1), 149–155.
Essa, M.M., Subash, S., Akbar, M., Al-Adawi, S., Guillemin, G.J., 2015. Long-term dietary supplementation of pomegranates, figs and dates alleviate neuroinflammation in a transgenic mouse model of Alzheimer's disease. PLoS ONE 10 (3), e0120964.
Fagbohun, E., Egbebi, A., Lawal, O., 2012. Phytochemical screening, proximate analysis and in-vitro antimicrobial activities of methanolic extract of Cnidoscolus aconitifolius leaves. Int. J. Pharm. Sci. Rev. Res. 13 (1), 28–33.
Farombi, E.O., 2003. African indigenous plants with chemotherapeutic potentials and biotechnological approach to the production of bioactive prophylactic agents. Afr. J. Biotechnol. 2 (12), 662–671.
Farombi, E.O., Akanni, O.O., Emerole, G.O., 2002. Antioxidant and scavenging activities of flavonoid extract (Kolaviron) of *Garcinia kola* seeds. Pharm. Biol. 40, 107–116.
Farombi, E.O., Owoeye, O., 2011. Antioxidative and chemopreventive properties of *Vernonia amygdalina* and Garcinia biflavonoid. Int. J. Environ. Res. Public Health 8, 2533–2555.
Farombi, E.O., Tahnteng, J.G., Agboola, A.O., Nwankwo, J.O., Emerole, G.O., 2000. Chemoprevention of 2-acetylaminofluorene-induced hepatotoxicity and lipid peroxidation in rats by kolaviron—a *Garcinia kola* seed extract. Food Chem. Toxicol. 38, 535–541.
FedibaluChukwu, E.I., Aparoop, D., Kamaruz, Z., 2020. Antidiabetic, anthelmintic and antioxidation properties of novel and new phytocompounds isolated from the methanolic stem-bark of *Vernonia amygdalina* Delile (Asteraceae). Sci. Afr. 10, e00578. doi:10.1016/j.sciaf.2020.e00578.
Feigin, V.L., Vos, T., Nichols, E., Owolabi, M.O., Carroll, W.M., Dichgans, M., et al., 2020. The global burden of neurological disorders: translating evidence into policy. Lancet Neurol. 19 (3), 255–265. doi:10.1016/S1474-4422(19)30411-9.
Flieger, J., Forma, A., Flieger, W., Flieger, M., Gawlik, P.J., Dzierżyński, E., et al., 2024. Carotenoid supplementation for alleviating the symptoms of Alzheimer's disease. Int. J. Mol. Sci 25 (16), 8982. doi:10.3390/ijms25168982.

Fu, X., Yang, M., Cao, M., Li, D., Yang, X., Sun, J., et al., 2014. Strategy to suppress oxidative damage-induced neurotoxicity in PC12 cells by curcumin: the role of ROS-mediated DNA damage and the MAPK and AKT pathways. Mol. Neurobiol. 53 (1), 369–378. doi:10.1007/s12035-014-9021-1.

Gendelman, H.E., 2002. Neural immunity: friend or foe? J. Neurovirol. 8 (6), 474–479.

Goyal, A., Verma, A., Dubey, N., Raghav, J., Agrawal, A., 2022. Naringenin: a prospective therapeutic agent for Alzheimer's and Parkinson's disease. J. Food Biochem 46 (12), e14415. doi:10.1111/jfbc.14415, Epub 2022 Sep 15. PMID: 36106706.

Gul, K., Singh, A.K., Jabeen, R., 2016. Nutraceuticals and functional foods: the foods for the future world. Crit. Rev. Food Sci. Nutr. 56 (16), 2617–2627. doi:10.1080/10408398.2014.903384.

Gupta, R.C., Srivastava, A., Lall, R., 2018a. Toxicity potential of nutraceuticals. Methods Mol. Biol 1800, 367–394. doi:10.1007/978-1-4939-7899-1_18, PMID: 29934903.

Gupta, G., Tiwari, J., Dahiya, R., Sharma, R.K., Mishra, A., Dua, K., 2018b. Recent updates on neuropharmacological effects of luteolin. EXCLI J 17, 211–214.

Igado, O.O., Olopade, J.O., Oluwasanmi, A.A., Aina, O., Farombi, E.O., 2012. Morphological and biochemical investigation into the possible neuroprotective effects of kolaviron (Garcinia kola bioflavonoid) on the brains of rats exposed to vanadium. Drug and Chemical Toxicology 35 (4), 371–380.

Ijomone, O.M., Nwoha, P.U., Olaibi, O.K., Obi, A.U., Alese, M.O., 2012. Neuroprotective effects of kolaviron, a biflavonoid complex of *Garcinia kola*, on rats hippocampus against methamphetamine-induced neurotoxicity. Maced. J. Med. Sci. 5, 10–16.

Iwu, M.M., 1985. Antihepatoxic constituents of *Garcinia kola* seeds. Experientia 41, 699–700.

Iwu, M.M., Igboko, O.A., Okunji, C.O., Tempesta, M.S., 2011. Antidiabetic and aldose reductase activities of biflavanones of *Garcinia kola*. J. Pharm. Pharmacol. 42, 290–292.

Iyare, W.F., Bolanle, I.O., Akhigbemen, A.M., Uwaya, D.O., Oboigba, O.G., Gabriel, B.O., et al., 2024. Evaluation of the neuropharmacologic potentials of methanol leaf extract of *Cnidoscolus aconitifolius* in mice. Phytomed. Plus 4 (2), 100529. doi:10.1016/j.phyplu.2024.100529.

John, A., Bamidele, F., Ridwan, S., 2015. Neuroprotective effect of aqueous extract of *Garcinia kola* on monosodium glutamate – induced cerebellar cortical damage in adult Wistar rats. Eur. J. Med. Plants 5, 13–22.

John, R., Abolaji, A.O., Adedara, A.O., Ajayi, A.M., Aderibigbe, A.O., Umukoro, S., 2022. Jobelyn®® extends the life span and improves motor function in Drosophila melanogaster exposed to lipopolysaccharide via augmentation of antioxidant status. Metab. Brain Dis. 37 (4), 1031–1040.

Joshua, O., Okoro, I., Stephen Adelodun, T., Sunday yinka, O., Joshua Oladele, O., Iheanyichukwu, O., et al., 2020. *Garcinia kola* mitigates 3, 4-methylenedioxymethamphetamine-induced derangement in cellular and molecular characterization in the cerebellum of Wistar rat. Curr. Neurobiol. 11, 37–47.

Keber, U., Klietz, M., Carlsson, T., Oertel, W.H., Weihe, E., Schäfer, M.K.-H., et al., 2015. Striatal tyrosine hydroxylase-positive neurons are associated with l-DOPA-induced dyskinesia in hemiparkinsonian mice. Neuroscience 298, 302–317. doi:10.1016/j.neuroscience.2015.04.021.

Keservani, R.K., Kesharwani, R.K., Sharma, A.K., Gautam, S.P., Verma, S.K., 2017. Developing new functional food and nutraceutical products. Nutraceutical Formulations and Challenges 161–177. https://doi.org/10.1016/B978-0-12-802780-6.00009-2.

Khan, M.B., Khan, M.M., Khan, A., Ahmed, M.E., Ishrat, T., Tabassum, R., et al., 2012. Naringenin ameliorates Alzheimer's disease (AD)-type neurodegeneration with cognitive impairment (AD-TNDCI) caused by the intracerebroventricular-streptozotocin in rat model. Neurochem. Int. 61 (7), 1081–1093.

Kreiner, G., Bierhoff, H., Armentano, M., Rodriguez-Parkitna, J., Sowodniok, K., Naranjo, J.R., et al., 2013. A neuroprotective phase precedes striatal degeneration upon nucleolar stress. Cell Death Differ. 20 (11), 1455–1464.

Kudo, T., Ikeda, M., Nishikawa, M., Yang, Z., Ohno, K., Nakagawa, K., et al., 2012. The RASSF3 candidate tumor suppressor induces apoptosis and G1–S cell-cycle arrest via p53. Cancer Res. 72 (11), 2901–2911.

Kumar, A., Leinisch, F., Kadiiska, M.B., Corbett, J., Mason, R.P., 2015. Formation and implications of alpha-synuclein radical in Maneb- and Paraquat-induced models of Parkinson's disease. Mol. Neurobiol. 53 (5), 2983–2994. doi:10.1007/s12035-015-9179-1.

Makkar, R., Behl, T., Bungau, S., Zengin, G., Mehta, V., Kumar, A., et al., 2020. Nutraceuticals in neurological disorders. Int. J. Mol. Sci. 21 (12), E4424. doi:10.3390/ijms21124424.

Manzanero, S., Santro, T., Arumugam, T.V., 2013. Neuronal oxidative stress in acute ischemic stroke: sources and contribution to cell injury. Neurochem. Int. 62 (5), 712–718.

Mendrysa, S.M., Ghassemifar, S., Malek, R., 2011. p53 in the CNS: perspectives on development, stem cells, and cancer. Genes Cancer 2 (4), 431–442.

Michel, T.K., Ottoh, A.A., Chukwunonye, U.C.E., Obodoike, E.C., Christopher, O., Mmaduakolam, I.M., 2016. Bio-flavonoids and garcinoic acid from *Garcinia kola* seeds with promising anti-inflammatory potentials. Pharmacogn. J. 8, 56–58.

Mohammed, A., 2023. Hypoglycemic potential of African medicinal plants in diabetic and non-diabetic human subjects: a review. Clin. Complement. Med. Pharmacol. 3 (2), 100081. doi:10.1016/j.ccmp.2023.100081.

Mojab, F., Kamalinejad, M., Ghaderi, N., Vahidipour, H., 2003. Phytochemical screening of some Iranian plants. Iran. J. Pharm. Res. 4, 77–82.

Morrison, R.S., Kinoshita, Y., Johnson, M.D., Guo, W., Garden, G.A., 2003. p53-dependent cell death signaling in neurons. Neurochem. Res. 28, 15–27.

Moser, V., Li, A.A., 2020. Information Resources in Toxicology (Fifth Edition), Chapter 58 - Target Sites: Nervous System, Volume 1. Elsevier eBooks, pp. 569–574.

Moura, L., da Silva Neto, J.X., Lopes, T.D.P., Benjamin, S.R., Brito, F.C.R., Magalhães, F.E.A., et al., 2019. Ethnobotanic, phytochemical uses and ethnopharmacological profile of genus *Cnidoscolus* spp. (Euphorbiaceae): a comprehensive overview. Biomed. Pharmacother. 109, 1670–1679. doi:10.1016/j.biopha.2018.10.015.

Nazeruddin, G.M., Shirish, S.P., Samir, S.S., 2011. Pharmacological review of *Tridax procumbens* L. Der Pharmacia Sinica 2 (4), 172–175.

Nurjanah, R.N., Hermana, W., Retnani, Y., 2024. Chaya leaf infusion (*Cnidoscolus aconitifolius*) as a phytogenic for productivity and egg quality of Japanese quail (*Coturnix coturnix japonica*) of 17–20 weeks of age. Arch. Razi. Inst. 79 (1), 234–239. doi:10.32592/ari.2024.79.1.234.

Okochi, V.I., Okpuzor, J., Okubena, M.O., Awoyemi, A.K., 2003. The influence of African herbal formula on the haematological parameters of trypanosome infected rats. Afr. J. Biotechnol. 2 (9), 312.

Oladele, J.O., Oyeleke, O.M., Oladele, O.T., Olaniyan, M., 2020. Neuroprotective mechanism of *Vernonia amygdalina* in a rat model of neurodegenerative diseases. Toxicol. Rep. 7, 1223–1232. doi:10.1016/j.toxrep.2020.09.005.

Oladimeji, G.R., Fawole, A.O., Ekanola, Y.A., et al., 2009. Antibacterial activity and phytochemical composition of *Buccholzia coriacea* seeds. Polytechnic J. Sci. Tech. 5, 1.

Olanrewaju, J.A., Iheanyichukwu, O., Oladele, O.J., Yinka, O.S., Taiye, A.S., Oluseyi, F.S., 2022. *Garcina kola* neuroprotective effects on the prefrontal cortex cyto-architecture of MDMA-induced neuroinflammation in male Wistar rats model. Phytomed. Plus 2. doi:10.1016/j.phyplu.2021.100174.

Olatunji, S.Y., Okoro, I., Owolabi, J.O., Olanrewaju, J.A., Adelodun, S.T., Fabiyi, S.O., 2020. *Garcinia kola* attenuates MDMA-induced neuroinflammation in the CA1 region of the hippocampus in the rat model. Ann. Neurosci. 27, 104–113.

Onasanwo, A., Olaleye, S., 2011. Anti-ulcerogenic and proton pump (H+, K+ ATPase) inhibitory activity of kolaviron from *Garcinia kola* Heckel in rodents. Indian J. Exp. Biol 49 (6), 461–468.

Onasanwo, S.A., Oyagbemi, A.A., Saba, A.B., 2011. Anti-inflammatory and analgesic properties of the ethanolic extract of *Cnidoscolus aconitifolius* in rats and mice. J. Basic Clin. Physiol. Pharmacol. 22 (1–2), 37–41. doi:10.1515/jbcpp.2011.010.

Onasanwo, S.A., Adebimpe-John, O.E., Olopade, F.E., Olajide, O.O., 2021. Kolaviron Protects Rats from Cognitive Decline Induced by Lipopolysaccharide in Wistar Rat. Niger. J. Physiol. Sci. 36, 1–13.

Onasanwo, S.A., Osawe, S.O., Olubori, M.A., Faborode, O.S., Oyebanjo, O.T.. Evaluation of the possible mechanisms involved in the anti-inflammatory and anti-nociceptive properties of Jobelyn® in laboratory rodents.

Onasanwo, S.A., Oyebanjo, O.T., Ajayi, A.M., Olubori, MA., 2017. Anti-nociceptive and anti-inflammatory potentials of *Vernonia amygdalina* leaf extract via reductions of leucocyte migration and lipid peroxidation. J. Intercult. Ethnopharmacol. 6 (2), 192–198. doi:10.5455/jice.20170330010610.

Onasanwo, S.A., Rotu, R.A., 2016. Antinociceptive and anti-inflammatory potentials of kolaviron: mechanisms of action. J. Basic Clin. Physiol. Pharmacol. 27, 363–370.

Orji, O., Ibiam, U., Aja, P., Ugwu, P., Uraku, A., Aloke, C., et al., 2016. Evaluation of the phytochemical and nutritional profiles of *Cnidoscolus aconitifolius* leaf collected in Abakaliki South East Nigeria. World J. Med. Sci. 13 (3), 213–217.

Oyagbemi, A., Odetola, A., 2010. Hepatoprotective effects of ethanolic extract of *Cnidoscolus aconitifolius* on paracetamol-induced hepatic damage in rats. Pak. J. Biol. Sci. 13 (4), 164–169.

Oyagbemi, A., Odetola, A., Azeez, O., 2008. Ameliorative effects of *Cnidoscolus aconitifolius* on anaemia and osmotic fragility induced by proteinenergy malnutrition. Afr. J. Biotechnol. 7 (11).

Oyagbemi, A., Omobowale, T., Adedapo, A., Yakubu, M., 2016. Kolaviron, biflavonoid complex from the seed of *Garcinia kola* attenuated angiotensin II- and lypopolysaccharide-induced vascular smooth muscle cell proliferation and nitric oxide production. Pharmacogn. Res. 8, 50. doi:10.4103/0974-8490.178647.

Oyagbemi, A.A., Omobowale, T.O., Olopade, J.O., Farombi, E.O., 2018. Kolaviron and *Garcinia kola* attenuate doxorubicin-induced cardiotoxicity in Wistar rats. J. Complement. Integr. Med. 15. doi:10.1515/jcim-2016-0168.

Oyeyemi, I.T., Akinlabi, A.A., Adewumi, A., Aleshinloye, A.O., Oyeyemi, O.T., 2018. *vernonia amygdalina*: a folkloric herb with anthelminthic properties. Beni-Suef. Univ. J. Basic Appl. Sci. 7 (1), 43–49.

Oyinbo, C.A., Igbigbi, P.S., Avwioro, G.O., 2016. Landolphiaowariensis attenuates alcohol-induced cerebellar neurodegeneration: significance of neurofilament protein alteration in the Purkinje cells. Folia Med. 58 (4), 241–249.

Pan, T., Jankovic, J., Le, W., 2003. Potential therapeutic properties of green tea polyphenols in Parkinson's disease. Drugs Aging 20 (10), 711–721. doi:10.2165/00002512-200320100-00001, PMID: 12875608.

Pérez-González, M.Z., Jiménez-Arellanes, M.A., 2021. Biotechnological processes to obtain-bioactive secondary metabolites from some Mexican medicinal plants. Appl. Microbiol. Biotechnol. 105 (16 17), 6257–6274. doi:10.1007/s00253-021-11471-z.

Pham, V.C., Tran, T.H., Trieu, H.P., Le Hoang, T.A., Do Thi, T., 2021. Hypolipidemic and antidiabetic effects of methanolic leaf extract of *Vernonia amygdalina*: study on animal model. Internauka (12–3) 25–30.

Pringsheim, T., Jette, N., Frolkis, A., Steeves, T.D.L., 2014. The prevalence of Parkinson's disease: a systematic review and meta-analysis. Mov. Disord. 29, 1583–1590.

Puri, V., Nagpal, M., Singh, I., Singh, M., Dhingra, G.A., Huanbutta, K., et al., 2022. A comprehensive review on nutraceuticals: therapy support and formulation challenges. Nutrients 14 (21), 4637.

Qiaoa, H., Zhanga, X., Zhua, C., Donga, L., Wanga, L., Zhanga, X., Caoa, X., 2012. Luteolin downregulates TLR4, TLR5, NF-B and p-p38MAPK expression, upregulates the p-ERK expression, and protects rat brains against focal ischemia. Brain Res. 1448 (1), 71.

Paliwal, R., Sulakhiya, K., Rai Paliwal, S., Singh, V., Kenwat, R., Paramanik, D., 2022. Role of Nanoparticles in Neurotoxicity. Elsevier eBooks 165–182. doi:10.1016/b978-0-323-85544-0.00009-5.

Sachdeva, V., Roy, A., Bharadvaja, N., 2020. Current prospects of nutraceuticals: a review. Curr. Pharm. Biotechnol 21 (10), 884–896.

Samuel, I., Arthur, N., Jude, E., Henrietta, C., 2014. Antihyperglycaemic efficacy of *Cnidoscolus aconitifolius* compared with glibenclamide in alloxan-induced diabetic Wistar rats. Int. Res. J. Med. Sci. 2 (3), 1–4.

Singh, A., Kumar Singh, N., 2024. Pre-clinical evidence-based neuroprotective potential of naringin against Alzheimer's disease-like pathology: a comprehensive review. Curr. Pharm. Biotechnol 25 (9), 1112–1123. doi:10.2174/1389201024666230801095526, PMID: 37526460.

Spencer, P.S., Lein, P.J., 2023. In: Neurotoxicity, 6. Elsevier eBooks, pp. 727–740 online.

Tekou, F.A., Kuate, D., Nguekouo, P.T., Woumbo, C.Y., Oben, J.E., 2018. Effect of cooking treatments on the phytochemical composition and antidiabetic potential of *Vernonia amygdalina*. Food Sci. Nutr. 6 (6), 1684–1691.

Toyang, N.J., Verpoorte, R., 2013. A review of the medicinal potentials of plants of the genus Vernonia (Asteraceae). J. Ethnopharmacol. 146 (3), 681–723. doi:10.1016/j.jep.2013.01.040.

Trivedi, P.P., Jena, G.B., 2014. Mechanistic insight into beta-carotene-mediated protection against ulcerative colitis-associated local and systemic damage in mice. Eur. J. Nutr 54 (4), 639–652.

Trivedi, P.P., Jena, G.B., 2013. Role of α-lipoic acid in dextran sulfate sodium-induced ulcerative colitis in mice: studies on inflammation, oxidative stress, DNA damage and fibrosis. Food Chem. Toxicol. 59, 339–355.

Umukoro, S., Eduviere, A.T., Ogunbiyi, R., Adeoluwa, O.A., 2014. Pharmacological evaluation of the analgesic and anxiolytic activities of Jobelyn® in mice. J. Basic Clin. Physiol. Pharmacol. 25 (4), 439–444.

Umukoro, S., Oghwere, E.E., Ben-Azu, B., Owoeye, O., Ajayi, A.M., Omorogbe, O., Okubena, O., 2019. Jobelyn® ameliorates neurological deficits in rats with ischemic stroke through inhibition of release of pro-inflammatory cytokines and NF-κB signaling pathway. Pathophysiology 26 (1), 77–88.

Umukoro, S., Omogbiya, I.A., Eduviere, A.T., 2013. Evaluation of the effect of Jobelyn®® on chemoconvulsants-induced seizure in mice. Basic Clin. Neurosci. 4 (2), 125.

Valenzuela Soto, R., Morales Rubio, M.E., Verde Star, M.J., Oranday Cárdenas, A., PreciadoR-angel, P., Antonio González, J., et al., 2015. Cnidoscolus chayamansa organic hydroponic and its hypoglycemic capacity, nutraceutical quality and toxicity. Rev. Mex. de Cienc. Agríc. 6 (4), 815–825.

Williams, R.J., Mohanakumar, K.P., Beart, P.M., 2021. Neuro-nutraceuticals: natural products nourish the brain but be aware of contrary effects. Neurochem. Int 150, 105159. doi:10.1016/j.neuint.2021.105159, Epub 2021 Aug 14. PMID: 34400236.

Yi, L.T., Li, J., Li, H.C., Su, D.X., Quan, X.B., He, X.C., et al., 2012. Antidepressant-like behavioral, neurochemical and neuroendocrine effects of naringenin in the mouse repeated tail suspension test. Prog. Neuro-Psychopharmacol. Biol. Psychiatry 39 (1), 175–181.

Zhu, L., Bi, W., Lu, D., Zhang, C., Shu, X., Lu, D., 2014. Luteolin inhibits SH-SY5Y cell apoptosis through suppression of the nuclear transcription factor-κB, mitogen-activated protein kinase and protein kinase B pathways in lipopolysaccharide-stimulated co-cultured BV2 cells. Exp. Ther. Med 7, 10. doi:10.3892/etm.2014.1564.
Zindo, F.T., Joubert, J., Malan, S.F., 2015. Propargylamine as functional moiety in the design of multifunctional drugs for neurodegenerative disorders: MAO inhibition and beyond. Future Med. Chem. 7 (5), 609–629.

CHAPTER NINE

Neurotoxicity of heavy metals across the lifespan: The beneficial role of nutrition from fetus to the elderly

Titilayomi Ayomide Otenaike[a,b], Oluwabukola Mary Farodoye[a,b], Halimat Olaniyan[b], John Oluwafemi Teibo[c], Hammed Olawale Faleke[d], Israel Akinrotimi Akinola[e], Zainab Abiodun Molik[f], Michael Aschner[g], Amos Olalekan Abolaji[b,h,*]

[a]Programa de Pós-Graduação em Genética e Biologia Molecular, Universidade Federal do Rio Grande do Sul, Porto Alegre, RS, Brazil
[b]Drosophila Research and Training Centre, Basorun, Ibadan, Nigeria
[c]Department of Biochemistry and Immunology, Ribeirão Preto Medical School, University of São Paulo, Ribeirão Preto, São Paulo, Brazil
[d]Department of Chemistry and Biochemistry, Texas Tech University, Lubbock, Texas, USA
[e]Department of Biological Sciences, East Tennessee State University, Johnson City, Tennessee, USA
[f]Department of Pharmacognosy, Faculty of Pharmacy, University of Ibadan, Ibadan, Nigeria
[g]Department of Molecular Pharmacology, Albert Einstein College of Medicine, Bronx, New York, USA
[h]Drosophila Laboratory, Drug Metabolism and Toxicology Unit, Department of Biochemistry, Faculty of Basic Medical Science, College of Medicine, University of Ibadan, Ibadan, Oyo State, Nigeria
*Corresponding author. e-mail address: ao.abolaji@ui.edu.ng.

Contents

Advances in Neurotoxicology, Volume 16
ISSN 2468-7480
https://doi.org/10.1016/bs.ant.2026.03.003

Abstract

Heavy metals pose a profound threat to neurological health across all stages of human life, from prenatal development to old age. Heavy metals such as lead, mercury, cadmium, and arsenic are pervasive environmental pollutants that disrupt neural function through mechanisms including oxidative stress, inflammation, mitochondrial dysfunction, and neurotransmitter system imbalances. During critical developmental windows such as fetal growth and early childhood, exposure can impair neurogenesis, synaptic plasticity, and myelination, leading to lifelong cognitive deficits, behavioral disorders, and increased vulnerability to neurodegenerative diseases in later life. Even in adulthood and aging, chronic exposure exacerbates neurodegeneration, accelerating conditions like Alzheimer's and Parkinson's diseases through persistent oxidative damage and inflammatory cascades. This chapter underscores the dual role of nutrition as both a shield and a therapeutic tool against heavy metal neurotoxicity. Key nutrients, such as polyphenols, vitamins, and essential minerals, counteract heavy metal-induced damage by scavenging free radicals, enhancing antioxidant defenses, modulating inflammation, and promoting neuronal repair. By integrating evidence from epidemiological, preclinical, and clinical studies, this chapter emphasizes actionable strategies, such as fortified infant formulas, plant-based proteins, and micronutrient supplementation to reduce heavy metal bioavailability and bolster neurological resilience. Public health initiatives targeting vulnerable populations, alongside policies regulating environmental pollutants, are critical to curbing this silent epidemic. This chapter advocates for a proactive, nutrition-centered approach to safeguarding brain health, demonstrating that dietary interventions are not merely complementary but foundational in combating the pervasive threat of heavy metal neurotoxicity across generations.

Keywords: Antioxidant, Developmental neurotoxicity, Heavy metals, Lifespan health, Nutritional mitigation

1. Introduction

Neurotoxicity is a broad term which refers to the adverse effects of chemical, biological, or physical agents on the structure or function of the nervous system, leading to neurological impairments (Jadhav et al., 2022). Exposure to these agents can disrupt cognitive, motor, sensory, and behavioral functions, with effects ranging from acute to chronic. These effects may result in temporary or permanent damage to neural tissues depending on the nature and duration of exposure.

Heavy metals are naturally occurring compounds, but human activities highly influence exposure to these heavy metals. Their toxicity is often dose-dependent, although some compounds are toxic even at low doses. The severity of their effects also'depends on the duration of exposure, route of exposure, and the nature of the heavy metal (Jaishankar et al., 2014). Heavy metals such as lead, mercury, cadmium, and arsenic are particularly concerning due to their widespread presence in the environment and neurotoxic properties. Heavy metals enter the human body from different sources such as food, environment, water, soil and air. The primary ways individuals can be exposed include swallowing, breathing in, and skin contact. This exposure varies across different locations and is based on different factors, with occupation and lifestyle playing a role in the exposure (Neal and Guilarte, 2013). Heavy metals can disrupt the normal function of neurons and other cells in the nervous system through multiple mechanisms. Given the complex function and structure of the nervous system, the impact of neurotoxicity can vary widely. Exposure to heavy metals can result in physical, cognitive, and developmental defects in humans at different stages of life. They can induce oxidative stress, leading to cellular damage; interfere with the synthesis and function of neurotransmitters; impair energy production in neurons by disrupting mitochondrial function; and induce inflammatory responses that can lead to neuronal death (Koyama et al., 2024).

Heavy metals have significant neurotoxic impacts on human health, making them a significant public health challenge. The impacts occur in the long or short term, depending on various factors such as dosage, duration of exposure, and genetics. The effects vary across the lifespan, from childhood to adulthood, making it an important public health priority that requires prompt prevention and mitigation efforts. Understanding the role of heavy metals in neurotoxicity is important due to their significant effects on individuals and society. The prevention and mitigation strategies include strengthening regulations on industrial emissions and pollutants,

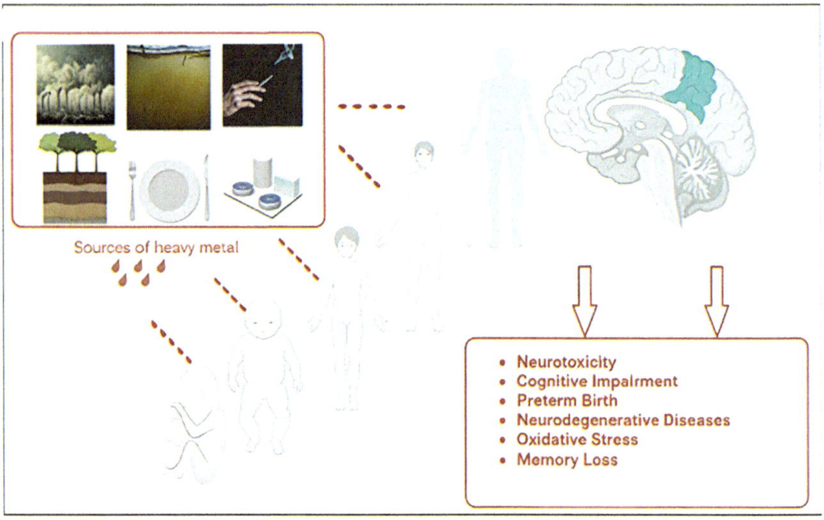

Fig. 1 Mechanisms by which heavy metals induce neurotoxicity.

improving workplace safety standards, promoting public awareness about the sources and risks of neurotoxic agents, and encouraging dietary and lifestyle modifications that enhance the body's natural defences against neurotoxicity (Ogbeide and Henry, 2024). Public health interventions should address disparities in exposure across different populations by prioritizing vulnerable groups to mitigate the long-term health impacts of neurotoxic exposure.

2. Mechanisms of metal neurotoxicity

The neurotoxic effects of these heavy metals are attributed to multiple molecular and cellular mechanisms, including oxidative stress, inflammation, mitochondrial dysfunction, and disruption of neurotransmitter systems (Teleanu et al., 2022). Understanding these mechanisms (Fig. 1) is crucial for developing strategies to mitigate heavy metal-induced neurotoxicity.

2.1 Oxidative stress

Oxidative stress is a primary mechanism through which heavy metals exert their neurotoxic effects, occurring through different means, including generating reactive oxygen species (ROS), depleting the antioxidant defense system, and lipid peroxidation. These free radicals damage cellular components such as lipids, proteins, and DNA, resulting in neuronal injury.

Lead induces oxidative stress by increasing ROS production while depleting antioxidants such as glutathione (GSH) (Virgolini and Aschner, 2021). Mercury, particularly in its organic form (methylmercury), strongly promotes oxidative stress by binding to thiol (-SH) groups in proteins and depleting GSH. Mercury exposure increases ROS levels, causing oxidative damage to neuronal membranes, disrupting calcium homeostasis, and leading to neurodegeneration (Farina et al., 2013). Arsenic exposure generates ROS through the redox cycling of arsenic metabolites, resulting in oxidative stress that damages cellular components, including lipids, proteins, and DNA and also impairs signaling pathways critical for neuronal survival and differentiation (Paithankar et al., 2021). Although cadmium is not redox-active, it indirectly induces oxidative stress by depleting GSH and other antioxidants, impairing the cell's ability to neutralize ROS. Cadmium also enhances NADPH oxidase activity, increasing ROS production and subsequent neurotoxicity (Hernández-Cruz et al., 2022).

2.2 Inflammation

Inflammation is another central mechanism in heavy metal neurotoxicity. Neuroinflammation is primarily driven by the activation of glial cells, particularly microglia and astrocytes, which release pro-inflammatory cytokines and chemokines (Smith et al., 2012). Lead exposure activates microglia and astrocytes, leading to the release of pro-inflammatory cytokines such as interleukin-1 beta (IL-1β), tumor necrosis factor-alpha (TNF-α), and interleukin-6 (IL-6). The resulting neuroinflammation disrupts neuronal signaling and promotes neuronal death, contributing to cognitive deficits and behavioral disorders (Gąssowska-Dobrowolska et al., 2023). Mercury also triggers microglial activation and the release of pro-inflammatory mediators, leading to increased levels of cytokines and chemokines that contribute to neurodegeneration. Mercury-induced inflammation is also linked to the activation of the nuclear factor-κB (NF-κB) pathway, which is a primary regulator of inflammatory responses in the brain (Novo et al., 2021). Chronic exposure to arsenic results in sustained neuroinflammation, characterized by glial cell activation and the production of pro-inflammatory cytokines (Escudero-Lourdes, 2016). Arsenic-induced inflammation disrupts the blood-brain barrier (BBB), allowing harmful substances to enter the brain and exacerbate neuronal damage (Wang et al., 2021). Cadmium induces neuroinflammation by activating microglia and astrocytes, leading to cytokine and ROS production. The activation of the NF-κB signaling pathway is central to cadmium-induced inflammation. The inflammatory response contributes to neuronal injury and is implicated in neurodegenerative disease pathogenesis (Vellingiri et al., 2022).

2.3 Mitochondrial dysfunction

Mitochondria are responsible for energy production, calcium homeostasis, and the regulation of apoptosis, but heavy metals disrupt these functions, leading to neuronal cell death (Sun et al., 2022). Lead impairs mitochondrial function by inhibiting the electron transport chain (ETC), leading to decreased ATP production and increased ROS generation. Additionally, lead disrupts mitochondrial membrane potential, leading to the release of pro-apoptotic factors like cytochrome c, which initiate cell death pathways (Sun et al., 2022). Mercury accumulates in mitochondria, where it interferes with the ETC and induces ROS production. Mitochondrial dysfunction in mercury toxicity is characterized by the loss of membrane potential, impaired ATP synthesis, and activation of apoptotic pathways, contributing to neurodegeneration (Kang et al., 2024). Arsenic disrupts mitochondrial function by impairing key enzymes in the ETC and inducing mitochondrial oxidative stress. Arsenic also promotes the opening of the mitochondrial permeability transition pore (mPTP), leading to the release of pro-apoptotic factors and subsequent neuronal apoptosis (Thakur et al., 2021). Cadmium disrupts mitochondrial function by inhibiting the ETC, leading to decreased ATP production and increased ROS generation. Cadmium exposure also disrupts mitochondrial membrane potential and triggers the opening of the mPTP, which facilitates cytochrome c release and initiates apoptosis (Choong et al., 2014).

2.4 Disruption of neurotransmitter systems

Heavy metals disrupt neurotransmitter systems, leading to imbalances in excitatory and inhibitory signaling in the brain. This disruption contributes to cognitive deficits, behavioral changes, and the development of neurological disorders (Rehman et al., 2021). Lead interferes with several neurotransmitter systems, including those involving glutamate, dopamine, and GABA. Lead exposure disrupts glutamatergic signaling by impairing NMDA receptor function, which is critical for learning and memory (Sanders et al., 2009). Additionally, lead alters dopamine synthesis and release, leading to deficits in motor control and reward processing (Speranza et al., 2021). Mercury, especially methylmercury, disrupts neurotransmitter systems by binding to and modifying neurotransmitter receptors. It inhibits glutamate uptake and enhances excitotoxicity, leading to excessive neuronal stimulation and cell death (Speranza et al., 2021). Mercury also impairs GABAergic transmission, contributing to anxiety and mood disorders (Teixeira et al., 2019). Arsenic exposure affects cholinergic, dopaminergic, and serotonergic systems by inhibiting acetylcholinesterase (AChE) activity, leading to acetylcholine accumulation and overstimulation of cholinergic receptors. Arsenic also disrupts dopamine and serotonin pathways, leading to mood disturbances, cognitive impairment, and motor dysfunction (Thakur et al., 2021). Cadmium disrupts neurotransmitter systems by altering neurotransmitter release, synthesis, and receptor function. It impairs dopaminergic signaling, leading to motor deficits and cognitive dysfunction. Cadmium also interferes with GABA and glutamate signaling, contributing to excitotoxicity and neuronal injury (Ojo et al., 2023).

3. Stages of life and vulnerability to heavy metal neurotoxicity

Heavy metals are widely distributed in nature, and their bioaccumulation frequently goes unrecognized until it manifests as a chronic health issue. These heavy metals in the environment have quite a ubiquitous presence. All age groups and organs are potentially vulnerable to heavy metal-induced toxicity (Tchounwou et al., 2012). A large body of evidence has demonstrated that exposure to neurotoxicants such as heavy metals at any stage of development, from childhood through adulthood and into old age, can have long-term adverse effects (Nabi and Tabassum, 2022). These effects includes impaired cognition, ataxia, incontinence (Chen et al., 2016), psychiatric and medical disease states (Akdağ and Erbas, 2023), and overall, a significant decline in the quality of life (Vellingiri et al., 2022). While studies have observed the effects of heavy metals, there is no comprehensive understanding of their impact from prenatal development through adulthood and into old age. Thus, this review gives an overview of how exposure to heavy metals, as observed in human and animal models, increases the risk of neurotoxicity, which may manifest in various forms across different life stages (Fig. 2).

3.1 Prenatal exposure to heavy metals

Evidence shows that heavy metals adversely impact brain function and behavior even at low doses, with critical periods of vulnerability spanning

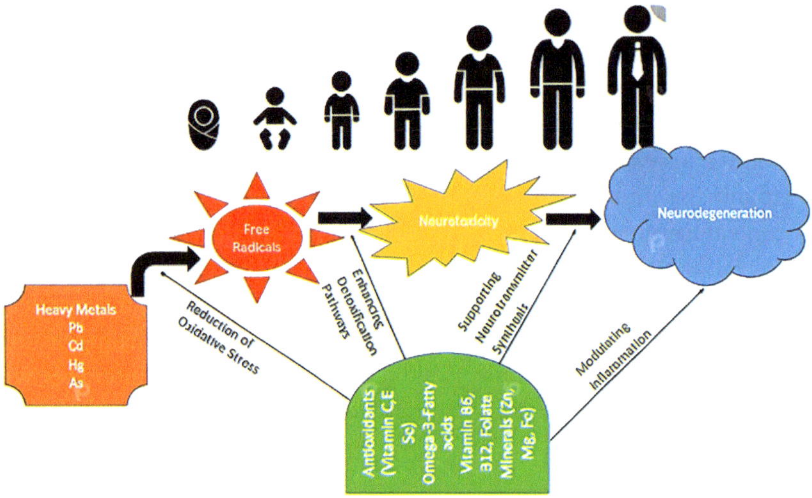

Fig. 2 Exposure to different sources of heavy metals at different stages of life results in neurotoxicological effects.

through pre-conception (in both sexes), pregnancy, and postnatal development (Zinia et al., 2023). Exposure during these periods can induce sex-specific neurotoxic effects, producing a greater risk after prenatal or postnatal exposure and propagating harm across subsequent generations.

Importantly, some heavy metals play essential roles in central nervous system (CNS) development as biometals, regulating structural organization and functional processes in small, physiologically relevant quantities (e.g., phosphorus, iron, copper). Conversely, non-essential metals (e.g., lead, mercury, chromium, silver) lack biological functions and exhibit pronounced neurotoxicity, even at trace levels (Hlatshwayo et al., 2024).

3.2 Mechanisms of placental transfer

During pregnancy, the placenta serves as a crucial selective barrier against harmful substances. Nevertheless, certain heavy metals (like nickel, lead, cadmium, arsenic, and mercury) can obstruct the placental transport system before passing through the placenta (Ding et al., 2021). The fetus is extremely sensitive to hazardous substances, although the environmental exposure level is far lower than the internal exposure. This is because the fetal physiological and biochemical levels differ from those of adults. Research has shown that most pregnant people have heavy metals such as lead, cadmium, arsenic, and mercury in their placentas, umbilical cord blood, and bloodstream. A birth cohort study conducted in Shanghai, China, between 2013 and 2016, reported that mean maternal blood levels of metals such as magnesium (Mg), calcium (Ca), and copper (Cu) were significantly higher during pregnancy compared to the preconception period (Zhang et al., 2023). These toxins can pass from mother to fetus via the placenta's passive diffusion, which may be responsible for abnormal placental function and inhibit the fetus's ability to receive the necessary nutrients and develop to its full potential (Habibian et al., 2022), ultimately resulting in preterm deliveries and adverse pregnancy effects. For example, in the hypothalamic-pituitary axis, heavy metals have been shown in certain animal studies to cause abnormalities by interfering with signal pathways within and between the pituitary gland and the hypothalamus at different doses and after different lengths of exposure (Asefi et al., 2022; Habibian et al., 2022). Furthermore, maternal exposure to toxic metals can cause several biochemical alterations during placental transfer, leading to the uncontrollably high production of free radicals in endothelial cells, including ROS and reactive nitrogen species (RNS). These radicals harm DNA, membrane lipids, and enzymes in placental tissues, and are often combated by antioxidant defenses in the placenta's endothelial cells (Singh et al., 2020). Oxidative stress is still observed, even in some pregnant populations during the second or third trimester, with low blood concentrations of environmental heavy metals (Kinjo et al., 2024). However, there is a notable increase in the progesterone-to-estrogen ratio and heavy metal-induced alteration of reproductive hormones in the placenta tissue, which may partially account for the increased uterine contractility. Therefore, it causes preterm birth following preterm labor and increases the spasm of uterine smooth muscle cells (Kinjo et al., 2024). Potential mechanisms for heavy metal exposure on placental development include reducing trophoblast cell proliferation, promoting apoptosis, inhibiting migration, and reducing the process of placental cell energy production (Zhao et al., 2020).

It is common to find heavy metals like cadmium in the environment, mostly in the diet and cigarette products. Contrary to mercury and lead, the placenta acts as an effective barrier, which causes the concentration of Cd to build up inside the placenta over time (Kinjo et al., 2024), impeding the healthy development of the fetus. After being absorbed, this heavy metal diffuses through the blood to several organs, including the placenta, and is consequently linked to reproductive, genotoxic, carcinogenic, and fetal toxicity.

3.3 Critical periods of brain development and effects on fetal brain development

How does heavy metal affect brain development in the fetus? This question needs to be especially considered during critical periods of growth, which may have the potential to harm the fetus and its brain development. The human brain undergoes a complex and prolonged developmental process, which can be significantly disrupted by environmental toxins, including heavy metals (Rock and Patisaul, 2018). The exposure time window is particularly important when considering the neurotoxic effects of the suggested heavy metal. In this regard, another important factor is the ability of toxic metals to bioaccumulate, which causes a further rise in their bioconcentration (Živančević et al., 2024).

Of the heavy metals studied for their effects on neurodevelopment, lead is one of the most extensively studied. Once inside the brain, it obstructs neurotransmitter release, synapse formation, and neurogenesis. It does this by crossing the blood-brain barrier (BBB). Reductions in brain volume can occur from lead exposure during critical periods of brain development, especially in the prefrontal cortex, hippocampus, and cerebellum (Kothapalli, 2021).

The brain is more vulnerable to harm during development. In this context, there is a need to understand that specific metals have specific detrimental effects. For example, disrupted neuronal migration, synaptogenesis, and neurotransmitter regulation, impaired cognitive function, motor skills, and sensory processing are associated with mercury exposure in the fetal brain (Milanković et al., 2024). Though there are various types of mercury, according to the National Research Council (NRC, US), the most hazardous is methylmercury (MeHg), which is mostly consumed through food, particularly fish. Methylmercury concentrates in the developing brain after it easily passes through the placenta and BBB (National Research Council (US) Committee on the Toxicological Effects of Methylmercury 2000). Although cadmium does not easily cross the BBB, with time, it can accumulate in the brain, causing oxidative stress, mitochondrial dysfunction, and disruption of calcium homeostasis (Arruebarrena et al., 2023). These processes lead to neuroinflammation, neuronal apoptosis, and deficits in learning and memory. In contrast, inorganic arsenic is particularly neurotoxic, affecting the brain by inducing oxidative stress, disrupting mitochondrial function, and interfering with neurotransmitter systems (Thakur et al., 2021). Chronic arsenic exposure has been linked to deficits in cognitive development, a reduced intelligence quotient (IQ), and increased risks of neurodevelopmental disorders (Rodríguez-Barranco et al., 2013).

A 2025 epidemiological study in humans by Kou et al. (2025) of 400 mother-infant pairs, which measured heavy metal levels in maternal urine at the 12th week of gestation and assessed infant neurodevelopment at 40 days using the Bayley Scales of Infant and Toddler Development, observe that Cadmium exposure causes significant linear relationships with cognitive scale and expressive language development (Kou et al., 2025). Also a study on maternal exposure to higher levels of manganese, nickel, lead, and aluminum correlated with lower Mental Development Index (MDI) scores in 2-year-old children, found that prenatal metal exposure impacts on MDI and these were more pronounced indicating greater susceptibility of the fetal nervous system to cognitive impairments during critical developmental periods (Xie et al., 2025).

In addition, experimental animal studies confirm that developmental heavy metal exposures cause neurobehavioral effects similar to those observed in humans. A landmark study examined the combinatorial effects of arsenic, cadmium, and lead on rat brain development administered to pregnant and lactating rats with pups, observed hyperlocomotion, increased grip strength, and learning-memory deficits, while biochemical changes revealed synergistic reduction in glial fibrillary acidic protein (GFAP) expression during brain development, increased astrocyte apoptosis in cerebral cortex and cerebellum, activation of extracellular signal-regulated kinase (ERK) signaling, downstream activation of Jun N-terminal kinase (JNK) pathway, and elevated intracellular calcium (Ca^{2+}) and reactive oxygen species generation at lethal concentration of 10–75 mm (Rai et al., 2010). The extensive experimental evidence reviewed here demonstrates that developmental heavy metal exposure causes neurobehavioral effects that closely mirror those observed in human populations. The early embryonic stage, encompassing the first trimester, is a critical period of organogenesis and rapid cellular differentiation, neurogenesis, synaptogenesis, and myelination (Ewool et al., 2024), which could in turn pose significant clinical challenges. Heavy metals can interfere with these processes through various mechanisms that cross the placenta and disrupt neural tube development, decrease steroidogenic enzymes, and induce oxidative stress and apoptosis in developing neural tissues (Liu et al., 2020). In utero, arsenic exposure affects neurogenesis and myelination, resulting in deficits in memory, attention, and executive function. Arsenic's neurotoxic effects are mediated through its interference with DNA methylation and histone modification (Vaidya et al., 2023). Likewise, a rat model of manganese poisoning used intraperitoneal injection of Manganese chloride ($MnCl_2$) at doses of 15.0, 25.0, and 50.0 mg/kg for three months revealed attention-deficit-like phenotypes in developing animals (Tian et al., 2018).

3.4 Long-term effects of exposure during adolescence and adulthood

During adolescence, significant cortical circuit reorganization occurs, influencing cognitive functions and emotional regulation. These neural changes are characteristics of this developmental stage and contribute to the cognitive functioning that is typical during this period (Pöpplau et al., 2024). Adolescents who are exposed to heavy metals may experience long-term consequences such as reduced IQ, attention deficit hyperactivity disorder (ADHD), learning disabilities, behavioral issues, and an increased risk of substance abuse and delinquency (Farmani et al., 2024). These developmental trajectories may be largely explained by changes in brain architecture and function.

Prenatal exposure to heavy metals can set the stage for neurodevelopmental challenges that manifest during adolescence. The developing fetal brain is highly susceptible to environmental toxins due to its rapid growth, the immaturity of the BBB, and the critical processes of neurogenesis, synaptogenesis, and myelination that occur in utero (Bell and O'Shaughnessy, 2022). In addition, it is important to emphasize that exposure to heavy metals during adolescence can have lasting effects that persist into adulthood, impacting cognitive, emotional, and behavioral outcomes (Rauh and Margolis, 2016).

Neuroimaging studies show functional and structural changes in front striatal regions, indicating toxic brain lesions at both early and delayed phases of disease progression (Rumboldt et al., 2019). The skeleton is thought to be a long-term storage location for heavy metal components due to the huge concentrations of these elements, which have lengthy half-lives (Akdağ and Erbas, 2023). As people get older, more senescent cells develop in their tissues and organs. Senescent cells can emit immune-stimulating chemicals, promote inflammation, and reduce the viability of surrounding tissues and cells (Cheng et al., 2021). Heavy metals like Cd accumulate in bone with increasing age, causing nephrotoxicity, while Mn is harmful to the CNS, contributing to cognitive impairment such as Alzheimer's disease and Parkinson's disease (Althomali et al., 2024).

Studies revealed that both single (Cd, Co, Cs, Pb, and Tl) and combined metals impact biological aging. Toxicological evidence suggests that heavy metals (e.g., Co and Cd) can cause senescence in stromal and vascular smooth muscle cells, as well as accelerate brain aging in animals, as evidenced by damaged molecules and impaired energy metabolism. Furthermore, a recent population study demonstrated the favorable relationships between heavy metals (e.g., Co) and biological aging markers (DNA methylation age) (Wu et al., 2024). As a by-product of their systemic interactions, heavy metals produce free radicals that can cause damage to cellular and molecular components. The sustained increased and prolonged generation of ROS and resulting chronic oxidative stress form the basis of carcinogenesis (Parida and Patel, 2023).

3.5 Neurodevelopmental changes and impact of early exposure on cognitive and behavioral outcomes

Previous literature has highlighted the mechanisms of molecular toxicity induced by substances such as arsenic, cadmium, chromium, nickel, lead, barium, and polychlorinated biphenyls (PCBs) (Petrosino et al., 2018). In animal studies, developmental lead exposure ranging from 25 to 600 ppm in drinking water or feed, administered during gestation or postnatal periods lasting up to several months results in notable neurobehavioral and biochemical alterations. For example, mice exposed to 25–100 ppm lead in drinking water from pre-mating through lactation exhibit impaired motor function, reduced social novelty preference, and sex-specific deficits in spatial learning and memory, as indicated by performance in the Morris water maze and social novelty tests (Dinçkol et al., 2022), confirmed with biochemical assay on dose-dependent lead accumulation in blood, brain, and bone, and parallel studies in monkeys (350–600 ppm dietary lead, yielding blood levels of 40–56 μg/dL) reveal increased reaction times and impaired executive function, with deficits worsening as task complexity rises (Parithathvi et al., 2024). In vitro experiments using mouse embryonic stem cell-derived neurons exposed to lead during differentiation mirror these findings, showing altered expression of neurodevelopmental genes (e.g., Ngn1, Bdnf, glutamate receptors) in key brain regions (Sánchez-Martín et al., 2013). During brain development, neuronal circuit shaping is a key event that relies on dendritic spine formation and pruning, which could be vulnerable to heavy metal toxicity. Moreover, the underlying molecular mechanisms of prenatal heavy metal toxicity partly overlap, hence the need to understand the path for placental transfer. Brain development and cognitive function are affected by multiple neurotoxicants. As a critical brain region involved in cognitive and mood-related processes, the hippocampus is particularly vulnerable to heavy metals (Chung et al., 2021).

Adult brain circuits are less conducive to learning new skills and more fixed than those of a developing brain. Both external sensory experiences and endogenous genetic factors influence how the brain develops; in the early stages of development, before many important neural circuits are fully formed, the brain is especially malleable. Because of this characteristic, early experience is essential for the development of language and musical skills. It is now evident that certain windows of early development, also referred to as critical periods, are when brain plasticity is at its highest. The brain grows in two somewhat overlapping phases following organogenesis: a phase of neuronal multiplication and a phase of glial proliferation, during and after which myelination and dendritic and axonal arborization take place (Hlatshwayo et al., 2024).

4. Accumulation over lifespan and interaction with age-related neurodegenerative processes

Neurodegeneration can be caused by oxidative stress and mitochondrial malfunction. Oxidative stress causes lipid peroxidation, which leads to DNA damage and eventually cell death in the brain (Misrani et al., 2021). In particular, exposure to heavy metals during early life can have a significant and long-lasting effect on behavioral and cognitive outcomes. Reducing exposure to these dangerous substances through efficient environmental regulations, pollution control strategies, and public health interventions is crucial for maintaining public health (Awewomom et al., 2024). Important neurodevelopmental processes may be disrupted during this time, which may cause long-term emotional, behavioral, and cognitive problems that last into adulthood (Dubovický, 2010). Heavy metals induce cellular senescence or physiological abnormalities that align with markers of aging (Vielee and Wise, 2023). Epidemiological and experimental studies consistently link chronic exposure to elevated risks of cognitive decline, Alzheimer's disease (AD), and Parkinson's disease (Bakulski et al., 2020). A longitudinal aging study shows that higher lifetime lead exposure predicts steeper declines in memory and executive function in older adults (Weisskopf et al., 2004). The brain's diminished capacity for detoxification with age makes it especially vulnerable to heavy metal-induced oxidative stress and inflammation. Animal models have been critical for clarifying mechanisms and dose–response relationships, for example APP/PS1 mice (an AD mice model) exposed to PbAc demonstrated more vascular amyloid deposition less neocortical myelination, and lower cognitive function at 50 mg/kg (Gu et al., 2024). Similarly, mercury chloride at doses of 0.5–2 mg/kg/day administered for several weeks impairs locomotor activity and induces anxiety-like behaviors in rodents (Bisen-Hersh et al., 2013). Biochemically, heavy metal exposure elevates lipid peroxidation (↑MDA), suppresses antioxidant enzymes (↓SOD, ↓GSH), and triggers pro-inflammatory cytokines (↑IL-1β, ↑TNF-α) (Kumar et al., 2021). In fact *Drosophila melanogaster* models fed cadmium chloride (0.1–1 mM) for 10–30 days displayed shortened lifespan, reduced climbing ability, and dopaminergic neuron loss (Bonilla-Ramirez et al., 2011).

Exposure to heavy metals reduces the quality of mitochondria (Althomali et al., 2024 ; Koyama et al., 2024). Neurodegenerative damage is caused by apoptosis in cells, which is triggered by a disturbance of calcium homeostasis, which is stabilized by calcium transporters in the mitochondrial membrane (Matuz-Mares et al., 2022). Changes in protein expression linked to exposure to heavy metals have been associated with PD and AD, according to recent system biology analyses (Pyatha et al., 2022). An adult may experience long-term cognitive, behavioral, and emotional difficulties because of the disruption of important neurodevelopmental processes that occur during this time. It is therefore crucial to comprehend the dangers of heavy metal exposure in adolescence to create preventative and intervention plans that will safeguard this susceptible group and encourage normal brain development.

5. Neuroprotective role of nutrition against neurotoxicity

The role of nutrition in neuroprotection cannot be overemphasized as it plays an essential role in various stages of life from fetus through adolescence, adulthood, and old age. Nutrients from various foods have roles in mitigating neurotoxicity and neurodegeneration induced by heavy metals and other toxic agents. Research in preclinical and clinical models has identified numerous functional foods and bioactive compounds with neuroprotective properties. These compounds exert their effects through multiple mechanisms, including reducing oxidative stress, enhancing detoxification pathways, supporting neurotransmitter synthesis, and modulating neuroinflammation. Two general principles support neuroprotective nutrition: preventing the generation of ROS and free radicals, and neutralizing free radicals generated as a result of neurotoxic exposure. Bioactive compounds, often electrophilic, donate electron pairs to unstable radicals, stabilizing them and facilitating their excretion from brain cells, neurons, and finally the kidneys (Lee et al., 2020). These functional foods and bioactive compounds serve as non-enzymatic antioxidants or support enzymatic antioxidants in their protective roles (Lobo et al., 2010).

An investigation by Oria et al. (2023), observed the effects of curcumin, a natural compound derived from turmeric, on memory deficits and oxidative damage induced by cobalt (known to cause neurotoxicity, leading to cognitive impairments and associated cellular damage, particularly in the hippocampus, a region critical for memory formation.) exposure in a model organism. The study highlights curcumin's ability to decrease astrocytosis, which refers to the proliferation of astrocytes in response to injury, and to enhance the expression of Nrf2, a transcription factor that promotes antioxidant defenses (Oria et al., 2023).

Diverse nutritional interventions have been shown to play a protective role in alleviating the neurotoxic effects of heavy metal exposure. Antioxidant vitamins such as vitamin C and vitamin E help counteract oxidative damage induced by metals like lead, mercury, and cadmium by scavenging reactive oxygen species and stabilizing neuronal membranes (Flora et al., 2012). While Selenium, a vital component of antioxidant enzymes like glutathione peroxidase, forms inert complexes with mercury, thereby mitigating neuroinflammation and cognitive deficits (Rayman, 2012).

In addition Zinc and iron has been evaluated to compete with lead and cadmium for absorption and cellular binding sites (Peng et al., 2023), thus limiting their neurotoxic uptake and supporting antioxidant enzyme activity (Kaur et al., 2014). Calcium also competes with lead at neuronal calcium channels, reducing lead accumulation in the brain and protecting synaptic function (Goyer, 1997). Omega-3 fatty acids, particularly EPA and DHA, are known to preserve synaptic integrity and reduce neuroinflammation triggered by methylmercury (Oken et al., 2005). Additionally, bioactive compounds such as curcumin exhibit both antioxidant and metal-chelating properties, which help restore neuronal health and cognitive functions after lead or arsenic exposure (Singh et al., 2013). The B-complex vitamins especially B6, B9, and B12aid in neural repair by maintaining proper DNA

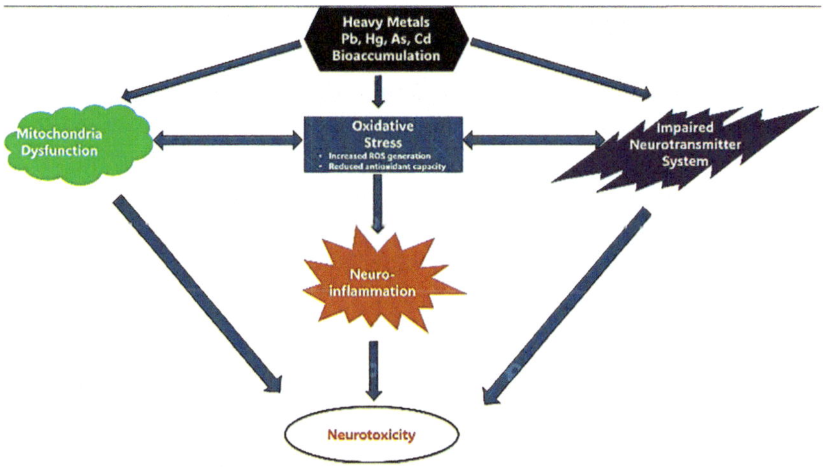

Fig. 3 General principle of Neuroprotective nutrition.

methylation and neurotransmitter synthesis (Zhang et al., 2016), while glutathione and its precursor cysteine enhance detoxification pathways by neutralizing metal-induced reactive intermediates (Flora et al., 2008). Collectively, these nutrients modulate various structural, physiological, and biochemical pathways to buffer the central nervous system against heavy metal-induced injury.

Also, livestocks can be beneficial sources of neuroprotection through the trace elements present within. As discussed by Suttle (2010), about the role of zinc, manganese and copper in rumen metabolism and how they are pivotal for immune condition through mineral nutrition preset in livestock present to meet animals' dietary needs and prevent deficiencies as well as enhance overall health, welfare, productivity, food safety, and reduce environmental impact when consumed through structural, physiological, catalytic, and regulatory benefits (Suttle, 2010).

5.1 Mechanisms of nutritional protection

Various mechanisms of nutritional protection are discussed below, viz a viz reducing oxidative stress, enhancing detoxification pathways, enhancing neurotransmitter synthesis and modulating neuroinflammation (Fig. 3).

5.1.1 Reducing oxidative stress

Increases in levels of ROS/RNS have been linked to several ailments in humans, including cancer, inflammatory illnesses, and neurological problems

(Qin, 2018). Therefore, antioxidant and detoxifying enzymes must function as part of the body's natural defense systems (Adwas et al., 2019). By stopping or slowing down the oxidation of macromolecules antioxidants are chemical compounds that shield the body from oxidative stress (Adwas et al., 2019). In particular, a variety of bioactive compounds that are oxidized themselves suppress oxidation reactions, hence mitigating the toxicity of ROS/RNS.

Several criteria can be used to categorize antioxidants: where they come from (natural and synthetic), how soluble they are (oil- and water-soluble), and how they work. Functionally, they are grouped as primary antioxidants or radical scavengers, secondary antioxidants or peroxide decomposers, and metal deactivators (Soleimani et al., 2018). Antioxidants can be classified as either endogenous or exogenous. The naturally occurring compounds known as endogenous antioxidants work against oxidative stress via many physiological mechanisms. Examples of enzymatic endogenous antioxidants include: Superoxide dismutase, glutathione peroxidase, catalase, and glutathione reductase, while non-enzymatic endogenous antioxidants include: albumin, ceruloplasmin, ferritin, glutathione, metallothionein, thioredoxin and transferrin. Additionally, enzyme co-factors such as coenzyme Q and alpha-lipoic acid, along with metabolites like bilirubin, melatonin, and uric acid, contributes to the body's antioxidant defense system (Horton and Török, 2018). Exogenous antioxidants are substances found in food and can be either synthetic or natural. Examples include synthetic antioxidants such as ethylenediaminetetraacetic acid (EDTA), propyl gallate, butylated hydroxytoluene (BHT), octyl gallate, butylated hydroxyanisole (BHA), and tert–butylhydroquinone. Naturally occurring antioxidants include essential vitamins like vitamins A, E, and C (Neha et al., 2019).

5.1.2 Enhancing detoxification pathways

A metalloprotein called superoxide dismutase (SOD) transfers an electron between superoxide molecules to catalyze the dismutation of superoxide into hydrogen peroxide. In particular, the recipient quickly reacts with two hydrogen ions to form hydrogen peroxide when the donor molecule turns dioxygen (Lumb, 2017). The human body contains three different isoforms of SOD: extracellular-, manganese- and copper/zinc-superoxide dismutase. These isoforms are primarily found in the cytosol and intermembrane space of the mitochondria. Catalase and glutathione peroxidase further convert hydrogen peroxide to water and molecular oxygen (Lumb, 2017).

Three kinds of intracellular antioxidant enzymes exist for catalase: pseudocatalase, manganese-catalase, catalase-peroxidase, and monofunctional catalase, often known as typical catalase (Velayati et al., 2019). Catalase is primarily found in peroxisomes and is a tetrameric hemoprotein according to its structural makeup. In contrast, glutathione peroxidase is an antioxidant enzyme found in the cytoplasm that functions against a broader range of peroxides because it is essential for preserving intracellular homeostasis, redox balance, and the prevention of lipid peroxidation (Lumb, 2017). The oxidized glutathione further oxidizes protein thiol groups after reducing hydrogen peroxide.

Glutathione (GSH) is present in nature and can also be enhanced by antioxidants found in food, such as glutamic acid, cysteine, and glycine. It is produced in the cytoplasm of the organism and is present in most bodily tissues, with the liver having the highest quantities. As the primary thiol-disulfide redox buffer in cells, GSH is essential to the body's detoxification and antioxidant processes (Ağardan and Torchilin, 2016). Glutathione-S-transferase catalyzes the reactions of glutathione, which include conjugation with halides or organic nitrates, aliphatic or aromatic epoxides. These organic nitrates are then subsequently reduced to nitrites. Moreover, GSH contributes to various protective mechanisms such as chelating transition metals, regenerating carotenoids and vitamins C and E, and preserving the redox balance of essential protein sulfhydryl groups needed for repairing DNA. Furthermore, research has demonstrated that the substantia nigra of PD patients has far lower glutathione levels (Simmons et al., 2018). Nuclear factor erythroid 2-related factor 2 (Nrf2) is known as a primary transcription factor in regulating antioxidant responses and maintaining redox homeostasis. Its mode of action involves either increasing the expression of these genes during stress conditions or influencing the basal expression of genes involved in the body's defense against oxidative damage (Paladino et al., 2018).

5.1.3 Supporting neurotransmitter synthesis

Iron plays a crucial role in the brain by supporting the activity of several enzymes involved in neurotransmitter synthesis and the myelination of motor neuron axons. Extracellular ascorbate may affect neurotransmission even if it doesn't function as a traditional neurotransmitter. Specifically, ascorbate affects the release of biogenic amines in the striatum and pituitary neuropeptides, reduces extracellular oxidants to regulate the redox status of catecholamines, and directly affects neural excitability by inhibiting T-type Ca^{2+} channels (Nelson et al., 2007).

5.1.4 Modulating inflammation

Phenolics are known for modulating neuro-inflammation through various mechanisms, including the regulation of toll-like receptor (TLR) pathways, such as TLR4, which play a key role in initiating inflammatory responses in the nervous system. By targeting these pathways, phenolics can reduce the production of pro-inflammatory cytokines and oxidative stress markers, thereby mitigating neuroinflammation and its contribution to neurodegenerative diseases. By reducing the levels of malondialdehyde and 8–hydroxy-2′-deoxyguanosine while enhancing glutathione and SOD activity, lycopene has been shown to alleviate cognitive impairment and oxidative stress. This has further avoided neuroinflammation and apoptosis (Lane and Bailey, 2005). Similar to this, astaxanthin, a xanthophyll carotenoid, has demonstrated neuroprotective effects through various mechanisms. It has been found to prevent hippocampal insulin resistance as well as AD-related complications in Wistar rats. Additionally, it inhibited lipopolysaccharide-induced neuroinflammation, oxidant activity, as well as prevented brain damage in offsprings exposed to prenatal maternal seizures (Shah et al., 2018). These substances have a wide range of characteristics and strong antioxidant activity despite their chemical heterogeneity.

6. Key nutrients and their mechanisms

6.1 Polyphenols

Polyphenols are a diverse class of non-nutritive phytochemicals widely distributed in plant-based foods such as grapes, green tea and berries exhibiting strong neuroprotective and antioxidant properties (Bhullar and Vasantha Rupasinghe, 2013) (Fig. 4). The neurobiological activity of polyphenols has been associated with their bioavailability in the CNS, ability to cross the BBB and reach brain cells, resulting in increased brain plasticity and improved cognition. Although there is limited information concerning their ability to enter the CNS, it has been suggested that polyphenols can accumulate in some tissues, including the brain, in a non-region-specific manner when consumed in low doses, regardless of their route of administration (Szpakowsk et al., 2023).

The neuroprotective effects of polyphenols are mediated through multiple mechanisms. Firstly, they act as metal chelators, binding to neurotoxic metals such as iron, lead, and mercury (Fig. 5). For example, epigallocatechin gallate (EGCG), a prominent polyphenol in green tea, reduces bioavailable metal ions and inhibits pro-oxidant reactions, thereby mitigating iron accumulation in AD (Zwolak, 2021). Secondly, their phenolic hydroxyl groups enable free radical scavenging, neutralizing ROS and preventing oxidative damage to lipids, proteins, and DNA. Thirdly, polyphenols modulate neuroinflammation by suppressing pro-inflammatory cytokines like TNF-α and IL-6 as well as inhibiting signaling pathways like NF-κB (Bhullar and Vasantha Rupasinghe, 2013).

Polyphenols also interact directly with neurotransmitter systems and intracellular signaling cascades. Resveratrol, abundant in grapes, enhances amyloid-beta (Aβ) clearance via an intracellular pathway involving proteasomal Aβ and inhibition of inducible nitric oxide synthase (Bhullar and Vasantha Rupasinghe, 2013). Similarly, grape seed proanthocyanidins extract reduced tau protein deposition by inhibiting phosphorylation, a key pathological event in tauopathies, highlighting their potential in neurodegenerative disease management. By regulating kinase pathways (e.g., MAPK, Akt/PKC), polyphenols further influence neuronal survival, apoptosis, and synaptic plasticity (Kennedy, 2014). Collectively, these mechanisms underscore their dual role as antioxidants and modulators of cellular signaling in neuroprotection.

6.2 Flavonoids

Among polyphenolic compounds, flavonoids are the most extensively studied, diverse and isolated class. To date, over 6000 flavonoids have been identified in plants; such as flavones, flavanols and anthocyanidins. Their core structure comprises of two phenyl rings (A and B) linked through a heterocyclic pyran C ring (C6-C3-C6) with functional groups that dictate their subclass and biological activities (Rodríguez-Arce and Saldías, 2021). Depending on the position 4 of the C ring and a double bind between C2 and C3 and the hydroxyl groups in ring B, flavonoids may be classified into 6 subclasses like flavones, flavanone, flavanols, flavonols, anthocyanidins and isoflavonoids (Mutha et al., 2021). The pharmacological effect of flavonoids is primarily attributed to their capacity to reduce the free radical loads within the biological environment. Additionally, their ability to interact with a variety of molecular targets, including membrane components, cytoplasmic proteins and nuclear factors (Selvaraj and Krishnan, 2021). Flavonoids play a crucial role in defending against oxidative damage. Their phenolic hydroxyl group enables potent free radical scavenging, neutralizing superoxide anions and lipid radicals while inhibiting ROS-generating enzymes like NADH

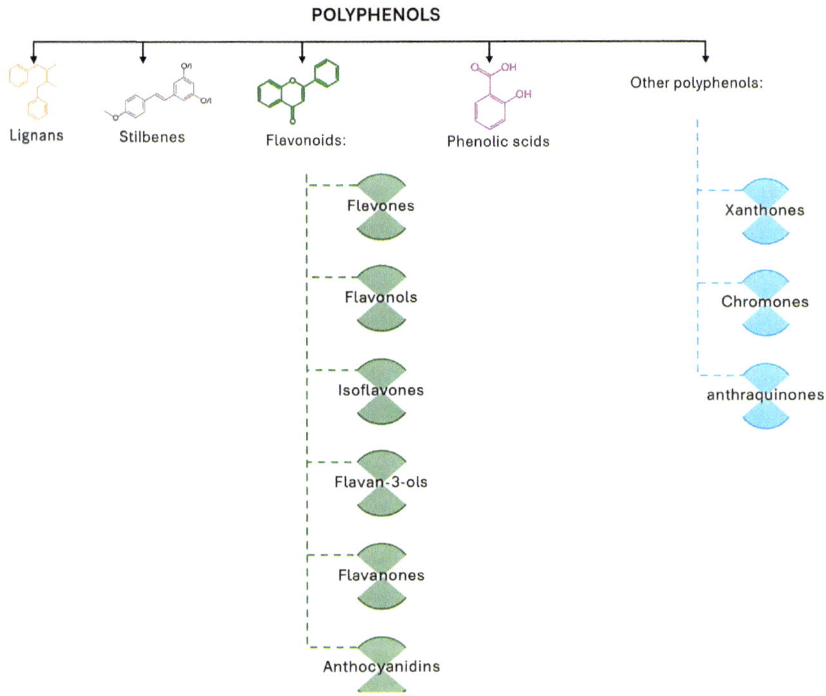

Fig. 4 Classification and chemical structure of major classes of polyphenols. Polyphenols are broadly classified according to the nature of action and structural components, they include phenolic acids, stilbenes, lignans and others including xanthones, chromones, anthraquinones.

oxidase (Mucha et al., 2021). Hesperidin, a flavanone subclass flavonoid, activates the ERK/Nrf2 pathway to enhance antioxidant defenses, reducing cadmium-induced neuronal degeneration and acetylcholinesterase inhibition by 50 percent (Famurewa et al., 2022). Several studies have examined the neuroprotective effect of flavonoids in both in vitro and in vivo models, particularly their role in attenuating neuroinflammation. During neuroinflammatory responses, activated glial cells, specifically microglia, release pro-inflammatory mediators such as cytokines and chemokines, leading to neuronal damage and impaired function. Flavonoids help counteract these effects by interacting with neuronal receptors and modulating gene/protein expression, kinase signaling pathways and transcription factors. This modulation suppresses microglial activation and minimizes the neurotoxicity caused by microglial-derived toxic species (Devi et al., 2021). For instance, apigenin preserves dopaminergic neurons in PD models by enhancing

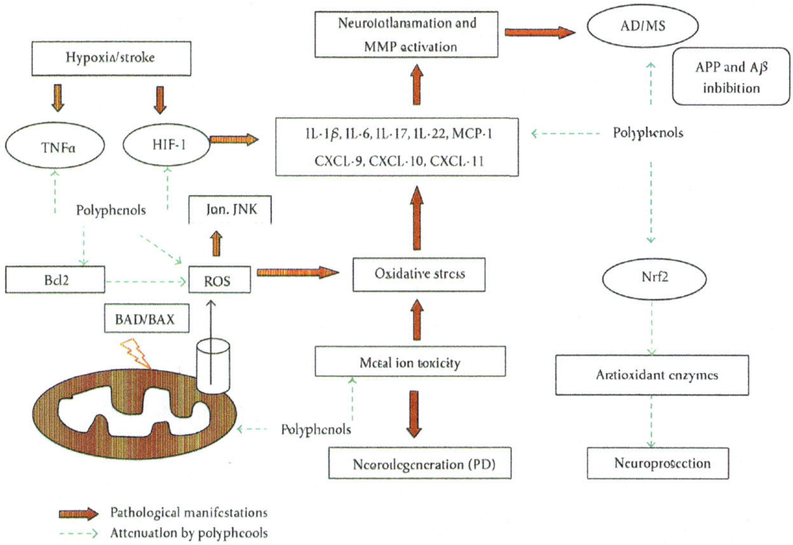

Fig. 5 Mechanistic pathway underlying the neuroprotective role of polyphenols in neurotoxicity. Neurodegenerative disorders are characterized by elevated levels of inflammatory cytokine concentrations such as tumor necrosis factor alpha (TNF-α), chemokines, interleukins, oxidative stress, mitochondrial dysfunction, prions and protein dysfunction. Due to their ability to cross blood-brain barriers, polyphenols may prevent, slow down or ameliorate these disorders by targeting the pathological pathway underlying these manifestations. One of such mechanisms includes the reduction of inflammatory cytokines and chemokines at the blood and mRNA levels, attenuation of oxidative stress through the Nrf2/HIF-1 pathway. As metal chelators, polyphenols can inhibit metal ion toxicity and accumulation in the brain. Additionally, they can prevent apoptosis through the upregulation of antiapoptotic Bcl-2 protein and downregulation of BAD/BAX protein expression (Bhullar and Vasantha Rupasinghe, 2013).

mitochondrial function and reducing inflammation (Anusha et al., 2017). Additionally, rutin downregulates mRNA expression of cytokines, caspase-1, and apoptosis-linked proteins while protecting dopaminergic neurons by suppressing microglial activation, thereby preventing inflammation-induced neurotoxicity (Devi et al., 2021).

6.3 Vitamins

Vitamins are essential micronutrients that play a key role in maintaining overall health. They catalyze nutrient utilization, act as a defense factor against cellular damage and ensure cellular metabolic functions (Guarnieri et al., 2024). Thirteen vitamins are required by humans for optimal physiological processes: vitamin A (retinoids and carotenoids), the

B complex group (B1, B2, B3, B5, B6, B7, B9, B12), vitamin C, vitamin D, vitamin E, and vitamin K (Godswill et al., 2020). These vitamins exert neuroprotective effects through diverse mechanisms, including antioxidant activity, modulation of neurotransmission, and regulation of gene expression. Heavy metal toxicity and oxidative stress increase the body's demand for vitamins, which can mitigate neuroinflammation, reverse oxidative damage, and enhance resilience against neurodegenerative diseases.

6.4 Vitamin C (Ascorbic acid)

A water-soluble antioxidant, vitamin C, is critical for brain function. It scavenges free radicals, reduces extracellular oxidants, and regenerates vitamin E, amplifying its neuroprotective effects (Lee et al., 2020). By clearing synaptic glutamate, it prevents excitotoxicity in stroke, hypoxia, and seizure models. Vitamin C also inhibits T-type Ca^{2+} channels to stabilize neuronal excitability while protecting against ethanol-induced apoptosis in hippocampal neurons (Nelson et al., 2007). Studies highlight its role in mitigating oxidative stress caused by heavy metals. For instance, in *Cyprinus carpio* exposed to metal toxicity, vitamin C significantly reduced accumulated heavy metal levels, demonstrating chelating properties (Sahiti et al., 2019).

6.5 Vitamin E (α-Tocopherol)

This lipid-soluble antioxidant prevents lipid peroxidation in brain cell membranes. Because vitamin E can cross the BBB, it has essential functions in the brain, including promoting neuroglial differentiation, safeguarding the hippocampal neurons, regulating neuroglial inflammatory pathways, and acting as a survival factor for Purkinje neurons. Notably, these protective roles help delay the onset of ataxia in individuals having vitamin E deficiency (Galli et al., 2017). However, vitamin E supplementation reduces ischemia-reperfusion injury and tau hyperphosphorylation in AD models (Yu et al., 2017). Although vitamin E is effective as an antioxidant and neuroprotective agent in neurodegeneration-induced chronic cerebral hypoperfusion, as shown by Sprague Dawley rat studies (Mohamed et al., 2018), its precise effectiveness is yet unknown. Despite prior studies indicating its potential, the prevention of AD by vitamin E and Selenium clinical trial results found that vitamin E alone was not effective in preventing dementia (Kryscio et al., 2017). Nevertheless, co-treatment of vitamin E with other antioxidants may enhance its neuroprotective effects. For example, vitamin C has been shown to regenerate the tocopheroxyl radical back to its reduced state, thereby amplifying vitamin E's antioxidant actions (González-Fuentes et al., 2018).

6.6 Vitamins B12, B6, and folate

Studies involving B12 supplements have demonstrated positive impacts on inflammation and cognitive function. After six months of folate and B12 supplementation, AD patients demonstrated improvement in cognition in a randomized, single-blind placebo study. This improvement was linked to lower serum homocysteine and TNF-α levels. Furthermore, when B12 supplements were administered alongside antipsychotic medications, they significantly reduced psychotic symptoms while increasing anti-inflammatory cytokines and decreasing pro-inflammatory cytokines (Vakilian et al., 2017), suggesting that B12 has an immune-modulatory function. When compared to their unaffected family relatives, patients with familial AD had noticeably lower serum B12 levels in a retrospective study (McCaddon et al., 1994). The most prevalent type of AD, sporadic, is likewise associated with similar outcomes in older people. A population-based longitudinal research in Sweden found that cognitively intact individuals aged 75 and older with lower baseline B12 levels were twice as likely to develop AD within three years compared to those with normal B12 levels (Wang et al., 2001). Similarly, a 10-year follow-up study indicated that subclinical B12 deficiency often precedes cognitive decline in older populations, further supporting the inverse relationship between B12 levels and cognitive function (Clarke et al., 2007). Elevated levels of TNF-α and IL-6, which have been demonstrated to promote Tau hyperphosphorylation and induce Aβ production, have also been linked to B12 deficiency in AD (Shah et al., 2023).

6.7 Vitamin B6 (pyridoxine)

Pyridoxine is commonly found in fruits, vegetables, and grains (or as a supplement). Nutrient metabolism, red blood cell production, maintenance of appropriate homocysteine levels, and support of the immune system and neurological system are all dependent on the biologically active form of this coenzyme, pyridoxal-5-phosphate (Parra et al., 2018). To evaluate whether systemic B6 levels influence cognitive performance in a healthy aging population, Jannusch and colleagues assessed blood vitamin B6 levels in relation to cognitive functions, cortical folding, and functional resting-state connectivity in 794 healthy individuals ages between 55 and 85 from the population-based 1000 BRAINS study (Jannusch et al., 2017). The study discovered no significant correlation between blood B6 levels and neuropsychological test scores or functional connectivity, but rather a strong positive correlation between local cortical folding throughout the brain. These findings shed new light on the potential function of vitamin B6 in maintaining brain structure during aging, since aging is associated with both brain atrophy and the reduction of folds and grooves within the cortical areas.

Vitamin A and its derivatives (retinoids) play a crucial role in neuronal differentiation, brain neurotransmitter expression and gene regulation through interaction with retinoic acid receptors (Guarnieri et al., 2024).

6.8 Minerals

Minerals are essential inorganic elements required by organisms for survival and proper functioning. They cannot be synthesized by living beings and must be obtained through diet or water. Calcium, sodium, magnesium, phosphorus, and potassium are the five most abundant minerals in the human body, while the remaining known as trace elements, include zinc, manganese, molybdenum, iodine, selenium, sulfur, iron, chlorine, cobalt and copper (Godswill et al., 2020). At balanced levels, minerals support neuroprotection by modulating synaptic activity, antioxidant defense, and neuronal plasticity. However, excess accumulation can trigger neurotoxicity, promoting oxidative stress, plaque formation, and neurodegeneration (Guarnieri et al., 2024).

6.8.1 Selenium (Se)

Selenium is an integral part of selenium-containing proteins and amino acids, as well as an enzymatic cofactor (Ellwanger et al., 2016). Se is necessary for healthy brain function and is involved in the pathophysiology of neurological conditions such as epilepsy, ALS, AD, and PD (Solovyev, 2015). Se has been shown through multiple experimental methodologies to have neuroprotective effects in addition to altering brain functioning and different CNS illnesses (Lu et al., 2014). More than half of the selenoproteins have antioxidant properties; the Trx reductase and GPx families are two of the most well-known families. Selenoproteins prevent a cytokine storm and the production of ROS/RNS by inhibiting the activation of NF-κB through redox signaling. Although Se is necessary for brain function, the brain has relatively low levels of Se expression in comparison to other tissues (Santamaría et al., 2005). The mechanisms of Se neuroprotection include inhibiting microglia invasion, promoting the biosynthesis of antioxidative selenoproteins, and controlling Ca2+ influx through ion channels

(Uğuz et al., 2012). Se homeostasis alterations have been noted in the HD model, and selenite supplementation has been shown to have a positive impact on mice that express mutant huntingtin. Furthermore, reduced Se content was found in both human and mouse HD brains (Lu et al., 2014). In the cases of traumatic brain damage, selenium injection in the cortex and hippocampus inhibits subsequent pathogenic events, which helps to reduce apoptotic cell death and avoid neuronal loss (Khalili et al., 2017).

6.8.2 Zinc (Zn)

Although zinc itself is redox inactive and does not directly interact with ROS, it is crucial for preserving the redox state of the antioxidant defense system in the cell in a number of ways. Zinc stabilizes antioxidant enzymes (SOD, GPx) and inhibits pro-oxidant enzymes like NADPH oxidase (Kloubert and Rink, 2015). It modulates synaptic activity by co-releasing with glutamate, influencing learning and memory. In AD models, zinc supplementation enhances mitochondrial function, reduces ROS, and improves cognitive deficits (Farbood et al., 2020). Clinically, zinc alleviates premenstrual symptoms by boosting antioxidant capacity and neurotrophic factors (Jafari et al., 2020). Copper and iron, two redox-active transition metals, compete with zinc for some binding sites. When zinc binds to these sites and forces copper and iron to hydrolytically polymerize into inert structures, blocking the generation of free radicals and the start of lipid peroxidation. In the brain, zinc is primarily expressed in the hippocampus, amygdala, cerebral cortex, thalamus, and olfactory cortex. Presynaptic glutamatergic neurons store zinc as free zinc ions ($Zn2+$). When synaptic activity occurs, zinc is liberated together with glutamate, acting as a strong extracellular modulator by interacting with several synaptic receptors (Kawahara et al., 2018).

6.8.3 Iron (Fe)

Iron can act as an electron donor or acceptor depending on whether it is reduced ($Fe2+$) or oxidized ($Fe3+$) in the cells. In particular, the ferrous form of iron, which produces hydroxyl radicals among other free radical species, can catalyze intensification of the toxicity of oxygen (Farina et al., 2013). Even though iron homeostasis is necessary for the brain's physiological processes, such as the synthesis of neurotransmitters and the myelination of motor neuron axons, the brain holds less than 2 percent of the body's total iron. A significant rise in iron levels in the brain is linked to the pathophysiology of neurodegenerative disorders and is associated with the generation of ROS (Farina et al., 2013; Carocci et al., 2018).

6.8.4 Magnesium (Mg)

Magnesium is necessary for cellular and hormonal signaling and stabilizes protein, nucleic acid, and lipid membranes. It also functions as a cofactor for several enzymes involved in the metabolism of carbohydrates and fats (Shah et al., 2023). Magnesium is essential for the function of neural transmission and has been connected to several neurological and neurodegenerative conditions (Yary and Kauhanen, 2019). Mg plays a crucial role in regulating N-methyl-D-aspartate (NMDA) receptors by blocking calcium channels. However, insufficient Mg levels can lead to increased excitotoxicity, which can cause neurotoxicity and the death of neuronal cells. Therefore, a magnesium deficit may result in decreased glutaminergic neuronal activity, which has been linked to neurological conditions like AD, PD, and epilepsy (Olloquequi et al., 2018). Studies have shown that systemic Mg levels of AD patients are lower than those of healthy controls. A recent meta-analysis confirmed that AD patients exhibit significant magnesium shortages, suggesting supplementation through diet could slow neurodegeneration (Du et al., 2022). Additionally, reduced magnesium levels have been found in key brain regions of AD patients, including the entorhinal cortex, Ammon's horn, and globus pallidus (Andrási et al., 2005).

7. Nutritional interventions across the lifespan: Pregnancy, infancy and childhood

Nutrition plays a vital continuum role across the entire lifespan, from preconception through gestation, birth, childhood, adolescence, adulthood and senescence (Fig. 6). The brain, comprising various anatomic regions and processes, has distinct developmental trajectories that begin from 2 to 3 weeks of fertilization and continue at a brisk pace until at least age 3 (Cusick and Georgieff, 2016). At this stage, known as the "critical period" or "sensitive period," the brain has heightened sensitivity to any perturbation from environmental factors. Maternal factors, including maternal diet and vitamin levels, including fetal factors are two factors that affect neurodevelopment (Hosseini et al., 2024). Perturbation during this sensitive period not only increases the risk of neurodevelopmental disorders after birth but may epigenetically program the brain, influencing susceptibility or resilience to neurological conditions later in life (Li et al., 2019). The Barker hypothesis, which suggests that alterations in fetal nutrition can have lasting effects on a person's health and disease predisposition later in adulthood, highlights the important position that maternal nutrition plays in the utero

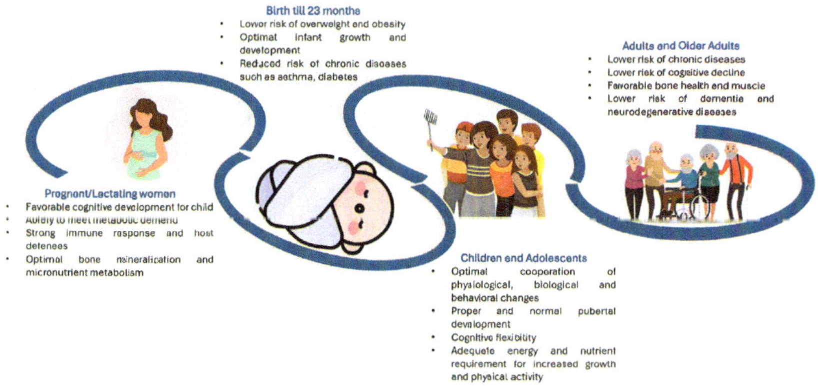

Fig. 6 Impact of optimal nutrition across the lifespan. Adequate maternal and early-life nutrition positively influences neurodevelopment, growth, immune function, and metabolic health from prenatal life through infancy, childhood, adolescence, adulthood, and old age, thereby reducing the risk of chronic diseases, cognitive decline, and neurodegenerative disorders later in life.

environment that programs the fetus for lifelong health (Muglia et al., 2022). Maternal nutritional disorder can have severe consequences on fetal neurodevelopment; thus, optimal pre- and peri–conception interventions are crucial for ensuring a healthy pregnancy and an ideal neurodevelopmental outcome for the developing fetus, which is entirely dependent on its mother for its nutrition (Camila and Paola, 2021).

During gestation, maternal metabolic demands increase significantly, leading to a rise in energy expenditure. This energy is used for new tissue growth, including the development of the fetus, amniotic fluid, the uterus and breast tissue, as well as the storage of fat in both the mother and fetus and overall maintenance of these newly formed tissues (Bowman et al., 2020). According to the WHO global guidance for antenatal care, key nutritional and health interventions recommended for a healthy pregnancy include multiple micronutrient supplementation with iron and folic acid, calcium supplementation to prevent preeclampsia particularly in regions with low calcium intake, and balanced energy and protein supplementation to reduce risk of low birth weight (WHO, 2020). While all nutrients contribute to brain growth and function, certain micronutrients have a particularly significant effect during early development (Cusick and Georgieff, 2016). Intrauterine growth restriction and postnatal growth failure caused by a deficiency in multiple nutrients can significantly affect neurodevelopment.

For example, linear growth rate before (but not after) 12 months of age and infant weight before 4 months of age are strong predictors of child IQ at 9 years. Moreover, intrauterine growth restriction can also alter the epigenetic landscape of the brain, providing a potential mechanism for long-term neurodevelopmental effects (Cusick and Georgieff, 2016). One of such is protein. Adequate protein intake during the prenatal period and early childhood is essential for optimal developmental trajectory. Research from preclinical models of early-life malnutrition demonstrates that protein or protein-energy restriction results in smaller brains with reduced RNA and DNA content, fewer neurons, simpler dendritic and synaptic head architecture, and decreased concentrations of neurotransmitters and growth factors (Pongcharoen et al., 2012). In addition to protein, carbohydrates, lipids, and essential nutrients are other key classes of nutrients. Some studies have linked supplementation with long-chain polyunsaturated fatty acids (LC-PUFAs) during gestation and early postnatal periods to improvements in cognition and attention. LC-PUFAs exert a profound effect on the neurotransmitter system and brain function. They can also modify the epigenetic landscape of the brain, potentially leading to long-term effects (Cusick and Georgieff, 2016). Consumption of a diet with balanced macronutrients before and during pregnancy provides the best chance for a healthy pregnancy and optimal perinatal outcomes. As such, diets that severely restrict any macronutrient class, such as the ketogenic diet that restricts consumption of dairy-based foods but excess saturated fats, should be avoided during and after pregnancy. A low-carbohydrate diet during pregnancy can promote starvation ketosis, increased lipolysis, which can have consequences on the fetus. Additionally, diets characterized by excessive saturated fat intake should also be discouraged (Marshal et al., 2022). For all women of reproductive age, regular supplementation with multivitamins and minerals, containing the optimal amount of folic acid, is recommended at least 2–3 months before conception and continuing through lactation or at least 4–6 weeks postpartum. Deficiencies in these nutritional supplements not only impede appropriate fetal development but can also result in the mother developing health complications and diseases (Santander et al., 2021).

Due to the physiological characteristics of pregnancy, the demand for vitamins substantially increases. For instance, decreased levels of serum retinol are often noticed during the third trimester. Vitamin A deficiency has been linked with adverse outcomes such as premature births, intrauterine growth retardation as well as increased maternal mortality (Ballestín et al., 2021). Apart from its important role in immune functions and host defense, Vitamin

A also plays a major role in cell division, differentiation, proliferation and development and maturation of organs. Vitamins C and E make up a group of antioxidants that act in a complementary and synergistic way to reduce oxidative stress during pregnancy. Oxidative stress has been hypothesized to contribute to endothelial dysfunction, a characteristic feature of pre-eclampsia and premature detachment of the placenta (Santander et al., 2021). The circulation of 1,25 dihydroxycholecalciferol, the biologically active form of vitamin D, in the plasma increases significantly during pregnancy. This vitamin is important for bone mineralization, calcium and phosphorus metabolism in both the mother and the fetus (Dominguez et al., 2010). The impact of vitamin D in maintaining glucose tolerance through insulin secretion and sensitivity has also been suggested (Hu et al., 2018). Deficiency of this vitamin may increase the risk of pregnancy-related complications like pre-eclampsia, gestational diabetes mellitus, caesarean delivery and bacterial vaginosis. Several studies by epidemiologists have highlighted a close link between folic acid deficiency and congenital malformations. Folic acids or folates are important nutrients involved in the synthesis of DNA and proteins, making them particularly important during periods of cell replication, such as pregnancy (Santander et al., 2021). Daily supplementation of folic acid (0.4 mg) for pregnant women is recommended by WHO to prevent maternal anemia, puerperal sepsis, low birth weight and preterm birth.

8. Tailoring nutrition for occupational exposures: A guide for adults

Adulthood, a life stage spanning approximately ages 19–59, is a pivotal stage in life characterized by increased independence and responsibilities. These responsibilities, coupled with limited time and financial resources, can pose significant challenges to maintaining a healthy lifestyle, including dietary habits. Many individuals enter adulthood with unhealthy dietary patterns established during childhood and adolescence. These patterns can contribute to the development of chronic diseases, with over half of adults living with at least one such condition. By adopting healthier dietary habits in adulthood, individuals can improve their overall health and reduce the risk of chronic diseases or slow their progression (DGA, n.d).

Beyond genetic and environmental factors, occupational exposures are also a critical risk factor for chronic diseases in these age groups. Meta-analyses have linked exposure to pesticides, metals, organic solvents and electromagnetic fields in workplaces to an increased risk for neurodegenerative diseases (Gunnarsson and Bodin, 2019). Nutritional status can impact neurodevelopment by modulating the effect of these occupational exposures and affect the body system's response to these environmental factors. Healthy dietary patterns must consider physical activity, body composition, occupation and presence of chronic diseases.

A study by Kachan et al. (2012) highlights the variations in dietary adherence across different occupational groups. It revealed that certain occupations exhibit distinct patterns of nutrient intake. For instance, farmers were more likely to meet the daily recommendation for fiber compared to service workers and blue-collar workers, while white-collar workers were adherent to the daily recommendation of sodium compared with all other occupational groups. These variations highlight the influence of occupational factors on dietary choices and subsequent health outcomes. Research has consistently demonstrated that higher intakes of certain nutrients, such as saturated fat, trans fat, and sodium, are associated with an increased risk of chronic conditions (Oh et al., 2005). Conversely, consuming adequate amounts of fiber, calcium, and vitamin D can contribute to improved health. Addressing these dietary disparities through targeted interventions can significantly reduce these disease risks. By encouraging increased consumption of fiber, calcium, and vitamin D, and decreasing consumption of added sugars, saturated fat, and sodium, individuals can improve overall health and well-being.

9. Conclusion and future directions

The neurotoxic consequences of heavy metals arise from their capacity to disrupt cellular redox equilibrium, triggering oxidative and nitrative stress that underlies a spectrum of neurological impairments. These metals, once introduced into the biological systems, catalyze the overproduction of reactive oxygen species and reactive nitrogen species, overwhelming endogenous antioxidant defenses such as glutathione, superoxide dismutase, and catalase. The resulting oxidative damage to lipids, proteins, and DNA, including the dysregulation of critical signaling pathways, manifests as cognitive deficits, behavioral abnormalities, and progressive neurodegeneration. Notably, developmental stages, prenatal, childhood, and adolescence, are particularly vulnerable, as heavy metal exposure during these periods disrupts neurogenesis, synaptic plasticity, and myelination, leaving indelible marks on lifelong brain health. The body's defense against heavy metal-induced toxicity hinges on both enzymatic and non-enzymatic antioxidant systems. While endogenous mechanisms like GSH synthesis and SOD activity form the first line of defense, exogenous dietary antioxidants; including polyphenols, vitamins, and trace minerals like zinc and selenium, play a pivotal role in scavenging free radicals, chelating toxic metals, and restoring cellular homeostasis. However, chronic exposure depletes these protective reserves, exacerbating oxidative damage and inflammation. Hence, there is a need for integrating nutritional strategies to strengthen antioxidant capacity, particularly in high-risk populations such as pregnant women, children, and aging adults. To translate these insights into actionable solutions, future research must prioritize elucidating the molecular mechanisms by which specific heavy metals modulate oxidative pathways and interact with genetic or epigenetic factors. Advanced studies on the synergistic effects of combinatorial antioxidant therapies, such as pairing polyphenols with essential minerals, could optimize neuroprotection. Additionally, identifying biomarkers for the early detection of oxidative stress and neurodegeneration will enable timely interventions. Finally, public health initiatives must also address environmental heavy metal contamination while promoting diets rich in antioxidants to mitigate exposure risks.

References

Adwas, A., Elsayed, A., Azab, A., 2019. Oxidative stress and antioxidant mechanisms in human body. J. Appl. Biotechnol. Bioeng. 6, 43–47.

Ağardan, N.B.M., Torchilin, V.P., 2016. Engineering of stimuli-sensitive nanopreparations to overcome physiological barriers and cancer multidrug resistance. In: Grumezescu, A.M. (Ed.), Engineering of Nanobiomaterials. William Andrew Publishing, Amsterdam, Netherlands, pp. 1–28.

Akdağ, Z., Erbas, O., 2023. Heavy metal toxicity: a potential risk factor for autism. J. Exp. Basic Med. Sci. 4, 192–196. doi:10.5606/jebms.2023.1065.

Althomali, R.H., Abbood, M.A., Saleh, E.A.M., Djuraeva, L., Abdullaeva, B.S., Habash, R.T., Alhassan, M.S., Alawady, A.H.R., Alsaalamy, A.H., Najafi, M.L., 2024. Exposure to heavy metals and neurocognitive function in adults: a systematic review. Environ. Sci. Eur. 36, 18. doi:10.1186/s12302-024-00843-7.

Andrási, E., Páli, N., Molnár, Z., Kösel, S., 2005. Brain aluminum, magnesium, and phosphorus contents of control and Alzheimer-diseased patients. J. Alzheimers Dis. 7, 273–284.

Anusha, C., Sumathi, T., Joseph, L.D., 2017. Protective role of apigenin on rotenone induced rat model of Parkinson's disease: suppression of neuroinflammation and oxidative stress mediated apoptosis. Chem. Biol. Interact 269, 67–79.

Arruebarrena, M.A., Hawe, C.T., Lee, Y.M., Branco, R.C., 2023. Mechanisms of cadmium neurotoxicity. Int. J. Mol. Sci. 24 (23), 16558. doi:10.3390/ijms242316558.

Asefi, Y., Gohari Mahmoudabad, A., Habibian Sezavar, A., Mirshahvaladi, S., Abyadeh, M., Abyareh, M., 2022. Association between maternal cadmium exposure and preterm birth: a meta-analysis. Int. J. Environ. Health Res. 32, 628–637. doi:10.1080/09603123.2020.1789947.

Awewomom, J., Dzeble, F., Takyi, Y.D., Ashie, W.B., Ettey, E.N.Y.O., Afua, P.E., Sackey, L.N.A., Opoku, F., Akoto, O., 2024. Addressing global environmental pollution using environmental control techniques: a focus on environmental policy and preventive environmental management. Discov. Environ. 2 (1), 8. doi:10.1007/s44274-024-00033-5.

Bakulski, K.M., Seo, Y.A., Hickman, R.C., Brandt, D., Vadari, H.S., Hu, H., Park, S.K., 2020. Heavy metals exposure and Alzheimer's disease and related dementias. J. Alzheimers Dis. 76 (4), 1215–1242. doi:10.3233/JAD-200282.

Ballestín, S.S., Campos, M.I.G., Ballestín, J.B., Bartolomé, M.J.L., 2021. Is supplementation with micronutrients still necessary during pregnancy? A review. Nutrients 13 (9), 3134. doi:10.3390/nu13093134.

Bell, K.S., O'Shaughnessy, K.L., 2022. The development and function of the brain barriers—an overlooked consideration for chemical toxicity. Front. Toxicol. 4, 1000212. doi:10.3389/ftox.2022.1000212.

Bhullar, K.S., Vasantha Rupasinghe, H.P., 2013. Polyphenols: multipotent therapeutic agents in neurodegenerative diseases. Oxidative Med. Cell. Longev. 2013 (1), 891748. doi:10.1155/2013/891748.

Bisen-Hersh, E.B., Farina, M., Barbosa, F., Rocha, J.B., Aschner, M., 2013. Behavioral effects of developmental methylmercury drinking water exposure in rodents. J. Trace Elem. Med. Biol. 28 (2), 117–124. doi:10.1016/j.jtemb.2013.09.008.

Bonilla-Ramirez, L., Jimenez-Del-Rio, M., Velez-Pardo, C., 2011. Acute and chronic metal exposure impairs locomotion activity in Drosophila melanogaster: a model to study Parkinsonism. BioMetals 24 (6), 1045–1057. doi:10.1007/s10534-011-9463-0.

Bowman, C.E., Arany, Z., Wolfgang, M.J., 2020. Regulation of maternal–fetal metabolic communication. Cell. Mol. Life Sci. 78 (4), 1455–1486. doi:10.1007/s00018-020-03674-w.

Camila, M., Paola, D., 2021. Maternal nutrition and neurodevelopment: a scoping review. Nutrients 13 (10), 3530. doi:10.3390/nu13103530.

Carocci, A., Catalano, A., Sinicropi, M.S., Genchi, G., 2018. Oxidative stress and neurodegeneration: the involvement of iron. Biometals 31, 715–735.

Chen, P., Miah, M.R., Aschner, M., 2016. Metals and neurodegeneration. F1000Research 5. doi:10.12688/f1000research.7431.1, F1000 Faculty Rev-366.

Choong, G., Liu, Y., Templeton, D.M., 2014. Interplay of calcium and cadmium in mediating cadmium toxicity. Chem. Biol. Interact. 211, 54–65. doi:10.1016/j.cbi.2014.01.007.

Chung, A., Jou, C., Grau-Perales, A., Levy, E.R.J., Dvorak, D., Hussain, N., Fenton, A.A., 2021. Cognitive control persistently enhances hippocampal information processing. Nature 600 (7889), 484–488. doi:10.1038/s41586-021-04070-5.

Clarke, R., Birks, J., Nexo, E., Ueland, P.M., Schneede, J., Scott, J., Molloy, A., Evans, J.G., 2007. Low vitamin B-12 status and risk of cognitive decline in older adults. Am. J. Clin. Nutr. 86, 1384–1391.

Cusick, S.E., Georgieff, M.K., 2016. The role of nutrition in brain development: the golden opportunity of the "First 1000 days". J. Pediatr. 175, 16–21. doi:10.1016/j.jpeds.2016.05.013.

Devi, S., Kumar, V., Singh, S.K., Dubey, A.K., Kim, J., 2021. Flavonoids: potential candidates for the treatment of neurodegenerative disorders. Biomedicines 9 (2), 99. doi:10.3390/biomedicines9020099.

DGA: Dietary guidelines for Americans, 2020-2025. n.d. https://www.dietaryguidelines.gov/sites/default/files/202103/Dietary_Guidelines_for_Americans-2020-2025.pdf. [accessed August 2024]

Dinçkol, Ö., Fuentes, B., Tartaglione, A.M., Pino, A., Calamandrei, G., Ricceri, L., 2022. Low-level lead exposure during development differentially affects neurobehavioral responses in male and female mouse offspring: a longitudinal study. Neurotoxicology 91, 188–199. doi:10.1016/j.neuro.2022.05.007.

Ding, A.L., Hu, H., Xu, F.P., Liu, L.Y., Peng, J., Dong, X.D., 2021. Pregnancy complications effect on the nickel content in maternal blood, placenta blood and umbilical cord blood during pregnancy. World J. Clin. Cases 9, 8340–8348. doi:10.12998/wjcc.v9.i28.8340.

Domínguez, M.C., Martínez, M.C., Méndez, J.I., Rodríguez, M.J., 2010. Supplements in pregnant women: controversies, evidence and recommendations. Nat. Health Syst 34, 1–12. https://www.msssi.gob.es/biblioPublic/publicaciones/recursos_propios/infMedic/docs/vol34n4_Suplementos.pdf.

Du, K., Zheng, X., Ma, Z.-T., Lv, J.Y., Jiang, W.J., Liu, M.Y., 2022. Association of circulating magnesium levels in patients with Alzheimer's disease from 1991 to 2021: a systematic review and meta-analysis. Front. Aging Neurosci. 13, 958.

Dubovický, M., 2010. Neurobehavioral manifestations of developmental impairment of the brain. Interdiscip. Toxicol. 3 (2), 59–67. doi:10.2478/v10102-010-0012-4.

Ellwanger, J.H., Franke, S.I.R., Bordin, D.L., Prá, D., Henriques, J.A.P., 2016. Biological functions of selenium and its potential influence on Parkinson's disease. An. Acad. Bras. Ciênc. 88 (3), 1655–1674.

Escudero-Lourdes, C., 2016. Toxicity mechanisms of arsenic that are shared with neurodegenerative diseases and cognitive impairment: role of oxidative stress and inflammatory responses. NeuroToxicology 53, 223–235. doi:10.1016/j.neuro.2016.02.002.

Ewool, J., Blankson, E.R., Quartey, J.K., Kyerematen, R., Gbogbo, F., 2024. Heavy metal concentrations in drinking water sources in two mining districts in Ghana. Heliyon 10, e33296. doi:10.1016/j.heliyon.2024.e33296.

Famurewa, A.C., Renu, K., Eladl, M.A., Chakraborty, R., Myakala, H., El-Sherbiny, M., Elsherbini, D.M.A., Vellingiri, B., Madhyastha, H., Ramesh Wanjari, U., Goutam Mukherjee, A., Valsala Gopalakrishnan, A., 2022. Hesperidin and hesperetin against heavy metal toxicity: insight on the molecular mechanism of mitigation. Biomed. Pharmacother. 149, 112914. doi:10.1016/j.biopha.2022.112914.

Farbood, Y., Sarkaki, A., Mahdavinia, M., Ghadiri, A., Teimoori, A., Seif, F., Dehghani, M.A., Navabi, S.P., 2020. Protective effects of co-administration of zinc and selenium against streptozotocin-induced Alzheimer's disease: behavioral, mitochondrial oxidative stress, and GPR39 expression alterations in rats. Neurotoxicol. Res. 38, 398–407.

Farina, M., Avila, D.S., Da Rocha, J.B.T., Aschner, M., 2013. Metals, oxidative stress and neurodegeneration: a focus on iron, manganese and mercury. Neurochem. Int. 62 (5), 575–594. doi:10.1016/j.neuint.2012.12.006.

Farmani, R., Mehrpour, O., Kooshki, A., Nakhaee, S., 2024. Exploring the link between toxic metal exposure and ADHD: a systematic review of pb and hg. J. Neurodev. Disord. 16, 44. doi:10.1186/s11689-024-09555-8.

Flora, S.J.S., Flora, G., Saxena, G., 2012. Environmental occurrence, health effects and management of lead poisoning. In: Flora, S.J.S. (Ed.), Lead: Chemistry, Analytical Aspects, Environmental Impact and Health Effects. Elsevier, Amsterdam, Netherlands and Boston, Massachusetts, USA, pp. 158–228.

Flora, S.J.S., Mittal, M., Mehta, A., 2008. Heavy metal-induced oxidative stress & its possible reversal by chelation therapy. Indian J. Med. Res. 128 (4), 501–523.

Galli, F., Azzi, A., Birringer, M., Cook-Mills, J.M., Eggersdorfer, M., Frank, J., Cruciani, G., Lorkowski, S., Özer, N.K., 2017. Vitamin E: emerging aspects and new directions. Free Radic. Biol. Med. 102, 16–36.

Gąssowska-Dobrowolska, M., Chlubek, M., Kolasa, A., Tomasiak, P., Korbecki, J., Skowrońska, K., Tarnowski, M., Masztalewicz, M., Baranowska-Bosiacka, I., 2023. Microglia and Astroglia—the potential role in neuroinflammation induced by pre- and neonatal exposure to lead (PB). Int. J. Mol. Sci. 24 (12), 9903. doi:10.3390/ijms24129903.

Godswill, A.G., Somtochukwu, I.V., Ikechukwu, A.O., Kate, E.C., 2020. Health benefits of micronutrients (vitamins and minerals) and their associated deficiency diseases: a systematic review. Int. J. Food Sci. 3 (1), 1–32.

González-Fuentes, J., Selva, J., Moya, C., Castro-Vázquez, L., Lozano, M.V., Marcos, P., Plaza--Oliver, M., Rodríguez-Robledo, V., Santander-Ortega, M.J., Villaseca-González, N., 2018. Neuroprotective natural molecules: from food to brain. Front. Neurosci. 12, 721.

Goyer, R.A., 1997. Toxic and essential metal interactions. Annu. Rev. Nutr. 17 (1), 37–50.

Gu, H., Liu, L.L., Wu, A., Yu, Y., Emir, U., Sawiak, S.J., Territo, P.R., Farlow, M.R., Zheng, W., Du, Y., 2024. Lead acetate exposure and cerebral amyloid accumulation: mechanistic evaluations in APP/PS1 Mice. Env. Health Perspect. 132 (10), 107004. doi:10.1289/EHP14384.

Guarnieri, L., Bosco, F., Leo, A., Citraro, R., Palma, E., De Sarro, G., Mollace, V., 2024. Impact of micronutrients and nutraceuticals on cognitive function and performance in Alzheimer's disease. Ageing Res. Rev. 95, 102210. doi:10.1016/j.arr.2024.102210.

Gunnarsson, L., Bodin, L., 2019. Occupational exposures and neurodegenerative diseases—a systematic literature review and meta-analyses. Int. J. Environ. Res. Public Health 16 (3), 337. doi:10.3390/ijerph16030337.

Habibian, A., Abyadeh, M., Abyareh, M., Rahimi Kakavandi, N., Habibian, A., Khakpash, M., Ghazi-Khansari, M., 2022. Association of maternal lead exposure with the risk of preterm: a meta-analysis. J. Matern. Fetal Neonatal Med. 35, 7222–7230. doi:10.1080/14767058.2021.1946780.

Hernández-Cruz, E.Y., Arancibia-Hernández, Y.L., Loyola-Mondragón, D.Y., Pedraza-Chaverri, J., 2022. Oxidative stress and its role in CD-induced epigenetic modifications: use of antioxidants as a possible preventive strategy. Oxygen 2 (2), 177–210. doi:10.3390/oxygen2020015.

Hlatshwayo, S.S.B., Bissessur, A., Selala, M.C., Takai, Y., Lebepe, J., 2024. Metal distribution in three organs and edibility assessment on Coptodon rendalli from the Umgeni River impacted by metallurgic industrial activities. Environ. Monit. Assess. 196 (8), 742. doi:10.1007/s10661-024-12875-w.

Horton, W., Török, M., 2018. Natural and nature-inspired synthetic small molecule antioxidants in the context of green chemistry. In: Török, B., Dransfield, T. (Eds.), Green Chemistry. Elsevier, Amsterdam, Netherlands, pp. 963–979.

Hosseini, S.M., Panahi-Azar, A., Sheybani-Arani, M., Morovatshoar, R., Mirzadeh, M., Asl, A.S., Mirsadeghi, M.N., Khajavi-Mayvan, F., 2024. Vitamins, minerals and their maternal levels' role in brain development: an updated literature-review. Clin. Nutr. ESPEN 63, 31–45. doi:10.1016/j.clnesp.2024.05.011.

Hu, L., Zhang, Y., Wang, X., You, L., Xu, P., Cui, X., Zhu, L., Ji, C., Guo, X., Wen, J., 2018. Maternal vitamin D status and risk of gestational diabetes: a meta-analysis. Cell. Physiol. Biochem. 45 (1), 291–300. doi:10.1159/000486810.

Jadhav, K., Raghuraj, R.E., Vaghasiya, K., Verma, R.K., 2022. Neurodegenerative disorders due to inhalation of various small particles. Nanomed. Drug Deliv. Neurodegener. Dis. 41–54. doi:10.1016/B978-0-323-85544-0.000101.

Jafari, F., Amani, R., Tarrahi, M.J., 2020. Effect of zinc supplementation on physical and psychological symptoms, biomarkers of inflammation, oxidative stress, and brain-derived neurotrophic factor in young women with premenstrual syndrome: a randomized, double-blind, placebo-controlled trial. Biol. Trace Elem. Res. 194, 89–95.

Jaishankar, M., Tseten, T., Anbalagan, N., Mathew, B.B., Beeregowda, K.N., 2014. Toxicity, mechanism and health effects of some heavy metals. In: Interdisciplinary Toxicology, 7. Slovak Toxicology Society, pp. 60–72. doi:10.2478/intox-2014-0009.

Jannusch, K., Jockwitz, C., Bidmon, H.J., Moebus, S., Amunts, K., Caspers, S., 2017. A complex interplay of vitamin B1 and B6 metabolism with cognition, brain structure, and functional connectivity in older adults. Front. Neurosci. 11, 596.

Kachan, D., Lewis, J.E., Davila, E.P., Arheart, K.L., LeBlanc, W.G., Fleming, L.E., Cabán-Martinez, A.J., Lee, D.J., 2012. Nutrient intake and adherence to dietary recommendations among US workers. J. Occup. Environ. Med. 54 (1), 101–105. doi:10.1097/jom.0b013e31823ccafa.

Kang, B., Wang, J., Guo, S., Yang, L., 2024. Mercury-induced toxicity: mechanisms, molecular pathways, and gene regulation. Sci. Total Environ. 943, 173577. doi:10.1016/j.scitotenv.2024.173577.

Kaur, N., Kumar, D., Mehta, D.S., 2014. Role of micronutrients in ameliorating heavy metal toxicity. Int. J. Adv. Pharm. Biol. Chem. 3 (2), 268–273.

Kawahara, M., Tanaka, K., Kato-Negishi, M., 2018. Zinc, carnosine, and neurodegenerative diseases. Nutrients 10 (2), 147.

Kennedy, D.O., 2014. Polyphenols and the human brain: plant "secondary metabolite" ecologic roles and endogenous signaling functions drive benefits. Adv. Nutr. 5 (5), 515–533. http://www.ncbi.nlm.nih.gov/pubmed/25469384.

Khalili, H., Ahl, R., Cao, Y., Paydar, S., Sjölin, G., Niakan, A., Dabiri, G., Mohseni, S., 2017. Early selenium treatment for traumatic brain injury: does it improve survival and functional outcome? Injury 48, 1922–1926.

Kinjo, Y., Shibata, E., Askew, D.J., Tanaka, R., Suga, R., Shimono, M., Sakuragi, T., Morokuma, S., Ogawa, M., Sanefuji, M., Hamada, N., Ochiai, M., Ohga, S., Tsuji, M., Kusuhara, K., Yoshino, K., 2024. Association of placental weight at birth with maternal whole blood concentration of heavy metals (cadmium, lead, mercury, selenium, and manganese): the Japan Environment and Children's Study (JECS). Environ. Int. 188, 108725. doi:10.1016/j.envint.2024.108725.

Kloubert, V., Rink, L., 2015. Zinc as a micronutrient and its preventive role in oxidative damage in cells. Food Funct. 6, 3195–3204.

Kothapalli, C.R., 2021. Differential impact of heavy metals on neurotoxicity during development and in aging central nervous system. Curr. Opin. Toxicol. 26, 33–38. doi:10.1016/j.cotox.2021.04.003.

Kou, X., Millán, M.P., Canals, J., Moreno, V.R., Renzetti, S., Arija, V., 2025. Effects of prenatal exposure to multiple heavy metals on infant neurodevelopment: a multi-statistical approach. Environ. Pollut. 367, 125647. doi:10.1016/j.envpol.2025.125647.

Koyama, H., Kamogashira, T., Yamasoba, T., 2024. Heavy metal exposure: molecular pathways, clinical implications, and protective strategies. Antioxidants 13 (1). doi:10.3390/antiox13010076.

Kryscio, R.J., Abner, E.L., Caban-Holt, A., Lovell, M., Goodman, P., Darke, A.K., Yee, M., Crowley, J., Schmitt, F.A., 2017. Association of antioxidant supplement use and dementia in the prevention of Alzheimer's disease by vitamin E and selenium trial (PREADVISE). JAMA Neurol. 74, 567–573.

Kumar, A., Siddiqi, N.J., Alrashood, S.T., Khan, H.A., Dubey, A., Sharma, B., 2021. Protective effect of eugenol on hepatic inflammation and oxidative stress induced by cadmium in male rats. Biomed. Pharmacother. 139, 111588. doi:10.1016/j.biopha.2021.111588.

Lane, M.A., Bailey, S.J., 2005. Role of retinoid signaling in the adult brain. Prog. Neurobiol. 75, 275–293.

Lee, K.H., Cha, M., Lee, B.H., 2020. Neuroprotective effect of antioxidants in the brain. Int. J. Mol. Sci. 21 (19), 7152.

Li, M., Francis, E., Hinkle, S.N., Ajjarapu, A.S., Zhang, C., 2019. Preconception and prenatal nutrition and neurodevelopmental disorders: a systematic review and meta-analysis. Nutrients 11 (7), 1628. doi:10.3390/nu11071628.

Liu, J., Zeng, L., Zhuang, S., Zhang, C., Li, Y., Zhu, J., Zhang, W., 2020. Cadmium exposure during prenatal development causes progesterone disruptors in multiple generations via steroidogenic enzymes in rat ovarian granulosa cells. Ecotoxicol. Env. Saf. 201, 110765. doi:10.1016/j.ecoenv.2020.110765.

Lobo, V., Patil, A., Phatak, A., Chandra, N., 2010. Free radicals, antioxidants and functional foods: impact on human health. Pharmacogn. Rev. 4, 118.

Lu, Z., Marks, E., Chen, J., Moline, J., Barrows, L., Raisbeck, M., Volitakis, I., Cherny, R.A., Chopra, V., Bush, A.I., 2014. Altered selenium status in Huntington's disease: neuroprotection by selenite in the N171-82Q mouse model. Neurobiol. Dis. 71, 34–42.

Lumb, A.B., 2017. Oxygen toxicity and hyperoxia. In: Nunn's Applied Respiratory Physiology. Elsevier, Edinburgh, New York, pp. 341–356.e342.

Marshall, N.E., Abrams, B., Barbour, L.A., Catalano, P., Christian, P., Friedman, J.E., Hay, W.W., Hernandez, T.L., Krebs, N.F., Oken, E., Purnell, J.Q., Roberts, J.M., Soltani, H., Wallace, J., Thornburg, K.L., 2022. The importance of nutrition in pregnancy and lactation: lifelong consequences. Am. J. Obstet. Gynecol. 226 (5), 607–632. doi:10.1016/j.ajog.2021.12.035.

Matuz-Mares, D., González-Andrade, M., Araiza-Villanueva, M.G., Vilchis-Landeros, M.M., Vázquez-Meza, H., 2022. Mitochondrial calcium: effects of its imbalance in disease. Antioxidants 11 (5), 801. doi:10.3390/antiox11050801.

McCaddon, A., Kelly, C.L., Tibblin, G., 1994. Familial Alzheimer's disease and vitamin B12 deficiency. Age Ageing 23, 334–337.

Milanković, V., Tasić, T., Leskovac, A., Petrović, S., Mitić, M., Lazarević-Pašti, T., Novković, M., Potkonjak, N., 2024. Metals on the menu-analyzing the presence, importance, and consequences. Foods 13, 1890. doi:10.3390/foods13121890.

Misrani, A., Tabassum, S., Yang, L., 2021. Mitochondrial dysfunction and oxidative stress in Alzheimer's disease. Front. Aging Neurosci. 13, 617588. doi:10.3389/fnagi.2021.617588.

Mohamed, W., Sayeed, S., Saxena, A., Oothuman, P., 2018. Oxidative stress status and neuroprotection of tocotrienols in chronic cerebral hypoperfusion-induced neurodegeneration rat animal model. Int. J. Nutr. Pharmacol. Neurol. Dis. 8, 47–52.

Mucha, P., Skoczyńska, A., Małecka, M., Hikisz, P., Budzisz, E., 2021. Overview of the antioxidant and anti-inflammatory activities of selected plant compounds and their metal ions complexes. Molecules 26 (16), 4886. doi:10.3390/molecules26164886.

Muglia, L.J., Benhalima, K., Tong, S., Ozanne, S., 2022. Maternal factors during pregnancy influencing maternal, fetal, and childhood outcomes. BMC Med 20, 418. doi:10.1186/s12916-022-02632-6.

Mutha, R.E., Tatiya, A.U., Surana, S.J., 2021. Flavonoids as natural phenolic compounds and their role in therapeutics: an overview. Future J. Pharm. Sci. 7 (1). doi:10.1186/s43094-020-00161-8.

Nabi, M., Tabassum, N., 2022. Role of environmental toxicants on neurodegenerative disorders. Front. Toxicol. 4, 837579. doi:10.3389/ftox.2022.837579.

National Research Council (US) Committee on the Toxicological Effects of Methylmercury, 2000. Toxicological Effects of Methylmercury. National Academies Press (US), Washington (DC):.

Neal, A.P., Guilarte, T.R., 2013. Mechanisms of lead and manganese neurotoxicity. Toxicol. Res. 2 (2), 99. doi:10.1039/C2TX20064C.

Neha, K., Haider, M.R., Pathak, A., Yar, M.S., 2019. Medicinal prospects of antioxidants: a review. Eur. J. Med. Chem. 178, 687–704.

Nelson, M.T., Joksovic, P.M., Su, P., Kang, H.W., Van Deusen, A., Baumgart, J.P., David, L.S., Snutch, T.P., Barrett, P.Q., Lee, J.H., 2007. Molecular mechanisms of subtype-specific inhibition of neuronal T-type calcium channels by ascorbate. J. Neurosci. 27, 12577–12583.

Oria, Rademene S., Godson E. Anyanwu, Emmanuel A. Esom, Johnson N. Nto, Amechi U. Katchy, Augustine U. Agu, and Omamuyovwi M. Ijomone. "Modulatory role of curcumin on cobalt-induced memory deficit, hippocampal oxidative damage, astrocytosis, and Nrf2 expression." *Neurotoxicity Research* 41, no. 3 (2023): 201-211.

Novo, J.P., Martins, B., Raposo, R.S., Pereira, F.C., Oriá, R.B., Malva, J.O., Fontes-Ribeiro, C., 2021. Cellular and molecular mechanisms mediating methylmercury neurotoxicity and neuroinflammation. Int. J. Mol. Sci. 22 (6), 3101. doi:10.3390/ijms22063101.

Ogbeide, O., Henry, B., 2024. Addressing heavy metal pollution in Nigeria: evaluating policies, assessing impacts, and enhancing remediation strategies. J. Appl. Sci. Environ. Manag. 28 (4), 1007–1051. doi:10.4314/jasem.v28i4.5.

Oh, K., Hu, F.B., Manson, J.E., Stampfer, M.J., Willett, W.C., 2005. Dietary fat intake and risk of coronary heart disease in women: 20 years of follow-up of the nurses' health study. Am. J. Epidemiol. 161 (7), 672–679. doi:10.1093/aje/kwi085.

Ojo, O.A., Rotimi, D.E., Ojo, A.B., Ogunlakin, A.D., Ajiboye, B.O., 2023. Gallic acid abates cadmium chloride toxicity via alteration of neurotransmitters and modulation of inflammatory markers in Wistar rats. Sci. Rep. 13 (1). doi:10.1038/s41598-023-28893-6.

Oken, E., Wright, R.O., Kleinman, K.P., 2005. Maternal fish consumption, hair mercury, and infant cognition in a U.S. cohort. Environ. Health Perspect. 113 (10), 1376–1380.

Olloquequi, J., Cornejo-Córdova, E., Verdaguer, E., Soriano, F.X., Binvignat, O., Auladell, C., Camins, A., 2018. Excitotoxicity in the pathogenesis of neurological and psychiatric disorders: therapeutic implications. J. Psychopharmacol. 32, 265–275.

Oria, R.S., Anyanwu, G.E., Esom, E.A., Nto, J.N., Katchy, A.U., Agu, A.U., Ijomone, O.M., 2023. Modulatory role of curcumin on cobalt-induced memory deficit, hippocampal oxidative damage, astrocytosis, and Nrf2 expression. Neurotox. Res. 41 (3), 201–211. doi:10.1007/s12640-023-00635-6.

Paithankar, J.G., Saini, S., Dwivedi, S., Sharma, A., Chowdhuri, D.K., 2021. Heavy metal associated health hazards: an interplay of oxidative stress and signal transduction. Chemosphere 262, 128350. doi:10.1016/j.chemosphere.2020.128350.

Paladino, S., Conte, A., Caggiano, R., Pierantoni, G.M., Faraonio, R., 2018. Nrf2 pathway in age-related neurological disorders: insights into microRNAs. Cell Physiol. Biochem. 47, 1951–1976.

Parida, L., Patel, T.N., 2023. Systemic impact of heavy metals and their role in cancer development: a review. Environ. Monit. Assess. 195 (6), 766. doi:10.1007/s10661-023-11399.

Parithathvi, A., Choudhari, N., Dsouza, H.S., 2024. Prenatal and early life lead exposure induced neurotoxicity. Hum. Exp. Toxicol. 43. doi:10.1177/09603271241285523.

Parra, M., Stahl, S., Hellmann, H., 2018. Vitamin B6 and its role in cell metabolism and physiology. Cells 7, 84.

Peng, X., Li, C., Zhao, D., Huang, L., 2023. Associations of micronutrients exposure with cadmium body burden among population: a systematic review. Ecotoxicol. Environ. Saf. 256, 114878.

Petrosino, V., Motta, G., Tenore, G., Coletta, M., Guariglia, A., Testa, D., 2018. The role of heavy metals and polychlorinated biphenyls (PCBs) in the oncogenesis of head and neck tumors and thyroid diseases: a pilot study. Biometals 31, 285–295. doi:10.1007/s10534-018-0091-9.

Pongcharoen, T., Ramakrishnan, U., DiGirolamo, A.M., Winichagoon, P., Flores, R., Singkhornard, J., Martorell, R., 2012. Influence of prenatal and postnatal growth on intellectual functioning in school-aged children. Arch. Pediatr. Adolesc. Med. 166 (5), 411–416. doi:10.1001/archpediatrics.2011.1413.

Pöpplau, J.A., Schwarze, T., Dorofeikova, M., Pochinok, I., Günther, A., Marquardt, A., Hanganu-Opatz, I.L., 2024. Reorganization of adolescent prefrontal cortex circuitry is required for mouse cognitive maturation. Neuron 112, 421–440.e7. doi:10.1016/j.neuron.2023.10.024.

Pyatha, S., Kim, H., Lee, D., Kim, K., 2022. Association between heavy metal exposure and Parkinson's disease: a review of the mechanisms related to oxidative stress. Antioxidants 11 (12), 2467. doi:10.3390/antiox11122467.

Qin, Y., 2018. Health benefits of bioactive seaweed substances. In: Bioactive Seaweeds for Food Applications. Academic Press, Kidlington, Oxford, United Kingdom San Diego, California, United States Cambridge, Massachusetts, United States London, United Kingdom, pp. 179–200.

Rai, A., Maurya, S.K., Khare, P., Srivastava, A., Bandyopadhyay, S., 2010. Characterization of developmental neurotoxicity of AS, CD, and PB mixture: synergistic action of metal mixture in glial and neuronal functions. Toxicol. Sci. 118 (2), 586–601. doi:10.1093/toxsci/kfq266.

Rauh, V.A., Margolis, A., 2016. Research review: environmental exposures, neurodevelopment and child mental health—new paradigms for the study of brain and behavioral effects. J. Child Psychol. Psychiatry Allied Discip. 57 (7), 775–793. doi:10.1111/jcpp.12537.

Rayman, M.P., 2012. Selenium and human health. Lancet 379 (9822), 1256–1268.

Rehman, Q., Rehman, K., Akash, M.S.H., 2021. Heavy metals and neurological disorders: from exposure to preventive interventions. In: Emerging Contaminants and Associated Treatment Technologies, pp. 69–87. doi:10.1007/978-3-030-66376-6_4.

Rock, K.D., Patisaul, H.B., 2018. Environmental mechanisms of neurodevelopmental toxicity. Curr Env. Health Rep. 5, 145–157. doi:10.1007/s40572-018-0185-0.

Rodríguez-Arce, E., Saldías, M., 2021. Antioxidant properties of flavonoid metal complexes and their potential inclusion in the development of novel strategies for the treatment against neurodegenerative diseases. Biomed. Pharmacother. 143, 112236. doi:10.1016/j.biopha.2021.112236.

Rodríguez-Barranco, M., Lacasaña, M., Aguilar-Garduño, C., Alguacil, J., Gil, F., González-Alzaga, B., Rojas-García, A., 2013. Association of arsenic, cadmium and manganese exposure with neurodevelopment and behavioural disorders in children: a systematic review and meta-analysis. Sci. Total Environ. 454–455, 562–577. doi:10.1016/j.scitotenv.2013.03.047.

Rumboldt, Z., Vavro, H., Špero, M., 2019. Exogenous toxins and CNS injuries: imaging techniques and diagnosis. In: Barkhof, F., Jäger, H.R., Thurnher, M.M., Rovira, À. (Eds.), Clinical Neuroradiology. Springer International Publishing, Cham, pp. 1365–1402. doi:10.1007/978-3-319-68536-6_66.

Sahiti, H., Bislimi, K., Rexhepi, A., Dalo, E., 2019. Metal accumulation and effect of vitamin C and E in accumulated heavy metals in different tissues in common carp (*Cyprinus carpio*) treated with heavy metals. Pol. J. Environ. Stud. 29 (1), 799–805. doi:10.15244/pjoes/103354.

Sánchez-Martín, F.J., Fan, Y., Lindquist, D.M., Xia, Y., Puga, A., 2013. Lead induces similar gene expression changes in brains of gestationally exposed adult mice and in neurons differentiated from mouse embryonic stem cells. PLoS One 8 (11), e80558. doi:10.1371/journal.pone.0080558.

Sanders, T., Liu, Y., Buchner, V., Tchounwou, P., 2009. Neurotoxic effects and biomarkers of lead exposure: a review. Rev. Environ. Health 24 (1). doi:10.1515/reveh.2009.24.1.15.

Santamaría, A., Vázquez-Román, B., La Cruz, V.P.D., González-Cortés, C., Trejo-Solís, M.C., Galván-Arzate, S., Jara-Prado, A., Guevara-Fonseca, J., Ali, S.F., 2005. Selenium reduces the proapoptotic signaling associated with NF-κB pathway and stimulates glutathione peroxidase activity during excitotoxic damage produced by quinolinate in rat corpus striatum. Synapse 58, 258–266.

Santander, B.S., Campos, M.I.G., Ballestín, J.B., Bartolomé, M.J.L., 2021. Is supplementation with micronutrients still necessary during pregnancy? A review. Nutrients 13 (9), 3134. doi:10.3390/nu13093134.

Selvaraj, S., Krishnan, U.M., 2021. Vanadium–Flavonoid complexes: a promising class of molecules for therapeutic applications. J. Med. Chem. 64 (17), 12435–12452. doi:10.1021/acs.jmedchem.1c00405.

Shah, H., Dehghani, F., Ramezan, M., Gannaban, R.B., Haque, Z.F., Rahimi, F., Abbasi, S., Shin, A.C., 2023. Revisiting the role of vitamins and minerals in Alzheimer's disease. Antioxidants 12, 415.

Shah, R., Shchepinov, M.S., Pratt, D.A., 2018. Resolving the role of lipoxygenases in the initiation and execution of ferroptosis. ACS Cent. Sci. 4, 387–396.

Simmons, A.D., 2018. Parkinson's disease. In: Rakel, D. (Ed.), Integrative Medicine. Elsevier, Philadelphia, PA, pp. 143–151.e143.
Singh, L., Anand, M., Singh, S., Taneja, A., 2020. Environmental toxic metals in placenta and their effects on preterm delivery-current opinion. Drug Chem. Toxicol. 43, 531–538. doi:10.1080/01480545.2018.1515216.
Singh, M., Murthy, V., Ramassamy, C., 2013. Beneficial effects of curcumin against heavy metal-induced neurotoxicity. Adv. Neurobiol. 6, 449–472.
Smith, J.A., Das, A., Ray, S.K., Banik, N.L., 2012. Role of pro-inflammatory cytokines released from microglia in neurodegenerative diseases. Brain Res. Bull. 87 (1), 10–20. doi:10.1016/j.brainresbull.2011.10.004N.
Soleimani, M., Dehabadi, L., Wilson, L.D., Tabil, L.G., 2018. Antioxidants classification and applications in lubricants. In: Johnson, D.W. (Ed.), Lubrication-Tribology, Lubricants and Additives. IntechOpen, Rijeka, Croatia doi:10.5772/intechopen.72621.
Solovyev, N.D., 2015. Importance of selenium and selenoprotein for brain function: from antioxidant protection to neuronal signaling. J. Inorg. Biochem. 153, 1–12.
Speranza, L., Di Porzio, U., Viggiano, D., De Donato, A., Volpicelli, F., 2021. Dopamine: the neuromodulator of long-term synaptic plasticity, reward and movement control. Cells 10 (4), 735. doi:10.3390/cells10040735.
Sun, Q., Li, Y., Shi, L., Hussain, R., Mehmood, K., Tang, Z., Zhang, H., 2022. Heavy metals induced mitochondrial dysfunction in animals: molecular mechanism of toxicity. Toxicology 469, 153136. doi:10.1016/j.tox.2022.153136.
Suttle, N.F., 2010. Mineral Nutrition of Livestock doi:10.1079/9781845934729.0000.
Szpakowski, P., Król, A., Kobyłecki, A., Głąbiński, A., Nowak, D., 2023. Polyphenols and their impact on the prevention of neurodegenerative diseases and development. Nutrients 15 (15), 3454. doi:10.3390/nu15153454.
Tchounwou, P.B., Yedjou, C.G., Patlolla, A.K., Sutton, D.J., 2012. Heavy metal toxicity and the environment. EXS 133–164. doi:10.1007/978-3-7643-8340-4_6.
Teixeira, F.B., Leão, L.K.R., Bittencourt, L.O., Aragão, W.a.B., Nascimento, P.C., Luz, D.A., Braga, D.V., Da Silva, M.C.F., Oliveira, K.R.M., Herculano, A.M., Maia, C.S.F., Lima, R.R., 2019. Neurochemical dysfunction in motor cortex and hippocampus impairs the behavioral performance of rats chronically exposed to inorganic mercury. J. Trace Elem. Med. Biol. 52, 143–150. doi:10.1016/j.jtemb.2018.12.008.
Teleanu, D.M., Niculescu, A., Lungu, I.I., Radu, C.I., Vladâcenco, O., Roza, E., Costăchescu, B., Grumezescu, A.M., Teleanu, R.I., 2022. An overview of oxidative stress, neuroinflammation, and neurodegenerative diseases. Int. J. Mol. Sci. 23 (11), 5938. doi:10.3390/ijms23115938.
Thakur, M., Rachamalla, M., Niyogi, S., Datusalia, A.K., Flora, S.J.S., 2021. Molecular mechanism of arsenic-induced neurotoxicity including neuronal dysfunctions. Int. J. Mol. Sci. 22 (18), 10077. doi:10.3390/ijms221810077.
Tian, Y., Chen, C., Guo, S., Zhao, L., Yan, Y., 2018. Exploration of the establishment of manganese poisoning rat model and analysis of discriminant methods. Toxicology 410, 193–198. doi:10.1016/j.tox.2018.08.006.
Uğuz, A.C., Nazıroğlu, M., 2012. Effects of selenium on calcium signaling and apoptosis in rat dorsal root ganglion neurons induced by oxidative stress. Neurochem. Res. 37, 1631–1638.
Vaidya, N., Holla, B., Heron, J., Sharma, E., Zhang, Y., Fernandes, G., Iyengar, U., Spiers, A., Yadav, A., Das, S., Roy, S., Ahuja, C.K., Barker, G.J., Basu, D., Bharath, R.D., Hickman, M., Jain, S., Kalyanram, K., Kartik, K., R, B.B., 2023. Neurocognitive analysis of low-level arsenic exposure and executive function mediated by brain anomalies among children, adolescents, and young adults in India. JAMA Netw. Open 6 (5), e2312810. doi:10.1001/jamanetworkopen.2023.12810.

Vakilian, A., Razavi-Nasab, S.M., Ravari, A., Mirzaei, T., Moghadam-Ahmadi, A., Jalali, N., Bahramabadi, R., Rezayati, M., Yazdanpanah-Ravari, A., Bahmaniar, F., 2017. Vitamin B12 in association with antipsychotic drugs can modulate the expression of pro-/anti-inflammatory cytokines in Alzheimer disease patients. Neuroimmunomodulation 24, 310–319.

Velayati, A.A., Farnia, P., 2019. Identification of nontuberculous mycobacterium: conventional versus rapid molecular tests. In: Velayati, A.A., Farnia, P. (Eds.), Nontuberculous Mycobacteria (NTM). Academic Press, Waltham, MA, pp. 11–59.

Vellingiri, B., Suriyanarayanan, A., Selvaraj, P., Abraham, K.S., Pasha, M.Y., Winster, H., Gopalakrishnan, A.V., G, S., Reddy, J.K., Ayyadurai, N., Kumar, N., Giridharan, B., P, S., Rao, K.S., Nachimuthu, S.K., Narayanasamy, A., Mahalaxmi, I., Venkatesan, D., 2022. Role of heavy metals (copper (Cu), arsenic (As), cadmium (Cd), iron (Fe) and lithium (Li)) induced neurotoxicity. Chemosphere 301, 134625. doi:10.1016/j.chemosphere.2022.134625.

Vielee, S.T., Wise, J.P., 2023. Among gerontogens, heavy metals are a class of their own: a review of the evidence for cellular senescence. Brain Sci. 13 (3), 500. doi:10.3390/brainsci13030500.

Virgolini, M.B., Aschner, M., 2021. Molecular mechanisms of lead neurotoxicity. Adv. Neurotoxicol. 159–213. doi:10.1016/bs.ant.2020.11.002.

Wang, H.X., Wahlin, Å., Basun, H., Fastbom, J., Winblad, B., Fratiglioni, L., 2001. Vitamin B12 and folate in relation to the development of Alzheimer's disease. Neurology 56, 1188–1194.

Wang, Y., Zhao, H., Liu, Y., Guo, M., Tian, Y., Huang, P., Xing, M., 2021. Arsenite induce neurotoxicity of common carp: Involvement of blood brain barrier, apoptosis and autophagy, and subsequently relieved by zinc (II) supplementation. Aquat. Toxicol. 232, 105765. doi:10.1016/j.aquatox.2021.105765.

Weisskopf, M.G., Wright, R.O., Schwartz, J., Spiro, A., Sparrow, D., Aro, A., Hu, H., 2004. Cumulative lead exposure and prospective change in cognition among elderly men: the VA Normative Aging study. Am. J. Epidemiol. 160 (12), 1184–1193. doi:10.1093/aje/kwh333.

World Health Organization, 2020. Nutritional Interventions Update Multiple Micronutrient Supplements During Pregnancy. WHO, Geneva.

Wu, L., Lu, X., Zhang, S., Zhong, Y., Gao, H., Tao, F.-B., Wu, X., 2024. Co-exposure effects of urinary polycyclic aromatic hydrocarbons and metals on lung function: mediating role of systematic inflammation. BMC Pulm. Med. 24 (1), 386. doi:10.1186/s12890-024-03173-9.

Xie, Y., Xiao, H., Zheng, D., Mahai, G., Li, Y., Xia, W., Xu, S., Zhou, A., 2025. Associations of prenatal metal exposure with child neurodevelopment and mediation by perturbation of metabolic pathways. Nat. Commun. 16 (1). doi:10.1038/s41467-025-57253-3.

Yary, T., Kauhanen, J., 2019. Dietary intake of magnesium and the risk of epilepsy in middle-aged and older Finnish men: a 22-year follow-up study in a general population. Nutrition 58, 36–39.

Yu, Z., Lin, L., Wang, X., 2017. Pathophysiology of ischemia-reperfusion injury and hemorrhagic transformation in the brain. In: Caplan, L.R., Biller, J., Leary, M.C., Lo, E.H., Thomas, A.J., Yenari, M., Zhang, J.H. (Eds.), Primer on Cerebrovascular Diseases. Academic Press, London, United Kingdom and San Diego, California, United States, pp. 121–124.

Zhang, C., Yu, W., Wang, L., 2016. Folic acid and vitamin B12 prevent DNA hypomethylation induced by lead exposure in rats. Environ. Toxicol. Pharmacol. 48, 61–67.

Zhang, T., Wang, X., Luo, Z., Liu, J., Chen, Y., Fan, P., Ma, R., Ma, J., Luo, K., Yan, C., Zhang, J., Ouyang, F., 2023. Maternal blood concentrations of toxic metal(loid)s and trace elements from preconception to pregnancy and transplacental passage to fetuses. Ecotoxicol. Environ. Saf. 264, 115394. doi:10.1016/j.ecoenv.2023.115394.

Zhao, H., Tang, J., Zhu, Q., He, H., Li, S., Jin, L., Zhang, X., Zhu, L., Guo, J., Zhang, D., Luo, Q., Chen, G., 2020. Associations of prenatal heavy metals exposure with placental characteristics and birth weight in Hangzhou Birth Cohort: multi-pollutant models based on elastic net regression. Sci. Total Environ. 742, 140613. doi:10.1016/j.scitotenv.2020.140613.

Zinia, S.S., Yang, K.-H., Lee, E.J., Lim, M.-N., Kim, J., Kim, W.J., 2023. Effects of heavy metal exposure during pregnancy on birth outcomes. Sci. Rep. 13, 18990. doi:10.1038/s41598-023-46271-0.

Živančević, K., Baralić, K., Vukelić, D., Marić, Đ., Kotur-Stevuljević, J., Ivanišević, J., Savić, M., Batinić, B., Janković, R., Djordjevic, A.B., Miljaković, E.A., Ćurčić, M., Bulat, Z., Antonijević, B., Đukić-Ćosić, D., 2024. Neurotoxic effects of low dose ranges of environmental metal mixture in a rat model: the benchmark approach. Environ. Res. 252, 118680. doi:10.1016/j.envres.2024.118680.

Zwolak, I., 2021. Epigallocatechin gallate for management of heavy metal-induced oxidative stress: mechanisms of action, efficacy, and concerns. Int. J. Mol. Sci. 22 (8), 4027. doi:10.3390/ijms22084027.

CHAPTER TEN

Dietary mycotoxin toxicity in *Caenorhabditis elegans* model: Insights into neurotoxicological and neuroendocrine effects

Abiola J. Oladepo[a], Rebecca P. Dickson[a], Adesina A. Babalola[b], Felix A.A. Soares[b], Michael Aschner[c], Isaac A. Adedara[a,d,*]

[a] *Nauphoeta cinerea* Research Laboratory (NaciReL), Drug Metabolism and Toxicology Unit, Department of Biochemistry, College of Medicine, University of Ibadan, Ibadan, Nigeria
[b] Graduate Program in Biological Sciences, Department of Biochemistry and Molecular Biology, Federal University of Santa Maria, Santa Maria, RS, Brazil
[c] Department of Molecular Pharmacology, Albert Einstein College of Medicine, Forchheimer 209, Bronx, NY, United States
[d] Department of Food Science and Technology, Center of Rural Sciences, Federal University of Santa Maria, Camobi, RS, Brazil
*Corresponding author. e-mail address: isaac.adedara@ufsm.br.

Contents

Advances in Neurotoxicology, Volume 16
ISSN 2468-7480
https://doi.org/10.1016/bs.ant.2026.02.007

Abstract

Mycotoxins pose a serious threat to global food safety, agricultural productivity, and public health. Food contamination with mycotoxins leads to physiological and biochemical abnormalities in humans and livestock. *C. elegans*, due to its unique characteristics, offers a dependable complementary model for elucidating the cellular and molecular basis of mycotoxin-mediated neurotoxicity and other general toxicological impacts. This chapter covers recent studies on mycotoxin toxicity in *C. elegans*, examining four classes of mycotoxins, namely aflatoxins (AFB1), fumonisins (FB1), deoxynivalenol (DON), and zearalenone (ZEA). Data in the literature reveal that mycotoxin exposure in *C. elegans* elicits oxidative stress, mitochondrial dysfunction, DNA damage, neurotoxicity, neuroendocrine alterations, and reproductive impairment. Specifically, FB1 promotes abnormal GABAergic and serotonergic neurotransmission, resulting in behavioral deficits in *C. elegans*. Mechanistic evidence reveals the central role of conserved transcription factors in stress-response and detoxification pathways, notably SKN-1/Nrf2 and DAF-16/FOXO signaling, in mediating mycotoxin toxicity. Collectively, this chapter highlights the utility of *C. elegans* in evaluating mycotoxin toxicity at environmentally relevant exposures. *C. elegans* represents an integrative and complementary model in understanding mechanisms associated with neurobehavioral toxicity resulting from mycotoxin exposure.

Keywords: Caenorhabditis elegans, Mechanisms, Mycotoxin, Neurobehavioral toxicity, Nutrition

1. Introduction

Approximately 400 mycotoxins have been discovered and classified. The common classes of mycotoxins include: aflatoxins, fumonisins, and ochratoxins (Ismaiel et al., 2015; Agriopoulou et al., 2020). Mycotoxins are metabolic byproducts of fungi that grow on animal feeds and agricultural products (El-Sayed et al., 2022; Shekhar et al., 2025), causing significant contamination of farm produce globally (Eskola et al., 2020). Being biologically stable, mycotoxins can persist throughout the processing, transport, and storage of farm produce, posing serious international trade concerns (Chilenga et al., 2024). Hence, they lower the profit margin of farmers and reduce foreign exchange earnings, especially in developing countries. Food contamination with mycotoxins starts before harvest and can occur at any stage of the food production (Pandey et al., 2023). Consumption of mycotoxin-contaminated food has been implicated in provoking the onset of adverse health effects including metabolic diseases, neurotoxicity, disability-adjusted life years, and death (Adedara et al., 2025). Economically, mycotoxins have been extensively demonstrated to negatively impact agricultural productivity, perturb human and animal health, promote food insecurity, and adversely affect international trade (Goda et al., 2025; Da Rocha et al., 2025).

In agriculture, pathogenic fungi such as Fusarium spp. (*F. graminearum, F. verticillioides, F. proliferatum*, etc.) infect food crops, causing diseases such as head blights in wheat and barley (Teli et al., 2020; Burgess and Bryden, 2012), and ear rot in maize (Zhou et al., 2018a; Bryła et al., 2022), significantly reducing the yield and quality of crops. Aspergillus spp. (*A. flavus and A.* parasiticus) infect food crops, especially in warm and humid climates, causing the spoilage and rot of affected crops (Zakaria, 2024). Similarly, mycotoxins-producing Fusarium spp. infects stored agricultural products such as fruits, grains, juices, etc., rapidly reducing their quality, and contaminating them with mycotoxins and their metabolites (Habschied et al., 2021; Pandey et al., 2023). Moreover, contamination of livestock feeds by mycotoxins often result in reduced food intake, acute weight loss, and depleted production of milk and eggs in cows and poultry (Bryden, 2012; Akinmoladun et al., 2025). Mycotoxins have been extensively demonstrated to induce physiological and biochemical abnormalities promoting carcinogenesis, endocrine disruptions, neurotoxicity, hepatotoxicity, immune suppression, nephrotoxicity, to mention a few (Gacem et al., 2019; Shekhar et al., 2025).

Caenorhabditis elegans was first used in 1965 by Sydney Brenner to study neurobiology and animal development (Riddle et al., 1997). Since then, *C. elegans* has been used as a biological model in various fields of study (Leung et al., 2008; O'Reilly et al., 2013). *C. elegans*, unlike vertebrate models, is highly cost-effective, easy to manage, possesses a short life span of approximately 2 to 3 weeks, and is easily cultured on *E. coli* supplemented agar (Ha et al., 2022; Roussos et al., 2023). Also, the model possesses a simple and well-characterized anatomy and physiology, with a completely sequenced genome (Adedara et al., 2025). *C. elegans* is has approximately 20,000 genes (a significant proportion of which are homologous to human genes), with a simple and functionally conserved neurotransmission system, making it a suitable in vivo model for neurotoxicology assessment (Lai et al., 2000). *C. elegans'* simple anatomy and transparent cuticle facilitate easy microscopic visualization and real-time imaging of its cells, and enable the tracking of fluorescent reporter (Breimann et al., 2018; Shah et al., 2022).

Currently, *C. elegans* is not covered under many strict animal welfare regulations; hence, experimental research with *C. elegans* requires fewer ethical approvals. The model is also highly sensitive and responds rapidly to awide range of environmental contaminants, even at low concentrations (Li et al., 2024). The lifestyle of *C. elegans* as a soil-dwelling organism predisposes it naturally to environmental toxicants, highlighting their relevance as a biological sensor and as a model in toxicological studies (Leung et al., 2008; Long et al., 2023). Due to its numerous advantages over vertebrate models, *C. elegans* is increasingly used to assess the toxicological impact of toxic contaminants in foods and the environment (Avila et al., 2012; Adedara et al., 2025). This chapter presents recent experimental reports on the use of *C. elegans* as a biological model to elucidate the neurotoxicology and neuroendocrine outcomes associated with mycotoxin exposure. It highlights the use of nematodes in assessing behavioral, developmental, reproductive, molecular, and cellular impacts of different categories of mycotoxins.

2. General toxicological mechanisms of mycotoxins in vertebrates

The routes of mycotoxin exposure include oral ingestion, inhalation, and dermal contact with contaminated materials (Nawaf, 2023; Awuchi et al., 2022; Coppock and Jacobsen, 2009). After exposure, mycotoxins are absorbed and redistributed to various organs and tissues in the body. Some mycotoxins have been observed to exhibit organ-specific accumulation patterns; notably AFB1, which has been reported to preferentially accumulate in liver tissues (Tolosa et al., 2021; Xu et al., 2021). Some plants have been documented with the potential to metabolize parent mycotoxins, releasing less detectable derivatives. This complicates and limits effective regulatory control of such toxins, while enhancing their spread and toxicological consequences (Freire and Sant'Ana, 2017). The toxicological impact of mycotoxins on organisms is dependent on the mycotoxin involved.

The molecular mechanism underlying the toxicological impact of mycotoxins is dependent on the mycotoxin involved and the toxicological endpoint. Primarily, mycotoxins in biological systems accelerate the formation and accumulation of ROS and free radicals. Accumulation of ROS is associated with the reduced activities of SOD, CAT, GPx, GST, and glutathione reductase (GR), resulting in oxidative stress (Da Silva et al., 2018; Gusti et al., 2021). As free radicals and ROS accumulate in the tissues of the exposed animals, cellular mechanisms are disrupted, and oxidative damage to membranes, DNA, and proteins increases (Oke et al., 2024). The cellular metabolic activation of mycotoxins in humans and animals reportedly alters the conformation and activities of antioxidant enzymes, resulting in their inhibition (Grintzalis et al., 2014; Fouad et al., 2019). Ochratoxin A chelates and depletes Fe^{2+} and NADPH cofactors necessary for the activities of thioredoxin system and GR (Kőszegi et al., 2016). Mycotoxins also deregulate Nrf2/Keap1 signaling pathways, suppressing the Nrf2-mediated transcription and limiting the synthesis of the antioxidant enzymes (Kozieł et al., 2021). Cellular detoxification of certain mycotoxins consumes GSH, limiting the antioxidant functions of GPx and GST (Guilford et al., 2014; Mavrommatis et al., 2021). Also, mycotoxins damage mitochondria, depleting ATP and NADPH supply and limiting GR function (Janik-Karpinska et al., 2023).

Dermal toxicological impacts of mycotoxins have also been widely reported (Doi and Uetsuka, 2014). Trichothecene (T-2) toxin induces histopathological changes in the skin of exposed animals. This epidermal damage is linked to secondary ischemia mediated by the degeneration of microvessels in mycotoxin-affected dermis. The observed degradation is related to the apoptosis of keratinocytes, mediated by inflammation transcripts namely TNF-α, IL-1β, and TGF-β1 mRNA. C-jun and C-fos were also reported to elevate the sensitivity of keratinocytes to apoptosis (Doi and Uetsuka, 2014). Mycotoxins have been shown to alter gastrointestinal integrity and gut microbiota leading to gastrointestinal disturbances and mycotoxicosis. Liew and Mohd-Redzwan (2018) reported that trichothecenes and ochratoxin A significantly reduce the population of the gut microbiota and compromise the integrity of the GIT epithelial cells lining. Certain mycotoxins have been established to be genotoxic and are recognized as carcinogenic agents. Aflatoxin B1 and sterigmatocystin are metabolized in the liver by cytochrome P450 into a more toxic epoxide intermediate, which forms DNA adducts, causing the transversion of GC to TA (Thongorarm et al., 2025). This transversion is implicated in the occurrence of hepatic carcinoma, especially when tumor suppressor genes like tp5*3, rb1, brca1, brca2,* and *pten* are affected (Thongorarm et al., 2025). Likewise, OTA, patulin, zearalenone, DON, T-2 toxin, and alternaria toxins have been implicated in oxidative-mediated mutagenesis and carcinogenesis. OTA and patulin have also been reported to inhibit DNA repair mechanisms, while patulin is a direct inhibitor of DNA topoisomerase I. However, the molecular mechanism of OTA DNA repair inhibition has not been established (Tumini et al., 2025). Other known mycotoxins that inhibit topoisomerase are alternariol and alternariol monomethyl-ether, which similarly mediate the breakage of the DNA strand via oxidative stress (Islam et al., 2023; Fehr et al., 2025).

Mycotoxins are established neurotoxicants and have been established to cross the blood-brain barrier (Song et al., 2024; Shekhar et al., 2025). In the brain, they have been demonstrated to induce several neurotoxic effects. Notably, AFB1 and OTA exert their neuronal damage via the induction of oxidative stress, DNA damage, S-phase cell arrest, and apoptosis. Aflatoxin B1 alters the level of neurotransmitters in the brain (Dai et al., 2023). Fumonisin B1 inhibits ceramide synthase, minimizing sphingolipid synthesis (causing overactivation of PARP-1) and apoptosis (Boppana et al., 2014).

3. Known effects of mycotoxins in simple model organisms

Lower organisms such as *Caenorhabditis elegans,* zebrafish (*Danio rerio*), fruit fly (*Drosophila melanogaster*), and cockroaches (*Nauphoeta cinerea* and *Periplaneta americana*) are increasingly explored as a valuable model in toxicological research (Strange et al., 2016; Adedara et al., 2023). This increase is driven by challenges associated with the use of higher models, ethical concerns, high cost of maintenance, handling challenges, low throughput, and long life span. Moreover, they possess better environmental relevance, with less complexity associated with higher organisms. Other advantages of lower organisms include the presence of conserved genes and signaling pathways that are similar to those in humans; assurance of high-throughput toxicological testing for toxicants; sensitivity to environmental contaminants and amenability to genetic manipulation using tools such as CRISPR, RNAi, and transgenes. Several experimental studies have assessed the toxicological effects of mycotoxins in lower organisms. In fruit flies, mycotoxins such as AFB1, rubratoxin B, and diacetoxyscirpenol induce lethality at varying concentrations (Reiss, 1976). In zebrafish, mycotoxins (zearalenol, T-2 toxin, and DON) at sub-lethal doses induced behavioral and developmental defects, as well as estrogenic potency (Khezri et al., 2018; Bakos et al., 2019). Furthermore, mycotoxins at varying doses induce lethality, oxidative stress, and neurodegeneration in cockroaches (*Periplaneta americana*) (Doi et al., 2011; Chaurasia et al., 2016). Fig. 1 presents the general toxicological impacts of mycotoxins on *C. elegans.*

4. *C. elegans* as a model for toxicological research

C. elegans is a small, non-parasitic multicellular organism with 959 somantic cells in adult harmaphrodite while the males have 1031 cells. (Corsi, 2006). They are highly abundant in the environment and are saprophytic in nature. In the wild, they feed on bacteria from rotten fruit and

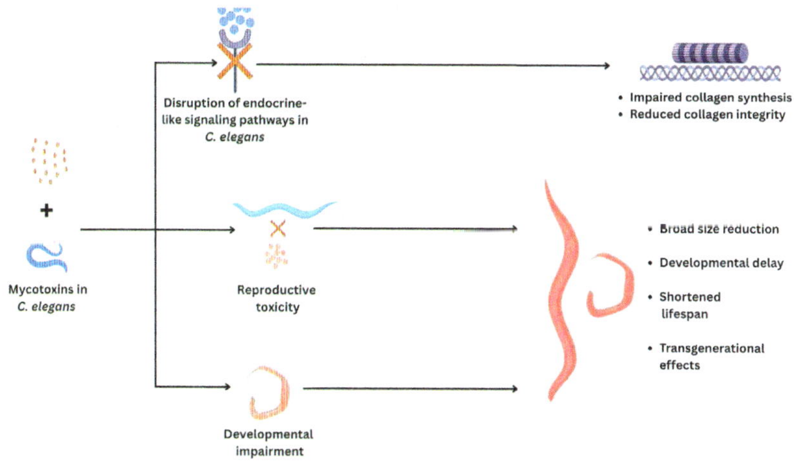

Fig. 1 Recapitulation of the common toxicological endpoints of mycotoxin exposures in *C. elegans*, revealing significant disruption of endocrine-like signaling pathways, and induction of developmental and reproductive toxicity.

dirt (Gives, 2022). *C. elegans* are around 1.0 mm long and are just barely visible to the human eye (Hunt, 2017). In the laboratory, *C. elegans* are cultured on media, e.g., solid agar plates (i.e., Nematode Growth Medium) or liquid media. Both media are seeded with *E. coli* bacteria as a food source for the nematodes (Espejo et al., 2022). They have a life cycle of approximately 12 to 35 days and a very high reproductive rate when cultivated at 20 °C. C. elgans possess four developmental stages termed L1, L2, L3, and L4 (Melnikov et al., 2023). The anatomy of *C. elegans* is simple, as it possesses a basic internal systems that parallel those obtainable in higher animals. Examples include the nervous, muscular, digestive, and excretory systems. Moreover, they possess a transparent body that allows easy assessment of internal processes through light microscopy (Corsi et al., 2018). The nematode has a completely sequenced genome, carrying more than 19,000 genes. Scientific reports show that these worms express gene homologues of around 60–80 percent of human genes. Therefore, some of their key biochemical pathways and biological functions are conserved in humans (Hillier et al., 2005). This allows extrapolation of findings in *C. elegans* to higher organisms. Also, the low cost of cultivating *C. elegans* and their high sensitivity to toxic substances position them as a suitable model in toxicological studies (Hunt, 2017; Xiong et al., 2017). This section explores various toxicological approaches using *C. elegans*, with emphasis on their

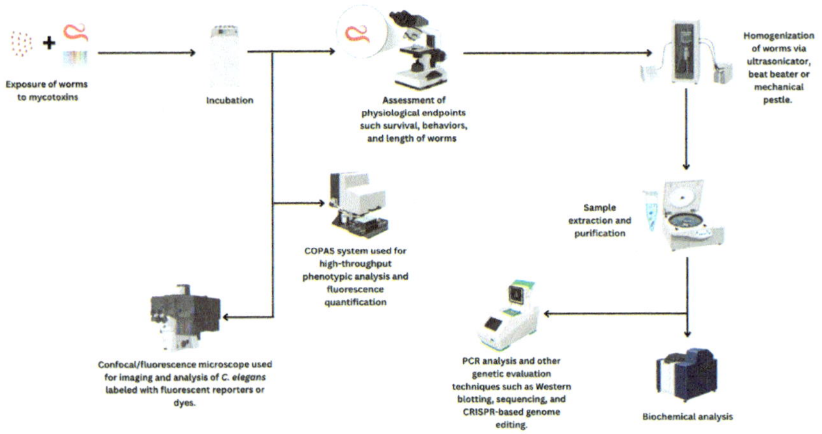

Fig. 2 Schematic overview of the simplified experimental workflow for investigating mycotoxin toxicity in the *C. elegans* model, highlighting toxicological exposure, incubation, physiological and phenotypic assessments, and mechanistic investigations of the biochemical and molecular basis of the endpoints.

methodology. Fig. 2 gives a schematic representation of mycotoxin toxicity assessments in *C. elegans*.

4.1 *C. elegans* as a model for neurotoxicity assessments

Although highly simplified, the *C. elegans* nervous system coordinates basic and complex neurobehavioral actions that are key to its survival. Some of these actions include locomotion, feeding, and defecation (Ruszkiewicz et al., 2018). In addition, complex actions, such as determining food sources, temperature, chemicals, oxygen levels, and pheromones, are controlled by their simple nervous system (Rodrigues et al., 2024). *C. elegans* has two sexes: hermaphrodites and the male sex. They manifest sex-specific behaviors like egg-laying in hermaphrodites mating, and locomotion patterns, making them useful in investigating sex-specific neurobehavioural patterns and dimorphism. Moreover, *C. elegans* has been shown to express behavioral plasticity facilitated by incremental learning and memory. Underlying these diverse behavioral functions are neuroendocrine signaling systems produced in specialized cells through proteolytic cleavage of propeptides (Li et al., 1999).

The versatility of *C. elegans* in characterizing the neurotoxicity and possible neuroprotective effects of food-related substances, including metals and organic compounds, has been previously reviewed (Adedara et al., 2025). Behaviors in *C. elegans* are regulated at the organismal, cellular, and subcellular levels (Schafer, 2005; Zhao et al., 2025). Similar to higher toxicological models, *C. elegans* is subject to behavioral disruptions caused by toxicants. These behavioral changes are characterized by the degree of exploration (crawling on a solid surface or swimming in liquid media) and exploration pattern (Wang and Xing, 2008; Melnikov et al., 2023). Worms respond to stimuli in their environment mainly through thermotaxis and chemotaxis. These behavioral responses enable them to seek food and avoid perceived harmful substances (Queirós et al., 2021). Through their sensory neurons, *C. elegans* are able to sense light, pH, oxygen level, osmolarity, volatile chemicals, mechanical stimuli, temperature, and water-soluble substances (Bargmann, 2006). Chemosensory neurons are associated with two types of glial support cells: sheath and socket cells. The signals are processed at the interneurons, sending instructions to another set of neurons (Oikonomou and Shaham, 2010; Emmons, 2024).

Moreover, *C. elegans* offers insight into the molecular basis of the observed behavioral changes (Zhao et al., 2025). The nematode genes are highly amenable to genetic manipulations. Transgenic worms expressing mutant proteins are readily available, and they provide molecular insights into neurotoxicological processes (Leung et al., 2008; Martinez-Finley et al., 2012). Investigation into intracellular pathways in *C. elegans* involves conducting forward and modifier genetic screenings following genetic mutations (Leung et al., 2008; Kimble and Nüsslein-Volhard, 2022). Also, *C. elegans* is susceptible to sex manipulation (possible generation of males, feminized males, masculinized hermaphrodites, or feminized hermaphrodites) (Morgan et al., 2012; Hu et al., 2018; Luo et al., 2024). This enables scientific research on sex-specific mechanisms of neurotoxicants or disorders and "rejuvenation" by forcing development through the quiescent dauer larval stage (Leung et al., 2008).

4.2 *C. elegans* as a model for assessing molecular and cellular responses

C. elegans offers a useful model in the assessment of molecular and cellular responses to toxicological exposure (Avila et al., 2012; Widmayer et al., 2022). They activate several cytoprotective pathways when exposed to toxicants. These pathways initiate an increase in the expression of antioxidative enzymes and heat shock proteins. The increased expressions of cytoprotective proteins are mediated by the activation of the corresponding genes (Jomova et al., 2023). Moreover, the transparency of *C. elegans*' external cuticle allows easy assessment of cellular, organelle, and lysosome-related organelles (LROs) changes that accompany toxicological exposures (Kucharíková et al., 2023). Likewise, these qualities facilitate the assessment of molecular responses through biological stains, dyes, probes, and markers. *C. elegans*' simple, small, and well-characterized genomic makeup enables the appropriate elucidation of genetic alterations associated with the toxicological effects of toxic substances (Lin et al., 2023). Also, the availability of a large spectrum of induced and natural genetic variation of *C. elegans* allows researchers to study genetic variability in toxicological responses, revealing the variation in susceptibility based on the genetic makeup of organisms (Hahnel et al., 2020; Widmayer et al., 2022). Summarized in Table 2 are the possible molecular and cellular response in *C. elegans* to toxicological exposures, and their assessment methods.

5. Mycotoxin toxicity in *Caenorhabditis elegans*: *Mechanistic insights*

C. elegans has been used extensively as an efficient complementary model in toxicology studies. It possesses ideal characteristics well-suited for studying toxicity mechanisms at different biological levels. Empirical studies have employed the model in elucidating the cellular and molecular basis central to the toxicity of mycotoxins (Yang et al., 2015; Keller et al., 2018). While mycotoxins differ in biochemistry and structure, *C. elegans*-based investigations show the convergence of their molecular mechanisms on specific set of conserved toxicological pathways. These toxicological pathways include oxidative stress, mitochondrial dysfunction, genomic instability, neurodegeneration, disruption of endocrine and reproductive processes, and dysregulation of stress-response signaling networks. This section presents a systematic integration of the mechanistic processes underlying mycotoxin toxicity in *C. elegans*. As a model, *C. elegans* possesses the ability to systematically integrate the mechanistic layers critical to a given biological process into a single experimental framework. This strongly strengthens its value for mycotoxin hazard assessment.

5.1 Oxidative stress is key to mycotoxin toxicity

Of all studied mycotoxins, AFB1 is the most explored mycotoxin in *C. elegans*. Exposure to mycotoxins consistently raises the level of intracellular reactive oxygen species (ROS). Consequently, it disrupts redox balance in exposed *C. elegans* and other experimental models. Oxidative stress has been shown to play a central role in mycotoxin toxicity. González-Hunt et al. (2014) reported that AFB1-exposure in *C. elegans* increases oxidative damage in exposed *C. elegans*. Subsequently, AFB1 interferes with key antioxidant signaling pathways, altering the expression of SKN-1 and DAF-16 (Chang et al., 2024; Wu et al., 2025). These establish oxidative stress as a key biochemical basis of key mycotoxin toxicological endpoints in *C. elegans*.

Like AFB1, FB1 has similarly been demonstrated to induce oxidative stress in *C. elegans* (Zhang et al., 2022). This possibly contributed to the observed neurobehavioral deficit and degeneration of neuronal cells in *C. elegans* following FB1 exposures. Likewise, deoxynivalenol (DON) induces oxidative stress indirectly through ribosomal damage and activation of inflammatory cascades. These observations explain the marked increase in growth retardation, reproductive failure, and transgenerational neurotoxicity following FB1 exposures (Zhou et al., 2018b; Di et al., 2018). In addition, oxidative stress was observed following the exposure of *C. elegans* to ZEA, which was further exacerbated by DAF-16 inhibition, a stress-response transcription factor, leading to significant decline in locomotor function (Huang et al., 2021). Overall, the review reveals that oxidative stress plays a central role in mycotoxin toxicity. Additionally, the observations indicates the significance of gene-level disruptions of antioxidant defense pathways on *C. elegans*' ability to withstand mycotoxin toxicity. This offers causal evidence for oxidative stress as a conserved toxicological hub across classes of mycotoxins.

5.2 Mitochondrial dysfunction and bioenergetic collapse

Empirical studies on mycotoxin toxicity in *C. elegans* have continuously shown mitochondria as the central target for the various classes of mycotoxins. González-Hunt et al. (2014) reported that AFB1 elicited considerable damage to mitochondrial DNA compared to nuclear DNA. Also, AFB1 exposure significantly altered mitochondrial functions, as revealed in the impaired oxidative phosphorylation, ATP production, and lifespan in *C. elegans*. Likewise, Luz and Meyer (2016) further strengthen González-Hunt et al. (2014)'s position as they uncovered the existence of a positive association between mtDNA copy number and sensitivity to AFB1 toxicity. Thereby highlighting the causal role of mitochondrial integrity in determining toxic outcomes.

Moreover, other classes of mycotoxins also weaken mitochondrial functions. For example, FB1-induced mitochondrial dysfunctions at environmentally relevant co-exposure conditions. FB1 and cadmium synergistically exacerbate oxidative stress, mitochondrial damage, and ferroptosis-mediated gene dysregulation (Wang et al., 2024). This also shows the cumulative effects of multiple exposures on the mitochondria, establishing the vulnerability of mitochondrial pathways to mixed toxicants. Likewise, ZEA exposure also damages mitochondrial homeostasis, inducing mitochondrial fragmentation and altered dynamics via DRP-1 signaling. The alterations were associated with mitochondrial alterations to dopaminergic neuron degeneration and progressive locomotor decline. Hence, mitochondrial dysfunction leads to oxidative stress and neurotoxicity. These studies, therefore, position mitochondrial impairment as a source and an amplifier of mycotoxin-induced cellular damage.

5.3 Genotoxicity and DNA damage responses

Another toxicological endpoint characterized for AFB1 toxicity in *C. elegans* is genomic instability. AFB1-exposures increase the formation of DNA adducts and strand breaks. These genetic modifications are dependent on the activation of cytochrome P450 enzymes similar to their action in mammalian systems. In support of that position, Feng et al. (2016) observed that xpa-1 mutant *C. elegans* that are deficient in excision repair are very sensitive to lethality, growth inhibition, and reproductive toxicity induced by AFB1. Hence, these findings confirm DNA damage as the primary toxic lesion. In contrast, trichothecenes and ZEA exhibit more indirect genotoxic effects. Trichothecenes and ZEA trigger ribosomal stress and activate compensatory immune and detoxification pathways, such as PMK-1 and UDP-glucuronosyltransferases (Di et al., 2018). Although direct DNA damage is less pronounced, sustained activation of stress-response pathways has a strong impact on developmental delay, reduced lifespan, and heritable toxicity effects (Zhou et al., 2018a). Genetic disturbances due to ZEA exposures are epigenetic in nature. In addition, Li and Wei (2024) observed that ZEA induces multigenerational reproductive and locomotor impairment through histone methylation via the mes-4 pathway. These findings highlight the capacity of mycotoxins to induce genetic reprogramming that persists across generations.

5.4 Neurotoxicity and behavioral impairment

Another established toxicological endpoint of mycotoxin exposure in *C. elegans* is behavioural toxicity. *C. elegans* exposure to AFB1 suppressed growth

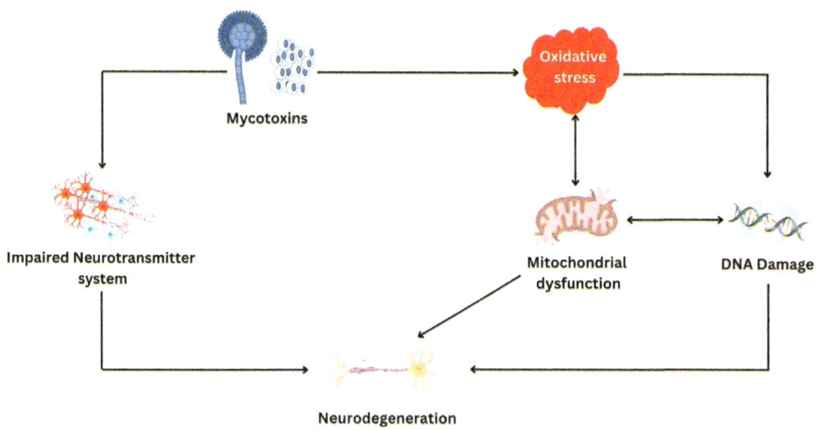

Fig. 3 Schematic overview of mycotoxins' neurotoxicological patterns in *C. elegans*, revealing the central role of oxidative stress in mediating mitochondrial dysfunction and DNA damage, which consequently alter normal behavioral patterns and contribute to neurodegeneration. Alternatively, mycotoxins induce neurotoxicity by directly altering *C. elegans'* neurotransmitter system.

performance and locomotion behavior with concomitant disruption of the structural integrity of the intestine (Wu et al., 2025). AFB1 was reported to induce selective dopaminergic neuron damage, resulting in significant locomotor dysfunction and observable neurodegenerative phenotypes (González-Hunt et al., 2014). Exposure of *C. elegans* to AFB1 resulted in a dose-dependent increase in the degeneration of dopaminergic CEP neurons possibly via the disruption of dlk-1 MAP kinase pathway (González-Hunt et al., 2014). These toxicological endpoints have been strongly associated with both oxidative stress and mitochondrial dysfunction, strongly reinforcing the interconnected basis of the pathways of mycotoxin toxicity. Likewise, FB1 alters the neurotransmitter system (mainly the serotonergic and GABAergic signalling), which is reflected in impaired locomotion and altered behavioral responses of the exposed worms (Zhang et al., 2022). ZEA causes dopaminergic neuron loss and progressive behavioral decline through oxidative and mitochondrial mechanisms (Huang et al., 2021). The conservation of these neurotoxic pathways in *C. elegans* and higher organisms enhances the translational relevance of the findings. This further establishes the capacity of *C. elegans* as a useful screening platform for neurotoxic hazards of mycotoxins. Fig. 3 presents the synthesized toxicological pattern of mycotoxin's toxicity in *C. elegans*.

5.5 Reproductive toxicity and neuroendocrine disruption

Reproductive impairment is a commonly reported toxicological endpoint of mycotoxin exposures in *C. elegans*. AFB1, FB1, and DON have been reported to induce significant brood size reduction, development delay, and shortened lifespan (Yang et al., 2015; Zhou et al., 2018a). These reflect the disruption of energy allocations, germline integrity, and hormonal signaling pathways. Moreover, ZEA elicits significant endocrine-disrupting activity. For example, Keller et al. (2018) showed that ZEA and its metabolites altered the lifespan and reproductive capacities of *C. elegans*. Also, Yang et al. (2018) associated ZEA exposure with disruptions of collagen synthesis and the integrity of cuticles. These underscore estrogenic mycotoxins' ability to interfere with developmental and reproductive processes, even in invertebrate systems lacking classical estrogen receptors. The discovery of transgenerational toxicity after ZEA exposure further broadens the scope of the toxicological significance of mycotoxins. Also, the inheritance of adverse phenotypic effects facilitated by epigenetics reveals the possible long-term effects of chronic low-dose exposure.

6. Integrative mechanistic perspective and model significance

While each mycotoxins possess different primary targets, their toxicological endpoints converge on a few sets of biological processes that regulate survival, reproduction, and aging. *C. elegans* offers the ability to link the molecular basis of mycotoxin actions to organismic outcomes within a single experimental system. Hence, *C. elegans* provides mechanistic clarity that complements mammalian models. It enhances the understanding of the potential of mycotoxin hazard. Consequently, this model is a valuable tool for mechanistic toxicology, mixture risk assessment, and the development of predictive frameworks for mycotoxin-induced health effects. Tables 1–4 summarizes the investigations on mycotoxin toxicity using *C. elegans as a* scientific model.

7. Methodological limitations

Although significant progressive strides have been made in elucidating the toxicological effects of mycotoxins in *C. elegans*, several methodological limitations are evident across the literature. For example, the exposure doses across studies often differ from realistic environmental levels. Several

Table 1 Summary of aflatoxin B1 toxicity studies using *Caenorhabditis elegans*.

Study	Toxin	Strain(s)	Exposure Conditions	Toxicological Endpoints	Key Findings
Yang et al. (2015)	AFB1	N2	Multiple doses; short-term exposure	Lethality, growth, reproduction, lifespan	AFB1 ranked second in lethality second to T-2 toxin; AFB1 significantly inhibited growth and reduced brood size; AFB1 shortened lifespan of exposed *C. elegans*
Feng et al. (2016)	AFB1	N2 and xpa-1	3–90 μM; larval exposure	Growth, reproduction, DNA damage, apoptosis	AFB1 induced dose-dependent growth and reproductive inhibition; AFB1 increased the occurrence of DNA lesions and germline apoptosis; the xpa-1 strain showed higher susceptibility
	AFB1	N2, emb-8, glp-1	30–100 μM	DNA damage, growth	AFB1 induced DNA damage required metabolic activation; CYP enzymes played key role in the observed metabolic disruption
González-Hunt et al. (2014)	AFB1	glp-1	3–100 μM	mtDNA damage, neurodegeneration	AFB1 mediated preferential mtDNA damage over nDNA; induced dopaminergic neurodegeneration
Luz and Meyer (2016)	AFB1	PE255, N2	5–50 μM; chronic exposure	mtDNA integrity, ATP, lifespan	Reduced mtDNA increased susceptibility to AFB1 toxicity; mitochondrial function is central to toxic response to AFB1 exposure
Chang et al. (2024)	AFB1	N2, skn-1	2.5–5 μM; long-term exposure	Aging, immunity, lifespan	AFB1 exposure delayed development, reduced lifespan and immune resistance; SKN-1/Nrf2 signaling involved was involved in the toxicological response
Wu et al. (2025)	AFB1	N2, TJ356	5–100 μM; 72 h exposure	Intestinal integrity, oxidative stress	AFB1 exposure resulted in intestinal barrier disruption, elevated ROS, and suppression of DAF-16/SKN-1 antioxidant pathways

Table 2 Summary of fumonisin toxicity studies using *Caenorhabditis elegans*.

Study	Toxin	Strain(s)	Exposure Conditions	Toxicological Endpoints	Key Findings
Yang et al. (2015)	FB1	N2	Multiple doses; short-term exposure	Lethality, growth, reproduction, lifespan	FB1 exhibited lower lethality of exposed *C. elegans* than AFB1 and T-2 toxin; FB1 significantly impaired growth and brood size; FB1 shortened lifespan the lifespan of exposed worms
Zhang et al. (2022)	FB1	N2; neurotransmission mutants	20–200 μg/mL; 24–48 h	Neurobehavior, neuronal integrity, neurotransmitters	FB1 impaired locomotion; FB1 damaged serotonergic and GABAergic neurons; FB1 exposure significntly reduced serotonin and GABA levels

Table 3 Summary of trichothecene toxicity studies using *Caenorhabditis elegans*.

Study	Toxin	Strain(s)	Exposure Conditions	Toxicological Endpoints	Key Findings
Gowrinathan et al. (2011)	DON, crude trichothecenes	N2, AU1	100–1000 mg/mL	Development, brood size	Purified DON significantly reduced larval development and brood size, while crude extracts inhibited egg hatching in exposed worms.
Zhou et al., 2018a	DON	N2	50–400 mg/mL	Growth, feeding, reproduction, transgenerational effects	DON exposure induced dose-dependent growth and fecundity reduction; DON exposure mediated toxic effects persisted across filial generations
Di et al. (2018)	DON	N2	125–500 μg/mL	Reproduction, lifespan, gene expression	DON exposure reduced egg production and lifespan. Exposure also upregulated immune and detoxification genes (PMK-1, UGTs)

Table 4 Summary of zearalenone toxicity studies using *Caenorhabditis elegans*.

Study	Toxin	Strain(s)	Exposure Conditions	Toxicological Endpoints	Key Findings
Keller et al. (2018)	ZEA, ZEA-14-S	N2	2.5–62.5 mg/L	Reproduction, lifespan, stress tolerance	ZEA exposure reduced *C. elegans'* reproductive output; However, ZEA-14-S exposure extended lifespan of exposed worms; Exposed worms metabolized ZEA into α- and β-zearalenol, *in vivo*
Yang et al. (2018)	ZEA	N2; mutants	10–40 μg/mL	Development, gene expression	ZEA exposure disrupted collagen synthesis pathways in exposed worms. Likewise, exposure to ZEA altered the expression of cuticle-related genes
Huang et al. (2021)	ZEA	N2	0.3–50 mM	Locomotion, aging, oxidative stress	Exposure to ZEA mediated a dose-dependent locomotor decline. Also, the exposure increased ROS level, and inhibited DAF-16 nuclear translocation
Li and Wei (2024)	ZEA	N2, mes-4	0.1–50 μM; multigenerational	Reproduction, locomotion, epigenetic inheritance	Toxic effects of ZEA was transmitted across generations. The multigenerational effect requires mes-4-mediated histone methylation.
Huang et al. (2021)	ZEA	N2	0.25–50 μM	Neurodegeneration, mitochondrial dynamics	ZEA exposure induced dopaminergic neuron loss and mitochondrial fragmentation via DRP-1 signaling

mycotoxin studies often use doses that fail to accurately represent chronic dietary exposure achievable in humans or animals. Although these doses are relevant for scientific investigations, they often limit ecological and translational relevance. Further, experimental design across studies often limits *C. elegans* cross-study comparisons. Differences in exposure stage, duration, and matrices often introduce variability in toxicokinetics and bioavailability. Also, many studies failed to standardize microbial interactions.

Moreover, studies of mycotoxins with *C. elegans* are overly reliant on phenotypic endpoints, which limits their mechanistic insights. Although these endpoints are sensitive and reproducible, they are insufficiently integrated with molecular, proteomic, or metabolomic data. Additionally, transgenerational and epigenetic studies are largely underexplored. While some mycotoxins have demonstrated the key role of histone modifications and altered energy metabolism in heritable effects, the majority of these investigations examined only a limited number of generations and focused on select epigenetic markers. Also, co-exposure studies are often few and oversimplified.

8. Research gaps in *C. elegans*-based mycotoxin neurotoxicology

Although the existing research on mycotoxin toxicity is broad, this chapter identifies some key research Gaps. Fumonisins, trichothecenes, and zearalenone have received limited research attention compared to aflatoxins. Only a few neurotoxicological indices and mechanistic bases have been explored. Across classes of mycotoxins, only a few studies combine transcriptomics with other omics concepts to construct unified toxicity networks. This limits the system-level understanding of mycotoxins' toxicity. While host–microbe–toxin interactions have been explored for AFB1, such an impact on fumonisins, trichothecenes, and zearalenone has not been investigated. Finally, translation of *C. elegans* for human risk assessment remains limited.

9. Conclusions

Mycotoxins have a serious toxicological impact on animals and humans. They pose a serious threat to global food safety, agricultural productivity, and public health. This chapter demonstrates the complementary roles of *C. elegans* in advancing the mechanistic understanding of mycotoxin neurotoxicity and other common toxic impacts. Because of its unique nature, *C. elegans* provides a suitable experimental system for connecting molecular processes to observable toxicological endpoints.

Evidence from literature revealed that the toxic effects of mycotoxins in *C. elegans* converge on a limited set of conserved pathways. Oxidative stress was shown as the central unifying mechanism that initiates several other toxicological endpoints. Stress-response regulators such as SKN-1/Nrf2 and DAF-16/FOXO were consistently implicated in mycotoxin toxicity. This reveals their key effects in redox imbalance and energy dysregulation in mycotoxin-induced toxicity. Notably, the mechanistic process of mycotoxins in *C. elegans* closely parallels those observed in mammalian systems, underscoring the translational relevance.

Although these fields of research have achieved major scientific milestones, major research gaps and methodological limitations have been identified. Given this, future research efforts should focus on more environmentally realistic doses and standardized experimental designs to enhance cross-study comparability and ecological relevance. More integration of multi-omics approaches must be employed to properly assess the system-level responses to mycotoxin exposure. Further, investigations must expand to cover the understudied mycotoxin classes and emerging or modified mycotoxins. In addition, increased host–microbiome–toxin interaction studies in *C. elegans* is key to elucidating the impact of bioavailability on mycotoxins' toxicity. Expanding host–microbiome–toxin interaction studies in *C. elegans* represents another critical future direction, as microbial metabolism can substantially modulate mycotoxin bioavailability and toxicity.

Acknowledgments

IAA gratefully acknowledge the financial support of CAPES (Coordenação de Aperfeiçoamento de Pessoal de Nível Superior; File No. 88887.568833/2020–00). MA was supported in part by grants from the National Institute of Environmental Health Sciences (NIEHS) R01ES10563 and R01ES07331.

References

Adedara, I.A., Mohammed, K.A., Canzian, J., Rosemberg, D.B., Aschner, M., Farombi, E.O., et al., 2023. *Nauphoeta cinerea* as an emerging model in neurotoxicology. In: Rocha, J.B.T., Aschner, M., Costa, L.G. (Eds.), Advances in Neurotoxicology. In: Advances in Neurotoxicology, 9. Elsevier Publishing Company, USA, pp. 181–196.

Adedara, I.A., Weis, G.C.C., Monteiro, C.S., Soares, F.A.A., Rocha, J.B.T., Schetinger, M.R.C., et al., 2025. Versatility of *Caenorhabditis elegans* as a model organism for evaluating foodborne neurotoxins and food bioactive compounds in nutritional neuroscience. Mol. Neurobiol 62 (6), 7205–7229. doi:10.1007/s12035-025-04705-y.

Agriopoulou, S., Stamatelopoulou, E., Varzakas, T., 2020. Advances in occurrence, importance, and mycotoxin control strategies: prevention and detoxification in foods. Foods 9 (2), 137. doi:10.3390/foods9020137.

Akinmoladun, O.F., Fon, F.N., Nji, Q., Adeniji, O.O., Tangni, E.K., Njobeh, P.B, 2025. Multiple mycotoxin contamination in livestock feed: implications for animal health, productivity, and food safety. Toxins 17 (8), 365. doi:10.3390/toxins17080365.

Avila, D., Helmcke, K., Aschner, M., 2012. The Caenorhabiditis elegans model as a reliable tool in neurotoxicology. Hum. Exp. Toxicol. 31 (3), 236–243. doi:10.1177/0960327110392084.

Awuchi, C.G., Ondari, E.N., Ogbonna, C.U., Upadhyay, A.K., Baran, K., Okpala, C.O.R., et al., 2021. Mycotoxins affecting animals, foods, humans, and plants: types, occurrence, toxicities, action mechanisms, prevention, and detoxification strategies-a revisit. Foods 10 (6), 1279. doi:10.3390/foods10061279.

Bakos, K., Kovács, R., Staszny, Á., Sipos, D.K., Urbányi, B., Müller, F., et al., 2013. Developmental toxicity and estrogenic potency of zearalenone in zebrafish (Danio rerio). Aquat. Toxicol 136–137, 13–21. doi:10.1016/j.aquatox.2013.03.004.

Bargmann, C.I, 2006. Chemosensation in *C. elegans*. In: WormBook: The Online Review of *C. elegans* Biology. WormBook, Pasadena (CA), pp. 2005–2018. https://www.ncbi.nlm.nih.gov/books/NBK19746/

Boppana, N.B., Kodiha, M., Stochaj, U., Lin, H.S., Haimovitz-Friedman, A., Bielawska, A., Bielawski, J., Divine, G.W., Boyd, J.A., Korbelik, M., Separovic, D., 2014. Ceramide synthase inhibitor fumonisin B1 inhibits apoptotic cell death in SCC17B human head and neck squamous carcinoma cells after Pc4 photosensitization. Photochem. Photobiol. Sci. 13 (11), 1621–1627. doi:10.1039/c4pp00292j.

Breimann, L., Preusser, F., Preibisch, S., 2018. Light-microscopy methods in *C. elegans* research. Curr. Opin. Syst. Biol 13, 82–92. doi:10.1016/j.coisb.2018.11.004.

Bryden, W.L., 2012. Mycotoxin contamination of the feed supply chain: implications for animal productivity and feed security. Anim. Feed Sci. Technol 173 (1–2), 134–158. doi:10.1016/j.anifeedsci.2011.12.014.

Bryła, M., Pierzgalski, A., Zapaśnik, A., Uwineza, P.A., Ksieniewicz-Woźniak, E., Modrzewska, M., et al., 2022. Recent research on *Fusarium* mycotoxins in maize-a review. Foods 11 (21), 3465. doi:10.3390/foods11213465.

Burgess, L., Bryden, W., 2012. Fusarium: a ubiquitous fungus of global significance. Microbiol. Aust 33 (1), 22. doi:10.1071/ma12022.

Chang, T.T., Chang, C.H., Liao, V.H.C, 2024. Early life long-term exposure to aflatoxin B1 induces aging and alters innate immunity associated with SKN-1/Nrf2 in *Caenorhabditis elegans*. Chem. Biol. Interact 406, 111349. doi:10.1016/j.cbi.2024.111349.

Chaurasia, S., Kaur, C., Rastogi, V., Poswal, A.K., Munda, D.S., Bhatia, R.K., et al., 2016. In situ measurement of ions parameters of laser produced ion source using high resolution Thomson parabola spectrometer. J. Instrum 11 (8), P08004. doi:10.1088/1748-0221/11/08/p08004.

Chilenga, C., Masamba, K., Kasapila, W., Ndhlovu, B., Munkhuwa, V., Rafoneke, L., et al., 2024. Mycotoxin management in Sub-Saharan Africa: a comprehensive systematic review of policies and strategies in Malawi. Toxicol. Rep 14, 101871. doi:10.1016/j.toxrep.2024.101871.

Coppock, R.W., Jacobsen, B.J, 2009. Mycotoxins in animal and human patients. Toxicol. Ind. Health 25 (9–10), 637–655. doi:10.1177/0748233709348263.

Corsi, A.K., Wightman, B., Chalfie, M. A transparent window into biology: a primer on *Caenorhabditis elegans*. In: WormBook: The Online Review of *C. Elegans* Biology. Pasadena (CA): WormBook; 2005–2018. https://www.ncbi.nlm.nih.gov/books/NBK299460/.

Corsi, A.K, 2006. A biochemist's guide to *Caenorhabditis elegans*. Anal. Biochem 359 (1), 1–17. doi:10.1016/j.ab.2006.07.033.

Da Rocha, M.E.B., Da Chagas Oliveira Freire, F., Maia, F.E.F., Guedes, M.I.F., Rondina, D., 2013. Mycotoxins and their effects on human and animal health. Food Control 36 (1), 159–165. doi:10.1016/j.foodcont.2013.08.021.

Da Silva, E.O., Bracarense, A.P.F.L., Oswald, I.P, 2018. Mycotoxins and oxidative stress: where are we? World Mycotoxin J 11 (1), 113–134. doi:10.3920/wmj2017.2267.

Dai, C., Tian, E., Li, H., Gupta, S.D., Hao, Z., Wang, Z., et al., 2023. Molecular mechanisms of aflatoxin neurotoxicity and potential neuroprotective agents. Food Sci. Hum. Wellness 13 (5), 2445–2455. doi:10.26599/fshw.2022.9250201.

Di, R., Zhang, H., Lawton, M.A, 2018. Transcriptome analysis of *C. elegans* reveals novel targets for DON cytotoxicity. Toxins 10 (7), 262. doi:10.3390/toxins10070262.

Doi, K., Uetsuka, K., 2014. Mechanisms of mycotoxin-induced dermal toxicity and tumorigenesis through oxidative stress-related pathways. J. Toxicol. Pathol 27 (1), 1–10. doi:10.1293/tox.2013-0062.

Doi, K., Uetsuka, K., 2011. Mechanisms of mycotoxin-induced neurotoxicity through oxidative stress-associated pathways. Int. J. Mol. Sci 12 (8), 5213–5237. doi:10.3390/ijms12085213.

El-Sayed, R.A., Jebur, A.B., Kang, W., El-Demerdash, F.M, 2022. An overview on the major mycotoxins in food products: characteristics, toxicity, and analysis. J. Future Foods 2 (2), 91–102. doi:10.1016/j.jfutfo.2022.03.002.

Emmons, S.W, 2024. Functions of *c. Elegans* neurons from synaptic connectivity. bioRxiv [Preprint] doi:10.1101/2024.03.08.584145.

Eskola, M., Kos, G., Elliott, C.T., Hajšlová, J., Mayar, S., Krska, R., 2020. Worldwide contamination of food-crops with mycotoxins: validity of the widely cited 'FAO estimate' of 25. Crit. Rev. Food Sci. Nutr 60 (16), 2773–2789. doi:10.1080/10408398.2019.1658570.

Espejo, L., Hull, B., Chang, L.M., DeNicola, D., Freitas, S., Silbar, V., et al., 2022. Long-term culture of individual *Caenorhabditis elegans* on solid media for longitudinal fluorescence monitoring and aversive interventions. J. Vis. Exp (190), e64682 doi:10.3791/64682.

Fehr, M., Pahlke, G., Fritz, J., Christensen, M.O., Boege, F., Altemöller, M., et al., 2009. Alternariol acts as a topoisomerase poison, preferentially affecting the IIalpha isoform. Mol. Nutr. Food Res. 53 (4), 441–451. doi:10.1002/mnfr.200700379.

Feng, W.H., Xue, K.S., Tang, L., Williams, P.L., Wang, J.S, 2016. Aflatoxin B_1-induced developmental and DNA damage in *Caenorhabditis elegans*. Toxins 9 (1), 9. doi:10.3390/toxins9010009.

Fouad, A.M., Ruan, D., El-Senousey, H.K., Chen, W., Jiang, S., Zheng, C., 2019. Harmful effects and control strategies of aflatoxin B_1 produced by *Aspergillus flavus* and *Aspergillus parasiticus* strains on poultry: review. Toxins (Basel) 11 (3), 176. doi:10.3390/toxins11030176.

Freire, L., Sant'Ana, A.S., 2017. Modified mycotoxins: an updated review on their formation, detection, occurrence, and toxic effects. Food Chem. Toxicol. 111, 189–205. doi:10.1016/j.fct.2017.11.021.

Gacem, M.A., Gacem, H., Telli, A., Khelil, A.O.E.H, 2019. Mycotoxin-Induced Toxicities and Diseases. Elsevier eBooks 117–154. doi:10.1016/b978-0-12-817998-7.00006-9.

Gives, P.M.D, 2022. Soil-borne nematodes: impact in agriculture and livestock and sustainable strategies of prevention and control with special reference to the use of nematode natural enemies. Pathogens 11 (6), 640. doi:10.3390/pathogens11060640.

Goda, A.A., Shi, J., Xu, J., Liu, X., Zhou, Y., Xiao, L., et al., 2025. Global health and economic impacts of mycotoxins: a comprehensive review. Environ. Sci. Eur 37 (1), 122. doi:10.1186/s12302-025-01166-x.

González-Hunt, C.P., Leung, M.C., Bodhicharla, R.K., McKeever, M.G., Arrant, A.E., Margillo, K.M., et al., 2014. Exposure to mitochondrial genotoxins and dopaminergic neurodegeneration in *Caenorhabditis elegans*. PLoS ONE 9 (12), e114459. doi:10.1371/journal.pone.0114459.

Gowrinathan, Y., Pacan, J.C., Hawke, A., Zhou, T., Sabour, P.M., 2011. Toxicity assay for deoxynivalenol using *Caenorhabditis elegans*. Food Addit. Contam. Part A Chem. Anal. Control. Expo. Risk Assess 28 (9), 1235–1241. doi:10.1080/19440049.2011.587836.

Grintzalis, K., Vernardis, S.I., Klapa, M.I., Georgiou, C.D, 2014. Role of oxidative stress in Sclerotial differentiation and aflatoxin B1 biosynthesis in Aspergillus flavus. Appl. Environ. Microb 80 (18), 5561–5571. doi:10.1128/AEM.01282-14.

Guilford, F.T., Hope, J., 2014. Deficient glutathione in the pathophysiology of mycotoxin-related illness. Toxins (Basel) 6 (2), 608–623. doi:10.3390/toxins6020608.

Gusti, A.M.T., Qusti, S.Y., Alshammari, E.M., Toraih, E.A., Fawzy, M.S, 2021. Antioxidants-related superoxide dismutase (*SOD*), catalase (*CAT*), glutathione peroxidase (*GPX*), glutathione-S-transferase (*GST*), and nitric oxide synthase (*NOS*) gene variants analysis in an obese population: a preliminary case-control study. Antioxidants (Basel) 10 (4), 595. doi:10.3390/antiox10040595.

Ha, N.M., Oh, J., Kim, C.S, 2022. *Caenorhabditis elegans* as a powerful tool in natural product bioactivity research. Appl. Biol. Chem 65 (1), 18.

Habschied, K., Krstanović, V., Zdunić, Z., Babić, J., Mastanjević, K., Šarić, G.K, 2021. Mycotoxins biocontrol methods for healthier crops and stored products. J. Fungi. (Basel) 7 (5), 348. doi:10.3390/jof7050348.

Hahnel, S.R., Dilks, C.M., Heisler, I., Andersen, E.C., Kulke, D., 2020. *Caenorhabditis elegans* in anthelmintic research – old model, new perspectives. Int. J. Parasitol. Drugs Drug Resist 14, 237–248. doi:10.1016/j.ijpddr.2020.09.005.

Hillier, L.W., Coulson, A., Murray, J.I., Bao, Z., Sulston, J.E., Waterston, R.H, 2005. Genomics in *C. elegans*: so many genes, such a little worm. Genome Res. 15 (12), 1651–1660. doi:10.1101/gr.3729105.

Hu, S., Skelly, L.E., Kaymak, E., Freeberg, L., Lo, T.W., Kuersten, S., et al., 2018. Multi-modal regulation of *C. elegans* hermaphrodite spermatogenesis by the GLD-1-FOG-2 complex. Dev. Biol 446 (2), 193–205. doi:10.1016/j.ydbio.2018.11.024.

Huang, C.W., Liao, W.R., How, C.M., Yen, P.L., Wei, C.C, 2021. Chronic exposure of zearalenone inhibits antioxidant defense and results in aging-related defects associated with DAF-16/FOXO in *Caenorhabditis elegans*. Environ. Pollut. 285, 117233. doi:10.1016/j.envpol.2021.117233.

Hunt, P.R, 2017. The *C. elegans* model in toxicity testing. J. Appl. Toxicol 37 (1), 50–59. doi:10.1002/jat.3357.

Islam, M.T., Martorell, M., González-Contreras, C., Villagran, M., Mardones, L., Tynybekov, B., et al., 2023. An updated overview of anticancer effects of alternariol and its derivatives: underlying molecular mechanisms. Front. Pharmacol. 14, 1099380. doi:10.3389/fphar.2023.1099380.

Ismaiel, A., Papenbrock, J., 2015. Mycotoxins: producing fungi and mechanisms of phytotoxicity. Agriculture 5 (3), 492–537. doi:10.3390/agriculture5030492.

Janik-Karpinska, E., Ceremuga, M., Niemcewicz, M., Synowiec, E., Sliwiński, T., Bijak, M., 2023. Mitochondrial damage induced by T-2 mycotoxin on human Skin—Fibroblast HS68 cell line. Molecules 28 (5), 2408. doi:10.3390/molecules28052408.

Jomova, K., Raptova, R., Alomar, S.Y., Alwasel, S.H., Nepovimova, E., Kuca, K., et al., 2023. Reactive oxygen species, toxicity, oxidative stress, and antioxidants: chronic diseases and aging. Arch. Toxicol. 97 (10), 2499–2574. doi:10.1007/s00204-023-03562-9.

Keller, J., Borzekowski, A., Haase, H., Menzel, R., Rueß, L., Koch, M., 2018. Toxicity assay for citrinin, zearalenone and zearalenone-14-sulfate using the nematode *Caenorhabditis elegans* as model organism. Toxins 10 (7), 284. doi:10.3390/toxins10070284.

Khezri, A., Herranz-Jusdado, J.G., Ropstad, E., Fraser, T.W.K, 2018. Mycotoxins induce developmental toxicity and behavioural aberrations in zebrafish larvae. Environ. Pollut. [Internet] 242 (Pt A), 500–506. doi:10.1016/j.envpol.2018.07.010.

Kimble, J., Nüsslein-Volhard, C., 2022. The great small organisms of developmental genetics: *Caenorhabditis elegans* and Drosophila melanogaster. Dev. Biol 485, 93–122. doi:10.1016/j.ydbio.2022.02.013.

Kőszegi, T., Poór, M., 2016. Ochratoxin A: molecular interactions, mechanisms of toxicity and prevention at the molecular level. Toxins (Basel) 8 (4), 111. doi:10.3390/toxins8040111.

Kozieł, M.J., Kowalska, K., Piastowska-Ciesielska, A.W, 2021. Nrf2: a main responsive element in cells to mycotoxin-induced toxicity. Arch. Toxicol 95 (5), 1521–1533. doi:10.1007/s00204-021-02995-4.

Kucharíková, S., Hockicková, P., Melnikov, K., Bárdyová, Z., Kaiglová, A., 2023. The *Caenorhabditis elegans* cuticle plays an important role against toxicity to bisphenol A and bisphenol S. Toxicol. Rep 10, 341–347. doi:10.1016/j.toxrep.2023.02.013.

Lai, C.H., Chou, C.Y., Ch'ang, L.Y., Liu, C.S., Lin, W., 2000. Identification of novel human genes evolutionarily conserved in *Caenorhabditis elegans* by comparative proteomics. Genome Res. 10 (5), 703–713. doi:10.1101/gr.10.5.703.

Leung, M.C., Williams, P.L., Benedetto, A., Au, C., Helmcke, K.J., Aschner, M., et al., 2008. *Caenorhabditis elegans*: an emerging model in biomedical and environmental toxicology. Toxicol. Sci 106 (1), 5–28. doi:10.1093/toxsci/kfn121.

Li, C., Nelson, L.S., Kim, K., Nathoo, A., Hart, A.C., 1999. Neuropeptide gene families in the nematode Caenorhabditis elegans. Acad. Sci. 897, 239–252. doi:10.1111/j.1749-6632.1999.tb07895.x.

Li, X., Shen, X., Jiang, W., Xi, Y., Li, S., 2024. Comprehensive review of emerging contaminants: detection technologies, environmental impact, and management strategies. Ecotoxicol. Environ. Saf 278, 116420. doi:10.1016/j.ecoenv.2024.116420.

Li, Y.S., Wei, C.C, 2024. Mycotoxin zearalenone induces multi-/trans-generational toxic effects and germline toxicity transmission via histone methyltransferase MES-4 in Caenorhabditis elegans. Environ. Pollut 361, 124787. doi:10.1016/j.envpol.2024.124787.

Liew, W.P., Mohd-Redzwan, S., 2018. Mycotoxin: its impact on gut health and microbiota. Front. Cell. Infect. Microbiol 8, 60. doi:10.3389/fcimb.2018.00060.

Lin, Y., Lin, C., Cao, Y., Chen, Y., 2023. *Caenorhabditis elegans* as an in vivo model for the identification of natural antioxidants with anti-aging actions. Biomed. Pharmacother 167, 115594. doi:10.1016/j.biopha.2023.115594.

Long, N.P., Kang, J.S., Kim, H.M, 2023. Caenorhabditis elegans: a model organism in the toxicity assessment of environmental pollutants. Environ. Sci. Pollut. Res. Int 30 (14), 39273–39287. doi:10.1007/s11356-023-25675-5.

Luo, J., Bainbridge, C., Miller, R.M., Barrios, A., Portman, D.S, 2024. *C. elegans* males optimize mate-preference decisions via sex-specific responses to multimodal sensory cues. Curr. Biol 34 (6), 1309–1323.e4. doi:10.1016/j.cub.2024.02.036.

Luz, A.L., Meyer, J.N, 2016. Effects of reduced mitochondrial DNA content on secondary mitochondrial toxicant exposure in *Caenorhabditis elegans*. Mitochondrion 30, 255–264. doi:10.1016/j.mito.2016.08.014.

Martinez-Finley, E.J., Chakraborty, S., Caito, S., Fretham, S., Aschner, M., 2012. *C. elegans* and neurodegeneration *in Caenorhabditis elegans: Anatomy, Life Cycles and Biological Functions*. Adv. Med. Biol 44, 1–46.

Mavrommatis, A., Giamouri, E., Tavrizelou, S., Zacharioudaki, M., Danezis, G., Simitzis, P.E., et al., 2021. Impact of mycotoxins on animals' oxidative status. Antioxidants 10 (2), 214. doi:10.3390/antiox10020214.

Melnikov, K., Kucharíková, S., Bárdyová, Z., Botek, N., Kaiglová, A., 2023. Applications of a powerful model organism *Caenorhabditis elegans* to study the neurotoxicity induced by heavy metals and pesticides. Physiol. Res 72 (2), 149–166. doi:10.33549/physiolres.934977.

Morgan, D.E., Crittenden, S.L., Kimble, J., 2010. The *C. elegans* adult male germline: stem cells and sexual dimorphism. Dev. Biol 346 (2), 204–214. doi:10.1016/j.ydbio.2010.07.022.

Nawaf, A., 2023. Mycotoxin source and its exposure causing mycotoxicoses. Bioinformation 19 (4), 348–357. doi:10.6026/97320630019348.

O'Reilly, L.P., Luke, C.J., Perlmutter, D.H., Silverman, G.A., Pak, S.C, 2013. *C. elegans* in high-throughput drug discovery. Adv. Drug. Deliv. Rev 69–70, 247–253. doi:10.1016/j.addr.2013.12.001.

Oikonomou, G., Shaham, S., 2011. The glia of *Caenorhabditis elegans*. Glia 59 (9), 1253–1263. doi:10.1002/glia.21084.

Oke, O.E., Akosile, O.A., Oni, A.I., Opowoye, I.O., Ishola, C.A., Adebiyi, J.O., et al., 2024. Oxidative stress in poultry production. Poult. Sci 103 (9), 104003. doi:10.1016/j.psj.2024.104003.

Pandey, A.K., Samota, M.K., Kumar, A., Silva, A.S., Dubey, N.K, 2023. Fungal mycotoxins in food commodities: present status and future concerns. Front. Sustain. Food Syst 7. doi:10.3389/fsufs.2023.1162595.

Queirós, L., Marques, C., Pereira, J.L., Gonçalves, F.J.M., Aschner, M., Pereira, P., 2021. Overview of Chemotaxis Behavior Assays in Caenorhabditis elegans. Curr. Protoc. 1 (5), e120. doi:10.1002/cpz1.120.

Reiss, J., 1975. Insecticidal and larvicidal activities of the mycotoxins aflatoxin B1 rubratoxin B, patulin and diacetoxyscirpenol towards Drosophila melanogaster. Chem. Biol. Interact 10 (5), 339–342. doi:10.1016/0009-2797(75)90055-1.

Riddle, D.L., Blumenthal, T., Meyer, B.J., Priess, J.R., 1997. Introduction to *C. elegans*. In: Riddle, D.L., Blumenthal, T., Meyer, B.J., Priess, J.R. (Eds.), C. *elegans* II, 2nd ed.. Cold Spring Harbor Laboratory Press, Cold Spring Harbor (NY) Chapter 1.

Rodrigues, D.T., Padilha, H.A., Soares, A.T.G., De Souza, M.E.O., Guerra, M.T., Ávila, D.S, 2024. The Caenorhabditis elegans neuroendocrine system and their modulators: an overview. Mol. Cell. Endocrinol 586, 112191.

Roussos, A., Kitopoulou, K., Borbolis, F., Palikaras, K., 2023. *Caenorhabditis elegans* as a model system to study human neurodegenerative disorders. Biomolecules 13 (3), 478. doi:10.3390/biom13030478.

Ruszkiewicz, J.A., Pinkas, A., Miah, M.R., Weitz, R.L., Lawes, M.J.A., Akinyemi, A.J., et al., 2018. C. elegans as a model in developmental neurotoxicology. Toxicol. Appl. Pharmacol 354, 126–135. doi:10.1016/j.taap.2018.03.016.

Schafer, W.R, 2005. Deciphering the neural and molecular mechanisms of *C. elegans* behavior. Curr. Biol 15 (17), R723–R729. doi:10.1016/j.cub.2005.08.020.

Shah, P., Bao, Z., Zaidel-Bar, R., 2022. Visualizing and quantifying molecular and cellular processes in *Caenorhabditis elegans* using light microscopy. Genetics 221 (4), iyac068. doi:10.1093/genetics/iyac068.

Shekhar, R., Raghavendra, V.B., Rachitha, P., 2025. A comprehensive review of mycotoxins, their toxicity, and innovative detoxification methods. Toxicol. Rep 14, 101952. doi:10.1016/j.toxrep.2025.101952.

Song, C., Wang, Z., Cao, J., Dong, Y., Chen, Y., 2024. Neurotoxic mechanisms of mycotoxins: focus on aflatoxin B1 and T-2 toxin. Environ. Pollut 356, 124359. doi:10.1016/j.envpol.2024.124359.

Strange, K., 2016. Drug discovery in fish, flies, and worms. ILAR J 57 (2), 133–143. doi:10.1093/ilar/ilw034.

Teli, B., Purohit, J., Rashid, M.M., Jailani, A.A.K., Chattopadhyay, A., 2020. Omics insight on Fusarium head blight of wheat for translational research perspective. Curr. Genomics 21 (6), 411–428. doi:10.2174/1389202921999200620222631.

Thongararm, P., Chancharoen, M., Suwanwong, N., Ruchirawat, S., Ruchirawat, M., Fedeles, B.I., et al., 2025. Structurally similar mycotoxins aflatoxin B_1 and sterigmatocystin trigger different and distinctive high-resolution mutational spectra in mammalian cells. Toxins (Basel) 17 (3), 112. doi:10.3390/toxins17030112.

Tolosa, J., Rodríguez-Carrasco, Y., Ruiz, M.J., Vila-Donat, P., 2021. Multi-mycotoxin occurrence in feed, metabolism and carry-over to animal-derived food products: a review. Food Chem. Toxicol 158, 112661. doi:10.1016/j.fct.2021.112661.

Tumini, E., Wellinger, R.E., Herrera-Moyano, E., Navarro-Cansino, P., García-Rubio, M., Salas-Lloret, D., et al., 2025. Patulin and Xestoquinol are inhibitors of DNA topoisomerase 1. Proc. Natl. Acad. Sci 122 (17), e2421167122. doi:10.1073/pnas.2421167122.

Wang, D., Xing, X., 2008. Assessment of locomotion behavioral defects induced by acute toxicity from heavy metal exposure in nematode *Caenorhabditis elegans*. J. Environ. Sci 20 (9), 1132–1137. doi:10.1016/s1001-0742(08)62160-9.

Wang, Y., Liu, Y., Huang, T., Chen, Y., Song, W., Chen, F., et al., 2023. Nrf2: a main responsive element of the toxicity effect caused by trichothecene (T-2) mycotoxin. Toxics 11 (4), 393. doi:10.3390/toxics11040393.

Widmayer, S.J., Crombie, T.A., Nyaanga, J.N., Evans, K.S., Andersen, E.C, 2022. *C. elegans* toxicant responses vary among genetically diverse individuals. Toxicology 479, 153292. doi:10.1016/j.tox.2022.153292.

Wu, Y., Wu, Z., Wen, J., Chen, M., Bao, W., Zhang, J., et al., 2025. Aflatoxin B1 disrupts intestinal barrier integrity: insights into DAF-16/SOD-3-mediated oxidative imbalance and autophagy in *Caenorhabditis elegans*. Ecotoxicol. Environ. Saf 303, 118894. doi:10.1016/j.ecoenv.2025.11889.

Xiong, H., Pears, C., Woollard, A., 2017. An enhanced *C. elegans* based platform for toxicity assessment. Sci. Rep 7 (1), 9839. doi:10.1038/s41598-017-10454-3.

Xu, F., Li, Y., Cao, Z., Zhang, J., Huang, W., 2021. AFB_1-induced mice liver injury involves mitochondrial dysfunction mediated by mitochondrial biogenesis inhibition. Ecotoxicol. Environ. Saf 216, 112213. doi:10.1016/j.ecoenv.2021.112213.

Yang, Z., Xue, K.S., Sun, X., Tang, L., Wang, J.S, 2015. Multi-toxic endpoints of the foodborne mycotoxins in nematode *Caenorhabditis elegans*. Toxins (Basel) 7 (12), 5224–5235. doi:10.3390/toxins7124876.

Yang, Z., Xue, K.S., Sun, X., Williams, P.L., Wang, J.S., Tang, L., 2018. Toxicogenomic responses to zearalenone in *Caenorhabditis elegans* reveal possible molecular mechanisms of reproductive toxicity. Food Chem. Toxicol 122, 49–58. doi:10.1016/j.fct.2018.09.040.

Zakaria, L., 2024. An overview of *Aspergillus* species associated with plant diseases. Pathogens 13 (9), 813. doi:10.3390/pathogens13090813.

Zhang, X., Ye, Y., Sun, J., Wang, J.S., Tang, L., Xu, Y., et al., 2022. Abnormal neurotransmission of GABA and serotonin in *Caenorhabditis elegans* induced by Fumonisin B1. Environ. Pollut 304, 119141. doi:10.1016/j.envpol.2022.119141.

Zhao, X., Li, X., Gao, J., Shen, S., Zou, W., 2025. Behavioral adaptations of *Caenorhabditis elegans* against pathogenic threats. PeerJ 13, e19294. doi:10.7717/peerj.19294.

Zhou, H., Tang, L., Xue, K.S., Qian, H., Sun, X., Williams, P.L., et al., 2018b. Trans-/multi-generational effects of deoxynivalenol on *Caenorhabditis elegans*. Chemosphere 201, 41–49. doi:10.1016/j.chemosphere.2018.02.173.

Zhou, D., Wang, X., Chen, G., Sun, S., Yang, Y., Zhu, Z., et al., 2018a. The major Fusarium species causing maize ear and kernel rot and their toxigenicity in Chongqing, China. Toxins (Basel) 10 (2), 90. doi:10.3390/toxins10020090.

Printed and bound by CPI Group (UK) Ltd, Croydon, CR0 4YY
14/09/2026
02216411-0001